Encyclopedia of Tropical Forests

Encyclopedia of Tropical Forests

Edited by **Lee Zieger**

R CALLISTO REFERENCE

New York

Published by Callisto Reference,
106 Park Avenue, Suite 200,
New York, NY 10016, USA
www.callistoreference.com

Encyclopedia of Tropical Forests
Edited by Lee Zieger

International Standard Book Number: 978-1-63239-306-7 (Hardback)

Contents

Preface

This book presents advanced in-depth information regarding tropical forests. The surprisingly rich and abundant biodiversity of tropical forests is declining at a fast rate. This has led to serious changes in important biogeochemical cycles of nitrogen, carbon, phosphorus, etc. and has resulted in an alteration in the global climate and pure natural ecosystems. This book broadly defines Tropical Forests in light of the following major aspects: tropical forest synthesis, synergy and structure; effect of anthropogenic pressure; tropical forest protection and process; tropical forest fragmentation; remote sensing and geographic information system. Advanced synthesis, descriptive recent reviews, latest experiences/experiments contributed by veteran scientists from across the globe, and several original data-rich researches have been included in this book. The aim of this book is to help in educating, generating knowledge, and understanding the role of "Tropical Forests" for the very survival of humans, variety of biota, and climate change across the planet. This book will serve as a valuable source of reference for ecologists, students, forest managers, scientists, and population and conservation biologists.

After months of intensive research and writing, this book is the end result of all who devoted their time and efforts in the initiation and progress of this book. It will surely be a source of reference in enhancing the required knowledge of the new developments in the area. During the course of developing this book, certain measures such as accuracy, authenticity and research focused analytical studies were given preference in order to produce a comprehensive book in the area of study.

This book would not have been possible without the efforts of the authors and the publisher. I extend my sincere thanks to them. Secondly, I express my gratitude to my family and well-wishers. And most importantly, I thank my students for constantly expressing their willingness and curiosity in enhancing their knowledge in the field, which encourages me to take up further research projects for the advancement of the area.

Editor

Part 1

Introduction

Structure, Diversity, Threats and Conservation of Tropical Forests

Madhugiri Nageswara-Rao[1,2,*], Jaya R. Soneji[2,3] and Padmini Sudarshana[4]

[1]*Department of Plant Sciences, University of Tennessee, Knoxville, TN,*
[2]*Polk State College, Department of Biological Sciences, Winter Haven, FL,*
[3]*University of Florida, IFAS, Citrus Research & Education Center, Lake Alfred, FL,*
[4]*Monsanto Research Center, Hebbal, Bangalore,*
[1,2,3]*USA*
[4]*India*

1. Introduction

In this elegant book, we have defined 'Tropical Forests' broadly, into five different themes: (1) tropical forest structure, synergy, synthesis, (2) tropical forest fragmentation, (3) impact of anthropogenic pressure, (4) Geographic Information System and remote sensing, and (5) tropical forest protection and process. The cutting-edge synthesis, detailed current reviews, several original data-rich case studies, recent experiments/experiences from leading scientists across the world (Fig.1) are presented as unique chapters. Though, the chapters differ noticeably in the geographic focus, diverse ecosystems, time and approach, they share these five important themes and help in understanding, educating, and creating awareness on the role of 'Tropical Forests' on the diversity of biota, impact of disturbances, climate change and the very survival of mankind.

2. Tropical forests - Structure, synergy, synthesis

Tropical forests are located in the 'tropics' which lie between the Tropic of Cancer and Capricorn, approximately between 23° N and 23° S latitudes (Thomas and Baltzer, 2002). They support vast biodiversity and are a source of wonderment, scientific curiosity, enormous complexity as well as a basic foundation for human welfare (Tilman, 2000). While occupying only one-tenth of the world's land area, tropical forests are economically, ecologically, environmentally (Fig. 2), culturally and aesthetically vital as they play crucial role in ensuring global food security, climate change, poverty eradication and improvement of human health (Rajora and Mosseler, 2001; Thomas and Baltzer, 2002; Nageswara Rao and Soneji, 2010a, 2011). They are important in terms of global biogeochemical cycles and are home to more than half of the world's species (Thomas and Baltzer, 2002).

It is estimated that more than 10 million species of plants, animals and insects live in the tropical rainforests (http://www.rain-tree.com). One-fifth of the world's fresh water is in

---

* Corresponding Author

the Amazon Basin and more tree species are found in 0.5 km² of some tropical forests than in all of North America or Europe (Burslem et al., 2001). These forests sustain the livelihoods of hundreds of millions of people globally (Nageswara Rao et al., 2008a; Uma Shaanker et al., 2001a) and studies estimate that at least 80% of the developed world's diet originated in the tropical rainforests. About 70% plants that are active against cancer cell lines found by the US National Cancer Institute (NCI) are found only in the tropical forests (http://www.rain-tree.com).

The dense leafy canopies of tropical forests make them highly productive plant communities storing almost 30% of the global soil carbon (Sayer et al., 2007). This makes tropical forests, with relatively high litterfall, a critical component of the global carbon cycle. To assess the tropical forest productivity, phenology, and turnover of biomass, litterfall collection is a standard non-destructive technique (Newbould, 1967; Lowman, 1988). The amount of leaf material falling reflects a forest's productivity and represents a major flux of carbon from vegetation to soil in the forest. Hence, changes in litter inputs are likely to have far-reaching consequences on the soil carbon dynamics (Proctor et al., 1983; Lowman, 1988; Sayer et al., 2007). In the chapter *"Comparing litterfall and standing vegetation: Assessing the footprint of litterfall traps"*, the authors have analyzed the correspondence between litterfall samples and standing vegetation at three different spatial scales. They examined the factors affecting the relative abundance of species in litterfall samples. To gain an insight for the scaling of litterfall data from the level of sampling plots up to the level of the forest stand, they compared the composition and relative abundance of species collected in litter traps. The authors' findings will prove instrumental for the improvement of methods in terrestrial and forest ecology, especially in the tropics where the high species diversity and structural complexity of forests impose tough challenges to the study of forest structure and their dynamics.

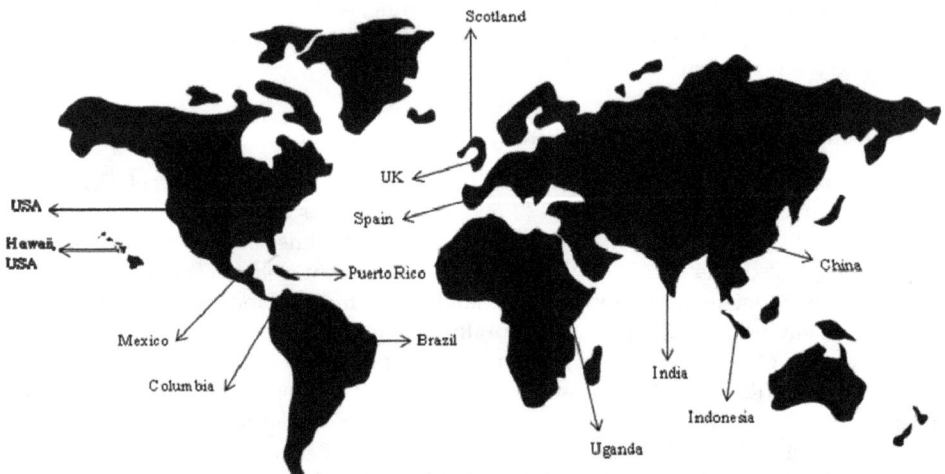

Fig. 1. Author geographic locations of studies in this book.

By regulating the microclimate, the litter layer helps to maintain favorable conditions for decomposition (Vasconcelos and Lawrence, 2005; Sayer et al., 2006) while the soil faunal activities can indirectly affect decomposition rates and the nutrient cycles (Moore and Walter, 1988). The interactions between the soil fauna and microbes can influence the microbial species composition (Visser, 1985), thus playing an important role in soil ecosystems (Lussenhop, 1992; Sayer et al., 2006) and creating habitats for arthropods (Arpin et al., 1995). Millipedes and other macroarthropods, as detritivores, affect the nutrient cycling by releasing chemical elements such as nitrogen and redistributing the organic material in the soil (Dangerfield and Milner, 1996). In the chapter *"Direct and indirect effects of millipedes on the decay of litter of varying lignin content"*, the authors have used a microcosm approach to answer what are the direct (leaf fragmentation) and indirect effects (microbial biomass) of millipedes on the decomposition of leaf litter and how these outcomes are influenced by the substrate (litter) quality and the density of millipedes. In the chapter *"Quantifying variation of soil arthropods using different sampling protocols: is diversity affected?"*, the authors have assessed how the diversity of extracted arthropods was affected by variations in the collection and extraction methodologies, and by variations in the duration of the extraction. This data will provide researchers with data to simplify the logistics of arthropod sampling and extraction, and to better choose a specific procedure for a given focal organism in a given habitat.

Fig. 2. A typical view of diverse tropical forest (see Chapter 20, by Canuto et al. in this book).

In the chapter *"Patterns of plant species richness within families and genera in lowland Neotropical forests: Are similarities related to ecological factors or to chance?"*, the authors have made a quantitative floristic comparison based on the patterns of species richness in families and genera for more than twenty five tropical areas, and correlated the floristic similarities with the ecological and stochastic factors (i.e. geographical distance). They also attempted to test the significance and relative roles of ecological and stochastic factors. Such studies can provide information on the present-day communities that have resulted from speciation, extinction, and migration (Leigh et al., 2004).

Compared to the wealth of botanical and ecological studies carried out in the tropical ecosystems, little is known about the status of mycorrhizae or the influence of mycorrhizal mutualisms on the tropical forest diversity and tree assemblages (Alexander and Lee, 2005; McGuire et al., 2008). In the chapter *"Dispersion, an important radiation mechanism for ectomycorrhiyzal fungi in Neotropical low land forest"*, the author has evaluated different hypothesis about the possible origin of ectomycorrhiyzal (EcM) fungi associated with *Pakaraimaea dipterocarpacea*. The EcM fungi diversity and community structure, and also the phylogenetic analysis have been carried out. This study is the first evidence of host sharing between both sympatric and allopatric tree species belonging to Dipterocarpaceae and EcM Fabaceae in the Neotropics.

3. Tropical forest fragmentation

Fragmentation, due to rapid economic growth and agricultural expansion, of the tropical forests and the natural habitats into smaller and non-contiguous patches is the most serious threat to the long-term survival of the biological diversity on earth (Myers, 1994; Chapin et al., 2000; Pimm and Raven, 2000; Cruse-Sanders and Hamrick, 2004; Nageswara Rao et al., 2008b).

Fig. 3. Shola forest fragments in the Western Ghats (one of the mega diversity 'hot-spot' in the world), India (see Rajanikanth et al., 2010).

As a consequence of fragmentation, natural or man-made, plant populations are isolated from their conspecific populations (Fig. 3), have reduced population size (Lamont et al., 1993; Hall et al., 1996; Risser, 1996; Rajanikanth et al., 2010) and have decreased fruit set or poor seed germination relative to large population (Menges, 1991; Byers and Meagher, 1992;

Hendrix, 1994; Heschel and Paige, 1995; Agren, 1996). These fragmented patches of forest are often embedded in a matrix of anthropogenically manipulated landscapes (such as pastures, agricultural fields or habitations; Fig. 4), behave as "islands" in a "sea" of pasture or agricultural ecosystem and may lead to distinct ecological, demographic and genetic consequences which result in the extinction of the native species (Tilman et al., 1994; Gilpin, 1988; Laurance, 2000; Nageswara Rao et al., 2001, 2007; Uma Shaanker et al., 2001b; Honnay et al., 2005). Fragmentation or conversion of forest into grassland or savanna due to forest harvesting, fertilization, atmospheric deposition, and climate change also affects the nitrogen mineralization of the tropical forests (Wang et al., 2004).

Anuran amphibians inhabit regions that have high moisture levels and moderate to warm temperatures owing to their skin permeability and dependence on aquatic and terrestrial habitats during their life cycles (Duellman and Trueb, 1994; Wells, 2007). Fragmentation and/or deforestation makes the environment drier and more seasonal, reduces the population size of anuran species, adversely affects the anuran richness in local assemblages that depend on breeding ponds for reproduction and sometimes eliminating those that depend on humid forest microhabitats (Haddad and Prado, 2005; Becker et al., 2007). In the chapter *"The role of environmental heterogeneity in maintenance of anuran amphibian diversity of the Brazilian mesophytic semideciduous forest"*, the authors employed tests of null hypotheses to assess whether patterns of spatial distribution of anuran assemblages differ from a random distribution among aquatic breeding sites monitored at Morro do Diabo State Park. They also verified the existence of indicator anuran species of environmental heterogeneity on a local scale.

The formation of treefall gaps and their influence on forest regeneration and dynamics have ecological consequences (Schnitzer et al., 2008). These canopy gaps, formed by death or injury to one or a few canopy trees, create sufficient resource heterogeneity to allow for resource partitioning and niche differentiation (Grinnell, 1917). They also release sufficient resources (e.g., light and nutrients) to permit the establishment or reproduction of plant species that would otherwise be excluded from the forest in the absence of gaps (Schnitzer et al., 2008). Such transitory events occur frequently in the tropical forests (Brokaw, 1985), where plant species of early successional stages (pioneers and secondary ones) take advantage of the gaps formed as they can tolerate higher micro-climate and ecological variations (Mulkey et al., 1996). In the chapter *"Gap area and tree community regeneration in a tropical semideciduous forest"*, the authors have identified ecological patterns related to richness and the potential of natural regeneration of tree species in natural gaps and have investigated whether the tree community responds to different levels of canopy openings represented by gaps of different sizes found in the tropical semideciduous forests.

4. Impact of anthropogenic pressure

In the tropical forests, where both species diversity and anthropogenic pressures on the natural environments are high, biodiversity is threatened by human-driven, land-use changes (Dirzo and Raven, 2003; Gibson et al., 2011). Rapid deforestation of tropical forests for agriculture (Fig. 4), timber production, pasture, firewood, construction of roads and dams, and other uses, have dire consequences on the tropical biodiversity along with the water sources and non-timber forest products (Sudarshana et al., 2001; Uma Shaanker et al., 2003, 2004; Foley et al., 2005; Lamb et al., 2005; Ravikanth et al., 2009; Gibson et al., 2011). The increasing rate of human population in the developing countries, where most of these forests are located, has triggered a greater demand for timber and other forest products, making sustainable management of these remnant forests a major challenge (Wright and Muller-Landau, 2006). Human disturbances often lead to altered environmental conditions, which influence the process that can both augment and erode species diversity in the tropical forest community (Kennard et al., 2002; Sapkota et al., 2010).

Fig. 4. Conversion of pristine tropical forests into agricultural lands, Western Ghats (one of the mega diversity 'hot-spot' in the world), India.

Changes in vertebrate assemblages in the tropical rain forests caused by anthropogenic disturbances affect the seed dispersal patterns and subsequent tree spatial recruitment patterns in the secondary tropical rain forests. Even though a variety of seed dispersal mechanisms are found within tropical forests, most plants produce fleshy fruits that are dispersed primarily by vertebrate frugivores (Jordano, 1992). Behavioral disparities among vertebrate seed dispersers could influence patterns of seed distribution and thus forest structure (Howe, 1990; Clark et al., 2001). In the chapter "*Seed dispersal and tree spatial recruitment patterns in secondary tropical rain forests*", the author examined the seed dispersal and tree spatial recruitment patterns in three tropical forests whose vertebrate populations have been altered differently over the past few decades. The changes in vertebrate seed disperser community on tree recruitment in the secondary forest landscapes in the wider context of the effectiveness of remnant vertebrate populations in seed dispersal and the possible consequences for tree demography are presented.

Forest fragmentation not only affects the plants but also the large predators that play an important role in regulating herbivore prey populations (Duffy, 2003). The ecological consequences of such fragmentation on the mesoherbivores remain largely undocumented.

In an effort to understand the magnitude of the effects of human-perturbed, mesoherbivore populations on the tropical forest plant communities, in the chapter *"Human altered mesoherbivore densities and cascading effects on plant and animal communities in fragmented tropical forests"*, the authors have reviewed substantial tropical literature on human overhunting of mesoherbivores and the consequences for the tropical forest plant communities. For the first time, the authors have synthesized the sparse and scattered literature involving tropical examples of mesoherbivore release, making a case that the cascading consequences of increased mesoherbivore abundance can be as widespread and ecologically destructive as those resulting from mesoherbivore decline. The authors have also addressed the most pressing conservation and management implications of the research on perturbed mesoherbivore populations and have identified topics in need of further investigation, in terms of both ecology and conservation.

Knowledge of forest structure and floristics are necessary for the study of forest dynamics, plant-animal interactions and nutrient cycling (Reddy and Pattanaik, 2009). In the chapter *"Floristic composition, diversity and status of threatened medicinal plants in tropical forests of Malyagiri hill ranges, Eastern Ghats, India"*, the authors have analyzed the diversity, distribution and population structure of tree species in a tropical deciduous forest stand, with special emphasis on the documentation of threatened medicinal plants. Medicinal plant species belonging to different threat categories were recorded from the forest and suggestive conservation measures for sustainable use of medicinal plant resources are presented.

Aboveground coarse necromass, a major component of the carbon cycle in the tropical forests, accounts for up to 20% of carbon stored above ground and for 14–19% of the annual aboveground carbon flux in the tropical forests (Palace et al., 2008). The dynamics of necromass production and loss through disturbance and decay are poorly understood and quantified in the tropical forests (Eaton and Lawrence, 2006). In the chapter *"A review of above ground necromass in tropical forests"*, the authors have examined literature pertaining to stocks or pools of above ground necromass, the disturbance and the episodic production of coarse necromass, and the slower process of decomposition in the tropical forests. They have described and defined important terms and components in necromass research and have various methodologies designed to measure these components and current literature involved with field based estimates of necromass.

5. Geographic information system and remote sensing

Combating deforestation requires factual information about the tropical forests which is not readily available (Ochego, 2003). Geographic Information System (GIS) and remote sensing provides a unique opportunity to assess (Fig. 5) and monitor deforestation, degradation, and fragmentation (Lyngdoh et al., 2005; Tejaswi, 2007). GIS integrates hardware, software and data for capturing, managing, analyzing and displaying all forms of geographically referenced information (http://www.gis.com/content/what-gis) and can be utilized for deciphering location, condition, trends, patterns and modeling of forests. Remote sensing utilizes the acquisition of information about an object, area or phenomenon through the analysis of data acquired by a device that is not in contact with the object, phenomenon or area under investigation (Lillesand and Kiefer, 1987). It has become a very powerful tool associated with the estimation of the interactions between earth's surface materials and electromagnetic energy reflected from them which are recorded by sensors aboard satellites in space (Ochego, 2003).

Remote sensing can work at multiple scales ranging from few meters to several kilometers, even in places where accessibility is an issue, and the data can be acquired periodically (e.g. daily, monthly) with measurements made in near real time basis (Tejaswi, 2007).

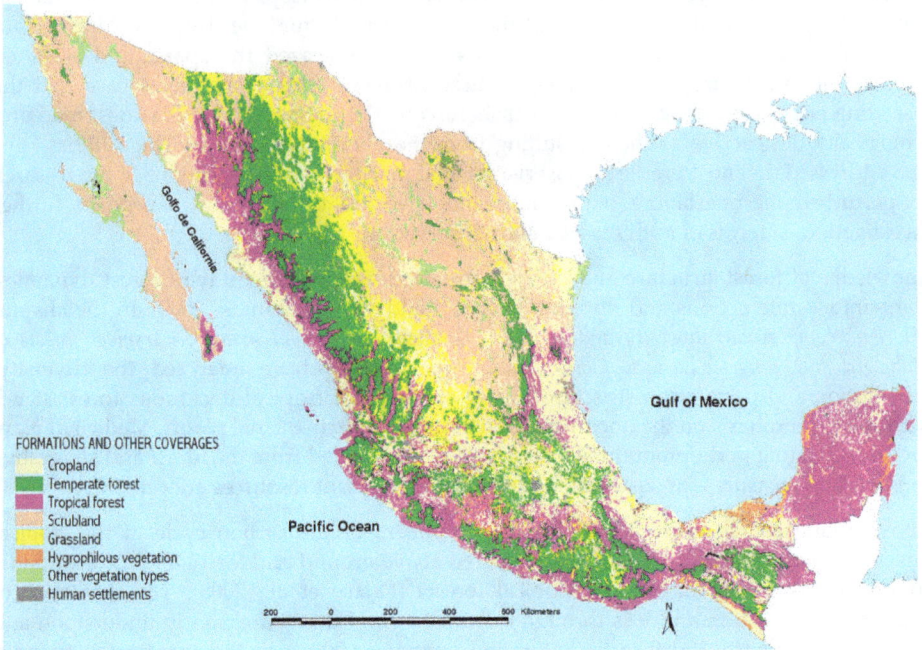

Fig. 5. National Forest Inventory map of Mexico (see Chapter 15, by Couturier et al. in this book).

Estimating the rate of change in tropical forest cover has become a crucial component of global change monitoring. In the chapter "*Seasonal pattern of vegetative cover from NDVI time-series*", the authors have employed seasonal adjustment of time-series statistical method to understand the phenology and detect disturbance on some woody vegetation utilizing the Normalized Difference Vegetation Index (NDVI) time-series of SPOT VEGETATION. In the chapter "*Measuring tropical deforestation with error margins: A method for REDD monitoring in South-eastern Mexico*", the authors present a methodological framework for the measurement of tropical deforestation in Southeast Mexico, based on the experience of accuracy assessment of regional land cover maps and on-site measurements of tropical forest cover in Mexico. In the chapter "*Natural forest change in Hainan, China, 1991-2008 and conservation suggestions*", the authors have analyzed the changes in natural forest and plantations on Hainan Island between 1991-2008 by using GIS and remote sensing, have tried to explore the driving factors of changes based on local policies, and have given suggestions for the future conservation plan. In the chapter "*Exchange of carbon between the atmosphere and the tropical Amazon rainforest*", the authors have examined the subcanopy flow dynamics and local micro-circulation features, how they relate to spatial and temporal distribution of CO_2 on the Manaus LBA Project site and have discussed the contribution of exchange of carbon between the atmosphere and the tropical Amazon Rainforest.

6. Tropical forest protection and process

During the last decade, a need to address conservation questions with a wider social, political and cultural framework was recognized (Hodgkin and Rao, 2002). With rapid vanishing of tropical forests and increasing extinction numbers, it is imperative to evolve holistic strategies to conserve the surviving populations. But launching of any such conservation program is contingent upon the knowledge of what, where and how to conserve (Ganeshaiah and Uma Shaanker, 1998). There is a general consensus among scientists and practitioners that no single conservation method is adequate and different methods should be applied in a complementary manner. In the recent past, approaches such as the *ex situ* conservation, *in situ* conservation, creating biosphere reserves, protected areas, etc. have been extended to address the conservation and restoration of tropical forest resources (Shands, 1991; Uma Shannker et al., 2001b,c; Nageswara Rao et al., 2007, 2011). In the chapter *"Direct sowing: an alternative to the restoration of ecosystems of tropical forests"*, the authors have analyzed ecological, technical, socio-economic and forestry aspects involved in the use of direct sowing to restore degraded ecosystems in the tropical forest regions (Fig. 6). The authors have also highlighted several experiments conducted in the tropical regions, which may contribute to broaden the perspective and enhance methodologies for ecological restoration and bring to light some experiences that may contribute to the decision making over the choice of direct sowing for restoration of degraded ecosystems.

Fig. 6. Seed variability and restoration efforts in tropical forests, Brazil (see Chapter 18, by Ferreira et al. in this book).

In the chapter *"Patterns of tree mortality in monodominant tropical forests"*, the author has used information from a long term study in permanent vegetation plots within 200 ha of monodominant Hakalau Forest National Wildlife Refuge, (Hawaiian, wet forest) to address basic ecological questions such as, how does tree mortality vary with respect to species, size, position in the canopy (crown class), and geographic location? What is the age of trees in this forest? To what extent can patterns of mortality provide evidence for succession in this forest?

Their results provide evidence that gap-phase dynamics may play a role in the succession, stand structure, and community composition in a large structured forest in Hawaii. Dead standing trees also provide important habitat for diverse wildlife, micro flora and fauna.

Protected areas are believed to be the corner stones for biodiversity conservation and the safest strongholds of wilderness around the globe (Pimm and Lawton, 1998; Bruner et al., 2001). With ever increasing threats to the tropical forests, protected areas and their networks offer the best possible approach to conserve the biological diversity (Hogbin et al., 2000; Bruner et al., 2001; Theilade et al., 2001). They harbor a greater level of biodiversity than the adjoining non-protected areas and may serve as *in situ* sites for the conservation of forest resources. In the chapter "*Conservation, management and expansion of protected and non-protected tropical forest remnants through population density estimation, ecology and natural history of top predators; case studies of birds of prey (Spizaetus taxon)*", the authors have described results of six studies conducted in Brazil by analyzing the incidence of specimens of the genus *Spizaetus* in areas with different fragmentation histories and considering the different population and reproductive ecological aspects of these taxons collected at each locality. The authors promote a reflection on the perspectives of local and punctual conservation of these species, according to their ecological requirements and have used these species as "flags" to point out the problems involving conservation of top predators, which present small density (Fig. 7) but demand a large area, in the fragmented and continuous areas. Protected areas could in fact be the last refugia for several tropical species (Ramesha et al., 2007; Nageswara Rao M et al., 2010). However, most protected areas may be too small to host viable populations. They may not allow for gene pool mixing across the population, due to their insular and isolated habitat. Efforts need to also be complemented by actions outside protected areas such as sustainable management and conservation of forests for multiple uses (FAO, 1993).

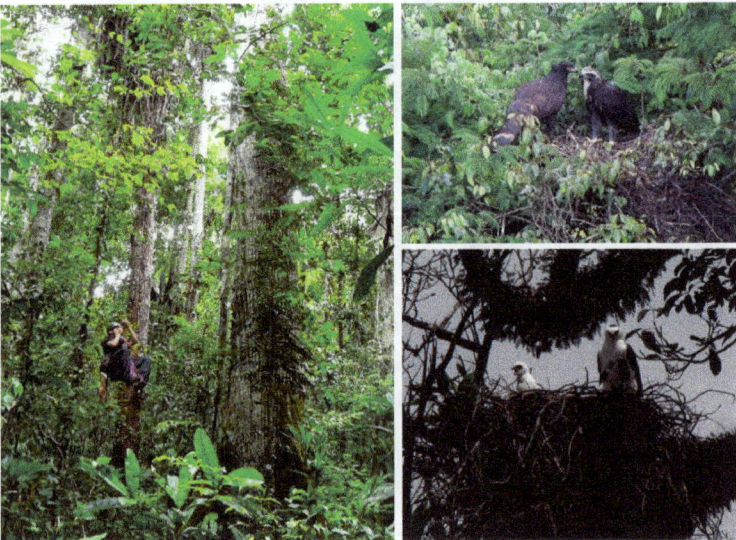

Fig. 7. Canopy survey, population estimation, conservation and management of Black Hawk, Black-and-White Hawk eagle in protected and non-protected tropical forests, Brazil (see Chapter 20, by Canuto et al. in this book).

The tropical forests, undoubtedly, are heritage for our future generations. They deal with the totality of gene, species, population and ecosystem on the basis of cellular, molecular, taxonomic and geographic criteria (Sharma, 1996) and face multiple threats. Although, monitoring and conserving the loss of forest biodiversity is crucial, there appears to be no single measure that can assess all the aspects of biodiversity. Consolidated efforts on the information on parameters such as the levels of threats, the spatial patterns of population/species richness, distribution, their interactions, genetic diversity, etc., are utmost needed for planning any effective conservation and sustainable utilization. First and foremost, ecosystems and landscapes with high concentration of endemic and useful species at risk need to be identified. Potential threats that these resources are facing should be highlighted. Species that are rare, endangered, highly threatened and economically important need to be selected and given highest priority (Uma Shaanker et al., 2001c), to study their effective population size, spatial structure, variability, and community interactions. Detailed data on all these parameters affected by native habitat loss, invasiveness, expansion of agriculture, and extraction patterns needs to be generated (Bawa et al., 2001). Mitigation strategies to counter the threats, restoration strategies, and understanding the local adaptive nature information should be an integral component of programs designed to conserve and manage the tropical forest resources (Nageswara Rao and Soneji, 2010b). Well organized national, as well as, international programs should be conducted to bridge the gap between local community, forest managers, policy-makers and the scientific community. They should be brought together through networking, training and public awareness programs. Thus, there is urgent need to consider, consolidate and complement research, policy making and on-field efforts to effectively conserve, efficiently utilize and sustainably manage the tropical forest resources before they are irrevocably lost (Nageswara Rao and Soneji, 2009).

7. References

Agren, J. (1996) Population size, pollinator limitation, and seed set in self incompatable herb *Lythrum salicaria. Ecology* 77: 1779-1790.

Alexander, I.J. & Lee, S.S. (2005) Mycorrhizas and ecosystem processes in tropical rain forest: implications for diversity. In: Burslem, D.F.R.P., Pinard, M.A. & Hartley, S.E. (eds) *Biotic interactions in the tropics: their role in the maintenance of species diversity.* Cambridge University Press, Cambridge, UK, pp 165–203.

Arpin, P., Ponge, J.F. & Vannier, G. (1995) Experimental modifications of litter supplies in a forest mull and reaction of the nematode fauna. *Fundamental & Applied Nematology* 18: 371-389.

Bawa, K.S., Ganeshaiah, K.N. & Uma Shaanker, R. (2001) Conserving tropical forest genetic resources: Threats and mitigation strategies. In: Uma Shaanker, R., Ganeshaiah, K.N. & Bawa, K.S. (eds.), *Forest genetic resources: Status, threats and conservation strategies.* Oxford and IBH Publications, New Delhi, pp. 303-308.

Becker, C.G., Fonseca, C.R., Haddad, C.F.B., Batista, R.F. & Prado, P.I. (2007) Habitat split and the global decline of amphibians. *Science* 318(5857): 1775-1777.

Brokaw, N.V.L. (1985) Gap-phase regeneration in a tropical forest. *Ecology* 66: 682-687.

Bruner, A.G., Gullison, R.E., Rice, E.R. & Fonseca, G.A.B. (2001) Effectiveness of parks in protecting tropical biodiversity. *Science* 291: 125-128.

Burslem, D.F.R.P., Garwood, N.C. & Thomas, S.C. (2001) Tropical forest diversity — The plot thickens. *Science* 291: 606-607

Byers, D.L. & Meagher, T.R. (1992) Mate vailability in the small populations of plant species with homomorphic sporophytic self-incompatibility. *Heredity* 68: 353-359.

Chapin, F.S., Zavaleta, E.S., Eviner, V.T., Naylor, R.L., Vitousek, P.M., Reynolds, H.L., Hooper, D.U., Lavore, S., Sala, O.E., Hobble, S.E., Mack, C.M. & Diaz, S. (2000) Consequences of changing biodiversity. *Nature* 405: 234-242.

Clark, C.J., Poulsen, J.R. & Parker, V.T. (2001) The role of arboreal seed dispersal groups on the seed rain of lowland tropical forest. *Biotropica* 33(4): 606-620.

Cruse-Sanders, J.M. & Hamrick, J.L. (2004) Genetic diversity in harvested and protected populations of wild American ginseng *Panax quinquefoliua* L. (Araliaceae). *American Journal of Botany* 91: 540-548.

Dangerfield, J.M. & Milner, A.E. (1996) Millipede faecal pellet production in selected natural and managed habitats of Southern Africa: implications for litter dynamics. *Biotropica* 28: 113-120.

Dirzo, R. & Raven, P.H. (2003) Global state of biodiversity and loss. *Annual Review of Environment & Resources* 28: 137-167.

Duellman, W.E. & Trueb, L. (eds) (1994) *Biology of amphibians*. McGraw-Hill, Baltimore, MD, USA, 670 pp.

Duffy, J.E. (2003) Biodiversity Loss, trophic skew, and ecosystem functioning. *Ecology Letters* 6(8): 680-687.

Eaton, J.M. & Lawrence, D. (2006) Woody debris stocks and fluxes during succession in a dry tropical forest. *Forest Ecology and Management* 232: 46-55.

Foley, J.A., DeFries, R., Asner, G.P., Barford, C., Bonan, G., Carpenter, S.R., Chapin, F.S., Coe, M.T., Daily, G.C., Gibbs, H.K., Helkowski, J.H., Holloway, T., Howard, E.A., Kucharik, C.J., Monfreda, C., Patz, J.A., Prentice, I.C., Ramankutty, N. & Snyder, P.K. (2005) Global consequences of land use. *Science* 309: 570-574.

Food and Agriculture Organization (1993) *Conservation of genetic resources in tropical forest management, Principles and concepts*. FAO, Rome, Italy.

Ganeshaiah, K.N. & Uma Shaanker, R. (1998) *Biligiri Ranganswamy temple wildlife sanctuary: Natural history, biodiversity and conservation*. Published by Ashoka Trust for Research in Ecology and the Environment and Vivekananda Girijana Kalyana Kendra (ATREE-VGKK), Bangalore, India, pp. 1-28.

Gibson, L., Lee, T.M., Koh, L.P., Brook, B.W., Gardner, T.A., Barlow, J., Peres, C.A., Bradshaw, C.J.A., Laurance, W.F., Lovejoy, T.E. & Sodhi, N.S. (2011) Primary forests are irreplaceable for sustaining tropical biodiversity. *Nature* 478: 378-383.

Gilpin, M.E. (1988) A comment on Quinn and Hastings: Extinction in subdivided habitats. *Conservation Biology* 2: 290-292.

Grinnell, J. (1917) Field tests and theories concerning distributional control. *American Naturalist* 51: 115-128.

Haddad, C.F.B. & Prado, C.P.A. (2005) Reproductive modes in frogs and their unexpected diversity in the Atlantic forest of Brazil. *Bioscience* 55(3): 207-217.

Hall, P., Walker, S. & Bawa, K.S. (1996) Effect of forest fragmentation on genetic diversity and mating system in a tropical tree, *Pithecellobium elegans*. *Conservation Biology* 10: 757-768.

Hendrix, H.D. (1994) Effects of population size on fertilization, seed production and seed predation in two prairie species. *Proceedings of the thirteenth North American Prairie conference: Spirit of land, our prairie legacy*. 13: 115-121.

Heschel, M. & Paige, K.N. (1995) Inbreeding depression, environmental stress and population size variation in scarlet gilia (*Ipomopsis aggregata*). *Conservation Biology* 9: 126-133.

Hodgkin, T. & Rao, R.V. (2002) People, plants and DNA: perspectives on the scientific and technological aspects of conserving and using plant genetic resources. In: Engels, J.M.M., Rao, V.R., Brown, A.H.D. & Jackson, M.T. (eds), *Managing plant genetic diversity*. CABI Publications, Wallingford, UK, pp. 469-480.

Hogbin, P.M., Peakall, R. & Sydes, M.A. (2000) Achieving practical outcomes from genetic studies of rare plants. *Australian Journal of Botany* 48: 375-382.

Honnay, O., Jacquemyn, H., Bossuyt, B. & Hermy, M. (2005) Forest fragmentation effects on patch occupancy and population viability of herbaceous plant species. *New Phytologist* 166: 723-736.

Howe, H.F. (1990) Seed dispersal by birds and mammals: implications for seedling demography. In: Bawa, K.S. & Hadley, M. (eds) *Reproductive ecology of tropical forest plants*. Parthenon Publishing Group, Paris, France, pp. 191-218.

Jordano, P. (1992) Fruits and frugivory. In: Fenner, M. (ed) Seeds: *The ecology of regeneration in plant communities*. CAB International, Wallingford, England, pp. 105-156.

Kennard, D.K., Gould, K., Putz, F.E., Fredericksen, T.S. & Morales, F. (2002) Effect of disturbance intensity on regeneration mechanisms in a tropical dry forest. *Forest Ecology and Management* 162: 197-208.

Lamb, D., Erskine, P.D. & Parrotta, J.A. (2005) Restoration of degraded tropical forest landscapes. *Science* 310: 1628-1632.

Lamont, B.B., Klinkhamer, P.G.L. & Witowaski, E.T.F. (1993) Population fragmentation may reduce fertility to zero in *Banksia goodii* – demonstration of the Allele effect. *Oecologia* 94: 446-450.

Laurance, W.F. (2000) Do edge effects occur over large spatial scales? *Trends in Ecology and Evolution* 15: 134-135.

Leigh, E.G., Davidar, P., Dick, C.W., Puyravaud, J.P., Terborgh, J., ter Steege, H. & Wright, S.J. (2004) Why do some tropical forests have so many species of trees? *Biotropica* 36: 447-473.

Lillesand, T.M. & Kiefer, R.W. (eds) (1987) *Remote sensing and image interpretation*. John Wiley & Sons, 768 pp.

Lowman MD (1988) Litterfall and leaf decay in three Australian rainforest formations. *Journal of Ecology* 76: 451-465.

Lussenhop, J. (1992) Mechanisms of microarthropod–microbial interactions in soil. *Advances in Ecological Research* 23: 1-33.

Lyngdoh, N., Hantode, S.S., Ramesha, B.T., Nageswara Rao, M., Ravikanth, G., Barve, N., Ganeshaiah, K.N. & Uma Shaanker, R. (2005) Rattan species richness and population genetic structure of *Calamus flagellum* in North-Eastern Himalayas, India. *Journal of Bamboo and Rattan* 4(3): 293-307.

McGuire, K.L., Henkel, T.W., Granzow de la Cerda, I., Villa, G., Edmund, F. & Andrew, C. (2008) Dual mycorrhizal colonization of forest-dominating tropical trees and the mycorrhizal status of non-dominant tree and liana species. *Mycorrhiza* 18(4): 217-222.

Menges, E.S. (1991) Seed germination percentage increases with population size in a fragmented prairie species. *Conservation Biology* 5: 158-164.

Moore, J.C. & Walter, D.E. (1988) Arthropod regulation of micro and mesobiota in below-ground detrital food webs. *Annual Review of Entomology* 33: 419-439.

Mulkey, S.S., Chazdon, R.L. & Smith, A.P. (eds) (1996). *Tropical forest plant ecophysiology*. Chapman & Hall, NY, 675 p.

Myers, N. (1994) Tropical forests and their species: going, going? In: Miller, G.T. (ed) *Living in the Environment*. International Thomson Publishing, Belmont, California, USA, pp 288-289.

Nageswara Rao, M. & Soneji, J.R. (2009) Threats to forest genetic resources and their conservation strategies. In: Aronoff, J.A. (ed) *Nature conservation: Global, environmental and economic issues.* Nova Science Publishers, Inc. New York, USA, pp 119-147.

Nageswara Rao, M. & Soneji, J.R. (2010a) Forest biodiversity: Issues and concerns, In: Nageswara Rao, M., Soneji, J.R. (eds) Tree and forest biodiversity. *Bioremediation, Biodiversity and Bioavailability* 4 (Special Issue 1): iv-v.

Nageswara Rao, M. & Soneji, J.R. (eds) (2010b) Tree and Forest Genetics. *Genes, Genomes and Genomics* 4 (Special Issue 1), 83 p.

Nageswara Rao, M. & Soneji, J.R. (eds) (2011) Tree Micropropagation and Tissue Culture. *Tree and Forestry Science and Biotechnology* 5 (Special Issue 1), 89p.

Nageswara Rao, M., Uma Shaanker, R. & Ganeshaiah, K.N. (2001) Mapping genetic diversity of sandal (*Santalum album* L.) in South India: Lessons for *in-situ* conservation of sandal genetic resources. In: Uma Shaanker, R., Ganeshaiah, K.N., Bawa, K.S. (eds) *Forest genetic resources: Status, threats and conservation strategies.* Oxford and IBH Publishing Company Private Limited, New Delhi, India, pp 49-67.

Nageswara Rao, M., Ganeshaiah, K.N. & Uma Shaanker, R. (2007) Assessing threats and mapping sandal (*Santalum album* L.) resources in peninsular India: Identification of genetic hot-spot for *in-situ* conservation. *Conservation Genetics* 8: 925-935.

Nageswara Rao, M., Sudarshana, P., Ganeshaiah, K.N. & Uma Shaanker, R. (2008a) Impacts of Human Disturbances on Medicinal NTFP species. *BioSpectrum Asia* (http://www.biospectrumasia.com/content/150908IND7092.asp)

Nageswara Rao, M., Sudarshana, P., Ganeshaiah, K.N., Uma Shaanker, R. & Soneji, J.R. (2008b), Indian Sandalwood Crisis. *Perfumer and Flavorist* 33(10):38-43.

Nageswara Rao, M., Ravikanth, G., Ganeshaiah, K.N. & Uma Shaanker, R. (2010) Role of protected area in conserving the population and genetic structure of economically important bamboo species, In: Nageswara Rao, M., Soneji, J.R. (eds) Tree and forest biodiversity. *Bioremediation, Biodiversity and Bioavailability* 4: 69-76.

Nageswara Rao, M., Soneji, J.R. & Sudarshana, P. (2011) Santalum. In: Kole C (ed) *Wealth of wild allied plant species, Volume 10: Allies of forest trees.* Springer – Heidelberg, Berlin, New York, Tokyo, pp 131-144.

Newbould, P.J. (1967) *Methods for estimating the primary production of forests.* IBP Handbook No. 2. Blackwell Scientific Publications, Oxford, 62 pp.

Ochego, H. (2003) Application of remote sensing in deforestation monitoring: A case study of the Aberdares (Kenya). Paper presented at the 2nd FIG regional conference, TS 11 - Management of water resources, Marrakech, Morocco.

Palace, M., Keller, M. & Silva, H. (2008) Necromass production: Studies in undisturbed and logged Amazon forests. *Ecological Applications* 18(4): 873-884.

Pimm, S.L. & Raven, P. (2000) Biodiversity: Extinction by numbers. *Nature* 403: 843-845.

Pimm, S.L. & Lawton, J.H. (1998) Planning for biodiversity. *Science* 279: 2068-2069.

Proctor, J., Anderson, J.M., Fogden, S.C.L. & Vallack, H.W. (1983) Ecological studies of four contrasting lowland rain forests in Gunung Mulu National Park, Sarawak: II. Litterfall, litter standing crop and preliminary observations on herbivory. *Journal of Ecology* 71: 261-283.

Rajanikanth, G., Nageswara Rao, M., Tambat, B., Uma Shaanker, R., Ganeshaiah, K.N. & Kushalappa, C.G. (2010) Are small forest fragments more heterogeneous among themselves than large fragments? In: Nageswara-Rao, M., Soneji, J.R. (eds) Tree and Forest Biodiversity. *Bioremediation, Biodiversity and Bioavailability* 4: 42-46.

Rajora, O.P. & Mosseler, A. (2001) Challenges and opportunities for conservation of forest genetic resources. *Euphytica* 118: 197-212.

Ramesha, B.T., Ravikanth, G., Nageswara Rao, M., Ganeshaiah, K.N. & Uma Shaanker, R. (2007) Genetic structure of rattan, *Calamus thwaitesii* in core, buffer and peripheral regions of three protected areas at central Western Ghats, India: Do protected areas serve as refugia for genetic resources of economically important plants? *Journal of Genetics* 86: 9-18.

Ravikanth, G., Nageswara Rao, M., Ganeshaiah, K.N. & Uma Shaanker, R. (2009) Impacts of harvesting on genetic diversity of NTFP species: Implications for conservation. In: Uma Shaanker, R., Joseph, G.C., Hiremath, A.J. (eds.) *Management, utilization, and conservation of non-timber forest products in the South Asia region.* Universities Press, Bangalore, India, pp 53-63.

Reddy, C.S. & Pattanaik, C. (2009) An assessment of floristic diversity of Gandhamardan hill range, Orissa, India. *Bangladesh Journal of Plant Taxonomy* 16(1): 29-36.

Risser, P.G. (1996) A new framework for prairie conservation. In: Samson FB, Knopf FL (eds) *Prairie Conservation: preserving North America's most endangered ecosystem.* Island press, Washington, D.C., USA, pp. 261-274.

Sapkota, I.P., Tigabu, M. & Oden, P.C. (2010) Changes in tree species diversity and dominance across a disturbance gradient in Nepalese Sal (*Shorea robusta* Gaertn. f.) forests. *Journal of Forestry Research* 21(1): 25-32.

Sayer, E.J., Powers, J.S. & Tanner, E.V.J. (2007) Increased litterfall in tropical forests boosts the transfer of soil CO_2 to the atmosphere. *PLoS ONE* 2(12): e1299. doi:10.1371/journal.pone.0001299

Sayer, E.J., Tanner, E.V.J. & Lacey, A.L. (2006) Effects of litter manipulation on early-stage decomposition and meso-arthropod abundance in a tropical moist forest. *Forest Ecology and Management* 229: 285-293.

Schnitzer, S.A., Mascaro, J. & Carson, W.P. (2008) Treefall gaps and the maintenance of species diversity in tropical forests. In: Carson, W.P., Schnitzer, S.A. (eds) *Tropical forest community ecology.* Blackwell Publishing, Oxford, pp 196-209.

Shands, H.L. (1991) Complementarity of *in situ* and *ex situ* germplasm conservation from the standpoint of the future user. *Israel Journal of Botany* 40(5-6): 521-528.

Sharma, A. (1996) Biodiversity, inventory, monitoring and conservation genetical aspects. In: *Conserving bio-diversity for sustainable development.* Published by Indian National Science Academy, New Delhi, India, pp. 27-33.

Sudarshana, P., Nageswara Rao, M., Ganeshaiah, K.N. & Uma Shaanker, R. (2001) Genetic diversity of *Phyllanthus emblica* in tropical forests of South India: Impact of anthropogenic pressures. *Journal of Tropical Forest Science* 13(2): 297-310.

Tejaswi, G. (2007) Manual on deforestation, degradation, and fragmentation using remote sensing and GIS: Strengthening monitoring, assessment and reporting on sustainable forest management in Asia (GCP/INT/988/JPN), Forestry Department, Food and Agriculture Organization of the United Nations, 49 pp

Theilade, I., Sekeli, P.M., Hald, S. & Graudal, L. (2001) Conservation plan for genetic resources of Zambezi teak (*Baikiaea plurijuga*) in Zambia Danida Forest Seed Center. DFSC Case study 2. Humlebaek, Denmark, pp. 1-32.

Thomas, S.C. & Baltzer, J.L. (2002) Tropical Forests. In: *Encyclopedia of life sciences.* Macmillan Reference Ltd, London, UK, Nature Publishing Group (http://www.els.net/), pp 1-8.

Tilman, D., May, R.M., Lehman, C.L. & Nowak, M.A. (1994) Habitat destruction and the extinction debt. *Nature* 371: 65-66.

Tilman, D. (2000) Causes, consequences and ethics of biodiversity. *Nature* 405: 208-210.

Uma Shaanker, R., Ganeshaiah, K.N. & Nageswara Rao, M. (2001a) Conservation of genetic resources of *Triphala* in South India: Identifications of hot-spots for *in-situ* conservation. In: *Medicinal plants, a global heritage*, IDRC and CRDI, pp 115-128.

Uma Shaanker, R., Ganeshaiah, K.N. & Nageswara Rao, M. (2001b) Genetic diversity of medicinal plant species in deciduous forests of India: Impacts of harvesting and other anthropogenic pressures. *Journal of Plant Biology* 28: 91-97.

Uma Shaanker, R., Ganeshaiah, K.N., Nageswara Rao, M. & Ravikanth, G. (2001c) Forest gene banks – a new integrated approach for the conservation of forest tree genetic resources. In: Engels, J.M.M., Rao, V.R., Brown, A.H.D. & Jackson, M.T. (eds), *Managing plant genetic resources*. Wallingford, UK, CABI Publications, pp. 229-235.

Uma Shaanker, R., Ganeshaiah, K.N., Nageswara Rao, M. & Ravikanth, G. (2003) Genetic diversity of NTFP species: Issues and implications. In: Hiremath, A.J., Joseph, G.C. & Uma Shaanker, R. (eds) *Proceedings of International workshop on 'Policies, management, utilization and conservation of non-timber forest products in the South Asia region'*, Ashoka Trust for Research in Ecology and Environment (ATREE) and Forest Research Support Programme for Asia and the Pacific, FAO, Bangkok, pp 40-44.

Uma Shaanker, R., Ganeshaiah, K.N., Nageswara Rao, M. & Aravind, N.A. (2004) Ecological consequences of forest use - from genes to ecosystem: a case study in the Biligiri Ranganswamy Temple Wildlife Sanctuary, South India, *Conservation and Society* 2: 347-363.

Vasconcelos, H.L. & Lawrence, W.F. (2005) Influence of habitat, litter type, and soil invertebrates on leaf-litter decomposition in a fragmented Amazonian landscape. *Oecologia* 144: 456-462.

Visser, S. (1985) Role of the soil invertebrates in determining the composition of soil microbial communities. In: Atkinson D, Fitter AH, Read DJ, Usher MB (eds) *Ecological interactions in soil*. British Ecological Society's Special Publication Service, Blackwell, Oxford, 4: 297-317.

Wang, C., Xing, X. & Han, X. (2004) Advances in study of factors affecting soil N mineralization in grassland ecosystems. *Ying Yong Sheng Tai Xue Bao (The Journal of Applied Ecology)* 15 (11): 2184-2188 (in Chinese).

Wells, K.D. (2007) The ecology and behavior of amphibians. The University of Chicago Press, Chicago, USA, 1148 pp.

Wright, S.J. & Muller-Landau, H.C. (2006) The future of tropical forest species. *Biotropica* 38: 287-301.

Part 2

Tropical Forest Structure, Synergy, Synthesis

2

Comparing Litterfall and Standing Vegetation: Assessing the Footprint of Litterfall Traps

Marcela Zalamea, Grizelle González and William A. Gould
International Institute of Tropical Forestry, USDA Forest Service,
Puerto Rico

1. Introduction

Litterfall traps could preferentially represent certain kinds of leaf litter. Several factors may cause bias while sampling litterfall leading to over- or under-representation of the species present in the surrounding vegetation. For example, species standing precisely above litterfall traps, having big and wide crowns, and/or with high leaf fall rate may be over-represented in litterfall samples. Additionally, species standing upslope or in the windward side of litterfall traps may be more likely to be collected in litterfall traps (Staelens et al., 2003). Conversely, species with big and/or heavy leaves or fronds such as palms or species from the *Cecropia* and *Heliconia* genera may be under-represented in litterfall traps (Clark et al., 2001). However, the few studies dealing with patterns of litterfall dispersal and collection have found contradictory results. For example, in Australian rainforests Lowman (1988) found that collected litterfall was not necessarily biased toward leaves coming from trees located precisely above traps. Similarly, in a dry forest in Costa Rica, Burnham (1997) found a low spatial correspondence between location of source stems and litterfall samples. In contrast, for a temperate mixed forest in northeastern Japan, Hirabuki (1991) found that estimated patterns of litterfall spatial distribution corresponded to the distribution of stems in the studied plot. In this chapter we report results from a study that takes advantage of an ongoing experiment in the Luquillo Experimental Forest, Puerto Rico, to examine the correspondence between litterfall samples and standing vegetation. Such correspondence was analyzed at three different spatial scales defined by the sampling units already in place: forest stand (10^6 m²), sampling blocks (4×10^4 m²), and plots (4×10^2 m²). Our first objective was to examine which factors, in addition to relative abundance of species in the vegetation, could affect the relative abundance of species in litterfall samples. Specifically, we evaluated the effect of tree size (measured as height and crown area), leaf size (measured as leaf area), and distance to litter traps using a stepwise regression procedure. We hypothesized that bigger trees (i.e., having high height and crown area) would produce more leaf litter and therefore would tend to occur more abundantly in litterfall samples; while trees with relatively big leaves would be in general under-estimated in litterfall samples because traps would fail to catch those leaves. Finally, if traps were capturing leaves from trees standing precisely above, then trees being closer to litter traps would tend to present higher relative abundances in litterfall samples. Additionally, we analyzed the similarity between litterfall and particular sub-sets of the whole vegetation community. Sub-sets were defined by tree height, crown area, and distance to traps, such that if litter traps were preferentially collecting leaves from any particular sub-set of the vegetation (e.g., bias toward either canopy or understory trees, wide-crowned trees, or trees located closer to traps), those sub-sets should bear a higher compositional similarity with the litterfall samples than the whole vegetation community.

The particular experimental set up used in this study (cf., Fig. 1), allowed us also to ask if litter traps located in the center of vegetation plots (i.e., surrounding plots. See Fig. 1) provided more representative samples of the surrounding vegetation than traps located adjacent to vegetation plots (i.e., adjacent plots. See Fig. 1). To address this second objective, we compared the composition and relative abundance of species collected in litter traps with the same parameters of the vegetation from the surrounding and adjacent plots (Fig. 1), using similarity indexes and parametric and non-parametric correlations. We hypothesized that if litter traps were collecting litterfall coming from all directions with the same likelihood, a higher similarity between litterfall samples and vegetation would be found for surrounding than for adjacent plots, both for the scale of the forest as for the scale of individual plots and for particular species.

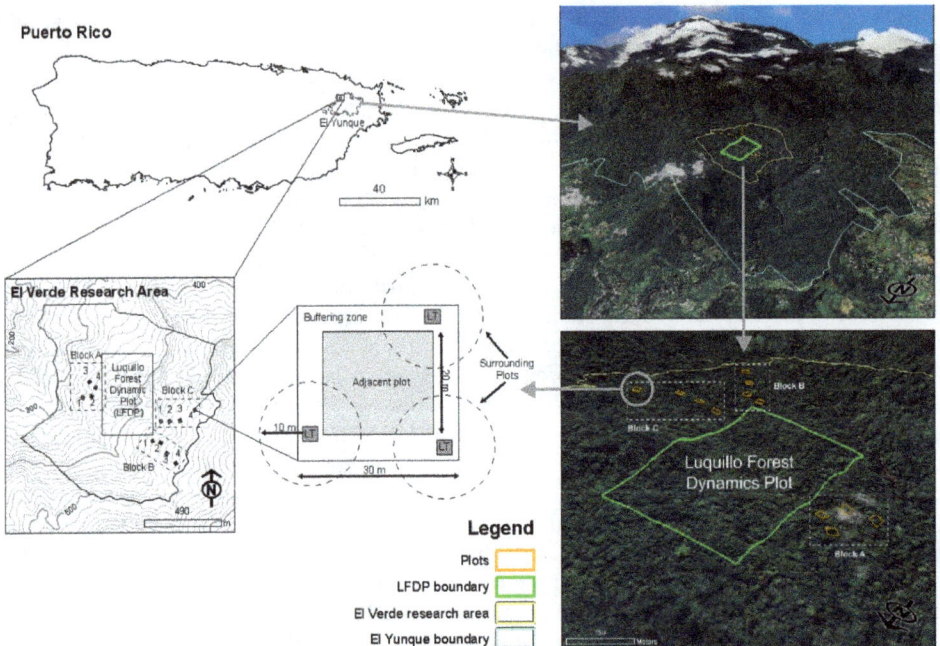

Fig. 1. Location of blocks, adjacent plots (square plots, numbered 1-4 within each block), surrounding plots (circular plots), and litter traps (LT) in El Verde research area within the Luquillo Experimental Forest. Inferred area covered by each block is 40,000 m² (broken lines). The complete study area covers around 10^6 m². The 16-ha Luquillo Forest Dynamic Plot (LFDP) is showed as reference; for more information about the LFDP please see Thompson et al. (2002)

Finally, our third objective was to gain insights for the scaling of litterfall data from the level of sampling plots up to the level of the forest stand. We addressed this by comparing the similarity between vegetation and litterfall across the three different scales mentioned before (i.e., plots, blocks and the forest stand; cf. Fig. 1) using similarity indexes, correlations, multivariate ordinations, and Mantel tests. An important aspect when examining the correspondence between litterfall and vegetation across different spatial scales is related to whether litter traps are capturing leaves from a wide range or only from the near vicinity around traps. On one hand, considering the potential far-ranged and random patterns of leaf dispersal (Jonard et al., 2006), a high compositional similarity between litterfall and vegetation at the scale of the forest type together with a low similarity at the smaller scales of sampling units might be expected. On the other hand, if litter traps are collecting leaf litter mainly from the vegetation in the near vicinity (for example, 10 m around traps), a high similarity between litterfall and vegetation at the scale of sampling units should be encountered as well. Particularly, the following outcomes could be expected: 1) high correlation between litterfall and vegetation dissimilarity matrices calculated for the smallest sampling units (i.e., plots), namely, pairs of plots with high dissimilarity in their vegetation should be also highly dissimilar in their litterfall; 2) litterfall and vegetation samples from the same plots should cluster together in an ordination space accurately representing compositional distances among sampling units; and 3) strong correlation between similarity among pairs of litterfall samples and the physical distance separating those samples (*i.e.*, distance among plots), namely, the more distant the plots were located, the higher the dissimilarity between them would be.

Litterfall collection using litter traps has become a ubiquitous method in terrestrial ecology. Thus it is important to understand the relevant variables behind the method and the implications of its limitations. We believe our findings will prove instrumental for the improvement of methods in terrestrial and forest ecology especially in the tropics were the high species diversity and structural complexity of forests impose tough challenges to the study of forest structure and dynamics.

2. Methods

2.1 Study site

The study was carried out in a subtropical wet forest in northeastern Puerto Rico (18o20′N, 65o49′W) in the Luquillo Experimental Forest. Mean monthly temperature is 23.03 oC and mean annual rainfall is 3592.3 mm (Zalamea & González, 2008). Soils are a complex of well- and poorly-drained ultisols and oxisols (Ruan et al., 2004). The forest type studied is dominated by *Dacryodes excelsa* Vahl., *Buchenavia tetraphila* (Vahl.) Eichl., *Homalium racemosum* Jacq., *Guarea guidonia* (L.) Sleumer, *Sloanea berteriana* Choisy, and *Prestoea montana* (Graham) Nicholson (Thompson et al., 2002). Mean canopy height is 21 m with some emergent trees reaching up to 30 m (Brokaw & Grear, 1991).

2.2 Sampling design

We followed the experimental design of a larger ongoing study in the Luquillo Long-Term Ecological Research program (LUQ- LTER) in Puerto Rico (See Richardson et al., 2010 for a description of the Canopy Trimming Experiment, CTE). Experimental layout of this experiment consists of three blocks (labeled A, B, and C) representing an area of around 4×10^4 m^2 each (Fig. 1). Block A is located between 340-360 m on a slight SW-facing slope;

block B is located between 450-485 m on a slight W-SW facing slope, while block C is at 435-480 m on a slight W-facing slope. Each block contains four 20 x 20 m square plots distanced at least 30 m from each other. In the buffering zone (cf. Fig. 1), three litter traps of 3 m² (1.73 m side length) were randomly installed adjacent to each square plot (around 1.6 m from plots border) at 1.3 m from ground level. Over this initial experimental set up, we installed 10 m radius circular plots around each litter trap (Fig. 1). Hereafter, we will refer to square plots as "adjacent" plots and to circular as "surrounding" plots to emphasize the respectively "lateral" and "central" position of litter traps relative to the two kinds of plots. Given the spatial distribution of blocks (cf. Fig.1), we assumed that data resulting from pooling all plots and blocks was a representative sample of the forest stand and corresponded to an area of 10^6 m².

2.3 Litterfall collection and vegetation inventories

Litter was collected every two weeks from November 2002 to November 2003. These samples served as base line data for planning a decomposition experiment as part of the CTE. The CTE was designed to experimentally disentangle the effects of canopy opening vs. debris deposition resulting from hurricane disturbance on organismal and ecosystem responses in a subtropical wet forest. Leaves were picked up from other litterfall components (such as reproductive parts, wood, and miscellaneous) and sorted out to species following Acevedo-Rodríguez (2003), and Little et al. (1974). Species belonging to the same genera and having similar leave morphology were pooled for data analysis. That was the case for *Miconia tetrandra* and *M. prasina* (pooled as *Miconia* spp.) and *Myrcia fallax*, and *M. leptoclada* (pooled as *Myrcia* spp.). Samples from all three traps around each square plot were pooled together, air-dried and weighed. Thus, the minimal sampling unit for litterfall was the plot. Species relative abundances in litterfall were calculated as % of annual litterfall per plot. Data for plots were then pooled to calculate % contribution of each species to total annual litterfall for each block and for the whole forest (i.e., after pooling all blocks).

Vegetation data for the adjacent plots was obtained from the Luquillo LTER web site (http://luq.lternet.edu, lterdb144) and corresponds to a vegetation survey carried out in April 2003 –as part of the CTE– in which all stems greater than 1 cm diameter at 1.3 m height (DBH) were tagged, identified to species and measured for DBH and height (see details about methods in http://luq.lternet.edu). In the surrounding plots, we carried out a vegetation inventory for all stems greater than 1 cm of DBH, for which we recorded: tree species, DBH, height, crown relative position (as canopy, sub-canopy, and understory), crown area, and distance of stem center to litter trap. Relative abundances were calculated as importance values: IV = (Relative density + Relative dominance)/2, where relative density is the % of total individuals per species per plot and relative dominance is the % of the total basal area per species per plot. We chose IV instead of just basal area to avoid big but non-numerous species to appear over-represented in the dataset (which in fact was the case for species such as *Ormosia krugii*). All the measured individuals in the surrounding plots were classified into three height, crown area, and distance classes. Distance and crown area classes were chosen arbitrarily to ensure that each class included roughly the same number of stems. Height classes, however, were chosen on the basis of the vertical structure of the forest (as described in Brokaw & Grear, 1991) and therefore the number of stems included in each class was not even. Height classes were: >10 m (447 stems), 10-5 m (979 stems), ≤5 m (529 stems), roughly corresponding to the crown relative position categories mentioned before (canopy, sub-canopy, and understory) which were obtained by visual examination of each tree in relation to neighboring trees. Crown area classes were: 0-1 m^2 (626 stems), 1-6.25 m^2 (788 stems), and >6.25 m^2 (541 stems). Distance classes were: 0-5.5 m (623 stems), 5.5-8 m (675 stems), and 8-10 m (657 stems).

2.4 Data analysis

A stepwise linear regression (SPSS 2002, version 11.5, Chicago, Illinois, USA) was carried out to determine the effect of: tree height, crown area, leaf size, and distance to litter trap, as the independent variables, over the % of annual litterfall per species, as the dependent variable. Values of height, crown area, and distance to trap per species were calculated as the corresponding importance values for each height, crown area and distance class. Leaf area (cm^2) for each species was calculated from digital images of herbaria and fresh specimens available in the following Internet sites: New York Botanical Garden Virtual Herbarium (http://sciweb.nybg.org/Science2/VirtualHerbarium.asp), Missouri Botanical Garden (http://mobot.mobot.org/W3T/Search/classicvast.html), Herbarium Berolinense (http://ww2.bgbm.org/herbarium/, Barcode B 10 0247501, ImageId 253751, accessed 28-May-08), La Selva Digital Flora (http://sura.ots.ac.cr/local/florula3/en/index.htm), and Biodiversity Information System for the Andes to Amazon Biodiversity Program –Atrium (http://atrium.andesamazon.org). Images were analyzed with Scion Image Software for Windows (Scion Corporation 2000-2001, version Alpha 4.0.3.2, Maryland, USA) by taking the area of 5 leaves from each herbaria specimen and calculating the average. Scale was set for each individual image before calculating leaf area. For some species images were not available. In those cases images from related species within the same genera and having similar leaf morphology were taken instead. Only species with >3 stems and IV > 0.1 were included in the regression analysis (44 out of the total pool of 91 species). Regression analysis was carried out in SPSS for Windows (SPSS 2002, version 11.5, Chicago, Illinois, USA) using arcsine-transformed data and standardized values (z-scores) in order to minimize the effect of collinearity among the independent variables included over the regression model (Rawlings et al., 1998, p. 370).

To determine how representative of the surrounding and adjacent vegetation were litterfall samples caught by central and lateral traps respectively, we used Spearman non-parametric correlation between the importance value of individual species and its correspondent % annual litterfall per plot (SPSS 2002, version 11.5, Chicago, Illinois, USA). Species with IV < 0.1, number of stems < 3, or frequency < 3 plots (either in litterfall, surrounding or adjacent plots) were excluded from this analysis. Out of 91 species in the combined data set of litterfall and vegetation plots, correlations were carried out for 41 species. Spearman's rank correlation was preferred over the parametric Pearson correlation because we were interested in accounting also for absences (i.e., zeros in the data set representing cases in which a given species occurred in the vegetation of a given plots but not in the litterfall or vice versa) and the Pearson correlation is known to be distorted by the presence of many zero values (Waite, 2000).

Differences in composition and relative abundance of species between vegetation and litterfall across spatial scales were explored by the Bray-Curtis index of dissimilarity, which is equivalent to the Sorensen index of similarity when subtracted from 1 (Waite, 2000). Similarities were calculated for the forest, blocks and plots scales by pooling data from plots into blocks and finally all blocks to get the complete forest type. Matrices for Mantel tests and multivariate ordinations were based on dissimilarities, while resemblances between litterfall and vegetation at different scales are hereafter presented as % of similarity for more clarity. Additionally, as another measure of similarity at the forest level we calculated the Pearson parametric correlation coefficient between % annual litterfall per species and the total abundance of each species in the forest (i.e., after pooling up data from all plots) for vegetation data obtained from the surrounding and adjacent plots separately. Data was arcsine-transformed before carrying out the correlations as recommended for relative values such as percentages (Waite, 2000).

Comparisons of litterfall and vegetation at the scale of blocks and plots were done using two methods: Mantel test and ordination analyses. Mantel test assesses correlation between two distance matrices (Lefkovitch, 1984). This method has been widely used in landscape ecology and population genetics where geographical distances are compared to genetic or ecological distances (e.g., Manel et al., 2003; Stehlik et al., 2001). In this case we compared: 1) the physical separation among plots (in meters) with the compositional dissimilarity among litterfall samples, and 2) the dissimilarity matrices for litterfall and vegetation (i.e., dissimilarity matrix of vegetation against vegetation vs. matrix for litter against litter in all plots). Physical distance matrix for plots was generated using a Geographic Distance Matrix Generator (Ersts P.J., version 1.2.2, American Museum of Natural History, Center for Biodiversity and Conservation. URL: http://biodiversityinformatics.amnh.org/open _source/gdmg. Accessed on 2008-4-3), based on State Plane Coordinates for Puerto Rico, which were obtained from the Luquillo LTER home page (http://luq.lternet.edu/data/, lterdb 144). Dissimilarity matrices and Mantel tests were done with XLSTAT (Addinsoft 2008, version 3.01, New York, USA).

Two ordination analyses were performed: Principal Component Analysis (PCA) and Multi-dimensional Scaling (MDS). PCA was used to explore underlying factors segregating litterfall and vegetation samples. As an explorative tool, PCA is an appropriate method despite no environmental variables were measured (Vervaet et al., 2002). MDS was used to visualize similarities among vegetation and litterfall samples plotted together and to evaluate if litter traps were preferentially collecting leaves from any of the height, crown or distance to trap classes. MDS ordinations were tested with Shepard diagrams using Kruskal's stress type 1 (Kruskal & Wish, 1978) to ensure that distances in the graph were proportional to calculated dissimilarities between plots. Ordination analyses were performed using the PC-ORD 5 software (McCune & Mefford, 2006).

3. Results

The final regression model obtained after the stepwise procedure explained 85% of the variability in the dataset ($r^2 = 0.854$) and included only two variables: tree height (height class >10 m: $r = 0.864$, $P < 0.001$) and crown area (crown area class >6.25 m^2: $r = 0.843$, $P < 0.001$), in contrast to 65% of variability accounted when using only the relative abundance of species in vegetation as the independent variable. Distance to traps was only marginally significant (distance class <5 m: $r = 0.626$, $P < 0.055$) meaning that a higher height and a wider crown were more important determining the presence and relative abundance of a species in litterfall samples than a closer location to litter traps. According to the regression model, some species were either over-estimated (e.g., *B. tetraphylla*, *H. rugosa*, and *H.*

racemosum) or under-estimated (*O. krugii* and *P. montana*) in litterfall samples, namely their abundance in litterfall was either higher or lower than expected from their relative abundance, height and crown size (Fig. 2).

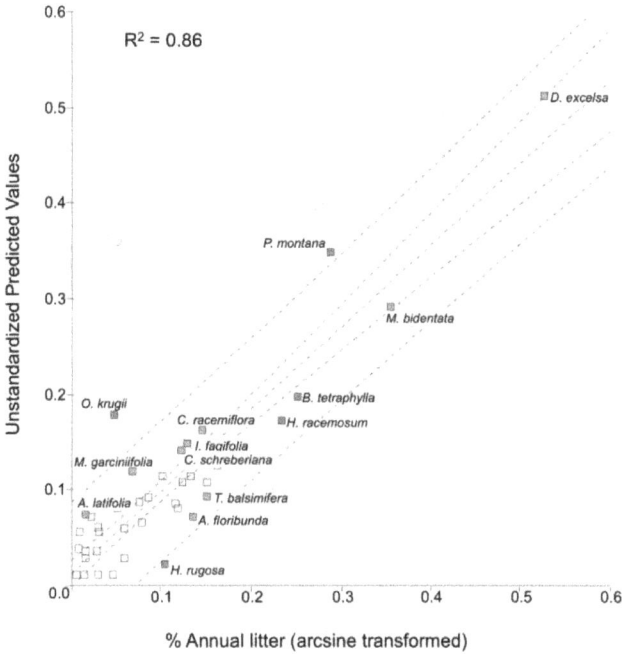

Fig. 2. Stepwise regression model for % annual litterfall (as the dependent variable) and importance values of vegetation for height class >10 m and crown area class >25 m^2 (as the independent variables). Unstandardized predicted values were obtained using SPSS 2002, version 11.5, Chicago, Illinois, USA. Refer to Table 1 for complete species names.

Although there was not any significant correlation between regression standardized residues (i.e., observed – predicted values, as a measure of over- or under- estimation) and leaf size, trees with small leaves tended to be over-estimated, while the two species that were under-estimated both have relatively big leaves (Fig. 3).

Comparisons of vegetation sub-sets and litterfall also suggested that distance to traps was not an important factor determining how representative of the vegetation were the litterfall samples, because sub-sets defined by distance to traps did not differ from the dissimilarity value calculated for the whole community (Fig. 4). In contrast, ordinations and similarity matrices for height, relative crown position, crown area, and distance to trap classes showed that vegetation sub-sets made up of the tallest trees (>10 m), occupying the canopy stratum, and having the biggest crowns (>6.25 m^2) presented the lowest dissimilarity between litterfall and vegetation (Fig. 4).

Similarity between litterfall and vegetation was 68% for surrounding and adjacent plots indicating that both types of plots provided equally representative samples of leaf litterfall at the scale of the forest. This result was corroborated by a high and positive correlation coefficient between vegetation relative abundance in both types of plots and % annual litterfall for the forest stand ($r_{Pearson}$ were 0.8 and 0.7 for surrounding and adjacent plots respectively; $P<0.0001$ in both cases). In contrast, correlations between % annual litterfall and relative abundance of vegetation species at the scale of plots showed that central traps provided better representative samples of leaf litter than lateral traps (Table 1). Percentage of annual litterfall for 20 out of the 41 species analyzed was positively correlated with relative abundance of vegetation species in surrounding plots, while only 6 species were correlated with vegetation in adjacent plots. For some species (*Cyathea arborea, Cyrilla racemiflora, Hirtella rugosa, Laetia procera, Manilkara bidentata, Matayba domingensis, Micropholis garciniifolia, Sapium laurocerasus, Tabebuia heterophylla,* and *Tetragastris balsamifera*) correlation coefficients were very high –especially between litterfall and vegetation in the surrounding plots. This might imply that leaf litter from these species has a relatively low range of horizontal mobility. However, correlation strength (measured as the Spearman coefficient magnitude) was not correlated with the species relative abundance, average height, average crown area, or leaf size.

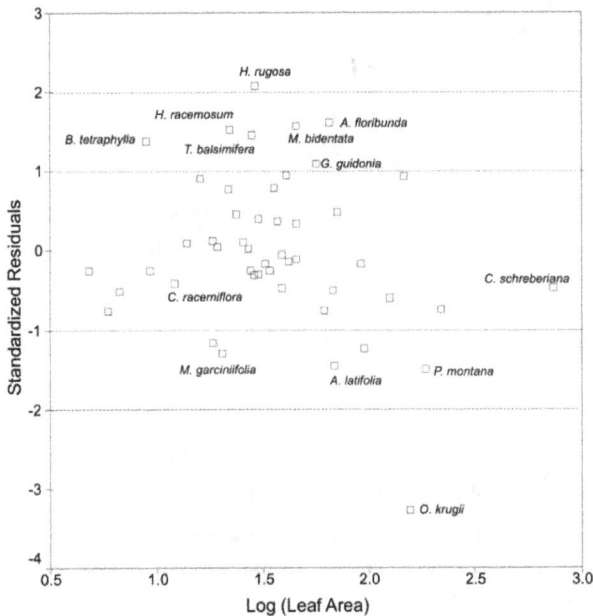

Fig. 3. Standardized regression residues as a function of log of leaf size. Analysis performed using SPSS 2002, version 11.5, Chicago, Illinois, USA. One and two standard deviations from the zero mean are indicated with dotted lines. Note that species with relatively big leaves such as *P. montana* and *O. krugii* are located in the lower right area of the graph, while species with relatively small leaves such as *H. rugosa* and *B. tetraphylla* are located in the upper left area.

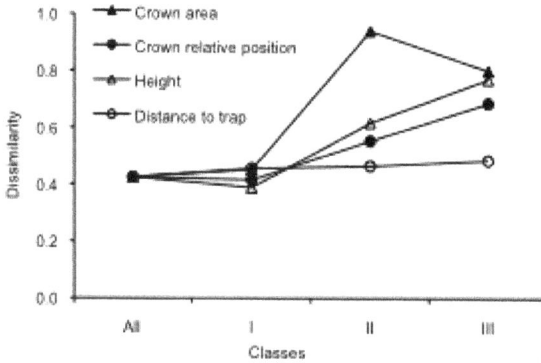

Fig. 4. Average dissimilarity between vegetation and litterfall from the same plot for the complete vegetation community and each of the three sub-sets corresponding to height (class I: >10m, II: 5-10 m, III: <5 m), crown relative position (I: canopy, II: sub-canopy, III: understory), crown area (I: >6.25 m², II: 1-6.25 m², III: <1 m²), and distance to traps (I: 0-5.5 m, II: 5.5-8 m, III: 8-10 m) classes. Differences between the dissimilarity for the whole community (all classes) and the sub-communities were significant for all factors except for distance to traps.

Species	$r_{Spearman}$ (*P*-value)			
	Litterfall vs. Surrounding vegetation		Litterfall vs. Adjacent vegetation	
Alchornea latifolia	0.391	(0.206)	0.389	(0.209)
Alchorneopsis floribunda	0.47	(0.123)	-0.213	(0.507)
Ardisia glauciflora	Not present in litterfall			
Buchenavia tetraphylla	0.599	(0.043)*	0.481	(0.114)
Byrsonima spicata	0.612	(0.035)*	0.393	(0.206)
Byrsonima wadsworthii	0.631	(0.028)*	0.631	(0.028)*
Casearia arborea	0.656	(0.020)*	-0.194	(0.546)
Casearia sylvestris	0.372	(0.234)	0.549	(0.065)
Cecropia schreberiana	0.688	(0.013)*	0.135	(0.676)
Coccoloba swartzii	0.264	(0.407)	-0.134	(0.677)
Cordia borinquensis	0.324	(0.304)	0.345	(0.272)
Cordia sulcata	0.054	(0.865)	0.322	(0.302)
Croton poecilanthus	0.492	(0.054)	-0.06	(0.828)
Cyathea arborea	1	(<0.001)***	Not present in plots	
Cyrilla racemiflora	0.825	(<0.001)***	0.778	(0.001)***
Dacryodes excelsa	0.21	(0.506)	0.524	(0.082)
Drypetes glauca	0.709	(0.003)**	0.372	(0.155)
Eugenia stahlii	0.448	(0.082)	-0.201	(0.455)
Faramea occidentalis	Not present in litterfall			
Garcinia portoricensis	-0.134	(0.677)	-0.243	(0.446)

Guarea glabra	-0.045	(0.895)	-0.28	(0.379)
Guarea guidonia	0.638	(0.009)*	0.49	(0.055)
Guettarda valenzuelana	0.326	(0.301)	-0.91	(0.779)
Hirtella rugosa	0.73	(0.002)**	0.799	(<0.001)***
Homalium racemosum	0.452	(0.08)	0.34	(0.195)
Inga fagifolia	0.285	(0.281)	-0.381	(0.146)
Laetia procera	0.844	(<0.001)***	0.183	(0.493)
Manilkara bidentata	0.791	(<0.001)***	0.7	(0.003)**
Matayba domingensis	0.854	(<0.001)***	0.776	(0.003)**
Miconia spp (*M. prasina* and *M. tetrandra*)	0.707	(0.01)*	-0.029	(0.930)
Micropholis garciniifolia	0.851	(<0.001)***	0.991	(<0.001)***
Myrcia spp (*M. fallax, M. splendens,* and *M. leptoclada*)	0.183	(0.569)	-0.118	(0.715)
Ocotea floribunda	-0.201	(0.530)	Not present in plots	
Ocotea leucoxylon	-0.102	(0.756)	-0.03	(0.934)
Ormosia krugii	0.789	(<0.002)**	0.24	(0.453)
Palicourea croceoides	Not present in litterfall			
Prestoea montana	0.403	(0.121)	0.344	(0.190)
Psychotria berteriana	-0.297	(0.349)	-0.087	(0.792)
Sapium laurocerasus	0.839	(<0.001)***	0.35	(0.182)
Schefflera morototoni	0.697	(0.004)**	0.021	(0.936)
Sloanea berteriana	0.059	(0.826)	-0.079	(0.771)
Tabebuia heterophylla	0.874	(<0.001)***	0.403	(0.121)
Tetragastris balsamifera	0.791	(<0.001)***	0.351	(0.181)
Trichilia pallida	0.42	(0.173)	-0.092	(0.780)

Table 1. Spearman's rank correlation ($r_{Spearman}$) between % annual litterfall and relative abundance of tree species in the surrounding and adjacent vegetation (n = 12 plots each). Degree of significance is indicated besides P-values as: * = $P \leq 0.05$, ** = $P \leq 0.005$, and *** = $P \leq 0.001$.

When the whole vegetation community was compared with litterfall across scales, there was a general trend of decreasing similarity from the scale of the forest to the scale of plots. When litterfall and vegetation from the same block were compared (i.e., A vs. A, B vs. B, and C vs. C), we found an average similarity of 72% (ranging from 63 to 82%); while among different blocks (i.e., A vs. B, A vs. C, and B vs. C) we found an average similarity of 56% (ranging from 42 to 68%). Litterfall and vegetation from the same plot had an average similarity of 58% (ranging from 34 to 70%). While, average similarity between litterfall and vegetation from different plots but within the same block was 45% (varying between 28 and 75%). Finally, average similarity among plots belonging to different block was 41% (ranging from 17 to 73%).

The relatively low similarity between litterfall and vegetation species composition at the scale of plots was also evident in other analyses: neither the Mantel test comparing vegetation and litterfall distance matrices, nor the one comparing matrices of physical against compositional distances were significant (standardized Mantel statistic = 0.23, P = 0.26 for vegetation vs. litterfall; Mantel statistic = 0.158, P = 0.202 for physical vs. compositional distances). According to the PCA (Fig. 5), the two main factors accounting for 28.7% of the variation among plots roughly corresponded to sample origin (i.e., litterfall vs. standing vegetation) and spatial distribution of vegetation (i.e., samples from blocks A, B, or C). It is interesting that block B appears as quite distinct from the other two blocks (check Fig. 1 for relative location of blocks over the study area). Finally, vegetation and litterfall samples from the same plots were not grouped together in the MDS ordination space. On the contrary, vegetation samples tended to cluster together while litterfall samples were scattered around (Fig. 5).

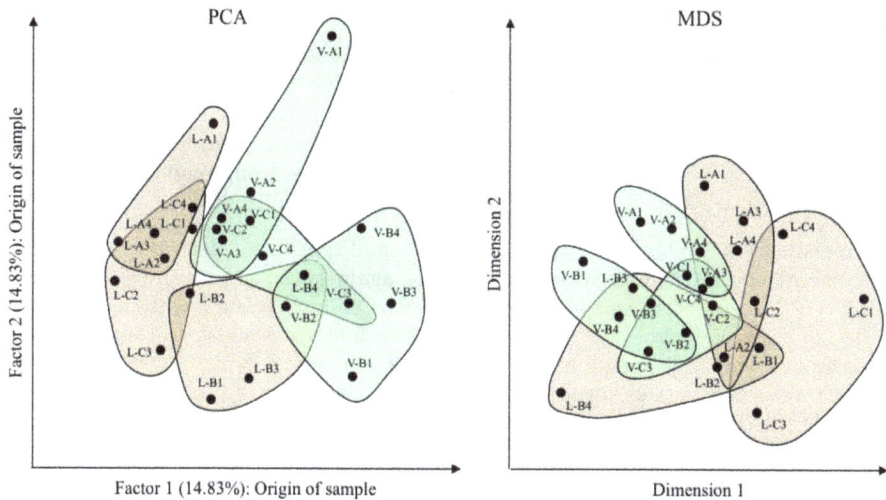

Fig. 5. PCA and MDS ordinations for vegetation in circular plots and litterfall collected in traps located at the center of plots. Codes refer to L = litterfall, V = vegetation, blocks (A, B, and C) and plots (1-4). Shading colors refer to vegetation (green) and litterfall (brown).

4. Discussion

Results from the regression analysis point to the significance of tree height and crown area as main factors determining litterfall composition and thus support ballistic models of leaf dispersal in which both wind and tree height are taken into account (e.g., Jonard et al., 2006). Ballistic models represent leaf dispersal as parabolic trajectories. This means that, even though distance is undoubtedly an important factor determining leaf dispersal, there might not be a simple inverse relation between leaf fall and distance from the source, as traditional models of leaf dispersal imply (e.g., Ferrari & Sugita, 1996). However, comparisons of litterfall and vegetation across scales indicated that patterns of leaf litter dispersal have an important random and wide-ranged character. In addition to the lack of correlation between distance matrices according to the Mantel test, the poor clustering of litterfall and vegetation samples in the MDS ordination space, and the separation of litterfall and vegetation samples along the first axis of the PCA (cf., Fig. 5), a case-based analysis of particular species provided further support for the thesis of a random and wide-ranged leaf dispersal in the forest studied. For example, the species *Clusia clusioides* and *Clusia rosea* were present in litterfall samples but not in any of the vegetation plots. These species of trees, however, have been recorded in the Luquillo Forest Dynamic Plot (cf., Fig. 1 for the location of this plot. See http://luq.lternet.edu/research/projects/forest_dynamics_description.html for detailed description of this plot) suggesting that litter traps could have collected leaves coming from much farther than 10 m.

Furthermore, the fact that height and crown size were the main factors determining composition of litterfall samples also suggests that small to medium size understory trees can be under-represented in litterfall samples. Such could have been the case for two species of small to medium size tree species –*Palicourea croceoides* and *Ixora ferrea*– which were absent in litterfall samples in spite of being well represented in both the surrounding and adjacent plots. Underestimation of certain components of vegetation in litterfall samples can have important implications for the estimation of forest and ecosystem parameters based on litterfall such as net primary productivity (Clark et al., 2001) and leaf area index (e.g., Vose et al., 1995). According to Clark et al. (2001) failure in the collection of big leaves in litterfall samples can lead to an up to 25% under-estimation of net primary productivity. Such under-estimation might be even greater if the failure to collect leaves from small understory trees is also accounted.

It is interesting that distance to traps was not a significant factor determining the relative abundance of species in litterfall samples. Here again specific cases are illustrative. For example, *Ardisia glauciflora* was absent in litterfall in spite of being mostly located close to the litterfall traps (the highest IV was recorded at distance class <5 m). In contrast, leaf litter of *B. tetraphylla* was higher than expected from the regression model given the height and crown size recorded for this species, despite the fact that most of the trees were located at distance class 8-10 m. These observations agree with previous studies in which proximity of big trees to litter traps was no warranty for catching leaf litter in the closest traps (e.g., Burnham, 1997; Lowman, 1988). Thus, if proximity to litter traps does not solve the problem of under-representation of small trees, it might be the case that in addition to a higher number of litter traps, litter traps should be placed at lower heights. Litter traps are commonly placed at 1 m above ground level (e.g., John, 1973; Kitayama & Aiba, 2002; Lam & Dudgeon, 1985; Martinez-Yrizar & Sarukhan, 1990; Newbould, 1970), although studies using traps at 0.4-0.8 m height (e.g., Kavvadias et al., 2001; Lowman, 1988; Rai & Proctor, 1986) and at 1.2-1.5 m height (e.g., Hirabuki, 1991; Hughes, 1971; Ukonmaanaho & Starr, 2001) are also quite common. However, there is rarely any explicit indication about the reason why a particular height is chosen. Results from this study suggest that at least the vertical structure of the forest should be considered to determine the optimal height at which traps are to be placed. We recommend a litterfall trap height lower than 1.3 m (the one used in this study), yet further studies are needed to establish exact canopy/litter trap height parametrization.

The comparison between litterfall samples and vegetation from the surrounding and adjacent plots proved to be informative for the experimental design of future studies employing litterfall traps. According to our results, studies in which the species composition of litterfall samples is a relevant parameter (e.g., decomposition and litterfall dynamics studies) should use litterfall traps located in the center of the studied plots, whereas studies focusing on general characterization of primary productivity at scales higher than 10^4 m^2 can use either central or lateral traps.

Even though we did not find a significant effect of leaf size over leaf litter samples composition, regression analysis suggested a trend in which small-leave trees were over-estimated whereas big-leave trees were under-estimated (cf., Fig. 3). We believe that further studies should investigate this trend by including either specific leaf area (SLA) or specific leaf weight (SLW) in the analyses. For example, *Alchornea latifolia* and *Micropholis garciniifolia* were slightly under-represented (cf. residues < -1 in Fig. 3) and both have relatively low SLA values (51 and 48.7 cm2/g respectively; data from Tanner & Kapos, 1982; Weaver & Murphy, 1990), suggesting that litter traps did less well at catching bigger and heavier leaves, and not simply bigger leaves. An approach based on SLA and/or SLW might also be promising in the study of specific patterns of horizontal mobility of leaf litter as were inferred in this study from the Spearman correlations. In addition, other factors such as wind pattern, animal distribution, and seasonal effects (Zalamea & González 2008) on litterfall patterns should be considered in future studies.

5. Conclusion

Higher height and a wider crown were more important determining relative abundance of species in litterfall samples. Trees with small leaves tended to be over-estimated. While distance to traps was not a significant factor determining the relative abundance of species in litterfall samples. The decreasing similarity between litterfall and vegetation from the scale of the forest stand down to the scale of sampling blocks and plots, plus the compositional differences among blocks inferred from the PCA indicates that sampling units at scales around 10^4 m^2 do not necessarily constitute proper replicates of units at bigger scales such as the forest stand (see for instance Williams et al., 2002 for the implications of scaling up highly spatially heterogeneous parameters), and that estimates at the forest stand scale should be calculated by pooling data coming from all the sampling units. Therefore, care must be taken when scaling up from small to intermediate sampling units such as plots and block respectively, due to the high variation of leaf litter dispersal at scales lower than 10^2 m^2.

6. Acknowledgment

This research was performed under grant DEB-0218039 from the National Science Foundation to the Institute of Tropical Ecosystem Studies, University of Puerto Rico (UPR), and the U.S. Department of Agriculture -Forest Service (USDA-FS), International Institute of Tropical Forestry (IITF) as part of the Long-Term Ecological Research Program in the Luquillo Experimental Forest. Additional support was provided by USDA-FS and UPR. We are thankful to the litterfall sorters: Yadira Ortíz, Omar Ortíz, Juan Ramírez, Marcos Rodríguez, Alberto Rodríguez, Vivian Vera, and Ivan Vicens. Aaron Shiels, Jill Thompson, and Eda Meléndez kindly provided unpublished and raw data on CTE plots vegetation and Luquillo Forest Dynamic Plot. El Verde technicians (especially John Bithorn) helped collecting litterfall samples. Maya Quiñones of the IITF GIS Remote Sensing Laboratory helped with figure 1. Finally, we are grateful to two anonymous reviewers whose comments helped to significantly improve the manuscript.

7. References

Acevedo-Rodríguez, P. (2003). *Bejucos y plantas trepadoras de Puerto Rico e Islas Vírgenes*, Smithsonian Institute, Washington, D.C.

Brokaw, N.V.L. & Grear, J.S. (1991). Forest Structure Before and After Hurricane Hugo at Three Elevations in the Luquillo Mountains, Puerto Rico. *Biotropica*, Vol. 23, No. 4A, pp. 386-392.

Burnham, R.J. (1997). Stand Characteristics and Leaf Litter Composition of a Dry Forest Hectare in Santa Rosa National Park, Costa Rica. *Biotropica*, Vol. 29, No. 4, pp. 384-395.

Clark, D.A., Brown, S., Kicklighter, D.W., Chambers, J.Q., Thomlinson, J.R. & Ni, J. (2001). Measuring Net Primary Production in Forests: Concepts and Field Methods. *Ecological Applications*, Vol. 11, No. 2, pp. 356-370.

Ersts,P.J.[Internet] Geographic Distance Matrix Generator (version 1.2.2). American Museum of Natural History, Center for Biodiversity and Conservation. Available from http://biodiversityinformatics.amnh.org/open_source/gdmg. Accessed on 2008-4-3.

Ferrari, J.B. & Sugita, S. (1996). A spatially explicit model of leaf litter fall in hemlock-hardwood forests. *Canadian Journal of Forest Research*, Vol. 26, No. 11, pp. 1905-1913.

Hirabuki, Y. (1991). Heterogeneous dispersal of tree litterfall corresponding with patchy canopy structure in a temperate mixed forest. *Plant Ecology (Vegetatio)*, Vol. 94, No. 1, pp. 69-79.

Hughes, M.K. (1971). Tree Biocontent, Net Production and Litter Fall in a Deciduous Woodland. *Oikos*, Vol. 22, No. 1, pp. 62-73.

John, D.M. (1973). Accumulation and Decay of Litter and Net Production of Forest in Tropical West Africa. *Oikos*, Vol. 24, No. 3, pp. 430-435.

Jonard, M., Andre, F. & Ponette, Q. (2006). Modeling leaf dispersal in mixed hardwood forests using a ballistic approach. *Ecology*, Vol. 87, No. 9, pp. 2306-2318.

Kavvadias, V.A., Alifragis, D., Tsiontsis, A., Brofas, G. & Stamatelos, G. (2001). Litterfall, litter accumulation and litter decomposition rates in four forest ecosystems in northern Greece. *Forest Ecology and Management*, Vol. 144, No. 1-3, pp. 113-127.

Kitayama, K. & Aiba, S.-I. (2002). Ecosystem Structure and Productivity of Tropical Rain Forests along Altitudinal Gradients with Contrasting Soil Phosphorus Pools on Mount Kinabalu, Borneo. *Journal of Ecology*, Vol. 90, No. 1, pp. 37-51.

Kruskal, J.B. & Wish, M. (1978). *Multidimensional Scaling*, SAGE publisher, University of California

Lam, P.K.S. & Dudgeon, D. (1985). Seasonal Effects on Litterfall in a Hong Kong Mixed Forest. *Journal of Tropical Ecology*, Vol. 1, No. 1, pp. 55-64.

Lefkovitch, L.P. (1984). A nonparametric method for comparing dissimilarity matrices, a general measure of biogeographical distance, and their application. *American Naturalist*, Vol. 123, No. 4, pp. 484-499.

Little, E.L., Woodbury, R.O. & Wadsworth, F.H. (1974). *Trees of Puerto Rico and the Virgin Islands*, USDA Forest Service, Washington, D.C.

Lowman, M. (1988). Litterfall and leaf decay in three Australian rainforest formations. *Journal of Ecology*, Vol. 76, No. 2, pp. 451-465.

Manel, S., Schwartz, M.K., Luikart, G. & Taberlet, P. (2003). Landscape genetics: combining landscape ecology and population genetics. *Trends in Ecology and Evolution*, Vol. 18, No. 4, pp. 189-197.

Martinez-Yrizar, A. & Sarukhan, J. (1990). Litterfall Patterns in a Tropical Deciduous Forest in Mexico Over a Five- Year Period. *Journal of Tropical Ecology*, Vol. 6, No. 4, pp. 433-444.

McCune, B. & Mefford, M.J. (2006). PC-ORD. Multivariate Analysis of Ecological Data. Version 5. MjM Software, Gleneden Beach, Oregon, USA

Newbould, P.J. (1970). *Methods for estimating the primary production of forests*, Blackwell, Oxford

Rai, S.N. & Proctor, J. (1986). Ecological Studies on Four Rainforests in Karnataka, India: II. Litterfall. *Journal of Ecology*, Vol. 74, No. 4, pp. 455-463.

Rawlings, J.O., Dickey, D.A. & Pantula, S.G. (1998). *Applied Regression Analysis: A Research Tool*, Springer-Verlag, Inc., New York

Richardson, B., Richardson, M., González, G., Shiels, A. & Srivastava, D. (2010). A Canopy Trimming Experiment in Puerto Rico: The Response of Litter Invertebrate Communities to Canopy Loss and Debris Deposition in a Tropical Forest Subject to Hurricanes. *Ecosystems*, Vol. 13, No. 2, pp. 286-301.

Ruan, H.H., Zou, X.M., Scatena, F.N. & Zimmerman, J.K. (2004). Asynchronous fluctuation of soil microbial biomass and plant litterfall in a tropical wet forest. *Plant and Soil*, Vol. 260, No. 1-2, pp. 147-154.

Staelens, J., Nachtergale, L., Luyssaert, S. & Lust, N. (2003). A model of wind-influenced leaf litterfall in a mixed hardwood forest. *Canadian Journal of Forest Research*, Vol. 33, No. 2, pp. 201-209.

Stehlik, I., Schneller, J.J. & Bachmann, K. (2001). Resistance or emigration: response of the high-alpine plant Eritrichium nanum (L.) Gaudin to the ice age within the Central Alps. *Molecular Ecology*, Vol. 10, No. 2, pp. 357-370.

Tanner, E.V.J. & Kapos, V. (1982). Leaf Structure of Jamaican Upper Montane Rain-Forest Trees. *Biotropica*, Vol. 14, No. 1, pp. 16-24.

Thompson, J., Brokaw, N., Zimmerman, J.K., Waide, R.B., Everham, E.M.I., Lodge, D.J., Taylor, C.M., García-Montiel, D. & Fluet, M. (2002). Land use history, environment, and tree composition in a tropical forest. *Ecological Applications*, Vol. 12, No. 5, pp. 1344-1363.

Ukonmaanaho, L. & Starr, M. (2001). The Importance of Leaching from Litter Collected in Litterfall Traps. *Environmental Monitoring and Assessment*, Vol. 66, No. 2, pp. 129-146.

Vervaet, H., Massart, B., Boeckx, P., Van Cleemput, O. & Hofman, G. (2002). Use of principal component analysis to assess factors controlling net N mineralization in deciduous and coniferous forest soils. *Biology and Fertility of Soils*, Vol. 36, No. 2, pp. 93-101.

Vose, J.M., Sullivan, N.H., Clinton, B.D. & Bolstad, P.V. (1995). Vertical leaf area distribution, light transmittance, and application of the Beer-Lambert Law in four mature hardwood stands in the southern Appalachians. *Canadian Journal of Forest Research*, Vol. 25, No. 6, pp. 1036-1043.

Waite, S. (2000). *Statistical Ecology in Practice. A Guide to Analysing Environmental and Ecological Field Data*, Prentice Hall, Pearson Education Limited, UK.

Weaver, P.L. & Murphy, P.G. (1990). Forest Structure and Productivity in Puerto Rico's Luquillo Mountains. *Biotropica*, Vol. 22, No. 1, pp. 69-82.

Williams, M., Shimabukuro, Y.E., Herbert, D.A., Lacruz, S.P., Renno, C. & Rastetter, E.B. (2002). Heterogeneity of Soils and Vegetation in an Eastern Amazonian Rain Forest: Implications for Scaling up Biomass and Production. *Ecosystems*, Vol. 5, No. pp. 692-704.

Zalamea, M. & González, G. (2008). Leaf Fall Phenology in a Subtropical Wet Forest in Puerto Rico: From Species to Community Patterns. *Biotropica*, Vol. 40, No. 3, pp. 295-304.

Quantifying Variation of Soil Arthropods Using Different Sampling Protocols: Is Diversity Affected?

María Fernanda Barberena-Arias[1], Grizelle González[2] and Elvira Cuevas[3]
[1]*Universidad del Turabo, School of Sciences and Technology, Gurabo,*
[2]*International Institute of Tropical Forestry/USDA Forest Service, Río Piedras,*
[3]*University of Puerto Rico, Department of Biology, San Juan,*
Puerto Rico

1. Introduction

In ecological studies, the use of different sampling methods for the same purpose influence data quality and thus the resulting conclusions (Coddington *et al.* 1996; Fisher 1999). For example, to collect arthropods from soil and litter samples a soil corer or a shovel may be used. Soil corers compact the soil (Meyer 1996) making difficult for organisms to leave the sample while shovels create a large disturbance (Longino *et al.* 2002) promoting mobile organisms to leave and reducing their apparent abundance in the sample. As a consequence the diversity of collected arthropods will vary between these two procedures, resulting in either an under- or overestimate of the diversity of the collected fauna (André *et al.* 2002). These different results will lead the researcher to infer different conclusions. Therefore it is essential to assess how different procedures affect the abundance, richness and species composition of the retrieved arthropods.

Arthropods are usually retrieved from soil/litter samples with Berlese-Tullgren funnels (Walter *et al.* 1987; Rohitha 1992; MacFadyen 1961; Bremner 1990; Lakly & Crossley 2000; MacFadyen 1953; Haarlov 1947). In these funnels, a source of heat (i.e. a light bulb) is placed above the sample, and a collecting vial filled with a killing solution (e.g. 70% ethanol) is placed below the sample. Light from the bulb has a double effect because light per se forces photophobic organisms to move away from the source, and light heats the sample. As the sample dries, a temperature and humidity gradient is created between the upper and lower surfaces of the sample (Haarlov 1947; Block 1966). As this gradient moves downwards, animals are forced down into the collecting liquid (Coleman *et al.* 2004). By increasing the temperature within the funnel, heat speeds drying (Coleman *et al.* 2004) but may also burn organisms before their collection and thus decreases estimates of their abundance (Walter *et al.* 1987). Alternatively, in remote field conditions, extractions without light are logistically more affordable and feasible, in which case the establishment of the gradient and the drying out of the sample depends on the room temperature in which the extractions are performed (Krell *et al.* 2005). Both, extractions with and without light, create different conditions within the sample, as a consequence, the use, or no use, of light during extractions, can result in different groups of arthropods being extracted, and thus a different set of data (Agosti *et al.* 2000).

The duration of arthropod extraction can also affect diversity estimates. Extraction periods reported in the literature vary from 2 d (Burgess *et al.* 1999), 3 d (Hasegawa 1997), to 4 d (Oliver & Beattie 1996; Bestelmeyer *et al.* 2000) and up to 7 d (Chen & Wise 1999; Walter *et al.* 1987). Long extraction periods are generally assumed to result in more complete extractions and higher abundance of the extracted fauna (Oliver & Beattie 1996) as organisms with low mobility require more time to exit the sample, but longer extraction periods may expose the samples to potential contamination with foreign organisms. On the other hand, to establish an adequate period of extraction, the environment of origin and the developmental stage should be taken into account (André *et al.* 2002). For example, organisms adapted to extreme environments, such as areas devoid of vegetation cover that have large temperature fluctuations, may require longer extraction periods than organisms adapted to less extreme environments. Furthermore, organisms from the same habitat but occurring in the dry or wet seasons (Oliver & Beattie 1996) or different developmental stages (Søvik & Leinaas 2002) may differ in the extraction period required to retrieve them. As a consequence, in order to collect reliable data, it is necessary to assess how an adequate duration of the extraction varies among environments of origin and developmental stages of the focal organism.

The present study was carried out in the Caribbean island of Puerto Rico, specifically in tropical dry and wet forests with contrasting environmental conditions (Ewel & Whitmore 1973). The objective of this study was to assess how the diversity of extracted arthropods was affected by variations in the collection and extraction methodologies, and by variations in the duration of the extraction. We present abundance, richness and composition of the collected fauna. The information presented here will provide researchers with data to simplify the logistics of arthropod sampling and extraction, and to better choose a specific procedure for a given focal organism in a given habitat.

2. Materials and methods

2.1 Study site

This study was carried out in north-eastern Puerto Rico in two forests of contrasting conditions. Samples from the litter and soil horizon (0-5 cm) were obtained in March 2003 from a wet forest site at the El Verde Field Station (Luquillo Experimental Forest, 18.33080, -65.82320, WGS 84), and from a dry forest site in the former Roosevelt Roads Military Base (Ceiba, 18.24800, -65.63290, WGS 84).

The wet forest site is located in the Luquillo Experimental Forest, where mean monthly temperature ranges from 23.5°C in January to 27°C in September (http://www.lternet.edu/sites/luq/fulldescription.php?site=LUQ), and total annual precipitation is 3524 mm yr^{-1} (García-Martinó *et al.* 1996) with a mild dry season from January to April (Schowalter & Ganio 1999). Soils are highly weathered; soil nutrients are 0.49% S, 0.35% N, 4.92% C, 0.30 P mg/g soil, and the C/N ratio is 14.2 (Gould *et al.* 2006); humus accumulation is low because there is rapid decomposition. Vegetation at the site is described as closed evergreen broad leaf forest that lays within the subtropical wet forest Holdridge life zone (Gould *et al.* 2006). The forest is dominated by *Dacryodes excelsa* and *Manilkara bidentata* (Schowalter & Ganio 1999; Gould *et al.* 2006).

The dry forest site is located within the former Roosevelt Roads Military Base, where mean monthly temperature is 27.5°C and annual precipitation is 1,262 mm yr-1 (Gould *et al.* 2006). It has a pronounced dry season that runs from November to April, and a wet season that usually runs from May to October (http://www.ceducapr.com/ceiba.htm). The soils are sandy or clayey with a developed organic matter (http://www.ceducapr.com/ceiba.htm). Soil nutrients are 0.06% S, 0.61% N, 6.34% C, 0.48 P mg/g soil, and the C/N ratio is 10.4 (Gould *et al.* 2006). This is a closed, mixed-evergreen deciduous, broad leaf forest that lays within the subtropical dry forest Holdridge life zone (Gould *et al.* 2006). This forest is dominated by *Bucida buceras* and *Guapira fragrans* (Gould *et al.* 2006).

In summary, these forests present contrasting conditions because, the wet forest has lower temperature and higher precipitation than the dry forest. In addition, the dry forest has a pronounced dry season while in the wet forest; the dry season is measured as number of days with no effective rain. The organic horizon is thin in the wet forest, and thick in the dry forest. As a consequence, the wet forest is warm and humid with thin litter and almost no humus, while the dry forest is hot and dry with deep litter and humus.

2.2 Data collection

In each forest, a 50 m x 50 m area was located, and within this area 40 litter samples were collected. Each sample was 100-cm^2 (10 cm x 10 cm), and was collected down to mineral soil. Litter depth was measured three times inside each of the 100-cm^2 areas. Inside the same 100-cm^2 area and after collecting the litter samples, two soil samples were collected: one using a soil corer (4.3 cm diameter and 5 cm height) and another one using shovels. For the shovel sampling, soil was collected with a shovel and served into a corer to assure that the soil volume in the shovelled sample was similar to that obtained with the soil corer. This sampling design resulted in 40 litter samples, 40 soil shovelled samples and 40 soil cored samples from each forest, giving a total of 120 samples in each forest type.

In the laboratory, the litter and soil shovelled samples were each placed in small Berlese-Tullgren funnels (Bioquip 2845) (10 cm height and 11 cm diameter) (Fig. 1A & 1B). The soil corer samples were placed in hand made funnels. For this, a wooden skeleton was built with basal holes covered with a mesh (Fig. 1C). Over each hole, a corer was placed and covered with a metallic funnel (11 cm x 5 cm) dia. The metallic funnels were then covered with a wooden ceiling (Fig. 1D). All funnels had the ceiling with an opening.

Samples were randomly assigned to two treatments in which extraction was done with or without light. For this, the 40 litter samples were split into two groups: 20 samples were extracted with light and 20 samples were extracted without light. The 40 soil shovelled samples and the 40 soil cored samples were similarly randomly assigned to one of these extraction treatments. When extraction was with light, a 20V-bulb was hanging through the funnel's opening, and was kept at maximum intensity during all the extraction period to control for the effect of changing light intensity during extractions. When extraction was without light, no bulb was placed over the funnel. All samples were located simultaneously in the same room where temperature and humidity were controlled.

Vials containing the killing solution (ethanol 70%) and collected arthropods were retrieved at 24, 48, 72, 144 and 168 hours after placement in the funnels. Thus each of the 120 samples

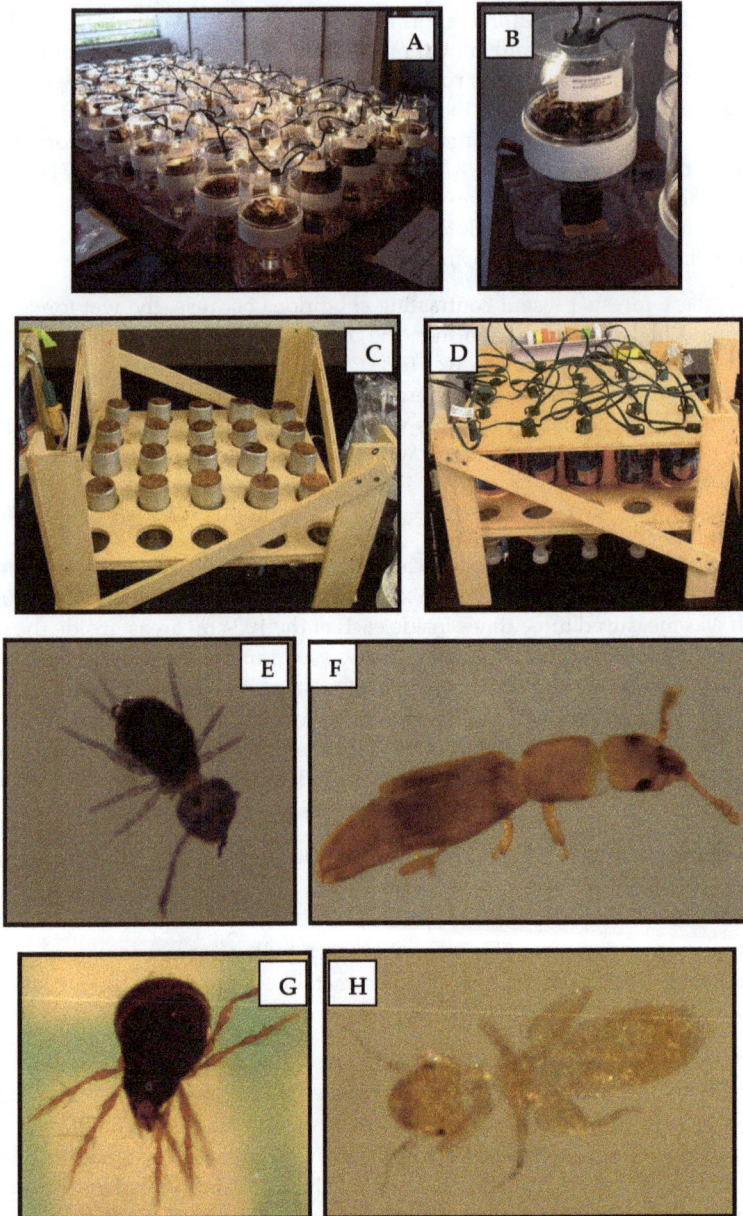

Fig. 1. A and B: Berlese-Tullgren funnels (Bioquip 2845) used for the litter and shovelled soil samples. (Photos A and B provided by M. F. Barberena-Arias, Universidad del Turabo). C and D: wooden skeleton used for the cored soil samples. E through H: examples of collected arthropods. E: Collembola Sminthurida, F: Coleoptera Corylophidae, G: Acari Oribatida, H: Psocoptera. (Photos C through H provided by G. González, Soil Ecology Program, IITF-US Forest Service).

per forest was retrieved at five sequential times giving a total of 600 samples per forest that were processed separately. For each sample, arthropod abundance was recorded, and arthropods were identified to the lowest category possible such as class, subclass, order or suborder, and classified as adult or immature (MacAlpine 1989; Triplehorn & Johnson 2004; Krantz 1978) (Fig. 1E through 1H). Collembola were not separated as adults or immature because it is difficult to differentiate among developmental stages.

All litter and soil samples were weighed before extraction, and after extraction all samples were oven dried at 65°C for a week and then weighed again. During the experiment, sample temperature and humidity were not measured but from reviewing literature, we assumed that the use of a light bulb during extractions increased sample temperature and dried the sample resulting in a gradient of temperature and humidity within the sample, while in the extractions without light, the establishment of the gradient would depend on the temperature and humidity of the room where extractions were performed (MacFadyen 1953; MacFadyen 1961; Haarlov 1947; Søvik & Leinaas 2002; Block 1966).

2.3 Data analyses

Data analyses were done using SigmaStat 3.0 (Systat Software Inc., www.sigmaplot.com) and PCORD 4.0 (PC ORD - Multivariate Analysis of Ecological Data, home.centurytel.net/~mjm/pcordwin.htm). Litter depth, and litter and soil dry weights were compared between forests using a Mann Whitney Rank Sum test (Sokal & Rohlf 1994). Arthropod abundance was standardized to individuals per square meter. Two way ANOVAs were used to establish the effect of light and duration of the extraction (time) on litter arthropod abundance and on the abundance of developmental stages. Three-way ANOVAs were used to establish the effect of corer vs shovel, light and duration of the extraction (time) on soil arthropod and developmental stage abundance. Non Metric Multidimensional Scaling (NMS) and a Multi Response Permutation Procedure (MRPP) were used to establish the identity and sequence in which arthropods were extracted. NMS was run with the Sorensen dissimilarity index based on presence/absence and a maximum of three axes were allowed. Non Metric Multidimensional Scaling places sampling units in space based on similarity such that close points have a similar composition, and MRPP establishes significant differences based on the dissimilarity matrix calculated with the Sorensen index (McCune & Grace 2002). The MRPP was used to establish statistical differences in composition due to collection, extraction and duration of the extraction (time), and the significant value was set at 0.05. Throughout the text, results are expressed as mean ± standard error.

3. Results

There were significant effects of forest type (dry vs wet) on litter depth (Mann Whitney, n=120, T= 9843.5, p<0.001), litter dry weight (Mann Whitney, n=40, T= 2410.0, p<0.001) and soil bulk density (Mann Whitney, n=80, T= 7206.0, p=0.009). Mean litter depth was higher in the dry forest, 3.7 cm (±0.1), than in the wet forest, 2.2 cm (±0.1), and mean litter dry weight was 5.7 kg m^{-2} (±458.8) in the dry forest and 0.8 kg m^{-2} (±54.5) in the wet forest. Mean soil bulk density was higher in the wet forest, 0.6 g cm^{-3} (±0.02), than in the dry forest, 0.5 g cm^{-3} (±0.02).

3.1 Litter arthropods

Acari and Collembola were dominant in both forests. Acari were significantly more abundant in the dry forest while, Collembola were more abundant in the wet forest (Table 1). In the dry forest, there were, on average, 655 Acari m^{-2} and 583 Collembola m^{-2}, while in the wet forest, there were 623 and 635, respectively.

	Dry forest			Wet forest		
	Mean(±s.e.)		%	Mean(±s.e.)		%
Acari	655 (±	106)a	30	623 (±	101)b	28
Collembola	583 (±	92)a	26	635 (±	92)b	29
Diptera	187 (±	33)a	8	179 (±	33)b	8
Hymenoptera	166 (±	65)a	7	152 (±	65)a	7
Araneae	155 (±	20)b	7	144 (±	19)a	7
Pseudoscorpiones	138 (±	27)a	6	142 (±	28)b	6
Homoptera	104 (±	31)a	5	101 (±	32)a	5
Isopoda	85 (±	17)b	4	93 (±	18)a	4
Coleoptera	70 (±	16)a	3	60 (±	11)b	3
Thysanoptera	58 (±	24)a	3	48 (±	24)a	2
Others	14 (±	4)		16 (±	5)	
Total	2214 (±	436)		2192 (±	428)	

Table 1. Mean abundance m^{-2} (±standard error) and percent dominance of litter arthropods in dry and wet forests. Alphabets indicate significant difference between forests for a particular group (Mann-Whitney Rank Sum Test, $\alpha=0.05$, n=40 for each forest).

In both forests, there was a significant effect of light on the abundance of litter arthropods, being higher when extractions were done with light. In the dry forest, there were significant effects of light and time on the abundance of the extracted arthropods while in the wet forest only time had a highly significant effect (Table 2). In the dry forest, >20,000 ind m^{-2} were extracted with light and in the wet forest <5,000 ind m^{-2} were extracted with light (Fig. 2). Through time, in both the dry and the wet forests, a high abundance was obtained in the first 24 h, with a slight increase at 48 h (Fig. 3). In the dry forest, some individuals were still recovered after 168 h, while in the wet forest all individuals were collected in the first 48 h. In general, in the dry forest, >90% of total extracted individuals was obtained after 144 h (6 d) of extraction with light, while in the wet forest >90% of total individuals was obtained after 48 h (2 d).

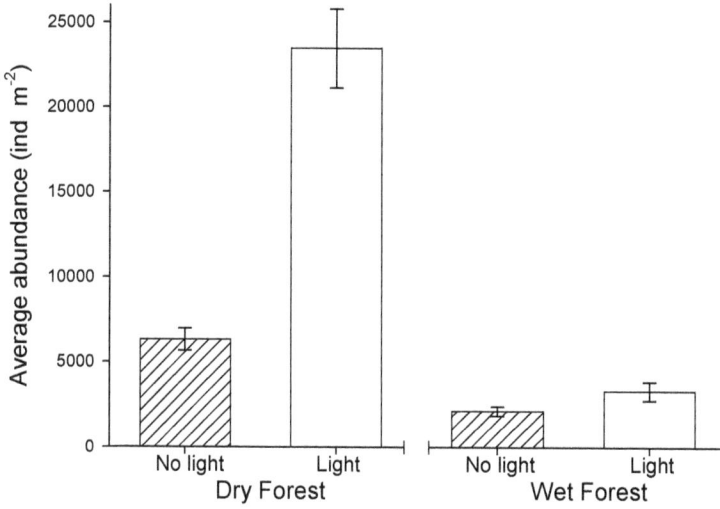

Fig. 2. Mean abundance of litter arthropods (ind m⁻²) extracted with and without light. Bars represent standard error.

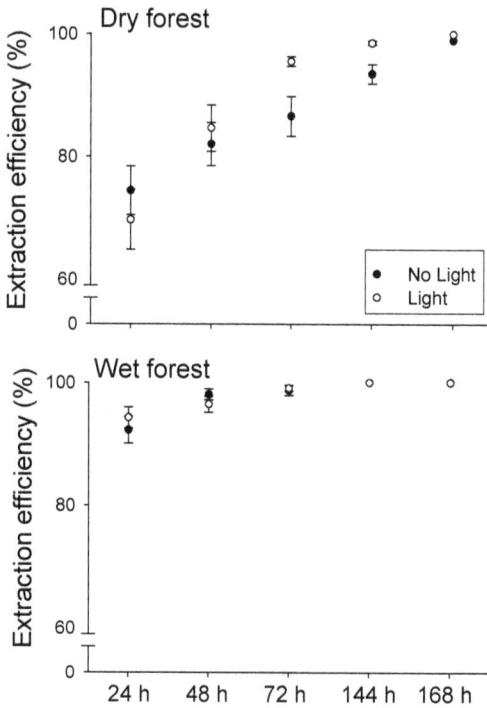

Fig. 3. Relative extraction efficiency (below) in dry and wet forests. Bars represent standard error.

In both forests, duration of the extraction (time) and light significantly influenced the identity and sequence in which arthropods were extracted (NMS, MRPP) (Fig. 4). In the dry forest and during the first 24 h, Blattodea and Protura were extracted mainly without light, while Acari (Fig. 1G), Collembola (Fig. 1E) and Pseudoscorpiones were extracted mainly with light. These groups represent organisms that are mostly considered detrivitivores, omnivores or predators, but microbivores (mainly Acari, Oribatida) were still recovered after 144 h of extraction. In the wet forest, all groups were extracted in the first 24 h, Chilopoda, Hymenoptera and Symphyla were extracted without light after the first 24 h, and Acari, Diplopoda and Diptera were extracted with light in the same extraction period.

| | | | | | Developmental stage | | | |
| | | | Abundance | | immatures | | adults | |
Forest	Source	df	F	P	F	P	F	P
Dry	Light	1	48.66	<0.001	41.72	<0.001	35.49	<0.001
	Time	4	57.42	<0.001	48.29	<0.001	42.40	<0.001
	Light x time	4	14.91	<0.001	13.88	<0.001	10.50	<0.001
	Residual	189						
	Total	198						
Wet	Light	1	3.98	0.05	6.48	0.01	2.80	0.10
	Time	4	73.26	<0.001	66.55	<0.001	60.06	<0.001
	Light x time	4	3.20	0.01	6.50	<0.001	2.09	0.08
	Residual	190						
	Total	199						

Table 2. Effect of use of light during extraction (with and without light) and duration of the extraction (time) (24 h, 48 h, 72 h, 144 h and 168 h) on litter arthropods in dry and wet forests (two 2-way AOV, $\alpha = 0.005$, n=199 for the dry forest and n=200 for the wet forest). The effects were evaluated for total abundance (ind m^{-2}) and abundance per developmental stage (ind m^{-2}).

Developmental stages. There were significant effects of light and time on the abundance of both immature and adults in the litter (Table 2). In the dry forest, both immature and adults were more abundant when extraction was done with light, and in the wet forest immature were more abundant when extraction was done without light, but adults were abundant in both extraction treatments (Table 3). Through time, both immature and adults followed the same pattern as established before: in the dry forest, >90% was obtained after 144h (6 d), while in the wet forest, >90% was obtained after 48h (2 d).

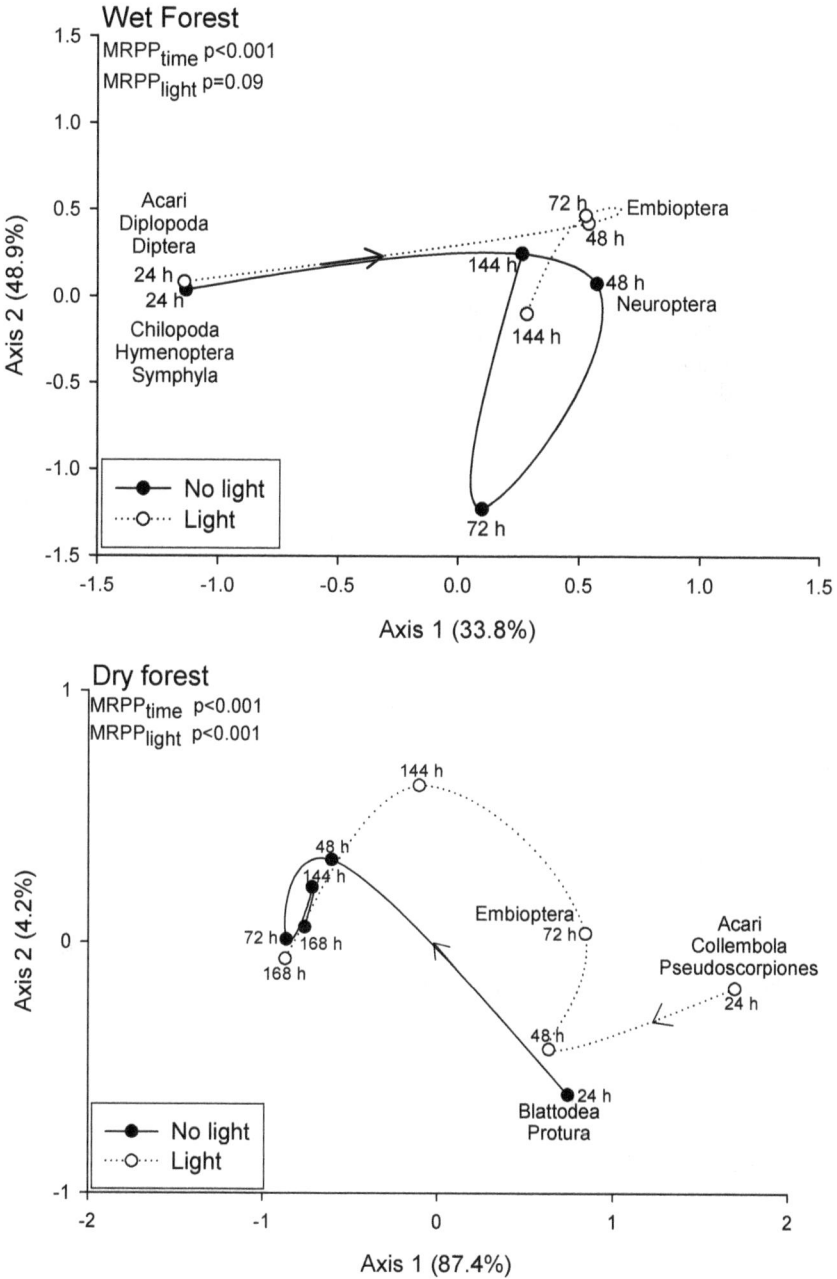

Fig. 4. Identity and sequence of extraction of litter arthropods extracted with and without light in dry and wet forests. The sequence of extraction within each type (with and without light) through time is connected by a line.

Sample	Stage	Collection	Dry forest		Wet forest	
			No light	Light	No light	Light
Litter	Immatures		2,115(±10113)b	8,120(±947)a	630(±90)a	345(±72)b
	Adults		4,205(±487)b	15,245(±1857)a	1,790(±239)a	2,690(±497)a
Soil	Immatures	Corer	11,877(±10113)a	10,924(13367)a	5,132(±5,091)a	1,466(±2,562)b
		Manual	19,062(±11,969)a	7,038(±7,724)b	2,346(±1,867)a	2,859(±3,027)a
	Adults	Corer	13,196(±17,395)a	5,792(±7,939)b	15,249(±8,968)a	4,179(±2,744)b
		Manual	11,070(±7,604)b	16,496(±8,015)a	9,310(±9,937)a	3,739(±2,912)b

Table 3. Mean abundance per square meter (+/- standard error) of litter and soil arthropods in dry and wet forests. Alphabets indicate significant differences between extraction methods (with and without light) for a particular developmental stage in a specific forest (Mann-Whitney Rank Sum Test, α=0.05, n=40 for litter arthropods in each forest, and n=80 for soil arthropods in each forest).

3.2 Soil arthropods

Overall, soil arthropod abundance was higher in the dry forest than in the wet forest (Table 4). Acari and Collembola were dominant in both forests, but both orders were significantly more abundant in the dry forest. There were, on average, 14,021 Acari m^{-2} and 3,904 Collembola m^{-2} in the dry forest, while in the wet forest, there were 2,511 and 2,786 ind m^{-2} respectively.

In both forests, there were significant effects of sampling technique, and of light and time, on the abundance of the extracted arthropods (Table 5). In both forests, extraction without light rendered higher abundance of soil microarthropods than extraction with light (Fig. 5). In the dry forest, collection with a corer and extraction without light rendered 5,015 ind m^{-2} while corer with light rendered 3,343, and shovel without light and shovel with light rendered 6,026 ind m^{-2} and 4,780, respectively. In the wet forest, corer without light had 4,076 ind m^{-2}, corer with light had 1,129, shovel without light had 2,331.4 and shovel with had 1,319. As established before, in the dry forest, arthropods continued to be collected after 144 h of extraction, and samples collected with shovel and extracted without light rendered the highest abundance. In the wet forest, arthropods were collected within the first 48 h except for the corer without light where arthropods were collected even after 144 h and this collection method rendered the highest abundance (Fig. 6).

In both forests, time significantly influenced the identity and sequence in which soil arthropods were extracted (NMS, MRPP) (Fig. 7). In the dry forest, Isoptera was extracted with corer with light in the first 24 h, Pseudoscorpiones with corer without light in the same time period 24 h, and Protura and Hemiptera with corer without light after 48 h. Also, Hymenoptera, Isopoda, Diplopoda and Chilopoda were extracted with shovel after 24 h (Fig. 7). In this forest, organisms representing several trophic groups (such as predators and omnivores) were extracted in the first 48 h, but those representing microbivores (mainly Oribatida) were still collected after 144 h. In the wet forest, Isoptera was extracted with corer with light after 24 h, Diplopoda and Chilopoda were extracted with shovel with light after 24 h, and Hymenoptera and Collembola were extracted with shovel without light after 24 h.

	Dry forest		Wet forest	
	Mean(±s.e.)	%	Mean(±s.e.)	%
Acari	14021 (± 1166)a	60	2511 (± 380)b	28
Collembola	3904 (± 538)a	17	2786 (± 343)b	29
Hymenoptera	1650 (± 813)a	7	1723 (± 531)a	8
Diplopoda	1100 (± 184)a	5	18 (± 18)b	7
Diptera	605 (± 390)a	3	642 (± 145)a	7
Coleoptera	440 (± 112)a	2	220 (± 59)b	6
Pseudoscorpiones	403 (± 111)a	2	220 (± 64)b	5
Protura	385 (± 135)a	2	861 (± 170)b	4
Araneae	183 (± 106)a	1	587 (± 124)b	3
Homoptera	183 (± 66)a	1	422 (± 111)b	2
Chilopoda	147 (± 96)a	1	73 (± 36)a	1
Hemiptera	128 (± 70)a	1	37 (± 26)b	0
Others	63 (± 9)a	0	92 (± 11)a	1
Total	23212 (± 1060)a		10189 (± 264)b	

Table 4. Mean abundance per square meter (+/-standard error) and percent dominance of soil arthropods in dry and wet forests. Alphabets indicate significant difference between forests for a particular group (Mann-Whitney Ran Sum Test, $\alpha=0.05$, n=80 for each forest).

Developmental stages. For immature arthropod abundance, there were significant effects of light and duration of the extraction (time) in both forests (Table 5). For adult arthropod abundance, there were significant effects of collection and duration of extraction (time) in the dry forest and, of collection, light and duration of the extraction (time) in the wet forest (Table 5). In the dry forest, immature were significantly more abundant when the sample was collected with shovels and extracted without light (Table 3). The abundance of adults depended on the collection method: for samples collected with a corer, a higher abundance was obtained when extracted without light; but for samples collected with shovel a higher abundance was obtained when extracted with light. In the wet forest, all soil samples, both corer and shovel that were extracted without light rendered a higher abundance than their counterparts extracted with light (Table 3).

Forest	Source	df	Abundance		Developmental stage			
					Immature		Adults	
			F	P	F	P	F	P
Dry	Collection	1	4.05	0.045	0.64	0.424	5.30	0.022
	Light	1	5.76	0.017	9.91	0.002	0.21	0.646
	Time	4	79.02	<0.001	39.01	<0.001	54.47	<0.001
	Collection x light	1	0.12	0.727	7.21	0.008	11.66	<0.001
	Collection x time	4	6.90	<0.001	3.90	0.004	4.08	0.003
	Light x time	4	0.42	0.797	0.42	0.792	1.14	0.337
	Collection x light x time	4	0.43	0.787	6.33	<0.001	3.99	0.003
	Residual	380						
	Total	399						
Wet	Collection	1	5.59	0.019	1.10	0.295	4.65	0.032
	Light	1	36.29	<0.001	5.64	0.018	31.62	<0.001
	Time	4	43.75	<0.001	15.25	<0.001	31.55	<0.001
	Collection x light	1	8.67	0.003	9.90	0.002	3.45	0.064
	Collection x time	4	12.80	<0.001	6.87	<0.001	8.23	<0.001
	Light x time	4	2.62	0.350	4.09	0.003	1.72	0.146
	Collection x light x time	4	8.52	<0.001	2.72	0.030	8.13	<0.001
	Residual	380						
	Total	399						

Table 5. Effect of collection (corer and shovel), extraction (with and without light) and duration of the extraction (time) (24 h, 48 h, 72 h, 144 h and 168 h) on the abundance of soil arthropods (two 3-way AOV, $\alpha = 0.005$, n=400 for each forest). The effects were evaluated for total abundance (ind m^{-2}) and abundance per developmental stage (ind m^{-2}).

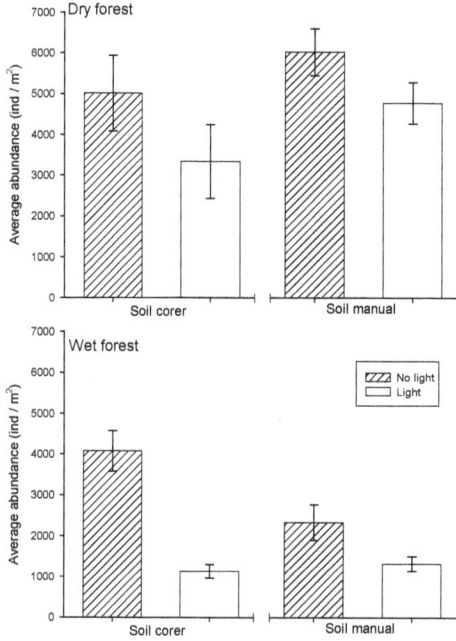

Fig. 5. Mean abundance of soil arthropods (ind m^{-2}) collected by corer or shovel, and extracted with and without light. Bars represent standard error.

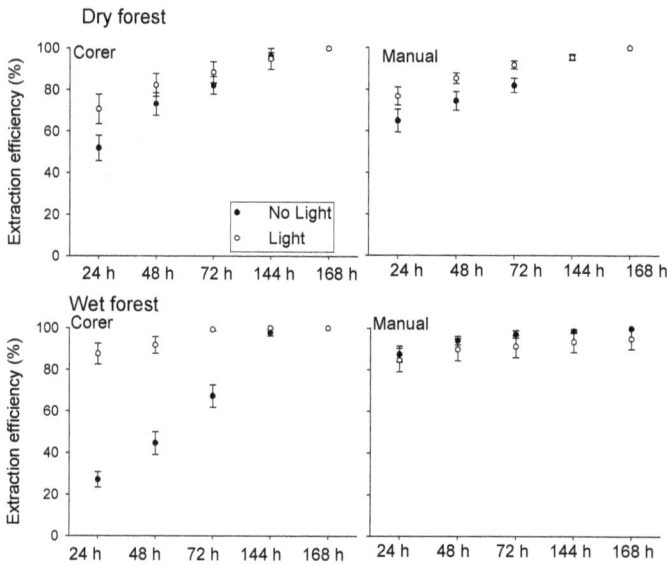

Fig. 6. Extraction efficiency in dry and wet forests. Bars represent standard error.

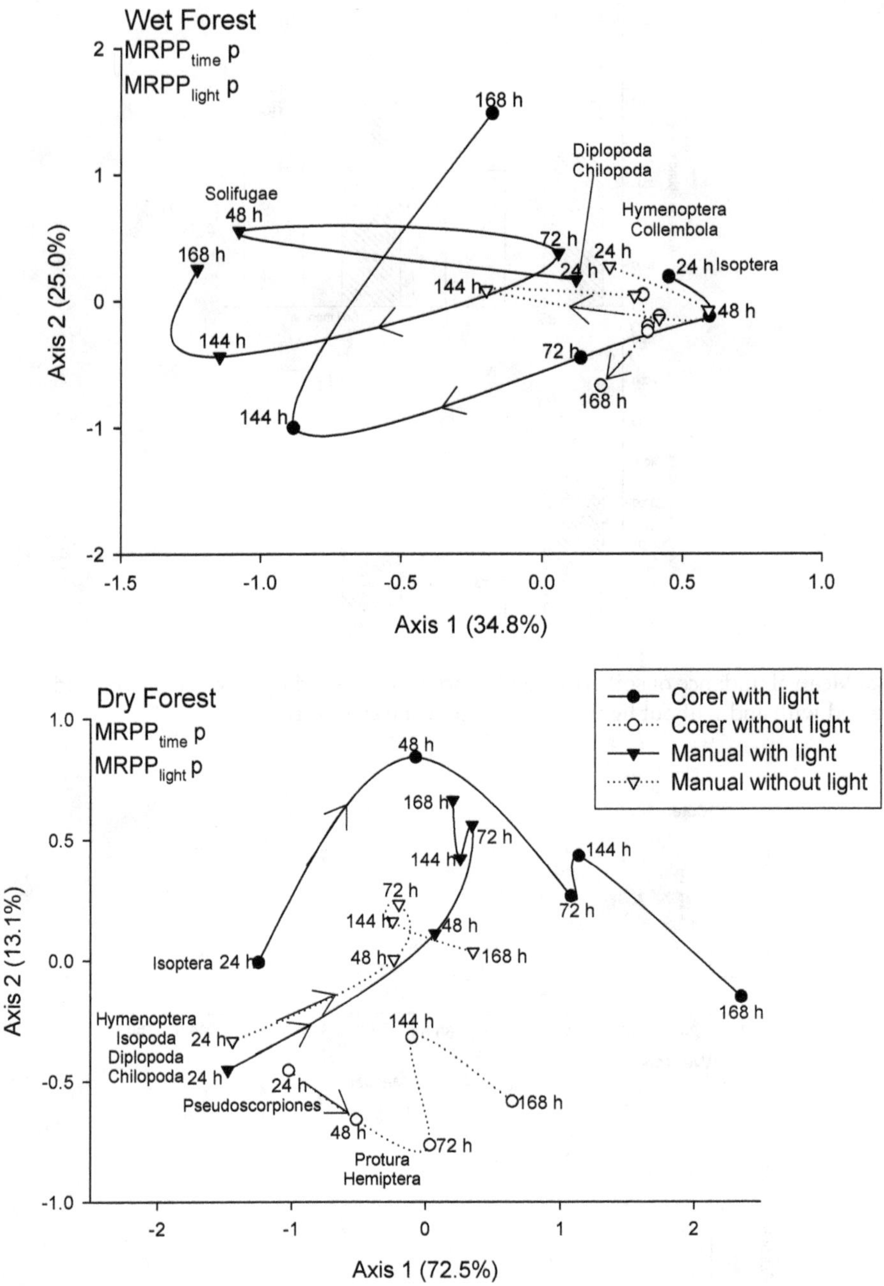

Fig. 7. Identity and sequence of soil arthropods collected by corer or shovel, and extracted with and without light in dry and wet forests. The sequence of extraction within each collecting technique with extraction type through time is connected by a line.

4. Discussion

The objective of this study was to determine how the diversity of retrieved arthropods was affected by collection technique, use of light during extractions in Berlese-Tullgren funnels and duration of the extraction. For use of light during extractions, we found that litter arthropod abundance was highest when extraction was done with light but soil arthropods were highest when extraction was done without light. We found that forest type (tropical wet vs dry forest) influenced the sampling technique that was best suited because, in the wet forest, soil arthropod abundance was highest when collection was done with soil corers, while in the dry forest soil arthropod abundance was highest when collection was done with shovels. Finally, we found that forest type also influenced duration of the extraction because in the wet forest, 90% of arthropods were recovered within the first 24 h while, in the dry forest the same percent was obtained after 144 h of extraction.

Litter and soil arthropods responded differently to the use of light during extractions. One explanation is that the use of light may have speeded the desiccation of the sample forcing more litter animals to exit the sample than in the extractions without light. On the contrary, soil arthropods are more sensitive to increasing temperature or decreasing humidity than litter arthropods, in which case the use of light during extractions would made arthropods inactive before leaving the sample and thus their apparent abundance would decrease. Furthermore, litter arthropods inhabit a clear and warm habitat (litter) (Eviner & Chapin 2003) and thus may require an increase in temperature and in light incidence to exit the samples. But soil arthropods inhabit a comparatively cooler and darker habitat (soil) (Eviner & Chapin 2003) and thus may be sensitive to increasing temperature and light incidence to the point that the use of light during extraction may have resulted in an underestimation of soil arthropod abundance. In both forests, we retrieved more adults than immature, possibly because the soft cuticle of immature makes them more susceptible to the decreasing humidity within the extraction funnel, and because immature organisms, such as mites (majority of the immatures retrieved in this study), undergo inactivity when moulting and thus cannot leave the sample (Søvik & Leinaas 2002). As a consequence another extraction methodology, such as flotation, should be more suitable for immature forms (Hale 1964; Walter et al. 1987; Lakly & Crossley 2000; Søvik & Leinaas 2002).

In both forests, soil arthropods left the shovelled samples faster than the cored samples. Also, abundance was highest from shovelled samples in the dry forest, and in the wet forest abundance was highest from the cored samples although the pace of retrieval was slow. One explanation is that the loose structure of shovelled samples retained less humidity and dried out faster allowing the temperature/humidity gradient to be established sooner than in the compact cored samples where more humidity could be retained (MacFadyen 1953). This would have a dual effect, in wet forest samples, the gradient resulting from drying out the cored sample at room temperature moved slowly downwards forcing arthropods to leave the sample but not being large enough to kill them (as would occur in the shovelled samples) resulting in higher arthropod abundance in cored than in shovelled samples. In the dry forest samples, the gradient resulting from drying the shovelled sample at room temperature reached higher critical levels and became larger than in cored samples, forcing arthropods outwards, resulting in higher estimates of arthropod abundance in the shovelled samples.

Arthropods from the wet forest were recovered faster than arthropods from the dry forest. Macfadyen (1961) proposed that humidity-loving animals require high levels of humidity to be active, as a consequence they respond quicker to changes in humidity than humidity-resistant animals. On the contrary, humidity-resistant animals require high temperatures and low humidity to be forced to exit the sample. During extractions, temperature of the sample begins to increase immediately with a significant increase after 16 – 24 h, and humidity within the sample drops simultaneously with the significant increase in temperature (Haarlov 1947; Block 1966). Following MacFadyen (1953), we can explain why arthropods from the wet forest exited the sample faster than those from the dry forest. Arthropods from the wet forest left the sample within the first 24 h in response to the increase in temperature and the decrease in humidity that occurred in the sample when extraction begins. As the sample became hotter and drier, any animal remaining in the sample could have become inactive (or killed) by low humidity and high temperature. On the other hand, arthropods from the dry forest required longer extraction times because for these humidity-resistant animals, the critical levels of humidity required for them to leave the sample, took longer to be established.

The environmental characteristics of the two forests studied here were different and contribute another explanation to our results. Dry periods in the wet forest and the concomitant response of the biota to these periods are based on the number of dry days, because monthly rainfall is always above 100mm (Cuevas *et al.* 1991; Cuevas & Lugo 1998). On the contrary, dry days in the dry forest are the common condition, an average of 200 dry days per year, with pulses of heavy rainfall occurring during the wet season (http://cirrus.dnr.state.sc.us/cgi-bin/sercc/cliMAIN.pl?pr8412). As a consequence, arthropods from the dry forest come from a habitat with higher temperatures and longer periods of drier conditions than do arthropods from wet forest. The more extreme conditions in the dry forest may have make arthropods less responsive to higher temperature and drier air within the extraction funnel, resulting in longer extraction times.

We found that arthropod abundance was significantly higher in the dry forest than in the wet forest. Litter depth and dry weight were higher in the dry forest, litter was 40% deeper and 86% heavier in the dry forest than in the wet forest, suggesting that habitat and resource availability significantly influenced arthropod abundance (Mulder *et al.* 1999). Several researchers have found that Berlese-Tullgren extractions underestimate arthropod abundance. Nevertheless, in this study we found that total abundances fall within similar ranges to those reported in the literature, such as in Mexico and Perú, where abundances are reported to be 4,303 – 6,409 ind/m^2 (Lavelle *et al.* 1992) respectively. In addition, Berlese-Tullgren funnels are the predominant methodology to collect arthropods from litter/soil samples, but care should be taken because some groups are sensitive to light and are not effectively recovered with funnel extraction, such as collembolans in the family Onychiuridae (Coleman *et al.* 2004), or in our case, Proturans that were much more abundant in extractions without light and almost absent when light was used.

The structure of the retrieved community was affected by duration of extraction. Other authors have also found that during extractions, different taxonomic groups leave the sample at different times (Krell *et al.* 2005; Block 1966). In an extraction that lasted three days (Block 1966), Mesostigmata mites left the sample during the first day, while the majority of Collembola and Cryptostigmata mites left the sample mainly during the second day, and

Prostigmata began to leave the sample at the third day. Also, Krell *et al.* (2005) using an alternative extraction method, Winkler bags, also found that duration of the extraction affected the structure of the retrieved community, for example 70% of adult beetles and ants were retrieved within three days of extraction but Chilopoda required 3 to 4 wk. By using Berlese-Tullgren funnel, we found that both Collembola and Acari began to leave the sample during the first 24 h, also we retrieved few adult Coleoptera (the majority of Coleoptera were larvae), and the majority of ants and centipedes left the samples within the first 24 h of extraction.

5. Conclusion

Krell *et al.* (2005) proposed that if the aim of the study is to rapidly assess the litter/soil fauna, short extraction times should be enough. On the contrary, if the aim of the study is to exhaustively assess this fauna, then the methodology should be standardized, such as assessing optimum extraction times and biases due to collection methods. Our results also suggest that methodology standardization is necessary because (1) to reach similar percents (90%), extraction periods were longer for samples from dry forests than from wet forests, (2) the use of light promoted litter arthropods to leave the sample producing high abundances, but for immature and soil arthropods the use of light resulted in low abundances, and (3) cored samples rendered higher abundances in wet forests than in dry forests where shovel samples rendered higher abundances. In addition, our data suggest that samples from dry environments should be extracted for longer periods than those coming from wet environments. Also, if the focal organisms are soil arthropods, then extraction without light should result in high abundances. Finally, the collection method best suited depends on the environment to be sampled: in this study for wet habitats cored soil samples resulted in higher abundances than shovelled samples which resulted in highest abundances when sampling dry habitats.

6. Acknowledgements

This research received partial support from Crest-Center for Applied Tropical Ecology and Conservation (Crest-Catec) of the University of Puerto Rico at Rio Piedras Campus, grant NSF-HRD-0206200 through a fellowship to MFBA, and from grant DEB-0218039 from the National Science Foundation to the Institute of Tropical Ecosystem Studies, University of Puerto Rico, and the USDA Forest Service, International Institute of Tropical Forestry as part of the Long Term Ecological Research Program in the Luquillo Experimental Forest. Logistic support was received from the United States Department of Agriculture Forest Service International Institute of Tropical Forestry (USDA FS IITF). We would like to thank anonymous reviewers for their comments on previous versions of this manuscript. We would like to thank Maria Rivera for her help during sample collection, and Elvin Rodriguez, Xiomara Cruz and Paola Santiago for their help during sample processing.

7. References

Agosti, D. & Alonso, L.E. (2000). The ALL protocol: a standard protocol for the collection of ground-dwelling ants. *Ants: Standard methods for measuring and monitoring biodiversity* (eds D. Agosti, J.D. Majer, L.E. Alonso & T.R. Schultz), pp. 204-206. Smithsonian Institution, Princeton Editorial Associates, Washington.

Andre, H.M., Duarme, X. & Lebrum, P. (2002) Soil biodiversity: myth, reality or conning? *Oikos*, 96, 3-24.

Bestelmeyer, B.T., Agosti, D., Alonso, L.E., Brandão, C.R.F., Brown, W.L. Jr., Delabie, J.H.C. & Silvestre, R. (2000) Field techniques for the study of ground-dwelling ants: an overview, description, and evaluation pages. *Ants: Standard methods for measuring and monitoring biodiversity* (eds D. Agosti, J.D. Majer, L.E. Alonso & T.R. Schultz), pp. 122-143. Smithsonian Institution, Princeton Editorial Associates, Washington.

Block, W. (1966) Some characteristics of the Macfadyen high gradient extractor for soil micro-arthropods. *Oikos*, 17, 1-9.

Burgess, N.D., Ponder, K.L. & Goddard, J. (1999) Surface and leaf-litter arthropods in the coastal forests of Tanzania. *African Journal of Ecology*, 37, 355-365.

Bremner, G. (1990) A Berlese funnel for the rapid extraction of grassland surface macro-arthropods. *New Zealand Entomologist*, 13, 76-80.

Chen, B. & Wise, D.H. (1999) Bottom-up limitation of predaceous arthropods in a detritus-based terrestrial food web. *Ecology*, 80, 761-772.

Coddington, J.A., Young, L. & Coyle, F.A. (1996) Estimating spider species richness in a southern Appalachian cove hardwood forest. *Journal of Arachnology*, 24, 111-128.

Coleman, D.C., Crossley, D.A. Jr. & Hendrix, P.F. (2004) *Fundamentals of soil ecology*. Elsevier Academic Press, New York.

Cuevas, E., Brown, S. & Lugo, A.E. (1991) Above and belowground organic matter storage and production in a tropical pine plantation and a paired broad leaved secondary forest. *Plant and Soil*, 135, 257-268.

Cuevas, E. & Lugo, A.E. (1998) Dynamics of organic matter and nutrient return from litter fall in stands of ten tropical plantation species. *Forest Ecology and Management*, 112, 263-279.

Eviner, V.T. & Chapin III, F.S. (2003) Functional matrix: a conceptual framework for predicting multiple plant effects on ecosystem processes. *Annual Review of Ecology, Evolution and Systematics*, 34, 455-485.

Ewel, J.J. & Whitmore, J.J. (1973) *The ecological life zones of Puerto Rico, U.S.* Forest Service Research Paper ITF-18, Rio Piedras, Puerto Rico.

Fisher, B.L. (1999) Improving inventory efficiency: a case study of leaf-litter ant diversity in Madagascar. *Ecological Applications*, 9, 714-731.

García-Martinó, A.R., Warner, G.S., Scatena, F.N. & Civco, D.L. (1996) Rainfall, runoff and elevation relationships in the Luquillo Mountains of Puerto Rico. *Caribbean Journal of Science*, 32, 413-424.

Gould, W.A., González, G. & Carrero-Rivera, G. (2006) Structure and composition of vegetation along an elevational gradient in Puerto Rico. *Journal of Vegetation Science*, 17, 653-664.

Hale, W.G. (1964) A flotation method for extracting Collembola from organic soils. *Journal of Animal Ecology*, 33, 363-369.

Haarlov, N. (1947) A new modification of the tullgren apparatus. *Journal of Animal Ecology*, 16, 115-121.

Hasegawa, M. (1997) Changes in Collembola and Cryptostigmata communities during the decomposition of pine needles. *Pedobiologia*, 41, 225-241.

Krantz, G.W. (1978) *A manual of acarology.* Oregon State University Book Stores, Corvallis, Oregon.

Krell, F.T., Chung, A.Y.C., DeBoise, E., Eggleton, P., Giusti, A., Inward, K. & Krell-Westerwalbesloh, S. (2005) Quantitative extraction of macro-invertebrates from temperate and tropical leaf litter and soil: efficiency and time-dependent taxonomic biases of the Winkler extraction. *Pedobiologia*, 49, 175-186.

Lakly, M.B. & Crossley, D.A. Jr. (2000) Tullgren extraction of soil mites (Acarina): Effect of refrigeration time on extraction efficiency. *Experimental and Applied Acarology*, 24, 135-140.

Lavelle, P., Blanchart, E., Martin, A., Spain, A.V. & Martin, S. (1992) Impact of soil fauna on the properties of soils in the humid tropics. *Myths and science of soils of the tropics* (eds R. Lal & P. A. Sanchez), pp. 157-185. Soil Science Society of America Special publication no. 29, Madison.

Longino, J.T., Coddington, J. & Colwell, R.K. (2002) The ant fauna of a tropical rain forest: estimating species richness three different ways. *Ecological Society of America Ecological Archives* E083-011-A1, *Ecology*, 83, 689-702.

MacAlpine, J.F. (1989) *Manual of Nearctic Diptera.* Vol. 1 - 2. Research Branch, Agriculture Canada, Canadian Government Publishing, Ottawa.

McCune, B. & Grace, J.B. (2002) *Analysis of ecological communities.* MjM software design, Gleneden Beach, Oregon.

Meyer, E. (1996) Methods in soil zoology, mesofauna. *Methods in soil biology* (eds F. Schinner, R. Öhlinger, E. Kandeler & R. Margesin), pp. 338-345. Springer, New York.

Macfadyen, A. (1961) Improved funnel-type extractors for soil arthropods. *Journal of Animal Ecology*, 30, 171-184.

Macfadyen, A. (1953) Notes on methods for the extraction of small soil arthropods. *Journal of Animal Ecology*, 22, 65-77.

Mulder, C.P.H., Koricheva, J., Huss-Danell, K., Högberg, P. & Joshi, J. (1999) Insects affect relationships between plant species richness and ecosystem processes. *Ecology Letters*, 2, 237-246.

Oliver, I. & Beattie A.J. (1996) Designing a cost-effective invertebrate survey: a test of methods for rapid assessment of biodiversity. *Ecological Applications*, 6, 594-607.

Rohitha, B.H. (1992) A simple separation system based on flotation for small samples of insects contaminated with soil. *New Zealand Entomologist*, 15, 81-83.

Schowalter, T.D. & Ganio, L.M. (1999) Invertebrate communities in a tropical rain forest canopy in Puerto Rico following Hurricane Hugo. *Ecological Entomology*, 24, 191-201.

Sokal, R.R. & Rohlf, F.J. (1994) *Biometry the principles and practices of statistics in biological research.* WH Freeman, New Jersey.

Søvik, G. & Leinaas, H.P. (2002) Variation in extraction efficiency between juvenile and adult oribatid mites: *Ameronothrus lineatus* (Oribatida, Acari) in a Macfadyen high-gradient canister extractor. *Pedobiologia*, 46, 34-41.

Triplehorn, C.A. & Johnson, N.F. (2004) *Borror and Delong's introduction to the study of insects.* Thomson Brooks/Cole, Florence, Kentucky.

Walter, D.E., Kethley, J. & Moore, J.C. (1987) A heptane flotation method for microarthropods from semiarid soils, with comparison to the Merchant-Crossley high-gradient extraction method and estimates of microarthropod biomass. *Pedobiologia*, 30, 221-232.

4

Direct and Indirect Effects of Millipedes on the Decay of Litter of Varying Lignin Content

Grizelle González[1], Christina M. Murphy[1] and Juliana Belén[2]
[1]USDA FS International Institute of Tropical Forestry (IITF), Río Piedras,
[2]University of Puerto Rico, Mayagüez,
Puerto Rico

1. Introduction

Millipedes are considered to be important organisms involved in decomposition, both for their direct feeding on detritus and their indirect effects on microbial activity. Hanlon (1981a, 1981b) suggested that fragmentation of leaf litter by soil fauna increases microbial biomass by increasing leaf surface area and diminishing pore sizes. The passage of litter through the gut of macroarthropods, such as millipedes, can help in the establishment of soil bacteria (Anderson & Bignell, 1980; Hanlon, 1981a, 1981b; Tajovsky et al., 1991; Maraun & Scheu, 1996). The presence of millipedes has been shown to increase the decomposition of litter as well as increase growth of seedlings (Cárcamo et al., 2001). In a beech forest, Bonkowski et al. (1998) also found that the presence of millipedes significantly increased the decomposition of litter, much more so than endogeic earthworms. The presence of millipedes has also been found to greatly increase the release of litter nutrients into the soil, especially calcium and nitrates (Pramanik et al., 2001). Millipedes are selective about what leaves they eat (Lyford, 1943; Kheirallah, 1979; Cárcamo et al., 2000). The chemical composition of leaf litter, especially the lignin and nitrogen content, can greatly affect soil fauna populations, although this effect is not clear for millipedes (Tian et al., 1993). Van der Drift (1975) estimated that in temperate areas millipedes are responsible for ingesting 5–10 percent of the annual leaf litter fall and Cárcamo et al. (2000) estimated that a single species of millipede consumed 36 percent of the annual leaf litter in a British Columbian Cedar-Hemlock forest. Tropical studies have also found a large influence of millipedes on decomposition (Tian et al., 1995). In a Tabonuco forest in Puerto Rico, Ruan et al. (2005) found that millipede density explained 40 percent of the variance in leaf litter decomposition rates, while soil microbial biomass explained only 19 percent of the variance.

Millipedes make up a large part of the arthropod community on the forest floor in the Tabonuco forests of Puerto Rico. Richardson et al. (2005, pers. com.) found that diplopods in El Verde (a Tabonuco forest) constituted about 11.4 percent (73.09 mg dry/m^2) of the microarthropod biomass, second only to Isoptera. In the same forest, we found that Stemmiulidae were the most abundant millipede with a density of ca. 22 individuals/m^2 (Murphy et al., 2008).

In this study, we use a microcosm approach to answer the direct (leaf fragmentation) and indirect (microbial biomass) effects of millipedes on the decomposition of leaf litter and how these outcomes are influenced by the substrate (litter) quality and the density of millipedes. We expect that higher the litter quality (lower lignin content) and the higher density of millipedes would result in more leaf area lost, decreased leaf mass remaining, and higher biomass of soil microbes. We used microcosms containing one of three litter species with varying lignin to nitrogen (L/N) ratios and three different densities of millipedes.

2. Methods

2.1 Site

Millipedes and soil for microcosms were collected from the Luquillo Experimental Forest, a subtropical wet forest located near the El Verde Field Station, (18°19′ N, 65°45′W) in Río Grande, Puerto Rico in June 2006 (Fig. 1). Vegetation at this elevation (420m) is called Tabonuco forest after the dominant plant species *Dacryodes excelsa*.

Fig. 1. Millipedes and soil for microcosms were collected from the Luquillo Experimental Forest, a subtropical wet forest.

The mean annual air temperature in the Luquillo Mountains is 22.3°C (Brown et al., 1983) and the mean annual precipitation is 3525 mm with rainfall distributed more or less evenly throughout the year (Garcia-Martinó et al., 1996). The dominant soil orders in the Luquillo Experimental Forest are Ultisols and Inceptisols (Brown et al., 1983). Soil series in the Tabonuco forest vary according to topography: *Humatas* on ridges, *Zarzal* or *Cristal* on slopes, and *Coloso* in valleys (Johnston, 1992; Soil Survey Staff, 1995).

2.2 Experimental design

The microcosms consisted of clear plastic containers, measuring 19 cm x 13 cm x 9 cm (area ca. 0.025m²), with mesh tops (1.5mm) to keep millipedes within and to exclude other organisms (Fig. 2). The microcosms were kept in a covered room at El Verde Field Station that was open to the environment laterally to keep conditions as close to field conditions as possible. One milliliter of distilled water was sprayed into the microcosms each day to mimic mean daily rainfall. Each microcosm contained 115g of sieved (1.19mm) uniform soil from the Tabonuco forest.

Fig. 2. The microcosms consisted of clear plastic containers, with mesh tops to keep millipedes within.

We used three different leaf species (*Dacryodes excelsa*, *Manilkara bidentata*, and *Rourea surinamensis*) (Fig. 3), one species per microcosm and three densities of millipedes (0, 2, and 5 individuals) with three replicates of each of these treatments, collected at two weeks after setup (July 13, 2006) and four weeks after setup (July 27, 2006). Additionally, we had an initial collection (June 29, 2006) of nine microcosms (3 litter species, 3 densities of millipedes) that were returned to the laboratory immediately after set up to establish handling loss and dry mass relationships (e.g., González & Seastedt, 2001). In total, there were 63 microcosms used in the experiment. Millipedes used in the microcosms were all from the Order Stemmiulida, Family Stemmiulidae (Fig. 4). Millipedes in the density of two individuals had a combined fresh weight of approximately 0.04g and those in the density of five individuals had a combined net weight of about 0.08g. Leaves were obtained from El Verde Tabonuco forest and air dried. They were then cut into 3 x 3 cm squares and 5 grams of leaf squares were added to each microcosm. The three plant species were chosen because of their frequent occurrence in the Tabonuco forest and their leaves having a range of lignin to nitrogen ratios, with similar percents of nitrogen and varying lignin amounts (Table 1). Other plant species in this forest have a range of L/N ratios from 6.2 to 62.5 (Zalamea & González, unpublished data).

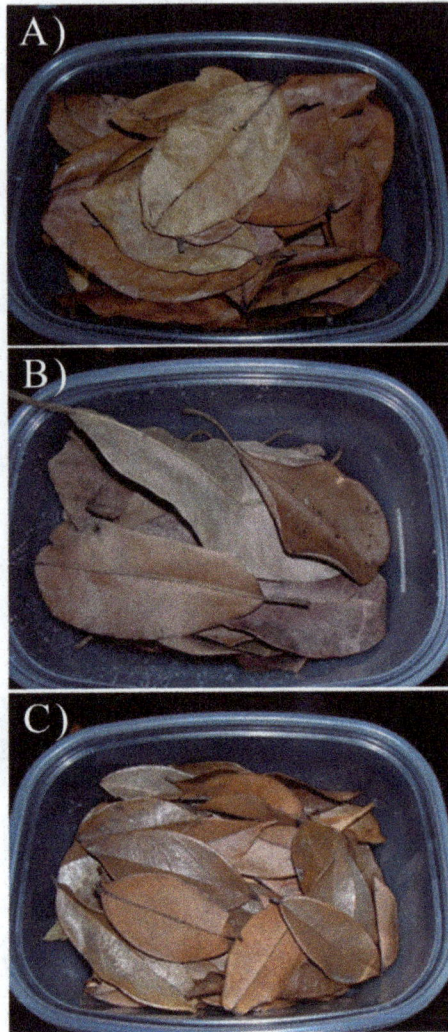

Fig. 3. Three leaf species with varying lignin to nitrogen (L/N) ratios were used in the experiment, A) *Dacryodes excelsa*, B) *Manilkara bidentata*, and C) *Rourea surinamensis*.

Plant Species	% Nitrogen	% Lignin	L/N ratio
Dacryodes excelsa	0.81	22.89	28.26
Manilkara bidentata	0.72	25.17	34.96
Rourea surinamensis	0.76	31.13	40.96

Table 1. Initial leaf chemistry of *Dacryodes excelsa, Manilkara bidentata,* and *Rourea surinamensis* (Zalamea & González, unpublished data).

Fig. 4. Millipedes used in the microcosms were all from the Order Stemmiulida, Family Stemmiulidae.

2.3 Measurements

At each collection, fifteen leaf squares were chosen at random and flattened in a plant press overnight. Leaf area for each of these 15 squares was then measured using an area scanner (LI-3100 Area Meter, Li-Cor, Inc.), taking three measurements of each square and averaging them. All leaves from the microcosms were oven-dried at 60° C and weighed. The fresh weights of millipedes were obtained initially and when microcosms were collected. The final oven dried weights of millipedes were also measured. In addition, when microcosms were collected all of the soil was removed, pH measured (with a 1:1 KCl to soil ratio) and inorganic nitrogen content (in the form of NO_3^- and NH_4^+) extracted with KCl and measured with a Rapid Flow Analyzer RFA-305 (Alpkem Corporation, Clackamas, Oregon). Soil microbial biomass C was estimated by the substrate induced respiration (SIR) method (Lin & Brookes, 1999) calibrated for the study area (Zalamea & González, 2007), using an ER-10 Columbus Instruments respirometer, and calculated from the CO_2 evolved (Anderson & Domsch, 1978).

2.4 Data analysis

The general linear model procedure was used for multivariate analyses of variance (MANOVA) to test the effect of litter species, millipede density, and collection time on the leaf area remaining, percent leaf mass remaining, millipede weight, soil pH, soil nitrate, ammonium concentrations, and soil microbial biomass (SPSS Inc., 2001, v. 11.0.1). Post-hoc tests were performed to determine significant differences among the litter species, millipede density, and collection time for each of the dependent variables, using Student-Newman-Keuls (SNK); α= 0.05. Additionally, Pearson's correlations (2-tailed) were calculated between the initial leaf chemistry (%N, %L, L/N ratio) and the leaf area remaining, percent leaf mass remaining, microbial biomass, and soil inorganic nitrogen. All statistical analyses were performed using SPSS Inc., 2001, v. 11.0.1.

3. Results

The leaf area remaining and the percent leaf mass remaining were significantly affected by all three independent variables (litter species, millipede density, and the time of collection) (Table 2). The percent of leaf mass remaining was significantly the lowest for *Dacryodes excelsa*, and the highest for *Manilkara bidentata* (Fig. 5A, Fig. 6).

The leaf area remaining was significantly different for all three litter types, following a pattern of significantly less leaf area remaining for the lower L/N ratio species (*D. excelsa*) and significantly the most area remaining for the highest L/N ratio species (*Rourea surinamensis*) (Fig. 5B). The highest density of millipedes (five individuals) had significantly less mass remaining than the litter from those microcosms without millipedes (Table 3).

The leaf area remaining was significantly less in the last collection, indicating that after four weeks the leaves had been significantly fragmented, while after two weeks the leaf area was not significantly different from the initial collection. The percent leaf mass remaining decreased significantly from one collection to the next (Fig. 6). At the two week collection Tabonuco leaves had significantly less percent of leaf mass remaining when millipedes were not present, and at the four week collection *R. surinamensis* leaves had significantly higher percent of leaf mass remaining when millipedes were not present (Fig. 6).

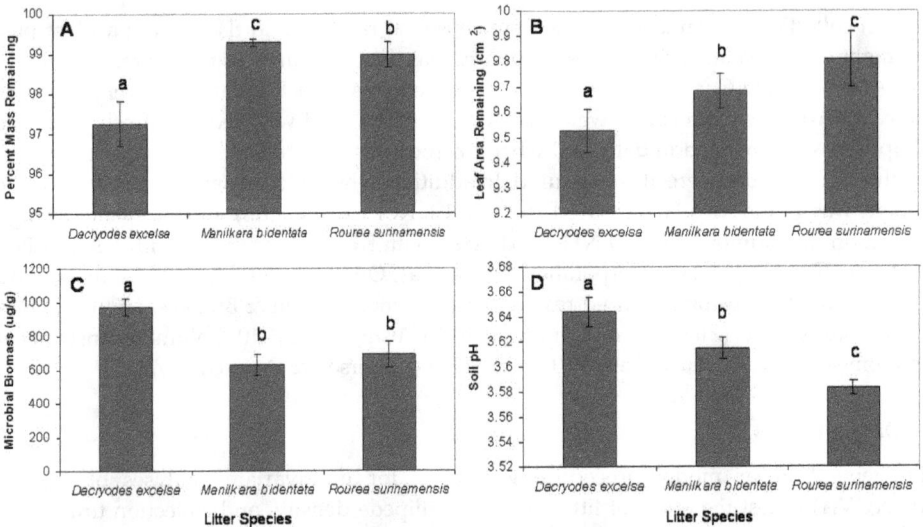

Fig. 5. A) The percent leaf mass remaining, B) leaf area remaining, C) microbial biomass, and D) the soil pH in microcosms containing *Dacryodes excelsa*, *Manilkara bidentata*, or *Rourea surinamensis* leaves. Significant differences are indicated by different letters (Student-Neuman-Keuls tests; α= 0.05). The standard error of the mean is indicated with error bars.

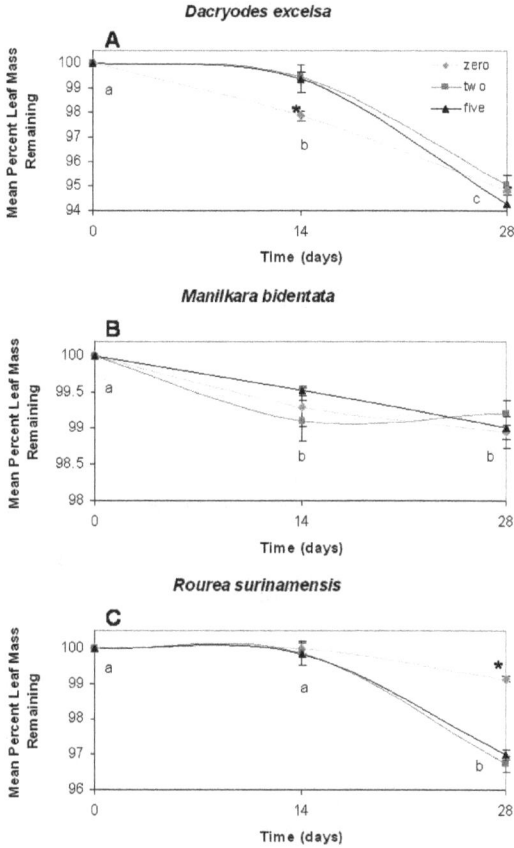

Fig. 6. The mean percent of mass remaining over time for each of the densities of millipedes (light gray for zero, dark grey for two, black for five) in microcosms with A) *Dacryodes excelsa*, B) *Manilkara bidentata*, or C) *Rourea surinamensis*. Different letters in the same graph represent significant differences among collection times. Asterisks (*) indicate a significantly different effect of the indicated millipede density at this collection time (Student-Neuman-Keuls tests; α= 0.05).

The initial leaf chemistry was significantly correlated with the leaf area remaining, percent leaf mass remaining, soil microbial biomass, and soil nitrate, but not soil ammonium. The percent nitrogen in leaves was negatively correlated with the percent leaf mass remaining (P<0.01) and positively correlated with the soil microbial biomass (P<0.01). The initial percent nitrogen in the leaves was not significantly correlated with either soil nitrate or soil ammonium. The initial percent lignin was positively correlated with the leaf area remaining (P<0.05) and the percent leaf mass remaining (P<0.05) and negatively correlated with soil microbial biomass (P<0.05) and soil nitrate (P<0.01). The L/N ratio was also positively correlated with the leaf area remaining (P<0.05) and the percent leaf mass remaining (P<0.01) and negatively correlated with soil microbial biomass (P<0.01) and soil nitrate (P<0.05).

Source	Variable	df	F	P	Power
Litter Species (L)	LAR	2	24.75	0.00	1.00
	PMR	2	48.13	0.00	1.00
	Soil pH	2	144.12	0.00	1.00
	Microbial Biomass	2	19.70	0.00	1.00
Millipede Density (M)	LAR	2	5.08	0.01	0.79
	PMR	2	8.62	0.00	0.95
	Soil pH	2	42.58	0.00	1.00
	Microbial Biomass	2	1.10	0.34	0.23
Time (CT)	LAR	2	115.86	0.00	1.00
	PMR	2	189.54	0.00	1.00
	Soil pH	2	220.18	0.00	1.00
	Microbial Biomass	2	130.46	0.00	1.00
L x M	LAR	4	5.02	0.03	0.94
	PMR	4	16.43	0.00	1.00
	Soil pH	4	8.32	0.00	1.00
	Microbial Biomass	4	1.04	0.40	0.29
L x CT	LAR	4	5.94	0.00	0.97
	PMR	4	43.99	0.00	1.00
	Soil pH	4	169.31	0.00	1.00
	Microbial Biomass	4	15.87	0.00	1.00
M x CT	LAR	4	6.35	0.00	0.98
	PMR	4	7.49	0.00	0.99
	Soil pH	4	18.93	0.00	1.00
	Microbial Biomass	4	0.62	0.65	0.18
L x M x CT	LAR	8	6.30	0.00	1.00
	PMR	8	5.67	0.00	1.00
	Soil pH	8	51.43	0.00	1.00
	Microbial Biomass	8	0.74	0.65	0.29

Table 2. Effect of litter species (*Dacryodes excelsa, Manilkara bidentata*, and *Rourea surinamensis*), millipede density (zero, two, and five), and the collection time (zero, fourteen, and twenty-eight days) on the leaf area remaining (LAR), percent leaf mass remaining (PMR), soil pH, and microbial biomass. Degrees of freedom (df), F and P values, and Power for MANOVA are presented. Statistics were performed using SPSS Inc., 2001, v. 11.0.1.

Density of Millipedes	Percent Leaf Area Remaining	Percent Leaf Mass Remaining	Millipede Weight lost (g)	Microbial Biomass (μg/g of soil)	pH
Zero	96.677[a]	98.692[b]	.000[b]	692.808[a]	3.633[b]
Two	97.125[a]	98.567[ab]	.003[b]	572.203[a]	3.605[a]
Five	97.345[a]	98.271[a]	.022[a]	615.774[a]	3.604[a]

Table 3. The mean percent leaf area, percent mass remaining, millipede weight lost, microbial biomass, and soil pH for each of the densities of millipedes in microcosms. Different letters in the same column signify significant differences among densities (Student-Neuman-Keuls tests; $\alpha = 0.05$).

Soil pH was significantly different among microcosms with different litter types, with *R. surinamensis* soil being the most acidic and Tabonuco (*D. excelsa*) soil being the least acidic (Fig. 5D). The presence of millipedes also caused the pH to slightly but significantly decline (Table 3). The soil pH at the final collection was significantly less acidic than in the first two collections (Table 2). Soil inorganic nitrogen was significantly lower in *R. surinamensis* soil than for the other two litter types (Figs. 7A-B). Soil nitrate did not vary through time (Figs. 7C). While, soil ammonium increased significantly from one collection to the next through time (Figs. 7D).

Neither form of inorganic nitrogen varied significantly with different densities of millipedes. Microbial biomass in the soil of Tabonuco microcosms was significantly higher than for the other two litter types (Fig. 5C). The microbial biomass did not differ depending on the density of millipedes. Additionally, microbial biomass significantly decreased throughout the experiment (Fig. 8). The millipedes lost weight as the experiment continued. Millipedes lost significantly more weight when they were in the highest density (Table 3).

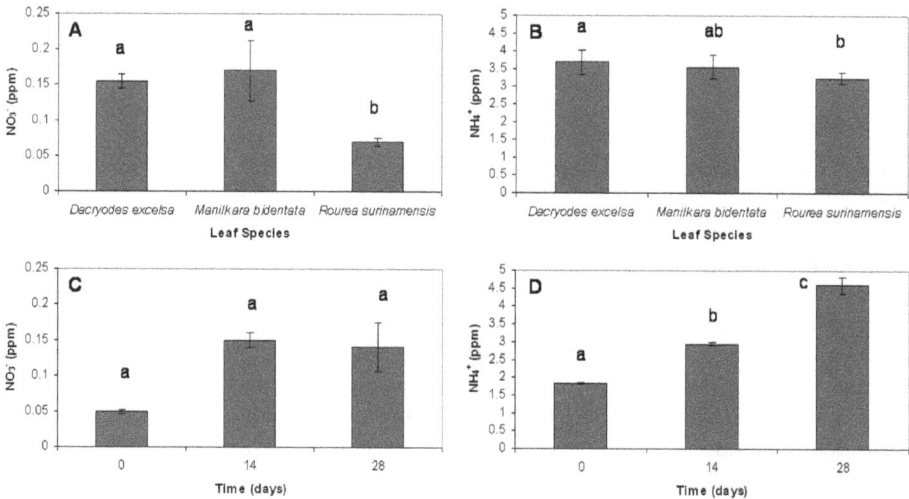

Fig. 7. (A) Soil nitrate and (B) ammonium content for microcosms containing leaves of *Dacryodes excelsa, Manilkara bidentata,* and *Rourea surinamensis.* (C) Soil nitrate and (D) ammonium measurements over the three collection times. Standard error of the mean is represented by error bars. Different letters in the same graph represent significant differences among collection times (Student-Neuman-Keuls tests; $\alpha = 0.05$).

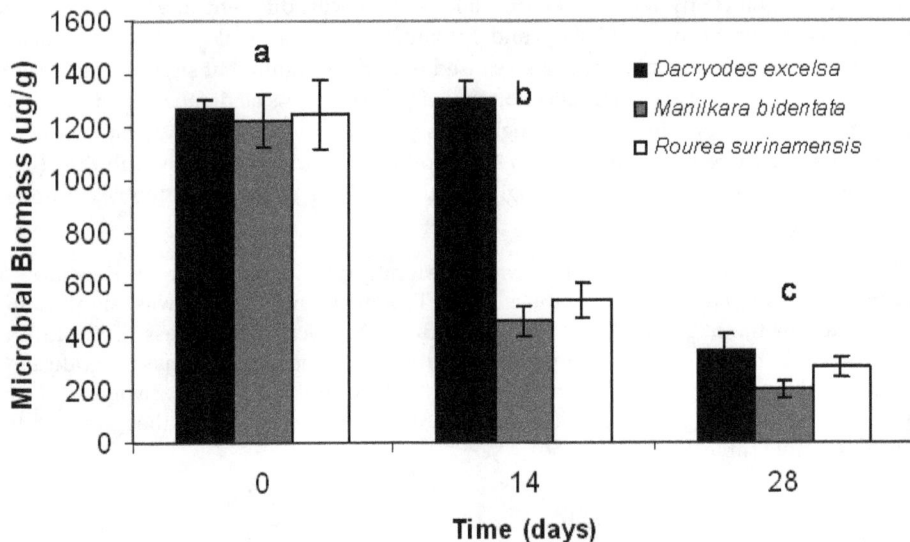

Fig. 8. The microbial biomass in microcosms containing leaves of *Dacryodes excelsa, Manilkara bidentata,* and *Rourea surinamensis* at three collection times. Different letters indicate significant differences among each of the collection times (Student-Neuman-Keuls tests, α= 0.05) of the overall microbial biomass.

4. Discussion

Millipedes directly impacted the decomposition of leaf litter in this microcosm study. The millipedes significantly fragmented the leaf litter by the fourth week. Direct fragmentation was affected by the substrate quality, with the higher lignin content (higher L/N ratio) leaf species having more leaf area remaining and percent mass remaining. The direct impact of millipedes also depended on the density of millipedes, with the higher density significantly decreasing the mass remaining of the leaves. Although this impact differed by leaf species: the low lignin content Tabonuco leaves lost more mass without millipedes at one of the collections, while the high lignin content *R. surinamensis* leaves lost more leaf mass with millipedes at the last collection (Fig. 6). These findings were similar to those of Tian et al. (1995), who found a decrease of percent mass remaining of leaves with very high lignin (47 percent) and C/N ratios by millipedes and earthworms in the field conditions. They concluded that microbes could independently breakdown the higher quality litter, so that the macrofauna influence was not as significant for higher quality leaves (<12 percent lignin content) (Tian et al., 1995), a probable explanation for our results as well. We did find that the high-quality *D. excelsa* soil had by far the greatest microbial biomass, suggesting that both the millipedes and microbes were influencing the breakdown of Tabonuco leaves. If we had continued this experiment for longer we might have seen more dramatic results in leaf fragmentation. For example, Coûteaux et al. (2002) found that after 198 days in a microcosm, a single millipede turned over on average nearly half of the five grams of pine litter given into frass when microcosms were kept at intermediate temperatures (15-32° C).

In the present study, leaf chemistry affected the fragmentation of leaves. Leaf species with a high percent of lignin and high L/N ratios were correlated with higher leaf area remaining and percent mass remaining. In the tropics, Aerts (1997) found that lignin-to-nitrogen ratios of leaf litter were the best chemical predictors of litter decomposition. In general, tropical leaf litter has lower L/N ratios with a mean of 24.2, while temperate regions average a value of 32.0, and 29.4 for Mediterranean regions (Aerts, 1997); our values ranged from 28 to 40 (Table 1).

Indirectly, the millipedes could have influenced the microbial biomass of the soil, though this influence is harder to tease apart than their direct effects. In this study, through time microbial biomass diminished, which could indicate that the millipedes had a negative effect on the microbes, although this argument is weakened by finding that adding millipedes to microcosms and increasing the density of millipedes had no significant effect on the microbial biomass. Our results suggest that millipedes had no effect on microbial biomass but that leaf chemistry did: percent leaf lignin and leaf L/N ratios were negatively correlated with soil microbial biomass. Kaneko (1999) found decreased microbial biomass when millipedes were present in a laboratory experiment, but when the same experiment was done in the field (an oak forest in Japan) millipedes had no effect on the microbial biomass. When Hanlon & Anderson (1980) introduced diplopods and isopods into microcosms they at first increased microbial biomass, but then returned to control levels after twelve days. We could have missed an initial increase in microbial biomass because our first collection was not until the fourteenth day. The decrease of microbes through time in this experiment could also be connected with the millipedes losing weight during the experiment, which seemed to be from the drier conditions of the microcosms because they did not hold water and humidity as well as their natural environment. We suggest that the effect of millipedes on microbial biomass be further tested under field conditions to confirm their impact.

The pattern of soil inorganic nitrogen could be explained in two ways. The first mechanism is by nitrogen leaching from the leaves. The soil under Tabonuco leaves, which had the highest percent nitrogen, also had the highest values of inorganic nitrogen while *R. surinamensis*, with the lowest percent nitrogen in their leaves, also had the lowest values of inorganic nitrogen in the soil. This pattern was not significant when a Pearson's correlation between the initial leaf chemistry and both soil inorganic nitrogen measures was calculated. Secondly, Tabonuco microcosms had the highest biomass of microbes, indicating a more active mineralization and faster turn over rate in the Tabonuco soil, which would release inorganic nitrogen into the soil. In this experiment, the latter seems to be the more probable source of inorganic nitrogen. Many authors have found significant increases in inorganic nitrogen from soil arthropods, due to their excrement, indirect efforts on microbes, and/or fragmentation of litter (Ineson et al., 1982; Persson, 1989; Setälä, et al., 1990; Teuben & Verhoef, 1992; Cárcamo et al., 2001; Pramanik et al., 2001; etc.), but this did not seem to be a factor in our results as millipede density had no significant effect on either soil ammonium or nitrate.

Both the leaf chemistry and the millipede density affected the soil pH: the less lignin and the absence of millipedes resulted in less acidic soil. Anderson & Domsch (1993) showed that microbial biomass decreases in more acidic soil, further influencing the decomposition. The distribution of microarthropods can also be influenced by soil pH (Hågvar, 1990;

Klironomos & Kendrick, 1995; Dlamini & Haynes, 2004). We found the highest microbial biomass in the less acidic Tabonuco soil (the least %L and L/N ratio) and this species also had significantly the least leaf area remaining and percent mass remaining, but the presence of millipedes also affected these results by decreasing the percent mass remaining.

5. Conclusion

We conclude that millipedes can impact leaf litter decomposition both directly and indirectly, but the extent of their effect depends on their density and the quality of the substrate (leaf lignin content). Directly, it is clear that millipedes fragment litter, which in some cases has been shown to indirectly increase microbial biomass (e.g. Hanlon, 1981a; 1981b). We did not find an increase but a decrease of soil microbial biomass over time, yet microbial biomass was not affected by the presence of millipedes, suggesting that other factors might have been driving this trend. Millipedes are not the only arthropods influencing decomposition, and their interactions with other organisms in their natural environment could affect the extent of their influence. Notwithstanding, we suggest that the results of this microcosm experiment be applied to millipedes in their natural setting, further confirming millipedes as important components of soil ecosystems and nutrient cycling.

6. Acknowledgements

This research was supported by grants #BSR-8811902, DEB-9411973, DEB-OO8538, and DEB-0218039 from the National Science Foundation to the Institute of Tropical Ecosystem Studies (IEET), University of Puerto Rico, and IITF as part of the Long-Term Ecological Research Program in the LEF. Additional support was provided by the Forest Service (U.S. Department of Agriculture) and done in cooperation with the University of Puerto Rico. Juliana Belén was supported through the REU Site program at El Verde Field Station, University of Puerto Rico, NSF-grant number DBI-0552567. We would like to thank María M. Rivera, Verónica Cruz, and Carmen Marrero for help with laboratory analyses. Special thanks to José Fumero and Benjamín Colón and all the REU students that were working at El Verde Field Station for assisting us with this project (Karla Campos, Danielle Knight, Johanne Fernández, Sharon Machín, Rebecca Clasen, and Miriam O'Neill). Drs. Barbara Richardson, Ariel E. Lugo, and D. Jean Lodge kindly provided comments on an earlier version of the manuscript.

7. References

Aerts, R. (1997). Climate, leaf litter chemistry and leaf litter decomposition in terrestrial ecosystems: a triangular relationship. *Oikos, 79*, 439-449.

Anderson, J. P. E., & Domsch, K. H. (1978). A physiological method for the quantitative measurement of microbial biomass in soils. *Soil Biology and Biochemistry, 10*, 215-221.

Anderson, J.M., & Bignell, D.E. (1980). Bacteria in the food, gut contents and faeces of the litter-feeding millipede *Glomeris marginata* Villers. *Soil Biology and Biochemistry, 12*, 251-254.

Anderson, T.H., & Domsch, K.H. (1993). The metabolic quotient for CO_2 (qCO_2) as a specific activity parameter to assess the effects of environmental conditions, such as pH, on the microbial biomass of forest soils. *Soil Biology and Biochemistry*, 25, 393-395.

Bonkowski, M., Scheu, S., & Schaefer, M. (1998). Interactions of earthworms (*Octolasion lacteum*), millipedes (*Glomeris marginata*) and plants (*Hordelymus europaeus*) in a beechwood on a basalt hill: implications for litter decomposition and soil formation. *Applied Soil Ecology*, 9, 161-166.

Brown, S., Lugo, A.E., Silander, S., & Liegel, L. (1983). Reseach history and opportunities in the Luquillo experimental forest. USDA Forest Service General Technical Report, New Orleans, LA.

Cárcamo, H.A., Abe, T.A., Prescott, C.E., Holl, F. B., & Chanway, C.P. (2000). Influence of millipedes on litter decomposition, N mineralization, and microbial communities in a coastal forest in British Columbia, Canada. *Canadian Journal of Forest Research*, 30, 817-826.

Cárcamo, H.A., Prescott, C.E., Chanway, C.P., & Abe, T.A. (2001). Do soil fauna increase rates of litter breakdown and nitrogen release in forests of British Columbia, Canada? *Journal of Forest Research*, 31, 1195-1204.

Coûteaux, M., Aloui, A., & Kurz-Besson, C. (2002). *Pinus halepensis* litter decomposition in laboratory microcosms influenced by temperature and a millipede, *Glomeris marginata*. *Applied Soil Ecology*. 20, 85-96.

Dlamini, T. C., & Haynes, R. J. (2004). Influence of agricultural land use on the size and composition of earthworm communities in northern KwaZulu-Natal, South Africa. *Applied Soil Ecology* 27, 77-88.

García-Martinó, A. R., Warner, G.S., Scatena, F.N., & Civco, D.L. (1996). Rainfall, runoff and elevation relationships in the Luquillo Mountains of Puerto Rico. *Caribbean Journal of Science*, 32, 413-424.

González, G., & Seastedt, T. R. (2001). Soil fauna and plant litter decomposition in tropical and subalpine forest. *Ecology*, 82, 955-964.

Hågvar, S. (1990). Reactions to soil acidification in microarthropods: Is competition a key factor? *Biology and Fertility of Soils*, 9, 178-181.

Hanlon, R.D., & Anderson, J.M. (1980) Influence of macroarthropod feeding activities on microflora in decomposition oak leaves. *Soil Biology and Biochemistry*, 12, 255-261.

Hanlon, R.D.G. (1981a). Influence of grazing by Collembola on the activity of senescent fungal colonies grown on media of different nutrient concentration. *Oikos*, 36, 362–367.

Hanlon, R.D.G. (1981b). Some factors influencing microbial growth on soil animal faeces II. Bacterial and fungal growth on soil animal faeces. *Pedobiologia*, 21, 264–270.

Ineson, P., Leonard, M.A., & Anderson, J.M. (1982). Effect of collembolan grazing upon nitrogen and cation leaching from decomposing leaf litter. *Soil Biology and Biochemistry*, 14, 601-605

Johnston, M.H. (1992). Soil-vegetation relationships in a Tabonuco forest community in the Luquillo mountains of Puerto Rico. *Journal of Tropical Ecology*, 8, 253-263.

Kaneko, N. (1999). Effect of millipede *Parafontaria tonominea* Attems (Diplopoda: Xystodesmidea) adults on soil biological activities: A microcosm experiment. *Ecological Research*, 14, 271-279.

Kheirallah, A.M. (1979). Behavioural preference of *Julus scandinavius* (Myriapoda) to different species of leaf litter. *Oikos*, 33, 466–471.

Klironomos, J.N., & Kendrick, B. (1995). Relationships among microarthropods, fungi, and their environment. *Plant and Soil*, 170, 183-197.

Lin, Q., & Brookes, P.C. (1999). Comparison of substrate induced respiration, selective inhibition and biovolume measurements of microbial biomass and its community structure in unamended, ryegrass-amended, fumigated and pesticide-treated soils. *Soil Biology and Biochemistry*, 31, 1999-2014.

Lyford Jr., W.H. (1943). Bottom-up control of the soil macrofauna community in a beechwood on limestone: manipulation of food resources. *Ecology*, 24, 252–261.

Maraun, M., & Scheu, S. (1996). Changes in microbial biomass, respiration and nutrient status of beech (*Fagus sylvatica*) leaf litter processed by millipedes (*Glomeris marginata*). *Oecologia*, 107, 131-140.

Murphy, C.M., González, G., & Belén, J. (2008). Ordinal abundance and richness of millipedes (Arthropoda: Diplopoda) in a subtropical wet forest in Puerto Rico. *Acta Científica 22, 1-3, 57-65*.

Persson, T. (1989). Role of soil animals in C and N mineralisation. *Plant and Soil*, 115, 241-245.

Pramanik, R., & Sarkar, K., Joy, V.C. (2001). Efficiency of detritivore soil arthropods in mobilizing nutrients of leaf litter. *Tropical Ecology*, 42, 51-58.

Richardson, B.A., Richardson, M.J., & Soto-Adames, F.N. (2005). Separating the effects of forest type and elevation on the diversity of litter invertebrate communities in a humid tropical forest in Puerto Rico. *Journal of Animal Ecology*, 74, 926-936.

Ruan, H., Li, Y., & Zou, X. (2005). Soil communities and plant litter decomposition as influenced by forest debris: Variation across tropical riparian and upland sites. *Pediobiologia*, 49, 529-538.

Setälä H., Martikainen, E., Tyynismaa, M., & Huhta, V. (1990). Effects of soil fauna on leaching of nitrogen and phosphorus from experimental systems simulating coniferous forest floor. *Biology and Fertility of Soils*, 10, 170-177.

Soil Survey Staff (1995). Order 1 soil survey of the Luquillo long-term ecological research grid, Puerto Rico. USDA, NRCS, Lincoln, Nebraska.

Tajovsky, K., Vilemin, G., & Toutain, F. (1991). Microstructural and ultrastructural changes of the oak leaf litter consumed by millipede *Glomeris hexasticha* (Diplopoda). *Revue D Ecologie Et De Biologie Du Sol*, 28, 287–302.

Teuben, A., & Verhoef, H.A. (1992). Direct contribution by soil arthropods to nutrient availability through body and faecal nutrient content. *Biology and Fertility of Soils*, 14, 71–75.

Tian, G., Brussaard, L., & Kang, B.T. (1993). Biological effects of plant residues with contrasting chemical compositions under humid tropical conditions: Effects on soil fauna. *Soil Biology and Biochemistry*, 25, 731-737.

Tian, G., Brussaard, L., & Kang, B.T. (1995). Breakdown of plant residues with contrasting chemical compositions under humid tropical conditions: Effects of earthworms and millipedes. *Soil Biology and Biochemistry*, 27, 277-280.

Van der Drift, J. (1975). The significance of the millipede *Glomeris marginata* (Villers) for oak-litter decomposition and an approach of its part in energy flow. In: Vanek, J. (Ed.), Progress in soil zoology. Academia, Prague, pp. 293–298.

Zalamea, M., & González, G. (2007). Substrate-induced respiration in puertorican soils: minimum glucose amendment. *Acta Científica*, 21, 1-3, 11-17.

5

Patterns of Plant Species Richness Within Families and Genera in Lowland Neotropical Forests: Are Similarities Related to Ecological Factors or to Chance?

Pablo R. Stevenson[1], María Clara Castellanos[2] and Juliana Agudelo T.[1]
[1]Departamento de Ciencias Biológicas, Universidad de los Andes, Bogotá,
[2]Consejo Superior de Investigaciones Científicas,
Centro de Investigaciones sobre Desertificación (CSIC-UV-GV), Valencia,
[1]Colombia
[2]España

1. Introduction

Present-day communities are the result of speciation, extinction, and migration (Leigh et al., 2004). Hence, the final outcome of these processes, the assemblage of species present at a site, has been sieved by both ecological and stochastic factors through time (Kristiansen et al., 2011; Stropp et al., 2009). Palinological evidence has demonstrated a correlation between Neotropical floral diversity and climatic change throughout the Cenozoic, showing the highest diversity in the Eocene during periods with high temperatures and an extensive area of tropical forests (Jaramillo et al., 2006). These results imply that present day communities are incomplete "museums" of plant diversity, which have suffered extinctions either through deterministic or stochastic effects. Then, findings from paleoecological studies suggest direct or indirect effects of macro-ecological factors, such as temperature and forested area, determining patterns of plant diversity in the tropics (Fine & Ree, 2006), and reveals patterns of species accumulation during hot and humid periods.

We still know little about the ecological and biogeographical factors that promote species diversification in plants. Plant-animal interactions, such as pollination and seed dispersal, have been proposed as mechanisms that promote plant diversification (Gentry, 1988). However, strict specializations are not common in pollination systems and the role of pollinators in generating diversification of plant species has not been supported (Gravendeel et al., 2004; Waser et al., 1996), although preferences may occur (Gong & Huang, 2011). Recent studies have suggested that habitat specialization to contrasting soil types, mediated by tradeoffs in strategies to avoid herbivory, has occurred in lowland Amazonian plants (Fine et al., 2006). Currently, among many theories to explain plant coexistence in diverse tropical forests at local scales, there are four hypotheses well supported which involve niche differentiation, infrequent competition among understory plants, host specific pests, and negative density dependent effects (Wright, 2002). According to Leigh et al. (2004), microhabitat specialization and disturbance appear insufficient to maintain alpha-diversity of trees in tropical forests, and there is a positive influence of larger areas covered by forests, where small populations have the chance to establish a new species.

Similarly, the disturbance and generation of new habitats that occurred during the uplift of the Andes ranges, was proposed as a major influence in the diversification of Neotropical plants. Gentry (1982) related the differences in familial composition to historical and ecological factors. He postulated the existence of two main centers of distribution in the Neotropics: Northern Andes and Central Amazonia. In each region, only certain families underwent processes of species diversification and, for this reason, present day floras in these two regions have an overrepresentation of certain families, in terms of species richness. Families with predominantly herb and shrub species tend to be diverse in Northwest South America (such as Costaceae, Gesneriaceae, Heliconiaceae, Zingiberaceae), while tree families are represented by many species in Central Amazonia (e.g. Burseraceae, Chrysobalanaceae, Lecythidaceae, Sapotaceae).

The most extreme extinction rates have been associated to drastic changes in global conditions (Benton & Twitchett, 2003), and local extinctions of particular plant species can also be caused by small changes in ecological conditions (Tilman & Lehman, 2001). Good examples of local extinctions come from studies of forest fragmentation, where changes in conditions, reductions in population numbers, and the extinction of mutualistic species tend to accelerate plant extinction rates and lower diversity (D'Angelo et al., 2004; Laurance et al., 2002). Therefore, even though fragmentation sets barriers to gene flow between populations, a process that might facilitate allopatric speciation and hence diversity (e.g. Haffer, 1969; Prance, 1982), evolutionary rates may be slow enough to allow differentiation of viable populations in such fragmented habitats.

Migration and colonization rates of trees have been estimated for temperate but not often for tropical regions. For instance, fossil pollen from northern latitudes has indicated rapid colonization rates unrelated with life history traits (i.e. dispersal kernels), which suggests that the colonization front was similarly limited to all plant species by climatic or geographic factors (Clark, 1998). Few studies have focused on colonization fronts in tropical plants, and these studies also suggest a large potential of rapid migration in these ecosystems (e.g. Charles-Dominique et al., 2003). In fact, the present geographic distributions of tropical plants show large variations in size, with some species restricted to particular sites, and other species with wide distributions (e.g. Henderson et al., 1995). This variation is consistent with the idea of rapid migration rates to areas with good climatic conditions, which may then be followed by local extinctions in less favorable periods, and low beta-diversity in western Amazonian forests (Condit et al., 2002). Although molecular analysis may provide information suggesting population dynamics of colonization, extinction and recolonization, this has been reported only a few times (e.g. Dutech et al., 2003), and it is difficult to be sure that local extinctions were driven by ecological or by stochastic population factors (e.g. random variations in population size).

In order to assess the relative importance of ecological vs. stochastic factors, several studies have quantified the proportion of the variation in floristic comparisons that can be attributed to ecological factors, then, the remaining variance in floristic patterns can be attributed to history or to chance (Tuomisto & Ruokolainen, 1997). On the other hand, neutral theories based on stochastic processes modeling community structure, predict a negative relationship between floristic affinity and geographical distance (Hubbell, 2001). In fact, some studies using abundance of individuals between plant groups showed that floristic affinities between sites decrease with geographical distance at some regional scales (Terborgh & Andressen, 1998; Tuomisto et al., 2003). However, these studies do not rule out the influence of ecological factors. In fact, several studies have found significant contributions of geographical distance and ecological factors in explaining patterns of floristic similarity (Chust et al., 2006; Plotkin et al., 2000; Pyke et al., 2001; Tuomisto et al., 2003).

All lowland rain forests in the Neotropics have floras with similar familial compositions (Gentry, 1988). This has been explained by the common origin of the most important families, which differentiated long before the separation of Gondwanaland (Gentry, 1982). Furthermore, lowland Neotropical forests have not been greatly influenced by the invasion of predominantly temperate families. Thus, plant families such as Leguminosae, Annonaceae, Lauraceae, Rubiaceae, Moraceae, Myristicaceae, Sapotaceae, Melicaceae, Palmae, Euphorbiaceae and Bignoniaceae are common in almost all Neotropical lowland forests (Gentry, 1988). However, there is variation in floristic composition among particular localities, and predominant plant families are not always the same among different regions. Gentry (1990) pointed out that macro-ecological factors (e.g. soil quality, rainfall patterns, pollination syndromes) are important in determining floristic composition and affinities of different areas. Based on a qualitative comparison of four florulas, he showed differences in families and habitat composition, which were explained mostly by broad categorization of ecological conditions. For instance, the richest taxa in hyper-humid sites corresponded to families with high representation of epiphytes such as Orchidaceae, Araceae, and Piperaceae, while Leguminosae dominated at other places. The aim of this study is to make a quantitative floristic comparison based on the patterns of species richness in families and genera for more than twenty tropical areas, and to correlate floristic similarities with ecological and stochastic factors (e.g. geographical distance). We attempted to test the significance and relative roles of ecological and stochastic factors from the following predictions. If floristic similarities are significantly affected by ecological variables we expected: 1) to find sites of similar conditions grouped together in ordination analyses, and 2) a significant correlation between matrices of floristic similarity and ecological factors in Mantel tests (Mantel, 1967). On the other hand, according to the hypothesis that floristic composition is determined by chance, we expected to find: 1) agglomeration of close-by sites in the ordination, and 2) a positive correlation between geographical distance and floristic dissimilarity.

2. Methods

2.1 Localities

We searched the literature of florulas in lowland Neotropical areas, and included all places with appropriate macro-ecological information and good collection effort. We obtained a database of 26 sites for families and 25 sites for genera (Fig 1). In order to avoid biased inventories due to small sampling effort, we only included humid forests with at least 1,000 species reported, and dry forest with more than 500 species. We searched for geographic coordinates of each site to estimate the geographic distance between sites (calculated as a distance along the earth curvature). When the florula corresponded to a large area, we used the centroid of the area to calculate the geographic distances to other places. The ecological conditions for each site were also extracted from the literature and from databases. We included information on average annual rainfall, average number of dry months (months with precipitation lower than 100 mm), and temperature for the 26 locations included in the analyses (Table 1).

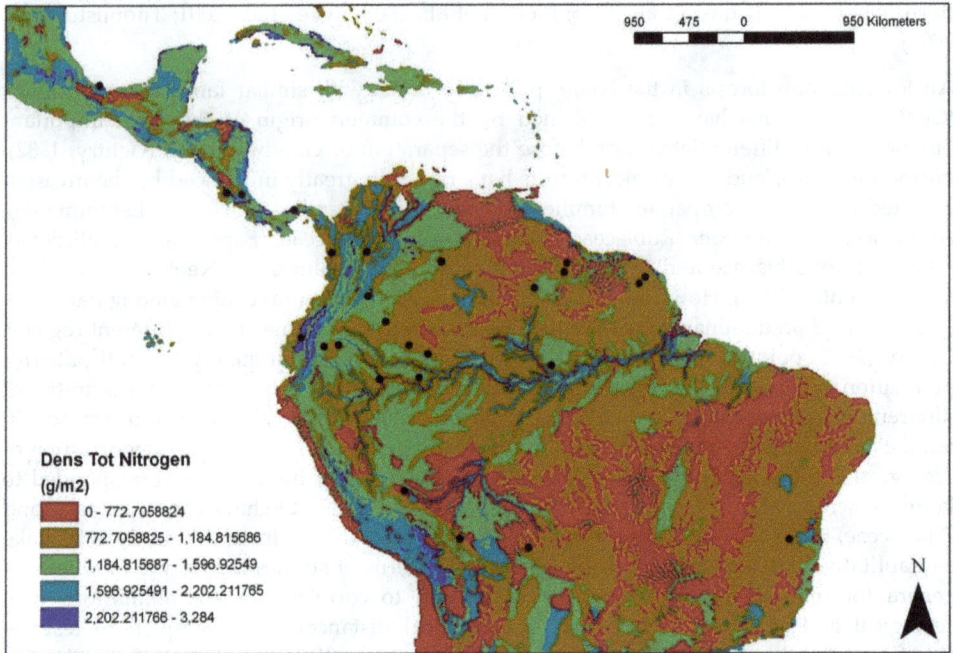

Fig. 1. Map of the neotropical region showing the location of the analyzed florulas, superimposed on a map of nitrogen content in the soils (from a geochemical and ecological database, ORNL DAAC, http://daac.ornl.gov/).

Site	Rainfall (mm)	Dry mo.	Temp (°C)	Reference	Coordinates		
					Long	Lat.	
1.Iquitos (Perú)	2949	0	25.9	Vásquez, 1997	73.30	3.50	S
2.Rio Palenque (Ecuador)	2650	0	23	Dodson & Gentry, 1978	79.40	0.40	S
3.La Selva (Costa Rica)	3962	0	25.8	Hammel & Grayum, 1982	84.00	10.40	N
4.Choco (Colombia)	6573	0	27	Forero & Gentry, 1989	77.00	6.00	N
5.Caqueta (Colombia)	3060	0	25.7	Duivenvoorden, 1996	71.00	1.00	S
6.Jatun Sacha (Ecuador)				http://www.mobot.org/MOBOT/research/ecuador/jatun/checklist.shtml	77.6	1.07	S

Site	Rainfall (mm)	Dry mo.	Temp (°C)	Reference	Coordinates		
					Long	Lat.	
7.Leticia (Colombia)	3215	0	25.8	Rudas & Prieto, 1998	70.20	3.40	S
8.Tuxtlas (Mexico)	4725	1	23.2	Ibarra-Manriquez & Sinaca-Colin, 1995	95.10	18.60	N
9.Tinigua (Colombia)	2702	2	26	Stevenson et al., 2000	74.20	2.70	N
10.Nouragues (French Guyana)	3124	2	27	Forget, 1994	52.70	4.10	N
11. Iwokrama (French Guyana)	2200	2		www.iwokrama.org	59.00	4.50	N
12.Mabura Hill (French Guyana)	2700	2	25.9	Renske & ter Steege, 1998	58.80	5.20	N
13.Cocha Cashu (Perú)	2028	3	24	Gentry, 1990	71.40	11.90	S
14.Bahia (Brazil)	1502	3.5	24.5	Mori et al., 1983	41.50	15.60	S
15.Ducke (Brazil)	2186	4	26.7	Gentry, 1990	60.00	2.50	S
16.BCI (Panama)	2656	4	27	Gentry, 1990	79.90	9.20	N
17. Maracá (Brazil)	2300	5	26.5	Thompson, 1992; Moskovits, 1985	61.30	3.30	N
18.Saul (French Guyana)	2413	6	27.1	Mori & Boom, 1987	53.20	3.60	N
19. Beni (Bolivia)	2550	2.5		Smith & Killeen, 1998	67.12	15.53	S
20.Las Quinchas (Colombia)	2654.7	0.5	27.8	Balcazar-Vargas et al., 2000	74.27	6.05	N
21. Chiribiquete (Colombia)	4000	0		Cortés-B & Franco-R,1997	72.80	0.80	N
22.Tuparro (Colombia)	2708	4		Mendoza et al., 2004	68.50	5.28	N
23. Yasuni (Ecuador)	2717	0		Valencia, 2004	76.5	0.95	S
24.Caparú (Colombia)	4000	0		Clavijo et al., 2009	69.52	1.65	S
25.Sta Rosa (Costa Rica)	1503	6	27.4	Enquist & Sullivan, 2001	85.20	10.50	N
26. Chiquitania (Bolivia)	1129	7	24.3	Killeen et al., 1998	61.80	16.20	S

Table 1. List of sites included in the study, their main climatic characteristics and location.

To facilitate comparison between localities, we used the floristic categories proposed by Foster and Hubbell (1990). Ferns are included as one group without differentiation of families, the three legume subfamilies (Papilionoidae, Mimosoidae and Caesalpinioidae) are presented as one (Fabaceae) and we did not treat Cecropiaceae within Urticaceae.

2.2 Climate and soil information

For each study site we obtained soil data from a geochemical and ecological database, ORNL DAAC (http://daac.ornl.gov/). Data were collected in the field, by satellite or generated by models. We chose 7 variables that describe soil properties: soil-carbon density (kg/m^2), profile available water capacity (mm), total nitrogen density (g/m^2), bulk density $(g/cm3)$, field capacity (mm) (PsiFC=-10 kPa), thermal capacity (J/m3/K) (Theta=0.00 %v/v), and wilting point (mm) (PsiWP=-1500 kPa).

We characterized each of our study sites with 25 climatic variables found in WorldClim (http://www.worldclim.org; Hijmans et al., 2005). These bioclimatic variables are derived from monthly rain values and temperature, and represent annual trends. We included limiting environmental factors such as the temperature of the coldest and the warmest month, the rainfall of the 3 rainiest, and driest months. Additionally, the information for the number of dry months was taken from the field database or from the literature, because this variable explains a large part of the variation on maximal diversity of plants in the Amazon (Ter Steege et al., 2003).

We characterized each of the study sites in terms of soil and climate variables by spatially locating their area using the ArcGis program (http://www.arcgis.com/). Then, we quantified the weighted average of each variable for each site, depending on the characteristics of each polygon. We used this average as the value for each variable to comparison between the sites.

2.3 Floristic analysis

We ranked families and genera within each locality because the collection efforts were dissimilar between florulas, thus, it was not possible to quantify the vegetation by the absolute number of species. We included the 20 families with the highest number of species for each location. We assigned ranks to the families in the list (i.e. the most species-rich family in the list got a value of 20, the second a value of 19, and so on). The same was done for the top 22 genera in each locality. Then, we calculated an index of floristic similarity between sites based on these ranks (Stevenson, 2004):

$$D_{ab} = \sum_i \left| \text{rank } i_a - \text{rank } i_b \right| / (n(n+1)) \qquad (1)$$

where, D_{ab} the floristic distance for each pair of localities (a and b) was the sum of the absolute differences between ranks. Then, 'i' was each of the families in the list included in both localities and 'n' is the number of families included (20 in this case). The division by $n(n+1)$ standardizes the index between 1 and 0. High values (close to 1) indicated higher floristic differences. For instance, shared families that have high ranks in both localities contribute little to the index, while a high-ranking family from one locality that is absent in the other contributes the most. This index was used for one of the matrices (floristic matrix) that we then used in Mantel Tests (Mantel, 1967). We also used this index for the 22 richest genera in each site. The cut off was generated as a tradeoff between increasing sampling size and avoiding sites with little information.

2.4 Statistical analysis

First, we ran a correlation analysis between the 26 environmental and soil variables to exclude redundant variables. Then we kept only variables which were not highly correlated with other independent variables (r > 0.6, Appendix 1).

We ordinated the localities according to the initial ranks using Nonmetric Multidimensional Scaling (NMS) in PCORD (Pc-Ord for Windows, Multivariate Analysis of Ecological Data. 5 version). We ran different analyses for family and genus information (Terborgh & Andresen, 1998). We did not make analyses at the species level because these comparisons could be compromised if species are misidentified, or if the same species is given a different name just because of its geographic location. To estimate floristic similarity, we used Euclidian distances from the rank matrix using PCORD. We allowed 20 runs for the NMS analysis, which used Euclidean distances. *A posteriori*, we determined the families (or genera) and the ecological factors that showed the highest correlation coefficients with the two main axes of the ordination, and according to the critical value of the Pearson's r (> 0.33) (Acton, 1966). Relevant families/genera and ecological factors can be graphically overlaid on the ordination.

To determine whether the observed patterns in the ordination were explained by geographical distances and/or ecological factors, we performed Mantel tests to evaluate the relationship between three different matrices for the 26 locations. The first matrix was a floristic distance matrix constructed from the ranks described above (either families or genera). The second matrix contained the values of climate and soil variables associated to each site. A third distance matrix included the Euclidean distances between the localities based on ecological factors and the distance from the young mountains (Andes and Central American ridges up to 1000m), as a proxy to relatively fertile sedimentation soils (Gentry, 1990).

The subsequent Mantel Tests were ran to evaluate the interdependency of the independent variables that we had chosen. Thus, we constructed distance matrices (differences between places for each variable), and we made comparisons between each pair of matrices to determine if the observed patterns in the ordination were correlated with geographic distances or with ecological factors. These correlations are an important step to identify general pattern variations, but they cannot estimate how important each variable is in relation to the effects of other ones. For this reason, from each distance matrix between variables, we first did Mantel Tests to evaluate the relation between the different variables and the floristic distance. We then identified the variables that better explained the floristic distance. We ran partial Mantel tests while holding the geographical distances constant to observe how much variation in the floristic distance was explained by it. The mantel Tests were made with Excel Mantel Stats (XLStats, Statistical Software for MS Excel), and the significance of Mantel coefficients was tested via permutation tests with 10000 iterations.

3. Results

3.1 Family level comparisons

We found a wide spread of the 26 localities in the ordination based on floristic similarities (Fig 2). The main ecological variables associated with the arrangement of sites were rainfall and temperature. In particular, we found that the two driest sites are located at the lower left corner of the ordination, and high values corresponded mainly to wet sites with short dry seasons. Therefore, the x axis of the ordination showed a high negative correlation with the number of dry months (Fig 2.b). Most sites in the upper right corner of the ordination corresponded to rainy Amazonian and Guyannan sites, relatively far from the Andean and Central American mountains. These regions usually have nutrient-poor soils; however, we did not find any significant correlation between ordination axes and soil traits (Table 1). The two sites with lowest values in the y-axis corresponded to hyper-humid forests in the Choco biogeographic region.

We found that the distribution of sites in the ordination was highly correlated with how rich are particular plant families in species (Table 2.a). For instance, the sites where Annonaceae was highly speciose, were located in the upper right corner of the floristic ordination (Figure 2c). We found a similar pattern for Lauraceae, Chrysobalanaceae, Clusiaceae and Sapotaceae. We also found that Bignoniaceae, Poaceae and Asteraceae were species rich families in the dry sites, showing a negative correlation with axis 1 (Table 2a). We also found a negative association between the number of species of Orchidaceae and Piperaceae and the axis 2 (Table 2a), mainly due to the richness in rainy forest in the Choco biogeographical region.

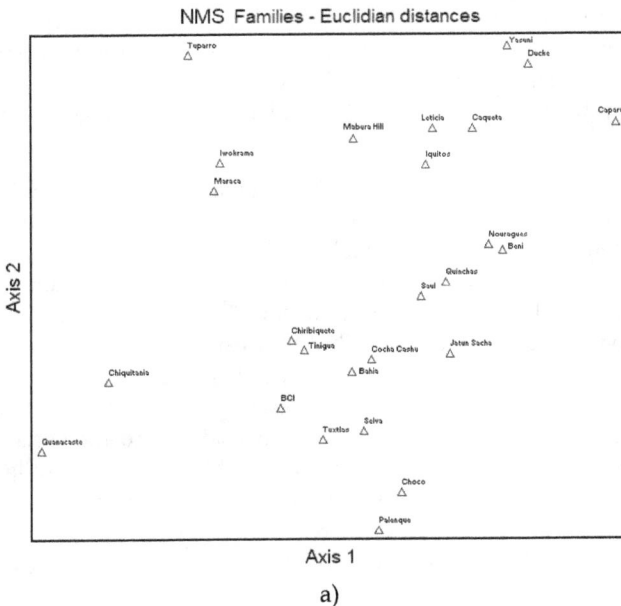

a)

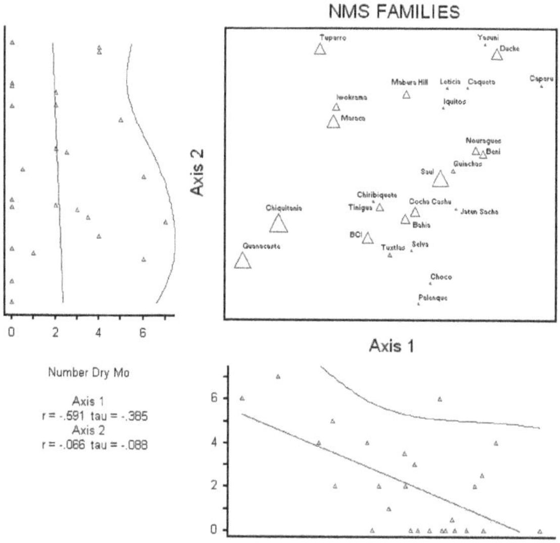

Fig. 2. NMS ordination of the 26 Neotropical localities, according to floristic affinities in terms of the number of species per family (a). In (b) the number of dry months for each site in the ordination is represented by the size of the triangle (a bigger size triangle indicates a longer dry season). The graphic also indicates the relation between the number of dry months and the two axes of the ordination. As an example, the number of species of the Annonaceae family is indicated in (c) again, by the size of the triangle. In this case the family is highly correlated with axis 2.

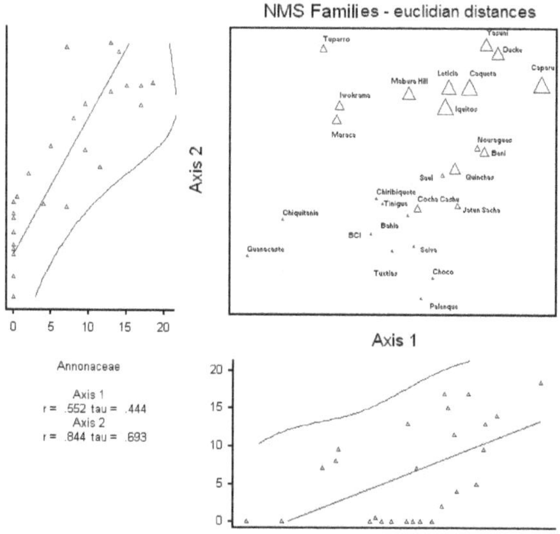

Even though we found that floristic dissimilarities among locations were correlated with ecological variables, we did not find that floristic affinities at the family level were associated with the geographical distances between them. Using Mantel tests, we obtained a significant association for three climatic variables (i.e. number of dry months, maximum temperature of the warmest month, and the mean day temperature; Table 3a). We obtained the same results in Mantel tests, even when the geographical distances were held constant, again showing a small effect of the geographic distance, and the largest effect from the number of dry months.

3.2 Genus level comparisons

In the ordination of the 25 localities using their similarities in genera composition, we found an atypical composition for the Costa Rican locality of Guanacaste (Fig 3.a), which was characterized by an extended dry season (Fig 3.b). The distribution of sites in the ordination was highly correlated with some plant genera (Table 2.b). The most influential genera showed a positive correlation with axis 1, and correspond to species-rich genera that are well represented in Guanacaste as *Desmodium* (Fig 3.c), *Hyptis, Ipomoea, Sida, Acacia, Capparis, Lonchocarpus and Mimosa* (Table 2.b.). *Eschweilera, Pouteria, Licania*, in contrast, showed negative correlations with axis 1, while *Trichomanes, Psychotria* and *Casseavia* were highly correlated with axis 2.

We found that floristic dissimilarities among the locations were correlated with macro ecological descriptors of the forests. Using Mantel tests, we found a significant correlation between the floristic and several climatic variables. In this case, the number of dry months again showed the highest correlation (e.g., Table 1b, Fig. 3b). The partial Mantel Tests showed that controlling for the geographic distance did not affect the effect of ecological variables.

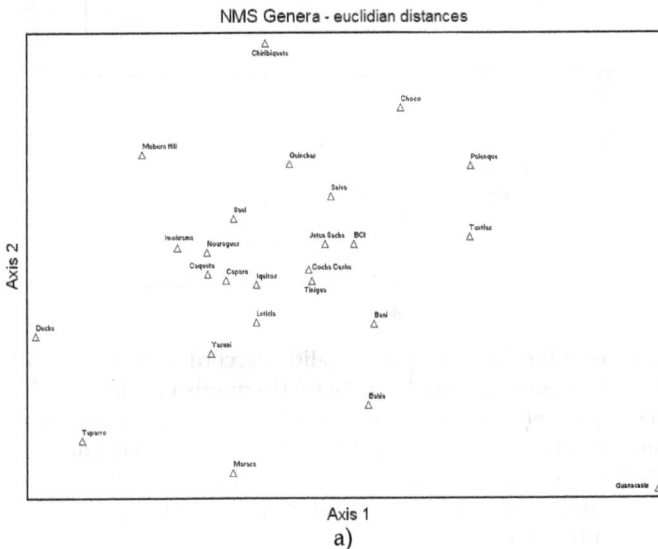

NMS Genera - euclidian distances

a)

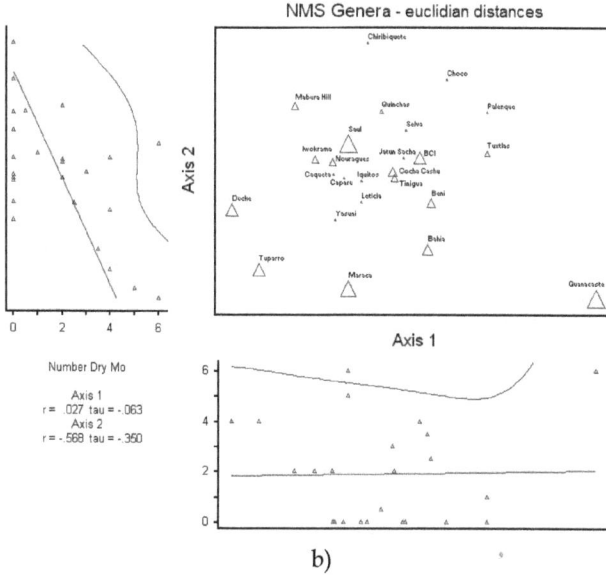

b)

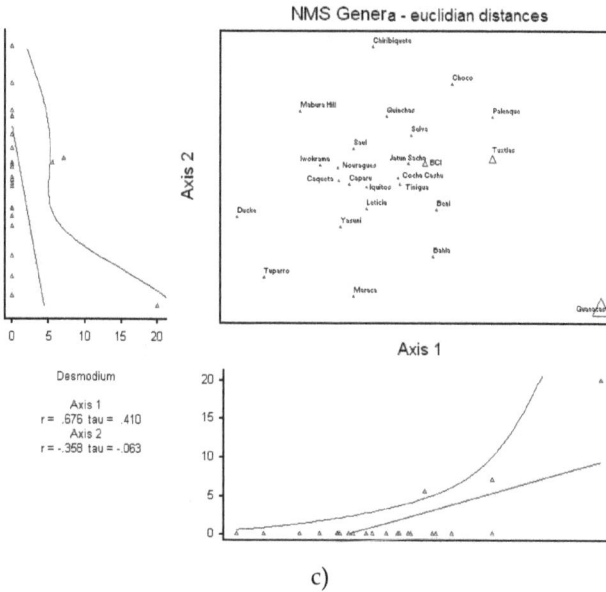

c)

Fig. 3. NMS ordination of the 25 Neotropical localities according to the number of species
per genera (a), and (b) the relation between the number of dry months in the ordination
from the floristic affinities, in terms of species richness per genera. A bigger size triangle
indicates longer duration of the dry season, and the graphics indicate a relation between the
number of dry months and the two ordination axis. One of the genera with the highest
correlation with axis 1 is *Desmodium* and it is mostly present in Guanacaste (c).

	Axis 1	Axis 2
	R	R
Annonaceae	0.552	0.844
Lauraceae	0.562	0.753
Chrysobalanaceae	0.252	0.627
Clusiaceae	0.422	0.607
Sapotaceae	0.619	0.523
Bignoniaceae	-0.619	0.011
Cyperaceae	-0.66	-0.002
Poaceae	-0.758	-0.392
Orchidaceae	-0.193	-0.626
Piperaceae	0.167	-0.722
Solanaceae	-0.209	-0.759
Asteraceae	-0.456	-0.832

a)

	Axis 1	Axis 2
	R	R
Desmodium	0.676	-0.358
Hyptis	0.665	-0.321
Ipomea	0.608	-0.416
Acacia	0.604	-0.430
Trichomanes	-0.069	0.636
Psychotria	-0.094	0.542
Solanum	0.592	-0.005
Acalypha	0.590	-0.420
Calliandra	0.590	-0.420
Capparis	0.590	-0.420
Cassia	0.590	-0.420
Lonchocarpus	0.590	-0.420
Mimosa	0.590	-0.420
Sida	0.590	-0.420
Piper	0.566	0.396
Croton	0.555	-0.396
Casearia	-0.248	-0.639
Pouteria	-0.525	0.009
Miconia	-0.601	0.004
Eschweilera	-0.604	0.09
Licania	-0.796	-0.238

b)

Table 2. Most dominant families (a) and genera (b), that are correlated with the sites distribution of the ordination.

Patterns of Plant Species Richness Within Families and Genera in Lowland Neotropical Forests. Are Similarities Related to Ecological Factors or to Chance?

83

	r(AB)	p-value
Number of dry months	0.245	< 0.001
Maximum Tº warmest month	0.179	0.002
Mean diurnal temperature	0.170	0.003
Distance to mountains	0.049	0.388
Temperature seasonality	0.106	0.053
Annual precipitation	0.077	0.166
Geographical distance	0.069	0.208
Total nitrogen density	0.016	0.768
Field capacity	-0.032	0.432
PAWC	-0.051	0.368

a)

	r(AB)	p-value
Number dry months	0.242	< 0.01
Maximum Tº warmest month	0.340	< 0.01
Geographical distance	0.269	< 0.01
Seasonality in temperature	0.268	< 0.01
Distance to mountains	0.210	< 0.01
Annual precipitation	0.167	< 0.01
Mean diurnal temperature	-0.032	0.437
Total nitrogen density	0.050	0.223
Field capacity	-0.053	0.200
PAWC	-0.014	0.740

b)

Table 3. Results of the bivariate Mantel Tests of each variable vs floristic distance for families (a) and for genera (b). The highlighted variables were the ones with a significant value.

4. Discussion

Our quantitative results support Gentry's ideas (1982, 1988, 1990) that floristic affinities in terms of species richness are determined mainly by ecological factors. Analyses at both the familial and generic levels showed significant correlations between ecological factors (i.e.

the number of dry months), and floristic affinities based on patterns of species richness. Interestingly, an analysis of plant diversity in Amazonia has also pointed to the duration of the dry season as the most important variable explaining maximum diversity (Ter Steege et al., 2003). It is possible that ecological variables related to temperature should have been important if highland forests were included, since temperature, altitude above sea level, and floristic affinities are well correlated (Gentry, 1988; 1995). Although the estimated amount of nitrogen in the soil was not highly correlated with floristic affinities in our analyses, we still think that nutrients in the soil may affect floristic composition. For instance, the fact that distance to young Andean and Central American mountains was correlated with the affinity of species richness within genera suggests that areas with similar sedimentation history (Latrubesse et al., 2010), show similar floristic composition. In addition, it is difficult to quantify soil nutrients at the spatial scale of florulas, and the spatial distribution of phosphorus, perhaps the most important nutrient in the soil for tropical plants, is not available at the spatial scales applied in this study.

The main families driving their placement in the first axis of the ordination differ in their dispersal systems and habit types. For instance, the most negatively correlated families were Poaceae, Cyperaceae, Bignoniaceae and Asteraceae, all characterized by abiotic seed dispersal and herbaceous and vine habits (Heywood et al., 2007). In contrast, the most positively correlated families were Sapotaceae, Lauraceae, Annonaceae, represented by trees with fleshy fruits dispersed by animals. The second axis shows the same dichotomy, but includes additional families of fleshy fruited species (Chrysobalanaceae) and wind dispersed species (Orchidaceae), both commonly found in humid forests (Gentry, 1995). These results suggest that particular families have functional traits that make them well suited for particular ecological settings and may coexist in places where they have reproductive advantages in comparison to other plant strategies. In fact, it is well established that large seeds are common in tree species (Foster & Janson, 1985), because large seeds have establishment advantages under closed canopy forests. In contrast, herbs and shrubs tend to have small seeds and are frequently represented in savannas, forest edges, and open canopy forests (Laurance et al., 2002; Stevenson & Rodriguez, 2008). However, biogeographical history might also influence the patters just described, since the large seeded, animal dispersed families represented by trees might have diversified in central Amazonia (Gentry, 1982; Stropp et al., 2009). Therefore, analyses of the plant traits promoting the establishment under particular conditions should control for phylogenetic and biogeographic history.

Similarly, at the genus level, the first axis of the ordination was negatively correlated with genera of large seeded animal dispersed seeds (*Licania* and *Eschweilera*), while it showed a high positive correlation with abiotically dispersed plants mainly represented by shrubs and vines (e.g., *Hyptis, Ipomea, Acacia, Acalypha* and *Calliandra*). However, it also includes genera dispersed by small animals such as birds and bats (*Miconia, Solanum* and *Capparis*), which are the most common seed dispersers in fragments and disturbed habitats (Pizo, 2004; Terborgh et al., 2008). The first axis was also positively correlated with the number of species of *Desmodium*, a genus dispersed in the fur of animals. These comparisons suggest that the patters of species richness might depend on ecological factors, such as the occurrence of dispersal agents and regeneration requirements.

The analyses at the genus level also showed a significant correlation between geographical distance and floristic affinity, as predicted by stochastic processes (Hubbell, 2001). Therefore, chance and mass-effects also play roles in the structure of plant communities in lowland Neotropical forests, and this effect seems to be more evident at low taxonomic scales. However, comparisons at the species level are more complicated, due to differences in the collection efforts and ambiguities in species determination, even though preliminary observations at the specific level also suggest a strong influence of ecological factors. For example, when the flora that we are more familiar with (Tinigua National Park, Stevenson et al., 2000) is compared to species lists of other Neotropical localities, we observed that the most similar sites correspond to western Amazonian localities with a dry period (i.e. Iquitos and Cocha Cashu). These sites are also very close to Tinigua in the generic ordination (Fig. 2), and not far away, though mixed, with other sites in the familial analysis (Fig. 1). These results could be better explained by ecological factors than by stochastic or historic events. Tinigua, Iquitos, and Cocha Cashu have 2-3 mo. dry seasons, and the three sites have relatively fertile soils because they have a greater influence of sedimentation from Andean soils. However, patches of forest on white sands are also common in the Iquitos area (Fine et al., 2006). Moreover, some of the same species are dominant at these localities, for instance, Foster's (1990) description of the floodplain at Cocha Cashu points to the dominance of *Guarea guidonia* and *Cecropia membranacea* in the early stages in riverine succession. These two species are not only the most important species in the flooded forests at Tinigua, but are also dominant in early succession processes (Stevenson et al., 2004), and *Heliconia marginata* dominates the understory in both places. Recent beaches are colonized mainly by *Tessaria integrifolia* and *Gynerium sagitatum* in both Cocha Cashu and Tinigua (Terborgh, 1983; Hirabuki et al., 1991).

The high floristic similarity between Tinigua and Cocha Cashu does not support the refuge theory, suggested originally by Haffer (1969) for Neotropical birds, and applied by Prance (1982) to the distribution of plant taxa. This theory proposed that in periods of increased aridity during the Pleistocene, populations were split into small patches of forests (refuges), where speciation occurred, followed by re-colonization of the forest. Haffer suggested that the actual distribution of species should therefore reflect the location of Pleistocene refuges. At least four refuges have been proposed for the upper Amazon basin, with Tinigua and Cocha Cashu near to two different refuges. If the present distribution of plant species originated in different refuges, then the flora at Tinigua and Cocha Cashu should be very different given the proximity to different refuges and the large geographical distance between the two places, but that is not the case. Interestingly, intermediate places such as Amazonian Ecuador, where there is no dry season, differ from Tinigua and Cocha Cashu in floristic composition. Thus our results provide futher evidence against the refuge theory (Colinvaux, 2005).

In summary, contrary to neutral theories, we can affirm that current and past macro-ecological factors have played significant roles determining the patterns of species richness in Neotropical lowland forests. Our analyses showed a minor effect of stochastic factors, but significant at some levels (i.e. genus). Does this mean that Neotropical plant communities are structured by niche differences? Although we did not address this question here, we think that this is not necessarily the case. For example, the fact that families with high representation of epiphytes dominate florulas in very humid sites does not imply that all species are partitioning the resources. On the contrary, the vast number of co-occurring species suggests that conditions are good enough to allow many species to coexist, in spite of using similar resources and ecological strategies (e.g., dispersal systems and establishment requirements). Thus, some degree of stochasticity in population dynamics and speciation patterns may occur nested within the controlling macro-ecological factors, as well as past ecological conditions (Stropp et al., 2009). Perhaps we should not be too worried about trying to understand how many species live in present-day tropical forests, since the fossil record shows that many more species can coexist under the appropriate climatic conditions (Jaramillo et al., 2006).

5. Conclusion

We compared information on 26 lowland Neotropical florulas, in order to assess which processes are correlated with the patterns of floristic similarities, based on plant species richness within families and genera. The results at the family level indicated that floristic similarity is significantly correlated with ecological factors (e.g., rainfall patterns, temperature and the distance to young mountains as a proxy of sedimentation processes), but is not correlated with geographical distance. At the genus level, again, ecological factors were highly correlated with floristic similarity. However, at this level geographical distance was also significantly correlated with floristic similarity. These quantitative results support Gentry's theory which states that floristic affinities, in terms of patterns of species richness, are determined mainly by ecological factors. However, stochastic processes seem to play a minor but significant role, given that the most species rich genera were similar between close-by areas, as predicted by neutral models. Our findings and an accumulating body of evidence show that forest composition does change along environmental gradients (e.g., Bohlman et al., 2008; Coronado et al., 2009; Engelbrecht et al., 2007; Pitman et al., 2008: Tuomisto, 2006), in spite of the occurrence of widely distributed species along Neotropical forests (Bohlman et al., 2008; Condit et al., 2002). This emphasizes the relevance of protecting in areas of high human preference, because they are usually located in particular ecological settings and floristic composition. It is clear that human impacts, such as deforestation, have been prevalent in areas of high crop and livestock productivity (Madriñan et al. 2007), that include a set of unique native species. Therefore, a holistic approach for biodiversity conservation should provide the protection of forest in all ecological settings, including sites with high quality soils and productivity.

6. Acknowledgment

We thank the department of Biological Sciences and the Faculty of Sciences at Universidad de Los Andes, for the support to conduct this investigation. We thank Diana Guzman for comments and corrections.

7. Appendix 1

Correlation analysis among the 26 environmental and soil variables. Excluded variables that were associated with a correlation index of more than 0.6 are shown in grey.

	1	2	3	4	5	6	7	8	9	10	11	12	13	14	15	16	17	18	19	20	21	22	23	24	25	26
1. Shriveling point	1																									
2. Total Nitrogen density	-0.548	1																								
3. Thermal conductivity	0.230	-0.813	1																							
4. Soil humidity	0.06	-0.24	0.00	1.00																						
5. Field capacity	0.87	-0.58	0.20	0.54	1.00																					
6. Apparent soil density	0.23	-0.81	1.00	0.00	0.19	1.00																				
7. Soil carbon density	-0.68	0.95	-0.75	-0.34	-0.74	-0.75	1.00																			
8. Mean annual temperature	-0.01	0.10	-0.06	-0.24	-0.13	-0.06	0.07	1.00																		
9. Daytime mean temperature	0.18	-0.46	0.37	0.37	0.34	0.37	-0.49	-0.43	1.00																	
10. Isothermality	0.24	-0.11	-0.06	0.01	0.21	-0.05	-0.14	0.42	-0.15	1.00																
11. Seasonal temperature	-0.21	-0.01	0.14	0.16	-0.10	0.14	0.02	-0.58	0.44	-0.87	1.00															
12. Temperature of hottest month	0.00	-0.07	0.15	-0.05	-0.02	0.15	-0.13	0.74	0.07	-0.09	-0.03	1.00														
13. Lowest temperature of coldest month	0.00	0.23	-0.21	-0.27	-0.14	-0.21	0.20	0.90	-0.71	0.56	-0.77	0.44	1.00													
14. Annual temperature range	0.00	-0.30	0.32	0.28	0.14	0.32	-0.29	-0.60	0.82	-0.68	0.84	0.06	-0.87	1.00												
15. Mean temperature of most humid quartile	-0.20	0.14	-0.06	-0.11	-0.22	-0.05	0.11	0.92	-0.32	0.27	-0.37	0.76	0.75	-0.42	1.00											
16. Mean temperature of driest quartile	0.06	0.12	-0.11	-0.18	-0.04	-0.10	0.08	0.94	-0.58	0.40	-0.62	0.65	0.94	-0.69	0.82	1.00										
17. Mean temperature of hottest quartile	-0.12	0.17	-0.06	-0.22	-0.21	-0.06	0.14	0.95	-0.39	0.16	-0.33	0.85	0.79	-0.42	0.93	0.89	1.00									
18. Mean temperature of coldest quartile	0.04	0.12	-0.12	-0.23	-0.08	-0.12	0.07	0.96	-0.50	0.61	-0.79	0.57	0.96	-0.75	0.83	0.93	0.84	1.00								
19. Annual precipitation	0.04	-0.06	-0.14	0.15	0.11	-0.14	-0.02	0.16	-0.37	0.42	-0.41	-0.14	0.36	-0.48	0.10	0.31	0.06	0.27	1.00							
20. Precipitation of most humid month	-0.22	0.24	-0.37	0.18	-0.09	-0.37	0.29	0.18	-0.47	0.07	-0.20	0.04	0.36	-0.38	0.18	0.38	0.19	0.23	0.81	1.00						
21. Precipitation of driest month	0.23	-0.25	0.05	0.18	0.28	0.06	-0.24	0.06	-0.19	0.63	-0.49	-0.33	0.25	-0.46	0.00	0.14	-0.12	0.20	0.83	0.41	1.00					
22. Seasonal precipitation	-0.43	0.32	-0.08	0.00	-0.37	-0.08	0.32	-0.19	0.23	-0.60	0.53	0.23	-0.32	0.49	-0.08	-0.25	0.00	-0.31	-0.60	-0.12	-0.82	1.00				
23. Precipitation of most humid quartile	-0.21	0.21	-0.38	0.27	-0.04	-0.38	0.24	0.15	-0.41	0.14	-0.21	-0.01	0.32	-0.36	0.15	0.32	0.14	0.21	0.86	0.98	0.51	-0.19	1.00			
24. Precipitation of driest quartile	0.24	-0.26	0.06	0.17	0.29	0.06	-0.24	0.10	-0.22	0.65	-0.52	-0.29	0.30	-0.49	0.03	0.20	-0.07	0.25	0.87	0.45	0.99	-0.84	0.54	1.00		
25. Precipitation of hottest quartile	-0.29	0.21	-0.36	0.44	-0.03	-0.36	0.21	-0.29	0.02	0.21	-0.02	-0.44	-0.15	-0.07	-0.15	-0.25	-0.33	-0.20	0.61	0.50	0.60	-0.29	0.61	0.57	1.00	
26. Precipitation of coldest quartile	0.33	-0.17	-0.04	-0.08	0.24	-0.04	-0.15	0.41	-0.38	0.51	-0.56	0.12	0.57	-0.56	0.20	0.54	0.29	0.50	0.81	0.61	0.65	-0.56	0.63	0.71	0.15	1.00

8. References

Acton, F. S. (1966). *Analysis of straight line data*. Dover, New York.

Balcázar-Vargas, M. P., Rangel-Ch., J.O. & Linares-C, E.L. (2000). Diversidad florística de la Serranía de Las Quinchas, Magdalena Medio (Colombia). *Caldasia*, 22, pp. (191-224)

Benton, M. J. & Twitchett, R. J. (2003). How to kill (almost) all life: the end-Permian extinction event. *Trends in Ecology & Evolution*, 18, pp. (358-365)

Bohlman, S. A., Laurance, W. F., Laurance, S. G., Nascimento, H. E. M., Fearnside, P. M., Ana, A. (2008). Importance of soils, topography and geographic distance in structuring central Amazonian tree communities. *Journal of Vegetation Science* 19, pp. (863-874)

Charles-Dominique, P., Chave, J., Dubois, M. A., De Granville, J. J., Riera, B. & Vezzoli, C. (2003). Colonization front of the understorey palm *Astrocaryum sciophilum* in a pristine rain forest of French Guiana. *Global Ecology and Biogeography*, 12, pp. (237-24)

Chust, G., Chave, J., Condit, R., Aguilar, S., Lao, S. & Perez, R. (2006). Determinants and spatial modeling of tree beta-diversity in a tropical forest landscape in Panama. *Journal of Vegetation Science*, 17, pp. (83-92)

Clark, J. S. (1998). Why trees migrate so fast: Confronting theory with dispersal biology and the paleorecord. *American Naturalist*, 152, pp. (204-224)

Clavijo, L., Betancur, J. & Cérdenas, D. (2009). Plantas con flores de la Estación Biológica Mosiro-Itajura-Caparú, Vaupés, Amazonia colombiana, In: *Estación Biológica Mosito Itajura-Caparú. Biodiversidad en el territorio del Yagojé-Apapporis*, G. Alarcón-Nieto & E. Palacios, pp. (55-97). Conservación Internacional Colombia, Bogotá

Colinvaux, P. (2005). The Pleistocene vector of neotropical diversity. In: *Tropical rainforests: past, present and future*, Bermingham, E. Dick C. W. Moritz C., pp (78-106). University of Chicago Press, Chicago.

Coronado, E. N., Baker, T. R., Phillips, O. L., Pitman, N. C. A., Pennington, R. T., Vasquez-Martinez, R., Monteagudo, A., Mogollon, H., Davila-Cardozo, N., Rios, M., Garcia-Villacorta, R., Valderrama, E., Ahuite, M., Huamantupa, I., Neill, D. A., Laurance, W. F., Nascimento, H. E. M., de Almeida, S. S., Killeen, T. J., Arroyo, L., Nunez, P., Freitas Alvarado, L. (2009). Multi-scale comparisons of tree composition in Amazonian terra firme forests. *Biogeosciences*, 6, pp. (2719-2731)

Condit, R., Pitman, N., Leigh, E. G., Chave, J., Terborgh, J., Foster, R. B., Nunez, P., Aguilar, S., Valencia, R., Villa, G., Muller-Landau, H. C., Losos, E. & Hubbell, S. P. (2002). Beta-diversity in tropical forest trees. *Science*, 295, pp. (666-669)

Cortes, R. & Franco, P. (1997). Análisis panbiogeografico de la flora de Chiribiquete, Colombia. *Caldasia*, 19, 3, pp. (465-478)

D'Angelo, S. A., Andrade, A. C. S., Laurance, S. G., Laurance, W. F. & Mesquita, R. C. G. (2004). Inferred causes of tree mortality in fragmented and intact Amazonian forests. *Journal of Tropical Ecology*, 20, pp. (243-246)

Dodson, C. & Gentry, A. H. (1978). Flora of the Rio Palenque Science Center. *Selbyana*, 4, pp. (1-623)

Duivenvoorden, J. F. (1996). Patterns of tree species richness in rain forests of the middle Caqueta area, Colombia, NW Amazonia. *Biotropica*, 28, 2, pp. (142-158)

Dutech, C., Maggia, L., Tardy, C., Joly, H. I. & Jarne, P. (2003). Tracking a genetic signal of extinction-recolonization events in a neotropical tree species: *Vouacapoua americana* aublet in French Guiana. *Evolution*, 57, pp. (2753-2764)

Engelbrecht, B. M. J., Comita, L. S., Condit, R., Kursar, T. A., Tyree, M. T., Turner, B. L., Hubbell, S. P. (2007). Drought sensitivity shapes species distribution patterns in tropical forests. *Nature* 447, pp. (80-82)

Enquist, B. J. & Sullivan, J. J. (2001). *Vegetative key and descriptions of tree species of the tropical dry forest of upland Sector Santa Rosa, Area de Coservación Guanacaste, Costa Rica*, Retrieved from: http://acguanacaste.ac.cr/paginas_especie/plantae_online/EnquistSullivanTreeK ey.pdf

Fine, P. V. A., Miller, Z. J., Mesones, I.,, Irazuzta, S., Appel, H. M., Stevens, M. H. H., Saaksjarvi, I., Schultz, L. C. & Coley, P. D. (2006). The growth-defense trade-off and habitat specialization by plants in Amazonian forests. *Ecology*, 87, pp. (150-162)

Fine, P. V. A. & Ree, R. H. (2006). Evidence for a time-integrated species-area effect on the latitudinal gradient in tree diversity. *American Naturalist* 168, 6, pp. (796-804)

Forero, E. & Gentry, A. H. (1989). *Lista Anotada de las plantas del departamento del Chocó, Colombia. Bogotá, Instituto de Ciencias Naturales - Museo de Historia Natural -,* Universidad Nacional de Colombia, Bogotá

Forget, P. M. (1994). Recruitment Pattern of *Vouacapoua americana* (Caesalpiniaceae), a Rodent-Dispersed Tree Species in French Guiana. *Biotropica*, 26, 4, pp. (408-419)

Foster, R. B. (1990). The Floristic Composition of the Rio Manu Floodplain Forest, In: *Four Neotropical Rainforests*, A. H. Gentry, pp. (99-111), Yale University Press, New Haven

Foster, R. B., & Hubbell, S. P. (1990). The Floristic Composition of the Barro Colorado Island Forest, In: *Four Neotropical Rainforests*, A. H. Gentry, pp. (85-89), Yale University Press, New Haven

Foster, S. A., & Janson, C. H. (1985). The relationship between seed size and establishment conditions in tropical woody-plants. *Ecology*, 66, pp. (773-780)

Gentry, A. H. (1982). Neotropical floristic diversity: phytogeographical connections between Central and South America, Pleistocene climatic fluctuations, or an accident of the Andean orogeny? *Annals of the Missouri Botanical Garden*, 69, pp. (557-593)

Gentry, A. H. (1988). Changes in plant community diversity and floristic composition on geographical and environmental gradients. *Annals of the Missouri Botanical Garden*, 75, pp. (1-34)

Gentry, A. H. (1990). Floristic similarities and differences between Southern Central America and Upper and Central Amazonia, In: *Four Neotropical Rainforests*, A. H. Gentry, pp. (141-157), Yale University Press, New Haven

Gentry, A. H. (1995). Patterns of diversity and floristic composition in neotropical montane forest. In: *Biodiversity and conservartion of neotropical montane forests*. Churchill S. P., Baslev, H., Forero E. & Lutyn, J. L., pp. (103-126). The New York Botanical Garden. New York

Gong, Y. B. & Huang, S. Q. (2011). Temporal stability of pollinator preference in an alpine plant community and its implications for the evolution of floral traits. *Oecologia*, 166, pp. (671-680)

Gravendeel, B., Smithson, A., Slik, F. J. W. & Schuiteman, A. (2004). Epiphytism and pollinator specialization: drivers for orchid diversity? *Philosophical Transactions of the Royal Society of London Series B-Biological Sciences*, 359, pp. (1523-1535)

Haffer, J. (1969). Speciation in Amazonian forest birds. *Science*, 165, pp. (131-137)

Hammel, B. E. & Grayum, M. H. (1982). Preliminary report on the flora project of La Selva Field Station, Costa Rica. *Annals of the Missouri Botanical Garden*, 69, 2, pp. (420-425)

Henderson, A., Galeano, G. & Bernal, R. (1995). *Field guide to the palms of the Americas*, Princeton University Press, Princeton, N.J, pp. (363)

Heywood, V, H., R. K. Brummitt, R. K., Culham A.& Seberg, O. (2007). *Flowering plant families of the world*. Royal Botanic Gardens, Kew, pp. (424)

Hijmans, R. J., S. E. Cameron, Parra, J. L., Jones, P. G. & Jarvis, A. (2005). Very high resolution interpolated climate surfaces for global land areas. *International Journal of Climatology*, 25, 15, pp. (1965-1978)

Hirabuki, Y., Takehara, A. & Hara, M. (1991) Some characteristics of fluvial soils along a riparian succession in the upper Colombian Amazon. *Field Studies of New World Monkeys La Macarena Colombia* 5, 17-24.

Hubbell, S. P. (2001). *The unified neutral theory of biodiversity and biogeography*, Princeton University Press, Princeton, N.J, pp. (375)

Ibarra-Manriquez, G. & Sinaca-Colin, S. (1995). Commented Checklist of Plants From the Los-Tuxtlas Biological Station, Veracruz, Mexico (Mimosaceae-Verbenaceae). *Revista De Biologia Tropical*, 44, 1, pp. (41-60)

Jaramillo, C., Rueda, M. J. & Mora, G. (2006). Cenozoic plant diversity in the Neotropics. *Science*, 311, pp. (1893-1896)

Killeen, T. J., Jardim, A., Mamani, F. & Rojas, N. (1998). Diversity, composition and structure of a tropical semideciduous forest in the Chiquitania region of Santa Cruz, Bolivia. *Journal of Tropical Ecology*, 14, pp. (803-827)

Kristiansen, T., J. C. Svenning, Pedersen, D., Eiserhardt, W. L., Grández, C. & Balslev, H. (2011). Local and regional palm (Arecaceae) species richness patterns and their cross-scale determinants in the western *Amazon. Journal of Ecology*, 99, pp. (1001-1015)

Latrubesse, E. M., Cozzuol, M., da Silva-Caminha, S. A. F., Rigsby, C. A., Absy, M. L. & Jaramillo, C. (2010). The Late Miocene paleogeography of the Amazon Basin and the evolution of the Amazon River system. *Earth-Science Reviews*, 99, pp. (99-124)

Laurance, W. F., Lovejoy, T. E., Vasconcelos, H. L., Bruna, E. M., Didham, R. K., Stouffer, P. C, Gascon, C., Bierregaard, R. O., Laurance, S. G. & Sampaio, E. (2002). Ecosystem decay of Amazonian forest fragments: A 22-year investigation. *Conservation Biology*, 16, pp. (605-618)

Leigh, E. G., Davidar, P., Dick, C. W., Puyravaud, J. P., Terborgh, J., ter Steege, H. & Wright, S.J. (2004). Why do some tropical forests have so many species of trees? *Biotropica*, 36, pp. (447-473)

Madriñan, L. F., Etter, A., Boxall, G. D., Ortega-Rubio, A. (2007). Tropical alluvial forest fragmentation in the eastern lowlands of Colombia (1939-1997). *Land Degradation & Development*, 18, pp. (199-208)

Mantel, N. (1967). The detection of disease clustering and a generalized regression approach. *Cancer Research*, 27, pp. (209–220).

Mendoza, H. (2004). *Caracterización florística y faunística del PNN El Tuparro*. Instituto Alexander von Humboldt, Bogotá

Mori, S. A. & Boom, B. M. (1987). The forest. *Memoirs of the New York Botanical Garden*, 44, pp. (9-29)

Mori, S. A., Boom, B. M., de Carvalho, A. M. & dos Santos, T. S. (1983). Southern Bahian Moist Forests. *Botanical Review*, 49, 2, pp. (155-232)

Moskovits, D. K. (1985). *The behavior and ecology of the two Amazonian tortoises, Geochelone carbonaria and Geochelone denticulata, in northwestern Brasil*. PhD thesis, The University of Chicago, Department of Biology, Chicago, pp. (328)

Pitman, N. C. A. Mogollon, H., Davila, N., Rios, M. Garcia-Villacorta, R., Guevara, J., Baker, T. R., Monteagudo, A., Phillips, O. L., Vasquez-Martinez, R., Ahuite, M., Aulestia, M., Cardenas, D., Ceron, C. E., Loizeau, P. A., Neill, D. A., Percy, N. V., Palacios, W. A., Spichiger, R, & Valderrama, E. (2008). Tree community change across 700 km of lowland Amazonian forest from the Andean foothills to Brazil. *Biotropica*, 40, pp. (525-535)

Pizo, M. A. (2004). Frugivory and habitat use by fruit-eating birds in a fragmented landscape of southeast Brazil. *Ornitologia Neotropical*, 15, pp. (117-126)

Plotkin, J. B., Potts, M. D., Leslie, N., Manokaran, N., LaFrankie, J. & Ashton, P. S. (2000). Species-area curves, spatial aggregation, and habitat specialization in tropical forests. *Journal of Theoretical Biology*, 207, pp. (81-99)

Prance, G. T. (1982) Forest refuges: evidence from woody angiosperms, In: *Biological divesification in the tropics*, G.T. Prance, pp. (137-158), Columbia University Press, New York.

Pyke, C. R., Condit, R., Aguilar, S. & Lao, S. (2001). Floristic composition across a climatic gradient in a neotropical lowland forest. *Journal of Vegetation Science*, 12, pp. (553-566)

Renske, C. & Ter Steege, H. (1998). Studies on the flora of the Guianas no. 89: The flora of the Mabura Hill area, Guyana. *Botanische Jahrbuecher fuer Systematik Pflanzengeschichte und Pflanzengeographie*, 120, 4, pp. (461-502)

Rudas, A. & Prieto, A. (1998). Analisis floristico del Parque Nacional Natural Amacayacu e Isla Mocagua, Amazonas (Colombia). *Caldasia*, 20, 2, pp. (142-172)

Smith, D. N. & Killeen, T. J. (1998). A comparison of the structure and composition of montane and lowland tropical forest in the Serrania Pilon Lajas, Beni, Bolivia, In: *Missouri Botanical Garden Webpage*, Available from: http://www.mobot.org/MOBOT/research/bolivia/pilonarticle/welcome.shtml

Stevenson, P. R. (2004). Phenological patterns of woody vegetation at Tinigua Park, Colombia: Methodological comparisons with emphasis on fruit production. *Caldasia*, 26, 1, pp. (125-150)

Stevenson, P. R., Quiñones, M. J. & Castellanos, M. C. (2000). *Guía de Frutos de los Bosques del Río Duda, Macarena, Colombia*, Asociación Para la Defensa de La Macarena – IUCN, Bogotá

Stevenson, P. R. & Rodríguez, M. E. (2008). Determinantes de la composición florística y efectos de borde en un fragmento de bosque en el Guaviare, Amazonia colombiana. *Colombia Forestal*, 11, pp. (5-17)

Stevenson, P. R., Suescun, M., and Quiñones, M. J. (2004). Characterization of forest types at the CIEM, Tinigua Park, Colombia. *Field Studies of Fauna and Flora Macarena Colombia*, 14, (1-20)

Stropp, J., Ter Steege, H. & Malhi, Y. (2009). Disentangling regional and local tree diversity in the Amazon. *Ecography*, 32, 1, pp. (46-54)

Ter Steege, H., Pitman, N., Sabatier, D., Castellanos, H., Van der Hout, P., Daly, D. C., Silveira, M., Phillips, O., Vasquez, R., Van Andel, T., Duivenvoorden, J., De Oliveira, A. A., Ek, R., Lilwah, R., Thomas, R., Van Essen, J., Baider, C., Maas, P., Mori, S., Terborgh, J., Vargas, P. N., Mogollon, H. & Morawetz, W. (2003). A spatial model of tree alpha-diversity and tree density for the Amazon. *Biodiversity and Conservation*, 12, pp. (2255-2277)

Terborgh, J. (1983) *Five new world primates*, Princeton University Press, Princeton, pp (260)

Terborgh, J. & Andresen, E. (1998) The composition of Amazonian forests: patterns at local and regional scales. *Journal of Tropical Ecology*, 14, pp. (645-664)

Terborgh, J., Nuñez-Iturri, G., Pitman, N. C. A., Valverde, F. H. C., Alvarez, P., Swamy, V., Pringle, E. G., Paine, C. E. T. (2008). Tree recruitment in an empty forest. Ecology, 89, pp (1757-1768)

Tilman, D. & Lehman, C. (2001). Human-caused environmental change: Impacts on plant diversity and evolution. *Proceedings of the National Academy of Sciences of the United States of America*, 98, pp. (5433-5440)

Thompson, J., Proctor, J., Viana, V., Milliken, W., Ratter, J. A. & Scott, D. A. (1992). Ecological studies on a lowland evergreen rain-forest on Maraca Island, Roraima, Brazil .1. Physical-environment, forest structure and leaf chemistry. *Journal of Ecology*, 80, pp. (689-703)

Tuomisto, H. (2006). Edaphic niche differentiation among Polybotrya ferns in western Amazonia: implications for coexistence and speciation. *Ecography*, 29, pp. (273-284)

Tuomisto, H. & Ruokolainen, K. (1997). The role of ecological knowledge in explaining biogeography and biodiversity in Amazonia. *Biodiversity and Conservation*, 6, pp. (347-357)

Tuomisto, H., Ruokolainen, K. & Yli-Halla, M. (2003). Dispersal, environment, and floristic variation of western Amazonian forests. *Science*, 299, pp. (241-244)

Valencia R. (2004). Yasuni forest dynamics plot, Ecuador. In: Losos EC, Leigh J, Giles E (eds). Tropical Forest Diversity and Dynamism: Findings from a Large-Scale Plot Network. University of Chicago Press: Chicago. pp (609–628)

Vásquez, R, Rudas LLeras, A. & Taylor, C. M. (1997). *Flórula de las reservas biológicas de Iquitos, Perú: Allpahuayo-Mishana, Explornapo Camp, Explorama Lodge,* Missouri Botanical Garden Press, St. Louis, pp. (1046)

Waser, N. M., Chittka L, Price, M. V., Williams, N. M. & Ollerton, J. (1996). Generalization in pollination systems, and why it matters. *Ecology*, 77, pp. (1043-1060)

Wright, J. S. (2002). Plant diversity in tropical forests: a review of mechanisms of species coexistence. *Oecologia*, 130, pp. (1-14)

WorldClim – Global Climate Data, Available from: http://www.worldclim.org

6

Dispersion, an Important Radiation Mechanism for Ectomycorrhizal Fungi in Neotropical Lowland Forests?

Bernard Moyersoen
University of Aberdeen, School of Biological Sciences,
UK

1. Introduction

Mycorrhizas are symbiotic associations between plant roots and fungi. They are divided into different categories depending on morphological characteristics and the identity of the fungi and plants (Smith & Read, 2008). In tropical rainforests, most tree species are associated with arbuscular mycorrhizas (AM) (Alexander, 1989). Neotropical tree species belonging to confirmed ectomycorrhizal (EcM) host genera are diverse and generally scattered in a wide range of vegetation types (Table 1). *Pakaraimaea dipterocarpacea,* an endemic tree species from Guayana region (Maguire et al., 1977; Maguire & Ashton, 1980; Moyersoen, 2006), is one of the few known locally dominant EcM tree species in the Neotropical lowland forests. This tree is phylogenetically related to the most important EcM tropical tree family in SE Asia: the Dipterocarpaceae. The disjunct distribution of *P. dipterocarpacea* suggests that the capacity to associate with EcM evolved in the ancestors of Dipterocarpaceae before the splitting of South America from Gondwana, c. 135 million year ago (Ma) (Moyersoen, 2006). There are alternative proposals to explain *P. dipterocarpacea* EcM status such as independent acquisition of EcM association in South America or a more recent transoceanic long-distance dispersal of EcM Dipterocarpaceae to the Neotropics (Alexander, 2006). Biotrophic fungi such as EcM can be an important source of information for improved understanding of disjunct distributions (Pirozynski, 1983). The phylogeny of *P. dipterocarpacea* EcM fungi might reflect the disjunct distribution of this endemic tree species. A pioneering survey on *P. dipterocarpacea* indicated that broad host range fungal lineages distributed across the tropics are associated with this host plant (Moyersoen, 2006), but no phylogeographic studies on *P. dipterocarpacea* EcM fungi are available to date.

Apart from human transport, there are three possible hypotheses to predict the phylogeography of EcM fungal community associated with *P. dipterocarpacea,* i.e. vicariance, dispersion and/or migration. Both migration and vicariance assume that there is a close and specialized relationship between the fungus and the plant partners in EcM symbiosis (Pirozynski, 1983; Halling et al., 2008). Any discovery of fungi with an old Gondwanan origin associated with *P. dipterocarpacea* would support the vicariance hypothesis. Radiations of EcM fungi across continents are explained by co-migrations of both the fungus and the host partners (Halling et al., 2008; Pirozynski, 1983) and the capacity of the fungi to switch hosts (Halling, 2001). The possible long term co-existence of Neotropical Dipterocarpaceae with other EcM host families in the same region (including the EcM Fabaceae *Dicymbe*) might have favored this radiation scenario (Moyersoen, 2006). Dispersion assumes that EcM fungi are able to cross environmental barriers, usually by spore dispersal. Long distance dispersion, presumably by *trans*-Tasman airflow, was proposed to explain the disjunct distribution of *Pisolithus* (Moyersoen et al., 2003) and *Inocybe* (Matheny et al., 2009) species between Australia and New Zealand. The biotrophic status of EcM fungi is an important constraint for the movement of these fungi and the relative importance of long distance dispersion versus migration and vicariance in global EcM fungi distribution is under debate (Halling et al., 2008).

Species in VG* (total species)	For each entire genus			
	Life forms	Vegetation range	Altitudinal range	Geographical range
Caesalpiniaceae Amherstieae				
Dicymbe 11 (19)	Small to large trees	Rainforests, shrublands, shrublands on rocky substrate, gallery forests	200-2700m	Southwestern Colombia, Southern Venezuela, Guyana, Surinam, Northwestern Brazil
Fabaceae s.str.				
Aldina 18 (22)	Small to large trees	Rainforests, shrublands, gallery forests, white sand shrub savannas, forest-savanna ecotone, savannas, riparian forests, swamp forests, shrublands on rocky substrate	100-1800m	Southwestern Colombia, Southern Venezuela, Guyana, Northwestern Brazil
Dipterocarpaceae				
Pakaraimaea 1 (1)	Trees,or shrubs	Forests on sandy soil, shrublands on rocky slopes	500-1100m	Southern Venezuela, Guyana
Pseudomonotes 0 (1)	Trees	Rainforests on clayey to sandy soil	200-300m	Southwestern Colombia
Pisonieae				
Guapira 13 (ca 50)	Trees or shrubs	Evergreen, semideciduous or deciduous forests, gallery forests, riparian forests, savanna-dry forest ecotone, white sand shrublands	50-1300m	Wide distribution from Mexico, West Indies to Brazil, Paraguay, and from Peru, Ecuador to French Guiana, Trinidad-Tobago
Neea 30 (ca 85)	Trees or shrubs	Rainforests, seasonally dry evergreen forests, semievergreen forests, riparian forests, savanna-dry forest ecotone, white sand savannas, white sand shrublands, shrublands	50-2000m	Neotropics

Species in VG* (total species)	Life forms	For each entire genus		
		Vegetation range	Altitudinal range	Geographical range
Pisonia 1 (ca 40)	Small trees, shrubs or climbers	on rocky substrate, gallery forests, flooded forests Varied	ca 300m	New World and Old World, subtropical and tropical
Coccolobeae				
Coccoloba 29 (400)	Shrubs, trees with scram-bling branches or lianas	Riparian forests, gallery forests, rainforests, savanna-forest ecotone, savannas, semideciduous forests, decidous forests, shrublands on rocky substrate, white sand shrublands, seasonnally flooded or flooded forests	1-2000m	Neotropics
Gnetaceae				
Gnetum 6 (ca 40)	Lianas	Riparian forests, gallery forests, flooded forest margins, rainforests, vegetation on rocky slopes or white sand soils, swamps, savanna-forest ecotone	50-1800m	New World and Old World, subtropical and tropical

Table 1. Richness and distribution of confirmed EcM plant genera in the Neotropical lowland forests. *Venezuelan Guayana. Source : Steyermark et al. (1995)

The objective of this study was to evaluate the different hypotheses about the possible origin of EcM fungi associated with *P. dipterocarpacea*. For comparisons between *P. dipterocarpacea* EcM fungi and other tropical tree species in the same region or elsewhere, EcM fungi diversity and community structure was evaluated in a plot, where clumps of *P. dipterocarpacea* are present together with individuals of the EcM tree *Aldina* sp., Fabaceae, Papilionoïdeae. *Inocybe* was selected for further phylogenetic analysis, because of its global importance as an EcM fungal lineage, the general knowledge about its biogeography and its hypothesized paleotropical origin (Kuyper, 1986; Matheny et al., 2009; Ryberg, 2009).

A great diversity of EcM fungi, comparable to other tropical or temperate EcM rich communities, was found in this study. The EcM community structure was similar to that found in another forest dominated by EcM Fabaceae *Dicymbe* sp. in the same region. This study was the first evidence that Neotropical host tree species belonging to the Dipterocarpaceae and the EcM Fabaceae share EcM fungi both within the same forest and across forest stands. The floristic similarities and phylogenetic relationships of the EcM fungi between *P. dipterocarpacea* and *Dicymbe* forests suggested that fungal dispersion is an important radiation mechanism for EcM fungi in the region. Close phylogenetic relationships between *Inocybe* species associated with *P. dipterocarpacea* and African strains confirmed EcM phylogeographic links between the two continents. This study should be extended to examine a new *P. dipterocarpacea* dominated forest. The possible importance of dispersion in EcM fungal radiation and the link with paleotropical fungi should be tested in additional fungal lineages.

2. Methods

2.1 Study site

The sampling site was as described by Moyersoen (2006), located at 4°20'N, 61°48'W, altitude 500 m, near Icabarú Village, in Gran Sabana, Estado Bolivar, Venezuela, (Fig. 1). The precise location of the 20 X 20 m plot was selected based on a previous report of a stand of *P. dipterocarpacea* ssp. *nitida* described by Maguire & Steyermark (1981) in the same area (Fig.2-a). A second EcM tree, *Aldina* sp., also occurs in the plot (Moyersoen, 2006). Four separate sampling expeditions were conducted, in November 2003 (one day), March 2006 (one day), July 2007 (5 days) and July 2008 (9 days), respectively.

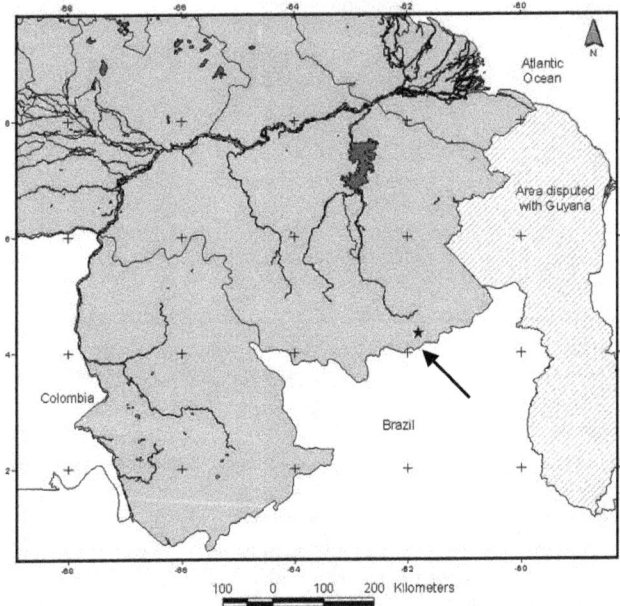

Fig. 1. Location of the field site

2.2 Fruit body and fine root sampling

All terricolous fruit bodies belonging to different morphotypes as well as resupinate fruit bodies growing under dead trunks and branches were collected inside the 20 X 20 m plot in 2007 and 2008 (Fig. 2-b, c, d). A brief description from fresh material was made in the field for each collection and a piece of mycelium from the cap was taken and fixed in cetyltrimethylammonium bromide (CTAB) for further molecular study. Time lapse between collection and fixation was less than one day. Microscopic observations of 3% KOH rehydrated material was carried out using a light microscope (Leica DM 2500) . Fruit body collections were assigned to families or genus using general keys and assigned to morphospecies. Fungi were considered to be putatively EcM on the basis of identification (e.g. Rinaldi et al., 2008; Singer, 1986; Smith & Read, 2008).

Fig. 2. Ectomycorrhizal (EcM) elements in *Pakaraimaea dipterocarpacea* forest plot. (a) Clump of *P. dipterocarpacea*; (b) *Cortinarius* sp. (BM08C30) surrounded by clavate *Clavulina* sp. (cf. BM07C9) fruit bodies; (c) *Austroboletus rostrupii* fruit body; (d) Resupinate *Tomentella* sp. (BM07C26, matching PD2_3) fruit body under dead trunk; (e) EcM system attached to *Aldina* sp. "B" secondary root (f) EcM system attached to *P. dipterocarpacea* secondary root ; (g) *Tomentella* sp. EcM (BM03M8, matching PD2_3) on *Aldina* sp.; *Cortinarius* sp. EcM (BM03M4, matching BM08C20) on *P. dipterocarpacea*. Bar, 0,4 mm.

Fine roots were sampled in each sampling expedition. They were traced from identified *P. dipterocarpacae* trees in the first three surveys and sampled in nine 10 X 10 X 5 cm deep soil cores scattered in the plot in the last harvest. The soil cores were located either in places where *P dipterocarpacea* and *Aldina* sp trees occurred at < 2 m distance from each other (5 cores) or in random locations (4 cores).

Traced root samples were morphotyped using a field dissecting microscope (Novex AP-7) or a 10X magnifying glass before fixation in CTAB. Time lapse between tracing and root fixation was always up to one day. Morphology of traced *P dipterocarpacea* and *Aldina* sp. fresh secondary roots was characterized for subsequent sorting in the soil cores. Color was rust brown to yellowish and silvery creamish in *P. dipterocarpacea* and *Aldina* sp., respectively (Fig. 2- e, f). Frequent longitudinal peels of bark were often observed on *P. dipterocarpacea* roots.

A 5 X 5 X 5 cm^3 subsample including top organic soil was taken from each soil core and most fine roots attached to secondary roots as well as loose root tips were carefully washed in the field Lab. EcM and non-EcM root systems were separated using a 10X magnifying glass. EcM roots were subsequently separated on *P. dipterocarpacea* and *Aldina* root systems using color and morphological features. Loose EcM tips were grouped into a third "unclassified" category. All roots from the three categories were fixed in CTAB for further morpho-anatomical study in the laboratory in Belgium. Time lapse between root sampling and fixation was less than one day.

For each soil core, all fine roots belonging to *P. dipterocarpacea*, *Aldina* and "unclassified" categories were screened separately under a dissecting microscope to assign them to EcM morphological categories (Fig. 2-g, h). Each morphological category, including several tips, was cross checked using anatomical features on mantle peels following Agerer's (1991) method with some modifications. Different morphotypes from each *P. dipterocarpacea*, *Aldina* and "unclassified" root category were stored separately in CTAB and kept as a representative sample for further molecular analysis.

2.3 DNA protocols

Genomic DNA was extracted using QIAGEN Dneasy™ Plant Mini Kit (Qiagen S.A. Courtaboeuf, France) from fruit body morphospecies, traced ECM and morphotypes from the soil cores. To test the accuracy of morphotypes in the soil cores, DNA was extracted from several replicates (between two and 6) in EcM morphotypes including more than one sample. Fungal internal transcript spacer (ITS) and partial large subunit (LSU, 25-28S) nuclear rDNA were PCR-amplified with forward primer ITS1f in combination with LR6 (Gardes & Bruns, 1993; Vilgalys & Hester, 1990). If multiple or no PCR products were obtained, DNA was extracted from another EcM tip or the same extract was reamplified using the following primers in different combinations: ITS1f, ITS4b, ITS4, 5,8SR, LR21, LROR, LR6 (Gardes & Bruns, 1993; Vilgalys & Hester, 1990; White et al., 1990; R Vilgalys Lab http://www.biology.duke.edu/fungi/mycolab/primers.htm). The PCR protocols were as described in Moyersoen (2006) with some modifications in cycling parameters for different primers. To check host identity of traced roots and morphotypes from the soil cores, *rbc*L DNA was amplified from 7 EcM samples of *P. dipterocarpacea* and *Aldina* sp. using the primers *rbc*LN and *rbc*LR (Käss & Wink, 1997), following the same PCR protocols as in Moyersoen (2006). ITS-LSU and *rbc*L amplification products were electrophoresed in

1% agarose gels stained with ethidium bromide and visualized under UV light. 1kb + DNA ladder (Invitrogen) was used as a marker. Controls with no DNA were included in every set of PCR amplifications. PCR products were purified using QIAquick protocol (Qiagen) or 96-well filtration system (Multiscreen-PCR plate, Millipore Corporation, MA, USA). Sequencing was performed by Genotranscriptomics Platform, GIGA, University of Liège , using the same primers as for PCR. Sequence editing was done using SEQUENCHER, version 4.0 (Gene Codes Corporation). ITS sequences of EcM specimens have been deposited at the National Center for Biotechnology Information (NCBI, GenBank: http://www.ncbi.nlm.nih.gov) under accession numbers JQ063044-JQ063063. Sequences of new fruit body species will be published separately.

2.4 Sequence analysis and phylogenetic analysis of *Inocybe*

ITS sequences were assigned to molecular species on the basis of arbitrary 3% similarity cut-off value (Nilsson et al., 2008). Sequence similarities were determined using the BLASTN sequence similarity tool (Altschul et al., 1997) in GenBank together with BLAST comparisons with recently published ITS sequences from a *Dicymbe* forest kindly supplied by M Smith (Smith et al., 2011).

A preliminary phylogenetic analysis was performed to place *Inocybe* sequences in Inocybaceae clades (Matheny et al., 2009). 5' end nLSU sequences of two *Inocybe* species from *P. dipterocarpacea* forest were aligned using Clustal X (V. 2.0.9) to Matheny et al. (2009) global Inocybaceae sequences data set kindly supplied by the first author. Maximum likelihood (ML) analysis was implemented in the program RAxML V. 7.0.6, using GTRCAT approximation. Taxon sampling was made on the basis of this analysis. All strains in *Inocybe* subgenus *Inocybe* sequences and a strain from *Nothocybe* lineage were selected in Matheny et al. (2009) data base together with sequences from recently published tropical strains Y01 (UDB 004238), Y02 (UDB004239) (Tedersoo et al., 2010a) and *I. tauensis* (GU97711, GU977212, GU977213) (Kropp & Albee-Scott, 2010). In total, the data set included 77 RNA polymerase II (RPB1), 68 RPB2 and 114 LSU sequences. The RPB1, RPB2 and nLSU sequences were aligned using Clustal X. For consistency, the same criteria as Matheny et al. (2005, 2009) were used for manual alignment using Bioedit V. Introns 1 of RPB1 and the intron of RPB2 were excluded from the analysis. Intron boundaries were inferred both by sequence comparison using published *I. lilacina* RPB1 intron 1 (AY351834), *I. sindonia* RPB1 intron 2 (AY351839) and by insertion between conserved amino acid and the canonical guanine-thymine and adenosine-guanine splice sites. Other positions, too ambiguous to align, were removed from the data set. RPB1 exon, RPB1 intron 2, RPB2 and nLSU were then concatenated in Bioedit. Data set partitioning was done following Matheny (2005). To test the phylogenetic relationships of *Inocybe* strains from *P. dipterocarpacea* forest with *Inocybe* data set, a ML Rapid Bootstrapping algorithm was implemented for 1000 replicates in RAxML, using GTRCAT approximation.

3. Results

3.1 Above and below ground EcM richness

A total of 64 EcM fruitbody samples were collected in the plot (Table 2). From these samples, 41 specimens (including replicates) were selected for molecular analysis and 38 (93%) were successfully sequenced. These sequences belonged to 26 molecular species. Descriptions of new fruit body species will be published separately.

From 15 adult trees and the soil cores, 150 EcM samples were recovered (Table 2). A total of 113 EcM samples were selected for molecular analysis and 97 EcM tips (86%) were successfully amplified and sequenced. These sequences belonged to 27 EcM fungal species. Only 12 of these species matched DNA from fruitbody surveys.

	Samples	PCR	Total sequences*	Success rate (%)	*Inocybe* sequences	Total species**	*Inocybe* species
Fruitbodies	64	41	38	0.93	2	26	1
EcM	150	113	97	0.86	8	27	2
Total	214	154	135	0.88	10	38***	2

*Including ITS1, 5.8S, ITS2, LSU DNA regions depending on species.
**Species were defined using 97% sequence similarity cutoff across the ITS1, 5.8S, ITS2 region as well as phylogenetic analysis of LSU on selected fungal groups.
***This figure does not include one unsequenced *Russula* and *Scleroderma* species.

Table 2. Observed fruitbody and EcM species in 400m² plot dominated by *Pakaraimaea dipterocarpacea*

EcM morphotypes accuracy was tested before further measurement of species density. A total of 96 EcM samples from soil cores were classified in 28 preliminary morphological categories. From these samples, 64 EcM were subject to molecular analysis, resulting in 56 (83%) successful sequences and 22 species. Some preliminary categories were lumped together and 20 morphotypes were defined after cross molecular analysis. Only two morphotypes belonging to *Cortinarius* and *Tomentella* were accurate at genus level. The remaining morphotypes were accurate at the species level. Between one and 12 different morphotypes could be retrieved from a single 125 cm³ soil core (Table 3). Species density per soil core was similar or above values reported in an highly diverse temperate *Picea abies* and *Tilia cordata* dominated EcM community where similar sampling strategy was used (Table 3).

Overall, 40 fungal species were recovered in the plot. Comparison with surveys in similar plot size in other tropical regions or in temperate regions showed that observed species richness was great and similar to or greater than values both in temperate and tropical areas (Table 4).

3.2 Host identity of ectomycorrhizal species

Host identity of all 7 *P. dipterocarpacea* EcM samples including traced roots (three samples) and morphotypes from the soil cores was confirmed after molecular cross checking (305/308, 99% similarity with *P. dipterocarpacea*, DQ406587). Among the 7 *Aldina* EcM samples (from soil cores), 5 tips matched *A. latifolia*, U74252 (99%, 594/600) *rbc*L. The remaining two samples belonged to *P. dipterocarpacea*. *Aldina* leaf morphology corresponded to "species B" in Flora of Venezuelan Guayana (1995) (G Aymard, pers. com.). Sequenced EcM singletons observed on "unclassified" roots belonged to *P. dipterocarpacea*.

In total, 22 EcM fungal species were associated with *P. dipterocarpacea*; 11 (50 %) of these species belonging to 7 fungal groups were also putative or molecularly confirmed associates of *Aldina*, and three singletons were associated with *Aldina* only (Table 5).

3.3 EcM community composition and comparisons with other tropical forest communities

The 40 fungal species were distributed in 7 fungal orders including Agaricales (11 species), Thelephorales (6 species), Cantharellales (6 species), Boletales (6 species), Russulales (5 species), Hymenochaetales (3 species) and Sebacinales (3 species) (Table 5). Agaricales were the richest fungal order including *Cortinarius* species (5 species), Amanitaceae (3 species), Inocybaceae (2 species) and one Tricholomataceae species (Table 5).

Eighty percent (80%) of the best aligned sequences were from the tropical areas including both the Paleotropics and the Neotropics (Table 5). Only 13 (33%) species matched (≥ 97% similarity) the published ITS sequences. These species belonged to *Amanita*, *Inocybe*, Boletaceae, *Clavulina*, Sebacinales and *Tomentella*. All but one of these species were reported from a *Dicymbe* dominated forest situated ca 240 kms apart in the Guayana region (Guyana, Upper Potaro River Basin). Confirmed tree hosts included *Dicymbe* and *Aldina* species (Smith *et al.* 2011).

	This study	Morris et al. (2009)	Tedersoo et al. (2010a)	Smith et al. (2011)	Tedersoo et al. (2003)	Morris et al. (2008)
Climate	Humid tropical	Humid subtropical	Humid tropical	Humid tropical	Temperate	Mediterranean
Host species	*P. dipterocarpacea*, *Aldina* sp.	*Quercus crassifolia*, *Quercus laurina*	*Coccoloba* sp., *Guapira* sp., *Neea* sp.	*Dicymbe* sp., *Aldina* sp.	*Picea abies*, *Tilia cordata*, *Betula pendula*, *Populus tremula*	*Quercus douglasii*, *Quercus wislizeni*
Soil core volume (cm³)	125	942	2250	1000	125	900
Number of root tips processed/ soil core	All root tips in the core	100	All root tips in the core	20	All root tips in the core	100
Replicates	9	80	60	57	108	64
EcM tips sampling strategy	Morphotyping*	Random selection**	Morphotyping	Morphotyping	Morphotyping	Random selection
EcM species density (species in soil core)	1- 6.6 ± 3.4 (M ± STD) - 12	6.2 (M)	1.42 (M)	9 - 18	3.65 ± 1.7 (M ± STD)	6.5 (M)

*root sorting in morphotypes before molecular analysis.
**molecular analysis of root tips randomly selected in soil core.
m=mean, STD= standard deviation

Table 3. Comparison of observed EcM species density between *Pakaraimaea dipterocarpacea* forest and recent tropical and temperate surveys

Stand type	Area	Survey	Species richness	Source
Tropical				
Pakaraimaea/Aldina	400 m^2	Mixed*	40	This study
Mixed Dipterocarpaceae forest in Borneo	400 m^2	EcM	4-26	Peay et al. (2010)
Temperate				
Arctostaphylos/Douglas fir	625 m^2	Mixed	> 40	From Horton &
Norway spruce	500 m^2	Mixed	25	Bruns (2001)
Bishop pine	625 m^2	EcM	20	
Bishop pine	625 m^2	EcM	7	

*mixed fruitbody and EcM survey.

Table 4. Observed EcM fungi richness in similar sized tropical and temperate forest plots

Fungal species (accession No) (host plant)[§]	Best aligned sequence (Accession No.) (overlapping base pairs/ total aligned bases)	Geographic origin	Host family (when tropical)
Agaricales			
Amanita BM08C22 (Pd*)	EcM 780 (JN168679) (627/628, 99%)	Neotropical, Guyana	Fabaceae (Caesalpiniaceae, Papilionoideae)
Amanita BM07C21	*Amanita vaginata* (AB458889) (433/508, 85%)	Paleotropical, Thailand	Dipterocarpaceae
Amanita BM07C1	*Amanita populiphila*[†] (HQ539724) (719/751, 96%)	North temperate	
Cortinarius BM08C20 (Pd*, A[δ])	*Cortinarius limonius* (GQ159869) (364/412, 88%)	North temperate	
Cortinarius BM07C2 (Pd*)	Uncultured ectomycorrhizal fungus L7789_Cort MAD02 (FR731470) (697/736, 95%)	Paleotropical, Madagascar	Not given
Cortinarius BM08C11 (Pd*)	Uncultured ectomycorrhizal fungus L7789_Cort MAD02 (FR731470) (724/761, 95%)	Paleotropical, Madagascar	Not given
Cortinarius BM08C30	Uncultured ectomycorrhiza clone SDL13 (FJ769528) (506/548, 92%)	North temperate	
Cortinarius 1PdM4 (JQ063044) (Pd°)	Uncultured ectomycorrhizal fungus isolate L5110_Cort_Cam02[†]	Paleotropical , Cameroon	Fabaceae (Caesalpiniaceae), Phyllantaceae

Fungal species (accession No) (host plant)§	Best aligned sequence (Accession No.) (overlapping base pairs/ total aligned bases)	Geographic origin	Host family (when tropical)
	(FR731783) (590/623, 95%)		
Inocybe BM08C27 (JQ063045) (Pd°, A°)	*Inocybe* EcM 434 (JN168723) (586/669, 88%)	Neotropical, Guyana	Fabaceae (Caesalpiniaceae, Papilionoideae)
Inocybe BM03M2 (JQ063046) (Pd°, A⁸)	*Inocybe* EcM 1091 (JN168720) (372/381, 98%)	Neotropical, Guyana	Fabaceae (Caesalpiniaceae, Papilionoideae)
Tricholomataceae BM08C2	*Catathelasma ventricosum*† (AM946418) (981/1088, 90%)	North temperate	
Boletales			
Boletaceae BM06M3 (JQ063047) (Pd*)	*Boletellus ananas* voucher TH6264 (JN168685) (198/198, 100%)	Neotropical, Guyana	Fabaceae (Caesalpiniaceae, Papilionoideae)
Boletaceae BM08C29	*Tylopilus ballouii* voucher R.E. Halling 8187† (EU430732) (647/673, 96%)	Neotropical, Costa Rica	Fagaceae
Boletaceae BM08C31	*Austroboletus rostrupii* TH8189 (JN168683) (537/537, 100%)	Neotropical, Guyana	Fabaceae (Caesalpiniaceae, Papilionoideae)
Boletaceae BM08C8	*Tylopilus felleus*† (AY586723) (734/772, 95%)	North temperate	
Boletaceae 7MM7 (JQ063048) (Pd°)	*Boletus* sp. MHM075 (EU569236) (252/280, 90%)	Neotropical, Mexico	Fagaceae
Scleroderma BM08C9	Unsequenced		
Cantharellales			
Cantharellus BM07C6	*Cantharellus cibarius* (AB453024) (236/251, 94%)	Paleotropical, Thailand	Dipterocarpaceae
Clavulina BM03sp1 (JQ063049) (Pd°, A⁸)	*Clavulina* voucher MCA 4022 (MCA4022) (541/545, 99%)	Neotropical, Guyana	Fabaceae (Caesalpiniaceae)
Clavulina BM07C10	*Clavulina amazonensis* isolate TH9191 (HQ680356) (596/650, 92%)	Neotropical, Guyana	Fabaceae (Caesalpiniaceae)
Clavulina BM07C9 (Pd⁸)	*Clavulina humicola* (DQ056368) (555/590, 94%)	Neotropical, Guyana	Fabaceae (Papilionoideae)

Fungal species (accession No) (host plant)§	Best aligned sequence (Accession No.) (overlapping base pairs/ total aligned bases)	Geographic origin	Host family (when tropical)
Clavulina BM08C24	*Clavulina monodiminutiva* (DQ056365) (570/643, 89%)	Neotropical, Guyana	Fabaceae (Caesalpiniaceae, Papilionoideae)
Hydnum BM03M19 (JQ063050) (Pd*)	*Hydnum repandum* (AY817136) (762/762, 100%)	North temperate	
Hymenochaetales			
Hymenochaetales 2PdM2 (JQ063051) (Pd^δ, A^δ)	*Coltriciella dependens* specimen voucher TU103611 (AM412254) (735/802, 92%)	Paleotropical, Seychelles	Fabaceae (Caesalpiniaceae), Dipterocarpaceae
Hymenochaetales 1MM1 (JQ063052)	*Coltriciella dependens* specimen voucher TU 103611 (AM412254) (715/799, 89%)	Paleotropical, Seychelles	Dipterocarpaceae
Hymenochaetales PD6.3 (JQ063053) (Pd*)	Uncultured *Coltricia* voucher EcM 731 (JN168708) (442/567, 78%)	Neotropical, Guyana	Fabaceae (Caesalpiniaceae)
Russulales			
Russula BM07C7	Uncultured ectomycorrhizal fungus isolate L6171_Russ_Y04 (FN557554) (550/570, 96%)	Neotropical, Ecuador	Nyctaginaceae
Russula BM07C18 (Pd*, A*)	Uncultured ectomycorrhiza voucher 403 (AY667427) (556/583, 95%)	Neotropical, Ecuador	Nyctaginaceae
Russulaceae 8PdM1 (JQ063054) (Pd°, A°)	Uncultured ectomycorrhizal fungus isolate L7612_Russ MAD41 (FR731334) (236/261, 90%)	Paleotropical, Madagascar	Not given
Russulaceae 1PdM3 (JQ063055) (Pd°)	Uncultured *Russula* voucher EcM1094 (JN168741) (225/245, 92%)	Neotropical, Guyana	Fabaceae (Caesalpiniaceae, Papilionoideae)
Russula "lilac" BM07C20	Unsequenced		
Sebacinales			
Sebacinales BM03M3 (JQ063056) (Pd°, A°)	Sebacinales EcM 17 (JN168754) (532/549, 97%)	Neotropical, Guyana	Fabaceae (Caesalpiniaceae, Papilionoideae)

Fungal species (accession No) (host plant)[§]	Best aligned sequence (Accession No.) (overlapping base pairs/ total aligned bases)	Geographic origin	Host family (when tropical)
Sebacinales PD10 (JQ063057) (Pd*, A[δ])	Uncultured ectomycorrhizal isolate L7604_Seb MAD02 (FR731328) (518/568, 91%)	Paleotropical, Madagascar	Not given
Sebacinales 6MM2 (JQ063058) (Pd°, A°) Thelephorales	Sebacinales voucher EcM84 (JN168756) (251/258, 97%)	Neotropical, Guyana	Fabaceae (Caesalpiniaceae)
Tomentella 7AM7 (JQ063059) (A°)	*Tomentella* EcM 963 (JN168770) (342/358, 96%)	Neotropical, Guyana	Fabaceae (Caesalpiniaceae)
Tomentella PD2_3 JQ063060 (Pd*, A°)	*Tomentella* EcM 963 (JN168770) (593/602, 99%)	Neotropical, Guyana	Fabaceae (Caesalpiniaceae, Papilionoideae)
Tomentella PD11 (JQ063061) (A°)	*Tomentella* EcM 712 (JN168764) (600/602, 99%)	Neotropical, Guyana	Fabaceae (Caesalpiniaceae, Papilionoideae)
Tomentella PD9_2 JQ063062 (A°)	*Tomentella* EcM712 (JN168764) (587/605, 97%)	Neotropical, Guyana	Fabaceae (Caesalpiniaceae, Papilionoideae)
Tomentella BM07C39 (JQ063063) (Pd[δ])	*Tomentella* MES348 (JN168772) (170/174, 98%)	Neotropical, Guyana	Fabaceae (Caesalpiniaceae, Papilionoideae)
Tomentella 2PdM6 (JQ063063) (Pd[δ])	*Tomentella* EcM 698 (JN168763) (318/329, 97%)	Neotropical, Guyana	Fabaceae (Caesalpiniaceae, Papilionoideae)

†LSU sequence. Unmarked aligned sequences include ITS region.
§Host for each fungal OTU was identified from EcM after *rbc*L sequencing°, root tracing* or root morphology[δ]. Fungal OTUs with no identified host were collected only from the fruiting bodies. Pd : *P. dipterocarpacea* ; A : *Aldina* sp. B.

Table 5. Sequence matches of ectomycorrhizas and ectomycorrhizal fruit bodies collected in *Pakaraimaea dipterocarpacea* forest

3.4 *Inocybe* EcM frequency and occurrence on host plants in the plot

Two *Inocybe* species were recovered in the plot. *Inocybe* EcM were characteristic and could be distinguished from other fungal genera on the basis of morphology and anatomy. They were both characterized by a smooth and whitish mantle with frequent rhizomorphs and a pinkish color at the apex (Fig. 2a). Hyphae in the rhizomorphs were uniformly shaped and agglutinated (type B in Agerer 2006). EcM habit corresponded to medium-distance smooth exploration type in Agerer's (2001) EcM mycelial system. Typical *Inocybe* prominent clamp connections (Fig. 2b) were observed in the rhizomorphs.

Despite the low number of *Inocybe* species in the total EcM fungal diversity, *Inocybe* EcMs were frequent and were observed in 55 % of the soil cores. Two *Inocybe* species were recovered from traced roots of *P. dipterocarpacea* and the host identity was confirmed molecularly. They were also recovered from *Aldina* sp. roots in soil cores and *Aldina rbc*L could be sequenced for one *Inocybe* EcM species.

3.5 Phylogeny and biogeography of *Inocybe* species associated with *P. dipterocarpacea*

One *Inocybe* species from *P. dipterocarpacea* forest (BM03M2) matched (372/381, 98% similarity) *Inocybe* EcM1091 ITS1/5.8S/ITS2 recovered from *Aldina insignis* and *Dicymbe alstonii* (Table 5). Phylogenetic analysis including LSU, together with fruit bodies' morphological comparisons, are necessary to confirm that EcM1091 and BM03M2 are conspecific. No significant match could be found for the second *Inocybe* species (Table 5). Best match with 25S LSU in GenBank was with *I. epidendron* (EU569840) for both species. Preliminary alignment of the two *Inocybe* species to Matheny et al. (2009) global Inocybaceae data set confirmed their position in *Inocybe* subgenus *Inocybe* (data not shown). These results were consistent with the characteristic *Inocybe* s. str. nodulose spores and pleurocystidia observed on *Inocybe* BM08C27 fruitbody species. LSU alignments with Matheny et al. (2009) global *Inocybe* subgenus *Inocybe* data set demonstrated that the two species were phylogenetically close to neotropical *Inocybe* species associated with *Dicymbe* and *Aldina* sp. in Guyana (Fig. 3). These species clustered within a tropical only, basal clade. A weakly supported node suggested a sister relationship with an African strain from Zambia. This clade was sister with a couple of related species from Zambia and Guyana.

4. Discussion

4.1 EcM community diversity and structure in *P. dipterocarpacea* forest

This is the first above and below ground EcM fungal diversity survey in a Neotropical forest dominated by a Dipterocarpaceae. Only few pioneering surveys have been reported on EcM fungal communities in Neotropical monospecific Dipterocarpaceae genera *Pseudomonotes* and *Pakaraimaea* (Franco-Molano et al., 2005; Moyersoen, 2006; Vasco-Palacios et al., 2005). Moyersoen (2006) study mostly focused on *P. dipterocarpacea* EcM status and only a small proportion of EcM community was described. The present study was necessary to get a more precise EcM diversity estimate in *Pakaraimaea* forest.

Despite the relatively small plot size, a great number of EcM fungal species was recorded. Total above and below ground species richness and species density in soil cores was comparable to highly diverse EcM temperate forest communities. It has been suggested that EcM diversity in tropical forests is lower than in temperate forests (Tedersoo & Nara, 2010; Tedersoo et al., 2011). In contrast, with boreal forests, where a niche differentiation of EcM fungi is observed at the level of soil horizons (Rosling et al., 2003; Tedersoo et al., 2003), a study by Tedersoo et al. (2011) suggested that EcM communities are homogeneous across soil horizons in the African forests. Recent studies indicate that trends in EcM fungal biodiversity are not simply driven by a latitudinal gradient (Smith et al., 2011). Biodiversity depends on the forest type under study (Brearly, 2011; Smith et al., 2011) and is also influenced by host identity (Haug et al., 2005; Tedersoo et al., 2010a). Amongst the EcM surveys in Dipterocarpaceae vegetation (see Brearly, 2011 review and references therein), comparisons could be made with a SE Asian Mixed-dipterocarp forest (Peay et al., 2010), where diversity was measured in plot size identical to this study. EcM richness (Table 2) was at the top end of observed values (Table 4) in SE Asia and total above and below ground richness (Table 4) was within the range of estimated richness (9.3-60.4) in Mixed-dipterocarp forest. This comparison still needs to be taken with caution since root sampling strategy differed between the two studies. Contrary to the relatively low species density observed on Neotropical Pisonieae (Tedersoo et al., 2010a), density in *P. dipterocarpacea* was within the range or greater than in the temperate forests (Table 3). Both total fungal diversity and species density indicated that EcM fungi occur at a fine scale in *P. dipterocarpacea* forest such as in boreal forests. A greater EcM fungal diversity might be expected by sampling both organic and mineral soil layers in *P. dipterocarpacea* forest. This study extends the findings of a previous survey in a Neotropical Fabaceae forest where great levels of EcM fungi biodiversity were observed (Smith et al. 2011).

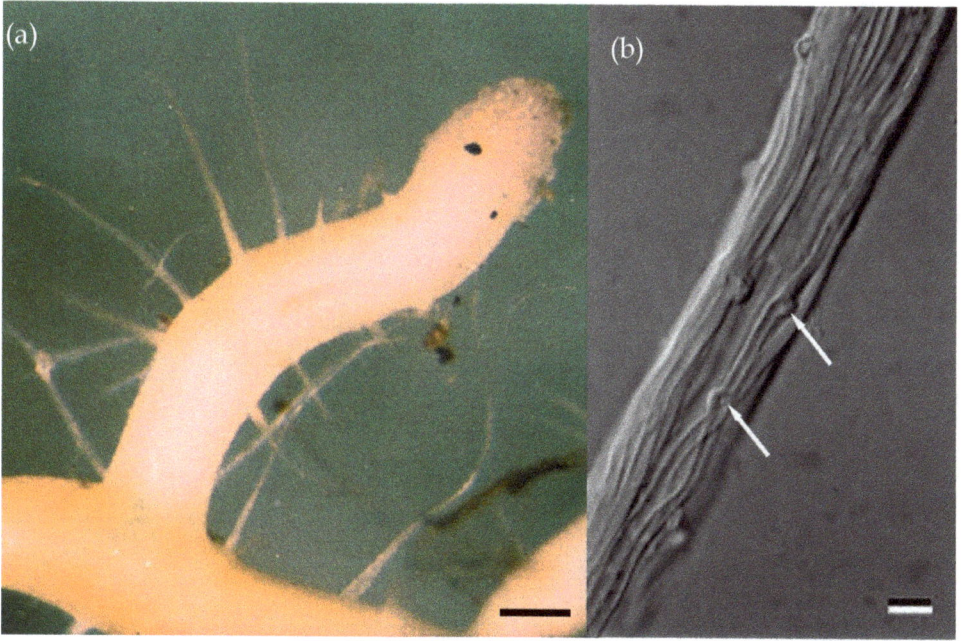

Fig. 2. (a): Habit of *Inocybe* sp. (BM03M2) ectomycorrhiza. Bar, 0.2 mm. (b): Detail of rhizomorph with clamp connections. Bar, 10μm.

EcM fungi in the *P. dipterocarpacea* forest belong to fungal orders generally observed in the tropical forests and elsewhere (e.g. Dickie & Moyersoen, 2008; Diédhiou et al., 2010; Lee et al., 2003; Peay et al., 2010; Riviere et al., 2007; Smith et al., 2011; Tedersoo et al., 2011). Species rich fungal orders included Agaricales, Thelephorales, Cantharellales, Boletales and Russulales. Particularly relevant was the number of *Clavulina* species. Many *Clavulina* species were also observed in *Dicymbe* forests and Smith et al. (2011) hypothesized that the Guayana region might be a centre of diversification for the *Clavulina* lineage. *Cortinarius* is usually considered to be rare in the tropical environments (Tedersoo et al., 2010b) and the great number of species identified in this study was striking. Interestingly, Peay et al. (2010) observed a greater number of *Cortinarius* in Mixed-dipterocarp forests on sandy soils. *P. dipterocarpacea* typically grows on white sands (Maguire & Steyermark, 1981). Whether preference for sandy soils is an important ecological character in these tropical *Cortinarius* species is worth investigating. Tricholomataceae fruit body species in this study did not match any RPB2/LSU/ITS sequences either deposited in GenBank or from a recent *Dicymbe* forest EcM survey. This fruit body species is currently under further investigation and is potentially a new endemic species. *Cortinarius*, *Amanita* and *Russula* are typical "late stage" fungi characteristic of undisturbed forest stands in the temperate forests (Deacon & Fleming, 1992; Nara et al., 2003).

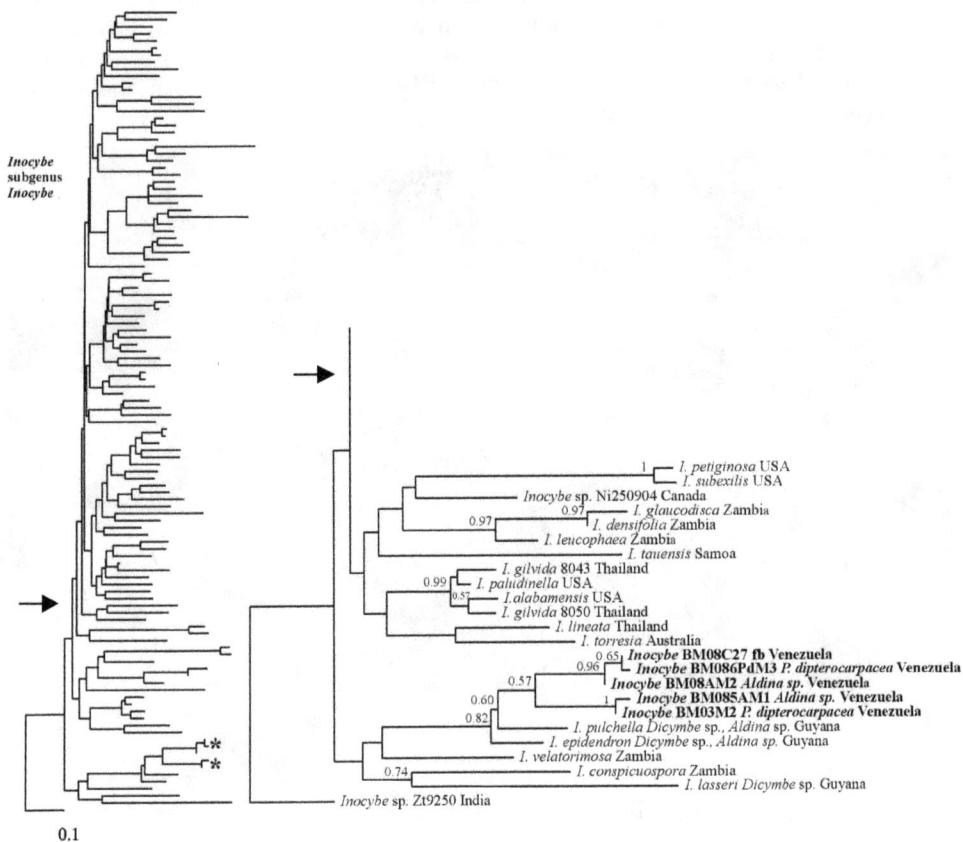

Fig. 3. Phylogenetic placement of two *P. dipterocarpacea* forest *Inocybe* species (*). Best ML tree based on combined analysis of RPB1, RPB2 and LSU sequences, resulting from 1000 replicates Rapid Bootstrapping algorithm and ML tree search in RAxML (v. 7.2.6) using GTRCAT approximation. The tree was rooted with *Inocybe* sp. ZT 9250. Bootstrap values < 50 % are not shown.

The great total diversity, together with the abundance of these late stage fungi and the presence of a possible novel endemic species, indicated that the plot is a well preserved *P. dipterocapacea* stand despite human activities in the region. This study also indicated that Neotropical *Pakaraimaea* and *Dicymbe* forests show similarities in EcM fungal community structure.

4.2 EcM fungal communities comparison between *P. dipterocarpacea* and other tropical tree host species

Until recently, biodiversity information on Neotropical lowland EcM fungi mostly relied on traditional fruit body surveys (e.g. Franco-Molano et al., 2005; Henkel et al., 2002, 2005, 2011; Matheny et al., 2003; Moyersoen, 1993; Singer et al., 1983; Vasco-Palacios et al., 2005) with inherent difficulties to confirm the host plant identity. Relatively few EcM surveys have included molecular host identifications (Haug et al., 2004; Moyersoen, 2006; Smith et al., 2011; Tedersoo et al., 2010a). This study in *P. dipterocarpacea* forest provides the first report of multi-host EcM fungal species associated both with a Dipterocarpaceae and EcM Fabaceae in the Neotropics.

With the exception of Pisonieae and *Gnetum* sp. where patterns of host specificity were reported (Bechem & Alexander, 2011; Haug et al., 2005; Tedersoo et al., 2010a), recent tropical EcM biodiversity surveys have shown that co-occurring (sympatric) tree species including different tree families can be colonized by identical dominant EcM fungi (Diédhiou et al., 2010; Smith et al., 2011; Tedersoo et al., 2011). These sympatric tree families included the Dipterocarpaceae and Caesalpiniaceae in the Paleotropics (Tedersoo et al., 2007, 2011). In addition to host overlap in the same forest, Tedersoo et al. (2011) also observed that some EcM fungal species are broadly distributed in tropical Africa. In contrast, host identity influences EcM fungal communities in the temperate forests (Ishida et al., 2007; Tedersoo et al., 2008). This study in *P. dipterocarpacea* forest indicated that several fungi from different fungal groups can colonise both *P. dipterocarpacea* and a sympatric Papilionoideae. Since EcM fractional colonisation and root length were not measured, sampling design did not allow to evaluate whether or not shared fungal species were true generalists (i.e. if they equally colonized the root system of both tree hosts).

The capacity of some EcM fungi to associate with *Pakaraimaea* and EcM Fabaceae was further demonstrated by the comparisons between *Pakaraimaea* forest fungal community and *Dicymbe* dominated forests. Around 15% of total species in *P. dipterocarpacea* forest (including undescribed *Amanita* sp. BM08C22, *Boletellus ananas*, *Austroboletus rostrupii*, *Clavulina MCA4022* and two undescribed *Tomentella* species) were conspecific (≥ 99% ITS similarity) with specimens collected in *Dicymbe* forest 240 km apart from the study site. *Cantharellus guianensis* might be an additional example of broadly distributed species but this hypothesis should be tested with complementary morphological and molecular study. Interestingly, *C. guianensis* was also described in *P. tropenbosii* dominated forest (Franco-Molano et al., 2005). Phylogenetics and morphological comparisons are necessary to confirm that the remaining 6 presumed shared species between *P. dipterocarpacea* and *Dicymbe* forests are conspecific. This study demonstrated that a proportion of EcM fungi associated with *P. dipterocarpacea* can also associate with sympatric and allopatric tree species belonging to EcM Fabaceae in the same region.

4.3 *Inocybe* ecology and phylogeography

Matheny et al. (2009) suggested that *Inocybe* flora in the Neotropics resulted from multiple recent immigration events in South America. The lower *Inocybe* diversity in the Neotropics than in the Paleotropics was attributed (apart from sampling effort) to a combination of factors including competition or extinction and paucity of EcM tree hosts.

The low proportion of *Inocybe* species in *P. dipterocarpacea* forest EcM fungal community was consistent with the general trend in the Neotropics. At least three different *Inocybe* lineages were observed in *Dicymbe* forests (Matheny et al., 2003; Matheny et al., 2009). *P. dipterocaracea* survey needs to be extended to test whether this tree species also associates with several *Inocybe* lineages.

Despite the few species recorded, the frequent occurrence of *Inocybe* EcM suggested that this fungal group is successful in *P. dipterocarpacea* plot. An interesting feature of morphotypes observed on two *P. dipterocarpacea Inocybe* species was the presence of rhizomorphs with agglutinated hyphae and ramified distal end in close contact with soil particles. A small proportion of *Inocybe* EcM morphotypes have been formally described (Agerer, 2006) and only one of them (*I. avellana* in association with *Shorea*, Dipterocarpaceae) (Ingleby, 1999) was tropical. One of the patterns in described morphotypes was the presence of emanating hyphae only as soil exploration structures (Agerer, 2006). Ryberg et al. (2010) demonstrated that soil preference reflects the evolutionary history of Inocybaceae. Preference for poor soil prevail in some phylogenetic groups. Agerer (2001) highlighted the possible ecological importance of rhizomorphs for soil exploration and plant nutrition. Peay et al. (2011) also suggested that differences in EcM exploration type influence EcM fungal dispersion ability. Rhizomorphs might contribute to the success of *Inocybe* in poor sandy soils associated with *P. dipterocarpacea* forest. Whether this morphological character is a conserved morphological characteristic in *Inocybe* subgenus *Inocybe* basal clade is worth investigating.

In contrast with the hypothesis that *Inocybe* species associated with *P. dipterocarpacea* represent an isolated immigration event (Matheny et al., 2009), both BLASTN and phylogenetic analysis demonstrated that *P. dipterocarpacea* shared identical or phylogenetically close *Inocybe* species with EcM Fabaceae. The capacity of two *Inocybe* species to associate with both *P. dipterocarpacea* and co-occurring *Aldina* species B is consistent with the host generalist status of Inocybaceae within angiosperms in the temperate forests (Ryberg et al., 2010; Kuyper, 1986). This host sharing ability was demonstrated for several *Inocybe* species on *Dicymbe* and co-occurring *Aldina* in the Guayana region (Smith et al., 2011). Most interestingly, *Inocybe* species in *P. dipterocarpacea* forest were phylogenetically close to species in *Dicymbe* dominated forest. Singer et al. (1983) described *I. amazoniensis* from Campinarana vegetation near Manaus, Northern Brazil. They did not identify *I. amazoniensis* host plant and putative EcM hosts in Campinarana include EcM Fabacaeae (*Aldina*), Pisonieae and Coccolobeae. Confirmation of *I. amazoniensis* phylogenetic position which is morphologically close to *I. epidendron* (Matheny et al., 2003) and *Inocybe* sp. BM08C27 (unpublished data) would further support the evidence that several closely related *Inocybe* species associate with diverse tree hosts including Dipterocarpaceae and EcM Fabaceae across the Guayana region.

The broad distribution of phylogenetically close *Inocybe* species might be the result of an efficient dispersion ability, the capacity to associate with different tree hosts and a long evolutionary history in the Guayana Region. A possible explanation for the disjunct distribution of *Inocybe* related species in the region is that extant forests with a dominance or codominance of one or several Neotropical EcM tree families including Dipterocarpaceae (Franco-Molano et al., 2005; Moyersoen, 2006; Vasco Palacios et al., 2005) and EcM Fabaceae (Henkel et al., 2002; Singer & Araujo, 1979, 1986) are remnants of more extensive EcM forests in the past. This hypothesis is difficult to evaluate since knowledge on Guayanan Region vegetation history is scanty (O Huber, pers. com). *Inocybe* is often mentioned in the literature as a pioneer and an effective coloniser by spore dispersal (Smith and Read, 2008). Although they are often scattered in AM dominated forests (Béreau et al., 1997; Lodge, 1987; Singer & Araujo, 1979; St John, 1980; Tedersoo et al., 2010a), species belonging to confirmed EcM tree hosts are diverse and widely distributed in the Guayana region (Table 1). These non-dominant Neotropical EcM plants might have contributed to the dispersion of EcM fungal groups such as *Inocybe* subgenus *Inocybe* by co-migration (Halling, 2001) or by providing "EcM tree islands" for fungus colonization (Peay et al., 2007). *Inocybe* species belonging to different clades in *Inocybe* subgenus *Inocybe* (data not shown) have been found on *Guapira* and *Neea* sp. from Ecuador, Belize and US Virgin Islands (Matheny et al., 2009; Tedersoo et al., 2010a) as well as on *Coccoloba* species in Ecuador (Tedersoo et al., 2010a). Fruit bodies belonging to 8 EcM lineages including Inocybaceae were reported by Moyersoen (1993) in Bana vegetation, Southern Venezuela, where Pisonieae (*Neea* and *Guapira*) are particularly conspicuous together with *A. kunhardtiana* and *C. excelsa*. Additional surveys are necessary to test the importance of non-dominant EcM tree hosts on EcM fungi dispersion and diversity in Guayana Region. Phylogeographic analyses similar to this *Inocybe* study should also be undertaken on more diverse fungal lineages such as *Clavulina* that are associated with different confirmed tree hosts from different families across the Guayana region (Henkel et al., 2005, 2011; Smith et al. 2011).

P. dipterocarpacea Inocybe clustering within a basal clade found only in the tropics demonstrated their tropical origin. The results are consistent with Matheny et al. (2009) in the species composition of the *Inocybe* subgenus *Inocybe* basal clade (tree node 84). These basal species were not included in the phylogenetic study by Kropp and Albee-Scott (2010), who selected a subset of Neotropical and Paleotropical *Inocybe* strains. According to molecular clock phylogeny, the estimated age of tree node 84 was 65 Myr (41-89 Myr, 95% CI) (Matheny et al. 2009). *Inocybe* species associated with *P. dipterocarpacea* are the recorded oldest origin on a Dipterocarpaceae. The present study highlights the need to get a more exhaustive Dipterocarpaceae *Inocybe* ITS/LSU/RPB1/RPB2 sequence data set across the tropics. Comparisons of EcM flora between *P. dipterocarpacea* and *P. tropenbosii* as well as Dipterocarpaceae from Africa or the related EcM family Sarcolanaceae (Ducousso et al., 2004) from Madagascar would be particularly appropriate. Basal clade topology fits with a vicariance hypothesis, but Matheny et al. (2009) excluded this hypothesis on the basis of an estimated divergence date. An alternative boreotropical migration route (Pennington & Dick, 2004) was proposed (Matheny et al., 2009) but this hypothesis is not reflected in the topology of *Inocybe* subgenus *Inocybe* phylogeny. Long distance dispersal between Africa and South America would be another alternative hypothesis. There is no clear cut biogeographic interpretation for the basal clade including Paleotropical and Neotropical *Inocybe* subgenus *Inocybe* strains. Observation of two of these basal species on *P. dipterocarpacea*, a tree host with possible Gondwanic ancestors, is intriguing.

5. Conclusion

This study demonstrated the importance of fungal dispersion and host sharing in understanding of the EcM community associated with *P. dipterocarpacea*. Host sharing between sympatric tree species belonging to different lineages within Fabaceae had already been described in a *Dicymbe* dominated forest (Smith et al. 2011). This study highlighted that

EcM fungal species can associate with allopatric tree hosts belonging to two typical EcM tree families in the Guayana region, Dipterocarpaceae and EcM Fabaceae. There is a need to extend *Inocybe* phylogeographic study to other fungal groups and to include non-dominant tree species including Pisoniae and Coccolobeae in the surveys in Guayana Region. These studies are necessary to evaluate the importance of host overlap in the area and the possible importance of non-dominant host tree species in EcM fungal dispersion and diversity.

The hypothesized origin of Inocybaceae is Paleotropical (Matheny et al., 2009). The old origin of *Inocybe* strains associated both with *P. dipterocarpacea* and EcM Fabaceae and their tropical root suggest a long evolutionary history of EcM symbiosis in the Neotropics. Present knowledge of Inocybaceae molecular clock phylogeny is insufficient to say with certainty which EcM radiation scenario could explain the disjunct distribution of the basal clade of *Inocybe* subgenus *Inocybe*. A more complete ITS/LSU/RPB1/RPB2 sequence data set including *Inocybe* strains associated with Dipterocarpaceae across the tropics is needed to improve the understanding of Dipterocarpaceae associated *Inocybe* phylogeography.

This study confirmed that forests dominated by typical EcM tree families host EcM rich fungal communities in the Guayana Region. EcM fungal species density was great and further studies are needed to evaluate the effect of soil substrate on EcM fungal communities. The presence of a possible tropical endemic Tricholomataceae species in the *P. dipterocarpaceae* forest suggests that EcM flora could include old relictual species, therefore confirming the great biodiversity value of EcM flora in the Guayana Region (Smith et al., 2011).

6. Acknowledgement

Several research institutions (Instituto Venezolano de Investigaciones Científicas, INRA Nancy, University of Liège, University of Tübingen, University of Aberdeen) provided support including Lab facilities at different stages of this research. FNRS sponsored two field trips and expenses for molecular analysis were partially funded by University of Liège and University of Aberdeen. Thanks to G. Cuenca, J. Rosales, T. Iturriaga, V. Demoulin and IJ. Alexander for their support in my research, M. Weiss for advises in phylogenetic analysis, O. Huber and K Scott for advice and comments on the manuscript and C. Rondon and K. Licea for help in the field.

7. References

Agerer, R. 1991. Characterisation of ectomycorrhiza. *Methods in Microbiology*, 23, pp. 25-73.

Agerer, R. 2001. Exploration types of ectomycorrhizae – A proposal to classify ectomycrrrhizal mycelial system according to their pattern of differentiation and putative ecological importance. *Mycorrhiza*, 11, pp. 107-114.

Agerer, R. 2006. Fungal relationships and structural identity of their ectomycorrhizae. *Mycological Progress*, 5, pp. 67-107.

Alexander, I.J. 1989. Mycorrhizas in tropical forests, In: *Mineral nutrients in tropical forest and savanna ecosystems*, Proctor, J, pp. 169-188, Blackwell Scientific Publications, Oxford.

Alexander, I.J. 2006. Ectomycorrhizas - out of Africa? *New Phytologist*, 172, pp. 589- 591.

Altschul, SF., Thomas, LM., Alejandro, AS., Jinghui, Z., Webb, M., David, JL. 1997. Gapped BLAST and PSI-BLAST: a new generation of protein database search programs. *Nucleic Acids Research*, 25, pp. 3389-3402.

Bechem EET., Alexander, IJ. 2011. Mycorrhizal status of *Gnetum* spp. in Cameroon: evaluating diversity with a view to ameliorating domestication effort. *Mycorrhiza*, doi 10.1007/s00572-011-0384-0.

Béreau, M., Gazel, M., Garbaye, J. 1997. Mycorrhizal symbiosis in trees of the tropical rainforest of French Guiana. *Canadian Journal of Botany*, 75, pp. 711-716.

Brearly, F. 2011. Ectomycorrhizal associations of the Dipterocarpaceae in tropical forests. *Biotropica, in Press*.

Deacon JW., Fleming, LV. 1992. Interactions of ectomycorrhizal fungi, In: *Mycorrhizal Functioning: An Integrative Plant-Fungal Process*, Allen, MF, pp. 249-300, Chapman & Hall, New York.

Dickie, IA., Moyersoen B. 2008. Towards a global view of ectomicorrhizal ecology. *New Phytologist*, 180, pp. 263-265.

Diédhiou, AG., Selosse, M-A., Galiana A., Diabaté M., Dreyfus B., Bâ AM., Miana de Faria S., Béna G. 2010. Multi-host ectomycorrhizal fungi are predominant in a Guinean tropical rainforest and shared between canopy trees and seedlings. *Environmental Microbiology*, 12(8), pp. 2219-2232.

Ducousso, M., Béna, G., Bourgeois, C., Buyck, B., Eyssartier, G., Vincelette, M., Rabevohitra, R., Randrihasipara, L., Dreyfus, B., Prin, Y. 2004. The last common ancestor of Sarcolanaceae and Asian dipterocarp trees was ectomycorrhizal before the India-Madagascar separation, about 88 million years ago. *Molecular Ecology*, 13, pp. 231-236.

Franco-Molano, AE., Vasco Palacios, AM., Quintero CL., Boekhout, T. 2005. *Macrohongos de la región del Medio Caquetá, Colombia. Guia de Campo Colombia*. Multimpresos Ltda, Colombia.

Gardes, M., Bruns, TD. 1993. ITS primers with enhanced specificity for basidiomycetes – application to the identification of mycorrhizae and rusts. *Molecular Ecology*, 2, pp. 113-118.

Halling, RE. 2001. Ectomycorrhizae: co-evolution, significance, and biogeography. *Annals of the Missouri Botanical Garden*, 88, pp. 5-13.

Halling, RE., Osmundson TW., Neves M-A. 2008. Pacific boletes: Implications for biogeograpic relationships. *Mycological Research*, 112, pp. 437-447.

Haug, I., Weiß, M., Momeier J., Oberwinkler, F., Kottke, I. 2005. Russulaceae and Thelephoraceae form ectomycorrhizas with members of the Nyctaginaceae (Caryophyllales) in the tropical mountain rainforest of southern Ecuador. *New Phytologist*, 165, pp. 923-936.

Henkel, TW., Terborgh, J., Vilgalys, RJ. 2002. Ectomycorrhizal fungi and their leguminous hosts in the Pakaraima mountains of Guyana. *Mycological Research*, 106, 5, pp. 515-531.

Henkel TW., Meszaros R., Aime MC., Kennedy A. 2005. New *Clavulina* species from the Pakaraima mountains of Guyana. *Mycological Progress*, 4, pp. 343-350.

Henkel TW., Aime MC., Ueling JK., Smith, ME. 2011. New species and distribribution records of *Clavulina* (*Basidiomycota*) from Guiana Shield. *Mycologia*, 103, 4, doi: 10.3852/10-355.

Horton, TR., Bruns, TD. 2001. The molecular revolution in ectomycorrhizal ecology: peeking into the black-box. *Molecular Ecology*, 10, pp. 1855-1871.

Ishida TA., Nara K., Hogetsu T. 2007. Host effects on ectomycorrhizal fungal communities: insight from eight host species in mixed conifer-broadleaf forests. *New Phytologist*, 174, pp. 430-440.

Ingleby K.1999. Inocybe avellana Horak + Shorea leprosila Miq. In: Description of
    Ectomycorrhizas 4 (Agerer R, Danielson RM, Egli S, Ingleby K, Luoma D, Treu R,
    eds.), pp 55-60, Einhorn-Verlag, Germany.
Käss, E., Wink, M. 1997. Phylogenetic relationships in the Papilionoideae (family
    Leguminosae) based on nucleotide sequences of cpDNA (*rbcL*) and ncDNA (ITS 1
    and 2). *Molecular Phylogenetics and Evolution*, 8, 1, pp. 65-88.
Kropp, BR., Albee-Scott S. 2010, *Inocybe tauensis*, a new species from the Samoan
    Archipelago with biogeographic evidence for a paleotropical origin. *Fungal Biology*,
    114, 9, pp. 790-796.
Kuyper, T. 1986. A revision of the genus *Inocybe* in Europe. I. Subgenus *Inosperma* and the
    smooth-spored species of subgenus *Inocybe. Persoonia*, supplement volume 3, pp. 1-247.
Lee, SS., Watling, R., Turnbull, E. 2003. Diversity of putative ectomycorrhizal fungi in Pasoh
    Forest Reserve, In: *Pasoh: Ecology of a lowland rainforest in Southeast Asia*, Okuda, T.,
    Manokaran, N., Matsumoto, Y., Niiyama, K., Thomas, S.C., Ashton, P.S., pp. 149-
    159, Springer, Tokyo.
Lodge, DJ. 1987. Resurvey of mycorrhizal associations in the El Verde rainforest in Puerto
    Rico, In: *Proceedings of the 7th North American Conference on Mycorrhizae*, Sylvia, DM.,
    Hung, LL., Graham, JH. pp. 127. Institute of Food and Agriculture Science,
    University of Florida, Gainsville.
Maguire, B., Ashton, PS., de Zeeuw, C., Giannasi, DE., Niklas, KJ. 1977. Pakaraimoideae,
    Dipterocarpaceae of the Western hemisphere. *Taxon*, 26, 4, pp. 341-385.
Maguire, B., Ashton, PS. 1980. *Pakaraimaea dipterocarpacaea* II. *Taxon*, 29(2/3), pp. 225-231.
Maguire, B., Steyermark, JA. 1981. *Pakaraimaea dipterocarpacea*. III. *Memoires of the New York
    Botanical Garden*, 32, pp. 306-309.
Matheny, PB., Aime MC., Bougher NL., Buyck B., Desjardin DE., Horak, E., Kropp BR.,
    Lodge DJ., Soytong K., Trappe JM., Hibbett DS. 2009. Out of the Paleotropics?
    Historical biogeography and diversification of the cosmopolitan ectomycorrhizal
    mushroom family Inocybaceae. *Journal of Biogeography*, 36, 4, pp. 577-592.
Matheny, PB. 2005. Improving phylogenetic inference of mushrooms with RPB1 and RPB2
    nucleotide sequences (*Inocybe*; Agaricales). *Molecular Phylogenetics and Evolution*, 35,
    pp. 1-20.
Matheny, PB., Aime, MC., Henkel, TW. 2003. New species of *Inocybe* from *Dicymbe* forests of
    Guyana. *Mycological Research*, 107(4), pp. 495-505.
Morris, MH., Pérez-Pérez, MA., Smith, ME., Bledsoe, CS. 2008. Multiple species of
    ectomycorrhizal fungi are frequently detected on individual oak root tips in a
    tropical cloud forest. *Mycorrhiza*, 18, pp. 375-383.
Morris, MH., Pérez-Pérez, MA., Smith, ME., Bledsoe, CS. 2009. Influence of host species on
    ectomycorrhizal communities associated with two co-occurring oaks (*Quercus spp.*)
    in a tropical cloud forest. *FEMS Microbiology Ecology*, 69, pp. 274-287.
Moyersoen, B. 1993. Ectomicorrizas y micorrizas vesículo-arbusculares en Caatinga
    Amazónica del Sur de Venezuela. *Scientia Guaianae*, 82 pp.
Moyersoen, B., Beever RE., Martin F. 2003. Genetic diversity of *Pisolithus* in New Zealand
    indicates multiple long-distance dispersal from Australia. *New Phytologist*, 160, pp.
    569-579.

Moyersoen, B. 2006. *Pakaraimea dipterocarpacea* is ectomycorrhizal, indicating an ancient Gondwanaland origin for the ectomycorrhizal habit in Dipterocarpaceae. *New Phytologist*, 172, pp. 753-762.

Nara K., Nakaya H., Wu, B., Zhou Z., Hogetsu, T. 2003. Underground primary succession of ectomycorrhizal fungi in a volcanic desert on Mout Fuji. *New Phytologist*, 159, pp. 743-756.

Nilsson RH., Kristiansson E., Ryberg M., Hallenberg N., Larsson K-H. 2008. Intraspecific *ITS* variability in the kingdom *fungi* as expressed in the international sequence databases and its implications for the molecular species identification. *Evolutionary Bioinformatics*, 4, pp. 193-201.

Peay KG., Bruns TD., Kennedy PG., Bergemann SE., Garbelotto, M. 2007. A strong species-area relationship for the eukaryotic soil microbes: island size matters for ectomycorrhizal fungi. *Ecology Letters*, 10, pp. 470-480.

Peay, KG., Kennedy PG., Davies SJ., Tan, S., Bruns, T. 2010. Potential link between plant and fungal distributions in a dipterocarp rainforest: community and phylogenetic structure of tropical ectomycorrhizal fungi across a plant and soil ecotone. *New Phytologist*, 185, pp. 529-542.

Peay, KG., Kennedy PG., Bruns T. 2011. Rethinking ectomycorrhizal succession: are root density and hyphal exploration types drivers of spatial and temporal zonation? *Fungal Ecology*, 4, pp. 233-240.

Pennington, RT., Dick, CW. 2004. The role of immigrants in the assembly of the South American rainforest tree flora. *Philosophical Transactions Royal Society London B*, 359, pp. 1611-1622.

Pirozynski KA. 1983. Pacific mycogeography: an appraisal. *Australian Journal of Botany*, Suppl. Ser. 10, pp. 137-159.

Rinaldi AC., Comandini O., Kuyper TW. 2008. Ectomycorrhizal fungal diversity: separating the wheat from the chaff. *Fungal Diversity*, 33, pp. 1-45.

Riviere T., Diedhiou AG., Diabate M., Senthilarasu G., Natarajan K., Verbeken A., Buyck B., Dreyfus B., Bena, G., Ba AM. 2007. Genetic diversity of ectomycorrhizal basidiomycetes from African and Indian tropical rainforests. *Mycorrhiza*, 17, pp. 415-428.

Rosling A., Landeweert R., Lindahl BD., Larsson K-H., Kuyper TW., Taylor AFS., Finlay RD. 2003. Vertical distribution of ectomycorrhizal fungal taxa in a podzol soil profile. *New Phytologist*, 159, pp. 775-783.

Ryberg, M., 2009. *An evolutionary view of the taxonomy and ecology of Inocybe (Agaricales) with new perspective gleaned fromGenBank metadata*, Ph.D. thesis, Department of Plant and Environment Sciences, University of Gothenburg.

Ryberg, M., Larsson, E, Jacobsson, S. 2010. An evolutionary perspective on morphological and ecological characters in the mushroom forming family *Inocybaceae* (*Agaricomycotina, Fungi*). *Molecular Phylogenetics and Evolution*, 55(2), pp. 431-442.

Singer, R. 1986. *The Agaricals in Modern Taxonomy*. Koeltz Scientific Books.

Singer, R., Araujo, I. 1979. Litter decomposition and ectomycorrhiza in Amazonian forests. 1. A comparison of litter decomposing and ectomycorrhizal basidiomycetes in latosol terra-firme rainforest and white sand podzol Campinarana. *Acta Amazonica*, 9, pp. 25-42.

Singer R, Araujo, I. 1986. Litter decomposing and ectomycorrhizal Basidiomycetes in an Igapó forest. *Plant Systematics and Evolution*, 153, pp. 107-117.

Singer, R., Araujo, I., Ivory, MH. 1983. The ectotrophically mycorrhizal fungi of the neotropical lowlands, especially Central Amazonia. *Beihefte zur Nova Hedwigia*, 77, pp. 1-352.

Smith SE., Read, DJ. 2008. *Mycorrhizal symbiosis*, Academic Press, 787 pp.

Smith ME, Henkel TW, Aime MC, Fremier AK, Vilgalys R. 2011. Ectomycorrhizal fungal diversity and community structure on three co-occurring leguminous canopy tree species in a Neotropical rainforest. New Phytologist doi: 10.1111/j.1469-8137.2011.03844.x.

Steyermark JA, Berry P.E., Holst BK. 1995. Flora of the Venezuelan Guayana. Volume 1. Missouri Botanical Garden. St. Louis. 320 pp.

St John, TV. 1980. A survey of mycorrhizal infection in an Amazonian rainforest. *Acta Amazonica*, 10(3), pp. 527-533.

Tedersoo, L., Kõljalg, U., Hallenberg, N., Larsson, K-H. 2003. Fine scale distribution of ectomycorrhizal fungi and roots across substrate layers including coarse woody debris in a mixed forest. *New Phytologist*, 159, pp. 153-165.

Tedersoo, L., Suvi T., Beaver K., Kõljalg, U. 2007. Ectomycorrhizal fungi of the Seychelles: diversity patterns and host shifts from the native *Vateriopsis seychellarum* (Dipterocarpaceae) and *Intsia bijuga* (Caesalpiniaceae) to the introduced *Eucalyptus robusta* (Myrtaceae), but not *Pinus caribea* (Pinaceae). *New phytologist*, 175, pp. 321-333.

Tedersoo, L., Nara K. 2010. General latitudinal gradient of biodiversity is reversed in ectomycorrhizal fungi. *New Phytologist*, 185, pp. 351-354.

Tedersoo, L., Jairus T., Horton BM., Abarenkov K., Suvi T., Saar I., Kõljalg, U. 2008. Strong host preference of ectomycorrhizal fungi in a Tasmanian wet sclerophyll forest as revealed by DNA barcoding and taxon-specific primers. *New Phytologist*, 180, pp. 479-490.

Tedersoo, L., Sadam A., Zambrano M., Valencia R., Bahram M. 2010a. Low diversity and high host preference of ectomycorrhizal fungi in Western Amazonia, a neotropical biodiversity hotspot. *The ISME journal*, 4, pp. 465-471.

Tedersoo, L., May TW., Smith ME. 2010b. Ectomycorrhizal lifestyle in fungi: global diversity, distribution, and evolution of phylogenetic lineages. Mycorrhiza, 20, pp. 217-263.

Tedersoo, L., Bahram M., Jairus T., Bechem E., Chinoya S., Mpumba R., Leal M., Randrianjohany E., Razafimandimbison S., Sadam A., Naaldel T., Kõljalg, U. 2011. Spatial structure and the effects of host and soil environments on communities of ectomycorrhizal fungi in wooded savannas and rainforest of Continental Africa and Madagascar. *Molecular Ecology*, 20, pp. 3071-3080.

Vasco Palacios, AM., Franco-Molano AE., López-Quintero CA., Boekhout T. 2005. Macromicetes (Ascomycota, Basidiomicota) de la Region del Medio Caquetá y Amazonas. *Biota Colombiana*, 6(1), pp. 127-140.

Vilgalys R., Hester, M. 1990. Rapid genetic identification and mapping of enzymatically amplified ribosomal DNA from several *Cryptococcus* species. *Journal of Bacteriology*, pp. 4238-4246.

White, TJ., Bruns, T., Lee, S., Taylor, J. 1990. Amplification and direct sequencing of fungal ribosomal RNA genes for phylogenetics, In: *PCR Protocols: a Guide to Methods and Applications*, Innis, MA., Gelfand, DH., Sninsky, JJ., White, TJ., pp. 315-322, Academic Press, New York.

Part 3

Tropical Forest Fragmentation

7

Gap Area and Tree Community Regeneration in a Tropical Semideciduous Forest

André R. Terra Nascimento[1], Glein M. Araújo[1],
Aelton B. Giroldo[2] and Pedro Paulo F. Silva[3]
[1]Instituto de Biologia, Universidade Federal de Uberlândia, Uberlândia, MG,
[2]Pós-Graduação em Ecologia, Universidade de Brasília, Brasília, DF,
[3]Pós-Graduação em Ecologia e Conservação de Recursos Naturais,
Universidade Federal de Uberlândia, Uberlândia, MG,
Brasil

1. Introduction

In tropical and temperate forests, the canopy gaps affect the architecture and the establishment of plants throughout their life-cycle. The gaps help to maintain the tree diversity by density effect and niche partitioning. The possibility of gap occurrence particularly interplays with recruitment limitation, allowing the coexistence of species that otherwise could not make it (Brokaw & Busing 2000; Lima, 2005; Gravel et al., 2010). A typical canopy gap goes through diverse stages, the initial one known as gap stage. In this stage, the invasion by lianas (Schnitzer & Carson, 2010) and bamboos (Larpkern et al., 2011) might occur. The next stage is the construction stage, in which the establishment of some species occurs, although they have not yet reached the canopy level. The last stage is the mature one, in which the canopy and the gap are virtually closed by true vegetation (Lima, 2005).

Natural gaps are caused by death (or injury) of one or more canopy trees (in some cases they are caused by the fall of large branches), and are defined as small openings on the canopy of forests, usually occupying a <0.1 ha (Yamamoto, 2000). This kind of transitory event is frequent in tropical forests (Brokaw, 1985), where plant species of early successional stages (pioneers and secondary ones) might take advantage of this kind of disturbance, since they can tolerate higher micro-climate and ecological variations (Mulkey et al., 1996).

The size distribution and frequency of gap occurrences in a forest is a function of local climate, topography, soil, bedrock and the composition and size distribution of trees (Deslow, 1980). All the trees, especially those in slopes, peaks, shallow or waterlogged soils, and the emerging ones, are subject to wind action, creating gaps proportional to their heights and crown sizes. Moreover, the gap-area is related to the number and orientation of falling trees. Each gap has specific geometry, climate and substratum (Lima, 2005; Deslow, 1980), which leads to important differences in spatial and temporal forest structure (Deslow 1980; Brokaw, 1982). The micro-climatic features of a canopy gap might change with its size from one season to another, and even with extreme climate events. These conditions may be optimal for certain species at a certain point of time, though they can change in a mid/long term (Brown, 1993). Some plant species can only regenerate in a narrow range of light availability (Barton et al., 1989; Whitmore, 1990). In addition to the marked variation in the canopy opening, the reduction of basal area and the increase in gap-area found in canopy gaps, can also interfere in the process of regeneration and in the growth of tree seedlings (Sapkota & Odén, 2009).

The colonization of gaps by species of different categories or successional groups is influenced by ecophysiological responses of species in the area by the seed bank and by seedlings and/or remnant individuals, as well as by post-disturbance migrant species via dispersal processes (Martins et al., 2009). Moreover, it will also depend on the time the opening has occurred, the opening size, the substratum conditions, the relationship with herbivores as well as on density dependent factors (Hartshorn, 1980). Understanding the dynamics of gaps in tropical forests is paramount for forest restoration, sustainable management and conservation of forest remnants (Martins et al., 2009).

In this context, the present study aims to: 1. Identify ecological patterns related to richness and the potential of natural regeneration of tree species in natural gaps. 2. Investigate whether the tree community responds or not to different levels of canopy openings represented by gaps of different sizes found in tropical semideciduous forests (TSF).

2. Material and methods

2.1 Study area

We conducted this study in a 30 ha tropical semideciduous forest (TSF) remnant showing a good conservation status. The fragment is situated in the southern region of Uberlândia municipality located in Minas Gerais state, at a region known as "Triângulo Mineiro". The study site lies within a matrix of planted pasture for cattle mixed with cerrado *sensu stricto* fragments and secondary riparian forest areas.

The dystrophic red latosol ranging from slight to high acidity levels predominates in the region. There are soils with a base saturation of 50 to 65% - Marília Formation, Bauru Group - in some scarce areas (Embrapa, 1982). In tropical semideciduous forest conditions at "Fazenda do Glória (18° 57′ S to 48° 12′ W) the dystrophic red latosol has a base saturation varying from 7.4 to 29% and low availability of N, P and Ca (Haridasan & Araújo, 2005).

2.2 Location and size-class distribution of canopy gaps

We calculated the gap area percentage in the canopy using the intercept method (Floyd & Anderson, 1987; Bullock, 1996), according to Kneeshaw & Bergeron methodology (1998).The gap areas in the canopy were calculated as following:

$$E(X) = \frac{1}{L} \sum_{j=1}^{n} \frac{Xj}{dj} \qquad (1)$$

Where: E (X) - estimation of canopy gaps proportion, L - transect length, X j - the gap area, dj - the gap diameter.

We marked each canopy gap center with a metal stake 90cm height.. Each stake was identified by a colorful tape bearing a numerical sequency. We recorded the geographic coordinates for each canopy gap using a Garmin GPS Etrex model. The geographic coordinates were obtained in the center of each canopy gap. For the purposes of analysis, each gap was regarded as a single sample. A total of 25 gaps were assessed and categorized into distinct size classes according to amplitude and size classes previously described by Brokaw (1985) for tropical rain forests. These classes were previously applied by Lima et al. (2008) who employed the same amplitude distribution in a tropical semideciduous forest at São Paulo state.

2.3 Canopy opening and gap-areas

We estimated the canopy opening and gap area for each canopy gap (N=25) using hemispherical photographs.

The hemispherical photographs were taken using a Sigma 8 mm fish-eye lens attached to a Nikon D80 camera (Figure 1). The camera was mounted on a leveled tripod 1.3 m above ground (Nascimento et al., 2007), with the fish eye focus setting into infinity, as previously described by Mitchell & Whitmore (1993).

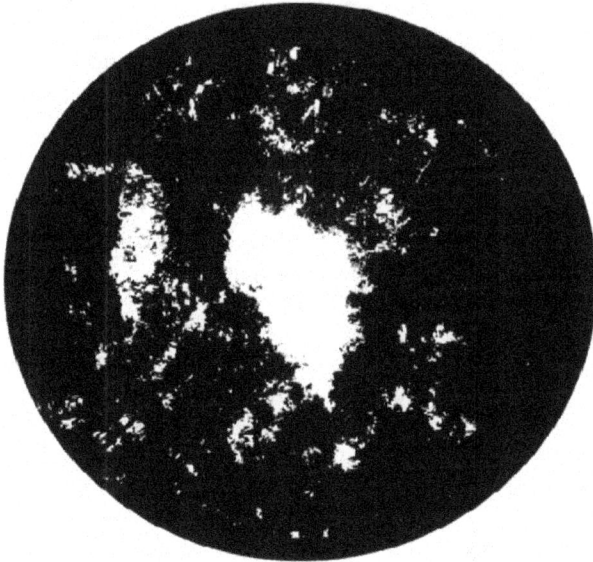

Fig. 1. Hemispherical photograph taken 1.3 m above ground in a canopy gap – open area in the center of the image – formed by the fall of one large sized tree.

The photographs were taken in full-color tones, always between 8-10 a.m., and 2-6 p.m., as recommended by Whitmore et al., (1993). Afterwards, we turned all the pictures taken into images in grey tones (8 bits) and reduced the resolution to 968 x 648 pixels. This is a necessary transformation in order to run the software Winphot 5.0 (Steege 1997), which we applied to analyze the set of hemispherical photographs and the area of each canopy gap. The function "gap area" is available as a tool of this software (Steege 1997). This kind of canopy gap area estimation had already been disseminated by other authors (Whitmore,

1990; Whitmore et al., 1993; van der Meer & Bongers, 1996; Eysenrode et al., 1998; Martins et al., 2002) and provides an accurate estimation of canopy opening as well as the occupied area by gaps in different vegetation types.

2.4 Natural regeneration in canopy gaps

We measured in each canopy gap the number of individuals (seedlings, saplings and some adults), height and number of all tree species higher than 10 cm and bearing a minimum 5 cm DBH (Diameter at Breast Height taken 1.3 cm above ground). The DBH is an inclusion criterion for tree species in forest formations at Cerrado biome (Felfili et al. 2005). When it was not possible to identify the species in the field, we collected botanical material (flowering or vegetative parts) of each plant and compared to specimens at reference herbariums (HUB, CENARGEN) both situated in Brazil. When the identification by comparison was not possible, we sent the material to specialists. All the species and families were identified and classified according to APG III (Chase & Reveal, 2009).

We studied the correlation between species richness, number of individuals and canopy-gap size using a linear correlation (Zar, 1999), investigating whether or not species show a response to size variation in canopy openings caused by gaps of different sizes. According to literature, larger gaps tend to hold a higher density of ruderals or invasive species (Whitmore, 1990). The presence of ruderal or invasive species within vegetation fragments is regarded as an important aspect of integrity of tropical plant communities (Tabarelli et al., 1999; Tabarelli et al., 2000; Tabanez & Viana, 2000).

The similarity of species present in canopy gaps was investigated using the Cluster Analysis (Hair et al. 2005) and applying the Jaccard similarity index. We used the medium distance from the group as the linkage method (UPGMA) (Kent & Coker, 1992). We ordinate the plots and the species within them, using a correspondence analysis by segments applying DECORANA (Detrended Correspondence *Analysis*) method (Hill, 1979). We created a matrix, species x plot, using the number of individuals in each sample (plot) as variable, following recommendations of Kent & Coker (1992). This type of multivariate analysis positions the species and plots along two ordination axis, permitting the investigation of ecological patterns in vegetational studies (ter Braack, 1995; Felfili et al., 2004). Both the multivariate techniques employed in this study (ordination and cluster analysis) were performed using the PC-ORD software version 5.10 (McCune & Mefford, 2006).

3. Results and discussion

The tree community species richness is represented by 80 species distributed among 65 genera and 37 families (Table 1). The families with the highest species richness were Fabaceae (13 species), Myrtaceae (8 species), Rubiaceae (8 species) and Annonaceae (4 species), totalling 40% of the total amount of species surveyed in this study. The genera with the highest species richness were *Myrcia* (4 species), *Myrsine* (3 species), *Trichilia* (3 species) and *Aspidosperma* (2 species). Most of the species found belong to initial colonization stages (successional and initial secondary), contributing 62.2% of the total amount surveyed. Most of the timber trees were late secondary species, and sometimes early secondary ones, which produce high or medium value timber.

Tree species	Family	Number of individuals	Number of gaps
Alibertia sessilis (Vell.) K. Schum.	Rubiaceae	11	7
Amaioua intermedia Mart.	Rubiaceae	2	2
Annona cacans Warm.	Annonaceae	2	1
Apuleia leiocarpa (Vogel) J.F. Macbr.	Fabaceae	4	1
Aspidosperma cylindrocarpon Müll. Arg.	Apocynaceae	10	7
Aspidosperma discolor A. DC.	Apocynaceae	7	6
Astronium nelson-rosae D. A. Santin	Anacardiaceae	25	17
Bauhinia rufa (Bong.) Steud.	Fabaceae	11	7
Bauhinia ungulata L.	Fabaceae	2	2
Campomanesia velutina (Cambess.) O. Berg	Myrtaceae	2	2
Cardiopetalum calophyllum Schltdl.	Annonaceae	21	15
Cariniana estrellensis (Raddi) Kuntze	Lecythidaceae	3	3
Casearia grandiflora Cambess.	Salicaceae	27	14
Casearia sylvestris Sw.	Salicaceae	2	2
Cedrella fissilis Vell.	Meliaceae	1	1
Celtis iguanaea (Jacq.) Sarg.	Cannabaceae	1	1
Cestrum sp.	Solanaceae	1	1
Cheiloclinium cognatum (Miers) A.C. Sm.	Celastraceae	9	4
Coccoloba mollis Casar.	Polygonaceae	2	2
Copaifera langsdorffii Desf.	Fabaceae	10	8
Cordia sellowiana Cham.	Boraginaceae	3	2
Cupania vernalis Cambess.	Sapindaceae	35	19
Dasyphyllum sp.	Asteraceae	2	1
Duguetia lanceolata A. St.-Hil.	Annonaceae	7	5
Dyospirus hispida A. DC.	Ebenaceae	2	2
Eriotheca candolleana (K. Schum.) A. Robyns	Malvaceae	1	1
Erythroxylum daphnites Mart.	Erythroxylaceae	3	3
Eugenia florida DC.	Myrtaceae	3	3
Faramea cyanea Müll. Arg.	Rubiaceae	2	1
Garcinia gardneriana (Planch. & Triana) Zappi	Clusiaceae	1	1
Genipa americana L.	Rubiaceae	4	3
Guapira areolata (Heimerl) Lundell	Nyctaginaceae	2	2
Handroanthus serratifolius (Vahl) S. O. Grose	Bignoniaceae	3	3
Hirtella gracilipes (Hook. f.) Prance	Chrysobalanaceae	10	4
Heisteria ovata Benth.	Olacaceae	9	7
Hymenaea courbaril L.	Fabaceae	1	1
Inga laurina (Sw.) Willd.	Fabaceae	13	5
Inga sessilis (Vell.) Mart.	Fabaceae	54	16
Ixora sp.	Rubiaceae	1	1
Ixora warmingii Müll. Arg.	Rubiaceae	3	2

Jacaranda cuspidifolia Mart. ex A. DC	Bignoniaceae	2	2
Lacistema aggregatum (P.J. Bergius) Rusby	Lacistemataceae	7	6
Maprounea guianensis Aubl.	Euphorbiaceae	2	2
Margaritaria nobilis L. f.	Euphorbiaceae	2	2
Matayba guianensis Aubl.	Sapindaceae	7	7
Miconia albicans (Sw.) Steud.	Melastomataceae	15	7
Miconia sellowiana Naudin	Melastomataceae	18	8
Myrcia splendens (Sw.) DC.	Myrtaceae	3	3
Myrcia sp.	Myrtaceae	4	4
Myrcia tomentosa (Aubl.) DC.	Myrtaceae	2	2
Myrcia variabilis DC.	Myrtaceae	1	1
Myrsine densiflora Scheff.	Primulaceae	1	1
Myrsine umbellata Mart.	Primulaceae	1	1
Myrsine sp.	Primulaceae	4	4
Nectandra megapotamica (Spreng.) Mez	Lauraceae	29	9
Ocotea corymbosa (Meisn.) Mez	Lauraceae	35	16
Ocotea spixiiana (Nees) Mez	Lauraceae	19	8
Ormosia arborea (Vell.) Harms	Fabaceae	1	1
Ormosia fastigiata Tul.	Fabaceae	1	1
Ouratea castaneifolia (DC.) Engl.	Ochnaceae	5	4
Platycyamus regnellii Benth.	Fabaceae	1	1
Pouteria torta (Mart.) Radlk.	Sapotaceae	2	2
Protium heptaphyllum (Aubl.) Marchand	Burseraceae	5	3
Pseudolmedia laevigata Trécul	Moraceae	2	2
Pterodon pubescens (Benth.) Benth.	Fabaceae	1	1
Rudgea viburnoides (Cham.) Benth.	Rubiaceae	1	1
Sclerolobium paniculatum Vogel	Fabaceae	2	2
Simira viridiflora (Allemao & Saldanha) Steyerm.	Rubiaceae	15	5
Siparuna guianensis Aubl.	Siparunaceae	106	20
Sorocea bonplandii (Baill.) W.C. Burger, Lanj. & Wess. Bôer	Moraceae	1	1
Sweetia fruticosa Spreng.	Fabaceae	13	9
Siphoneugena densiflora O. Berg	Myrtaceae	6	6
Tapirira obtusa (Benth.) J.D. Mitch.	Anacardiaceae	20	8
Terminalia glabrescens Mart.	Combretaceae	12	6
Trichilia catigua A. Juss.	Meliaceae	2	2
Trichilia pallida Sw.	Meliaceae	40	14
Trichilia elegans A. Juss.	Meliaceae	2	1
Virola sebifera Aubl.	Myristicaceae	3	2
Vitex polygama Cham.	Lamiaceae	1	1
Xylopia aromatica (Lam.) Mart.	Annonaceae	2	2

Table 1. Regenerating species of tree community sampled in 25 canopy gaps in tropical semideciduous forest (TSF), Uberlândia, Minas Gerais.

The species with highest number of individual values within the 25 surveyed gaps were *Siparuna guianensis* Aubl. (106 individuals), *Inga sessilis* (Vell.) Mart. (54 ind.), *Trichilia pallida* Sw. (40 ind.), *Ocotea corymbosa* (Meisn.) Mez (35 ind.), *Trichilia pallida* Sw. (40 ind.) and *Cupania vernalis* Cambess. (35 ind.). This cluster of species was found regenerating in gaps of different sizes, and they are probably favored by occurrence of canopy gaps in TSF conditions.

The cluster analysis identified 4 different canopy gap groups based on tree species composition (Figure 2), with a cut-off level of 0.5 or 50% of floristic similarity. The analysis categorized most part of the gaps into a huge group (C1, C11 up to C17). The other groups formed had shown a lower species similarity. These results underscore the heterogeneity in the species composition within gaps, separating the huge gaps (C4, C10 and C24) from the other, based on the occurrence of tree species that compound the regeneration in this environment. These three gaps are of large size ($\geq$ 1000.0 m²) and contribute to a greater number of individuals sampled over the small gaps.

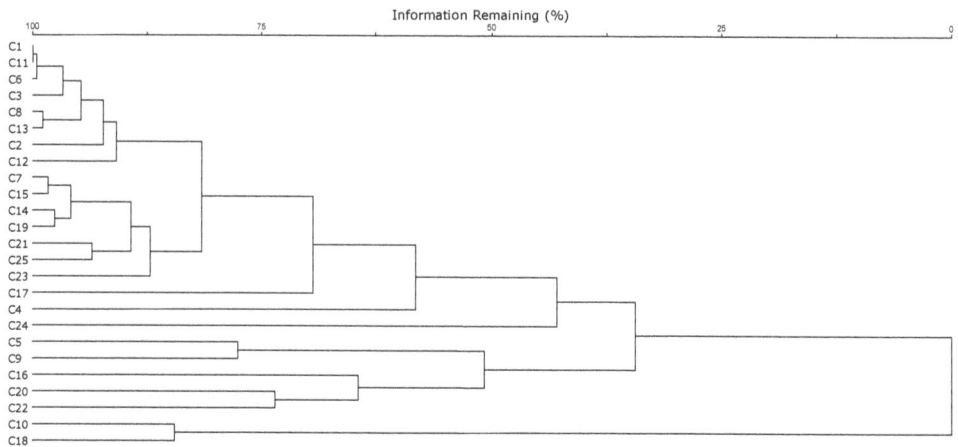

Fig. 2. Cluster analysis of 25 canopy gaps in TSF conditions, using UPGMA and the Jaccard index, Uberlândia, MG. The cut level for small groups was 0.5 or 50% of floristic similarity. The abbreviations C1-C25 reefer to the sequential number of each canopy gap sampled in this study.

On the other hand, the ordination by DECORANA (Figures 3 a and b) showed a density gradient, in which the species bearing the highest number of individuals, such as *Siparuna guianensis*, *Trichilia pallida* and *Ocotea corymbosa* formed a huge group, standing in the central portion of the ordination space (Figure 3b). The eigenvalue for the first value was significant (>0,3), denoting strong relationships from an ecological point of view (Kent & Coker, 1992).

The species *Virola sebifera* Aubl. and *Handroanthus serratifolius* (Vahl) S. O. Grose had the lowest number of individuals, separating from the main group formed by the most part of the species and occurring only in few gaps. The ordination also separated the largest gaps with different values for the two ordination axis (C4, C10 and C24), from the smaller ones based on density values, which in turn, had agreed with the cluster analysis.

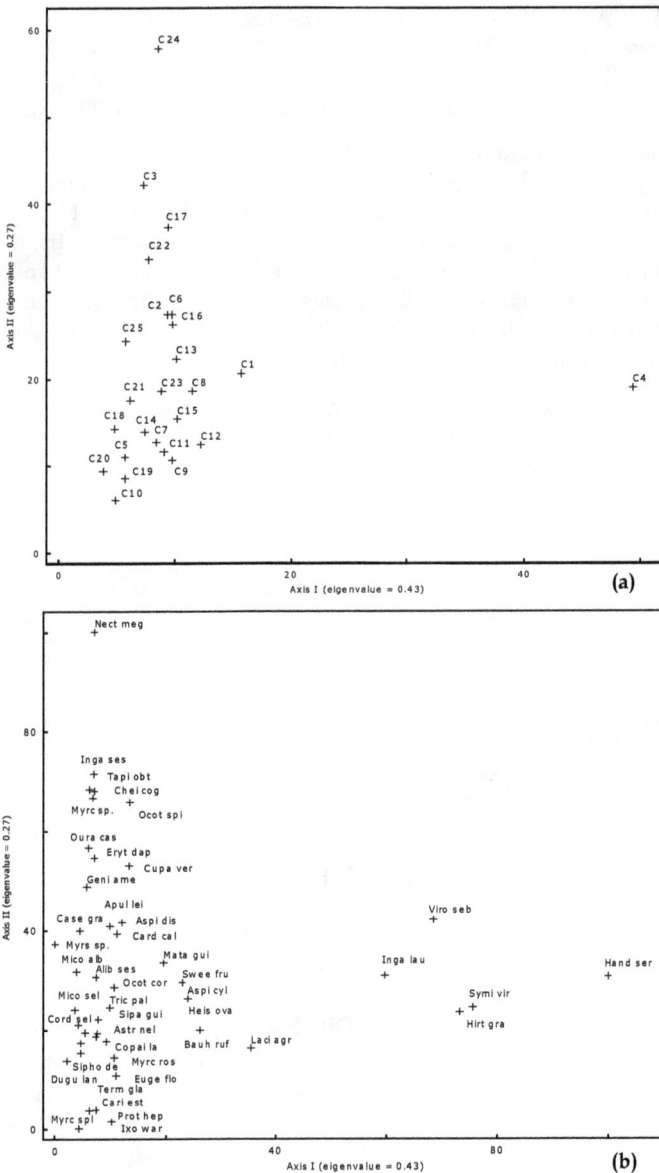

Fig. 3. Ordination of the 25 natural canopy gaps (a) and tree species (b) by DECORANA method in TSF forest, Uberlândia, MG. The first letters refer to the number of each gap.

More than half of the canopy gaps in the study area were formed by the fall of whole canopy trees (52% of the gaps) followed by break of branches and crowns (Table 2). The fall of more than one tree also played some role in the formation of large sized "linked gaps".

We could also observe, only one time, a whole collapsed tree, which probably was caused by strong winds. In this context, Nelson (2005) mentions that pioneer species which form the canopy are susceptible to wind damages, which can cause the fall of individuals in open areas and edges of tropical forest fragments.

The majority of the canopy gaps documented in the study area are small to medium sized ones (up to 500 m^2 area) (Figure 4). The average area of the surveyed gaps are 573.0 m^2 ranging from 52.6 m^2 for small gaps formed by the fall of a single tree to 3468.0 m^2 for the larger ones. We found a significant correlation (Figure 5) between the gap area and the number of tree species (r=0.68; p<0.001). When the number of sampled individuals were taken into account we found a less significant relationship (r=0.45; p<0.05). This parameter shows a higher variation rate (10-82 ind. per gap) when compared to the number of species (7-24 species per gap) therefore, the relationship found becomes weaker as the gap areas increase.

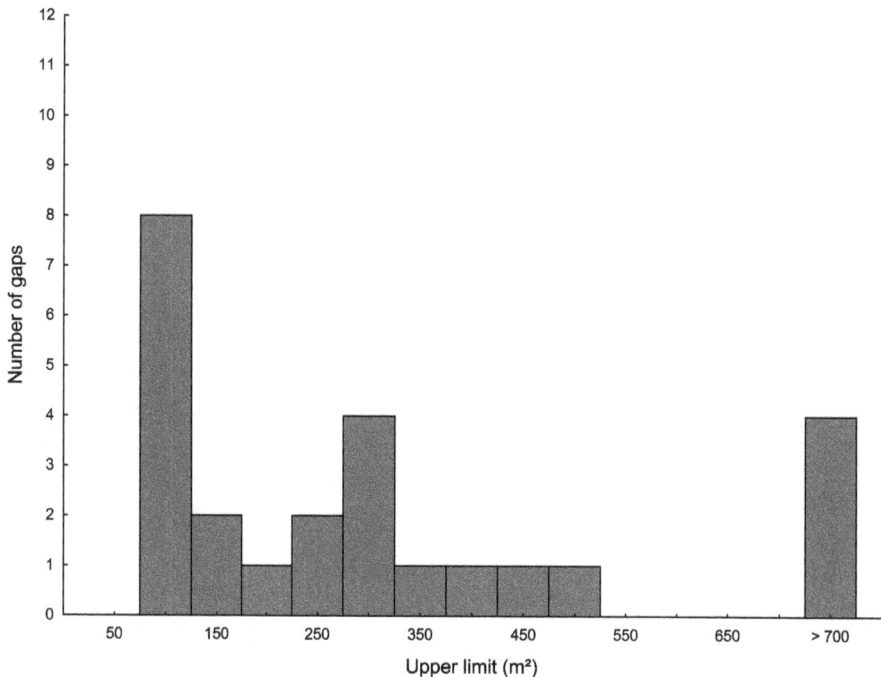

Fig. 4. Distribution by size-classes of natural canopy gaps (using hemispherical photographs) in a tropical semideciduous forest remnant (TSF), Uberlândia, MG.

Lima et al., (2008) found a high variation rate in the canopy gap sizes, ranging from 507.0 to 1109.0 m^2, in a tropical semideciduous forest, using the Runkle method for area estimation. The huge canopy gaps documented by the authors were, in many occasions, larger than the plots they had set. In our study area, the large sized gaps occupy areas over 1000 m^2, tending to hold a marked environmental variation (center, edges and limits with closed canopy) with higher light radiation rates reaching the soil surface (Figure 6).

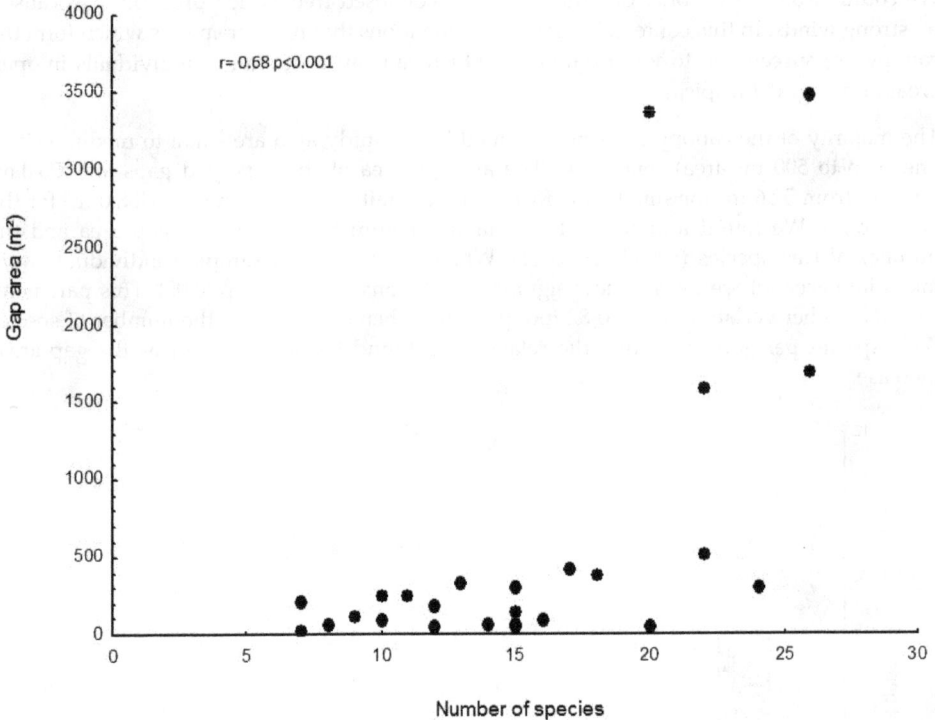

Fig. 5. Relationships between gap area and number of species, in a tropical semideciduous forest remnant (TSF), Uberlândia, MG.

The tropical plant species usually have their growth genetically determined by low or high light intensity rates (Lüttge, 1997). Therefore, some species are able to survive in the overstory or under the canopy in shaded environments. We can find in the same individual adapted leaves for sunny and shaded environments and these two distinct features related to different rates of light/radiation exposure are vital to understanding the ecophysiological variations of plants associated to their environments (Lüttge, 1997).

The estimation of gap areas in the canopy varied from 6.03 to 7.81% alongside the lines we had set for this purpose (Table 3). These are higher values than those mentioned by Van der Meer & Bongers (1996) for a tropical rain forest at Guiana and similar to the values of 7.1% found by Kneeshaw & Bergeron (1998) in a temperate boreal forest at secondary successional stage, in this case using the same methodology applied in this work.

Martins et al., (2004) described canopy gap size classes ranging from 35 to 454 m² in a tropical semideciduous forest (TSF) in Southeast Brazil. The canopy gaps formed by death of Bamboo groupings (*Merostachis riedeliana* Rupr.), had played some role in the structural and successional forest organization, creating successional environments for gaps colonization by tree and shrub species belonging to different ecological groups. In a TSF fragment situated at São Paulo state, Oliveira (1997), mentions that the majority of natural gaps were created by the fall of branches and dead trees, providing small sized gaps ($\cong 46{,}9$ m^2). This event, on a small scale, maintains a diversity of patches in the forest, forming renovation and degradation of eco-units surrounded by patches at later successional stages (Oldeman, 1990).

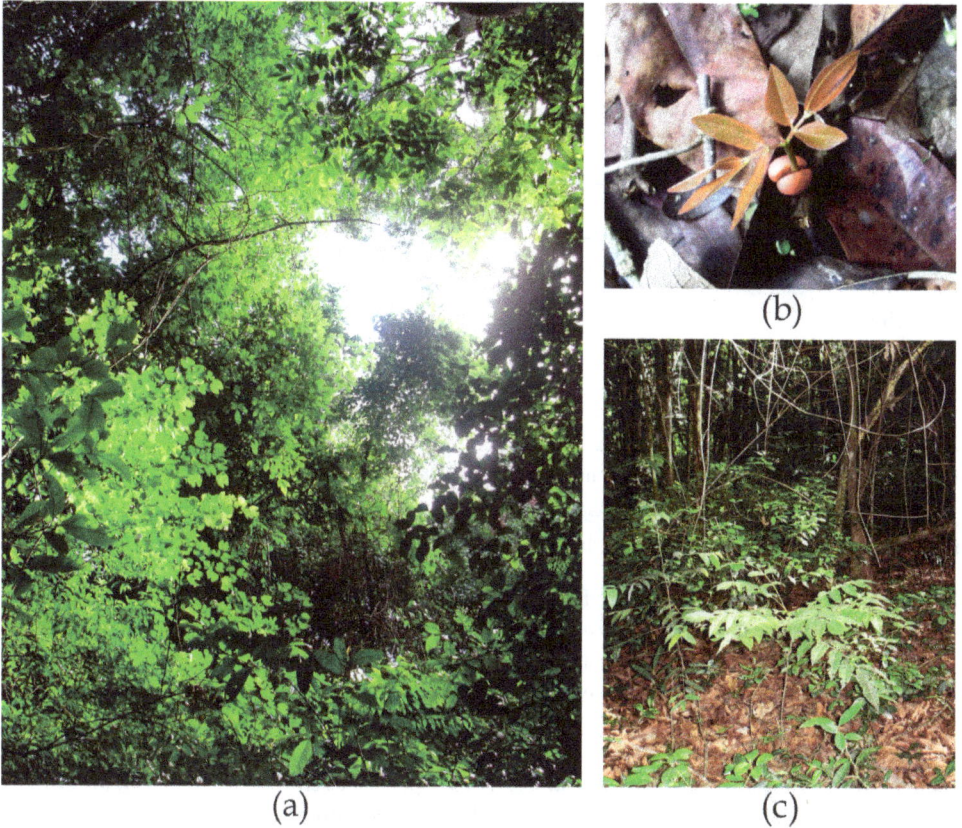

Fig. 6. Light input in a medium sized (a) canopy gap (500 m^2), seedling establishment in a recently formed canopy gap (b) and natural regeneration in advanced stage at an old gap (c) in TSF, Uberlândia-MG.

Formation cause	N	% of total	Type of event (*)
Tree Fall	13	52,0	Falling of a whole tree, creating small sized gaps (< 100 m²)
Fall of more than one tree	4	16,0	Falling of more than one tree creating large sized gaps and linked gaps (> 1000 m²)
Crown break	6	24,0	Breaking of branches, and usually, the whole crown of a tree
Collapsed tree	1	4,0	Collapsing of large sized tree caused by tension or weight
Unidentified	1	4,0	Old canopy gap where we could not identify its formation cause

Table 2. Identified causes of canopy gaps in a tropical semideciduous forest remnant (TSF), Uberlândia, Minas Gerais.

Lines (*)	Covered distance (m)	Occupied area (%)
Line 1	320	7,81
Line 2	320	6,03
Line 3	320	7,68
Line 4	320	6,45

(*) According to Kneeshaw & Bergeron methodology (1998).

Table 3. Estimations of occupied area by natural canopy gaps in a tropical semideciduous forest remnant (TSF), Uberlândia, MG.

In many canopy gaps at natural forests there is a strong regeneration of fast growing pioneer species, , as well as the regrowth of previously stablished species by vegetative growth of roots and stumps (Harstshorn, 1989; Putz & Brokaw, 1989; Whitmore, 1990; Riéra, 1995, Vieira & Scariot, 2006), improving the natural regeneration of tropical vegetation. We observed this fact in our study area, mainly at large sized gaps, where the overstory is subject to higher light radiation rates, showing a higher development of early secondary and pioneer species, besides ruderal and lianes. These species, even though were not accounted in our sample design, can surely interfere in the regeneration of tree species through either competition at ground level or reproducing vegetative in a pretty much shorter time scale than do the tree species.

4. Conclusions

The class-size distribution of canopy-gaps showed a trending towards a bimodal distribution, being the most part of gaps represented by short to medium sized classes (≤ 500 m²). The distribution of tree species is influenced by gap size, community composition and by the number of individuals of species present in this transitory event in STF conditions.

The environmental heterogeneity represented by canopy gaps of different sizes provides an important regeneration niche for pioneer and early secondary species, which we observed in colonizing gaps of different sizes and distinct micro-environments present in these locations. The variations in canopy openings within gaps of different sizes tend to increase the environmental variability in STF conditions improving thus, the natural regeneration in this type of vegetation.

5. Acknowledgments

We would like to thank FAPEMIG for financial support to the project CRA 2851. We are also grateful to researchers Bruno Machado T. Walter (Embrapa-Cenargen), Lúcia Helena Soares Silva (UNB), Ivan Schiavini, Rosana Romero and Jimi N. Nakajima (UFU) for their assistance in the identification of botanical material and to Biology students (UFU) Bárbara Valverde and Gastão V. de Pinho Júnior for their assistance in data gathering.

6. References

Barton, A. M., Fetcher, N., Redhead, S. (1989). The relationship between treefall gap size and light flux in a neotropical rain forest in Costa Rica, *Journal of Tropical Ecology 5*: 437-439.

Brokaw, N. V. (1982). Treefalls: frequency, timing and consequences. *In* Leigh Jr, E. G., Rand, A S. & Windsor, D. M. (eds) *The ecology of a tropical forest: seasonal rhythms and long term changes*, Smithsonian Institution Press, p.101-108.

Brokaw, N. V.L. (1985). Gap-phase regeneration in a tropical forest, *Ecology 66*: 682-687.

Brokaw, N. & Busing, R. T. (2000). Niche versus chance and tree diversity in forest gaps, *Trends in Ecology and Evolution 15*: 183-188.

Brown, N. (1993). The implications of climate and gap microclimate for seedling growth conditions in a bornean lowland rain forest, *Journal of Tropical Ecology 9*: 153-168.

Bullock J. (1996). Plants. *In* Sutherland W.J. (ed.) *Ecological census techniques- a handbook.* Cambridge University Press, p.111-138.

Chase, M.W. & Heveal, J.L. (2009). A phylogenetic classification of the land plants to accompany APG III, Botanical Journal of Linnean Society 161: 122-127.

Deslow, J. S. (1980). Gap partitioning among tropical rainforest trees, *Biotropica 12*: 47-55.

Embrapa. (1982). Serviço Nacional de Levantamento e Conservação dos Solos. *Levantamento de média intensidade dos solos e avaliação da aptidão agrícola das terras do Triângulo Mineiro*, Rio de Janeiro.

Eysenrod, S., Bogaert, J., Hecke, P.V., Impens, I. (1998). Influence of tree-fall orientation on canopy gap shape in an Ecuadorian rain forest, *Journal of Tropical Ecology 14*: 865-869.

Felfili, J.M., Carvalho, F.A., Haidar, R.F. (2005). *Manual para monitoramento de parcelas permanentes nos biomas Cerrado e Pantanal*, Universidade de Brasília 60p.

Felfili, J.M., Silva Junior, M.C., Sevilha, A.C., Fagg, C.W., Walter, B.M.T., Nogueira, P.E., Rezende, A.V. (2004). Diversity, floristic and structural patterns of cerrado in Central Brazil, *Plant Ecology 175 (1):* 37-45.

Floyd, D.A. & Anderson, J.A. (1987). A comparison of three methods for estimating plant cover, *Journal of Ecology 75:* 221-228.

Gravel, D., Canham, C. D., Beaudet, M., Messier, C. (2010). Shade tolerance, canopy gaps and mechanisms of coexistence of forest trees, *Oikos 119:* 475-484.

Hair, J.H., Anderson, R.E., Tatham, R.L., Black, W. (2005). *Análise multivariada de dados.* Bookman edições 593p.

Haridasan, M. & Araújo, G.M. (2005). Perfil nutricional de espécies lenhosas de duas florestas semideciduas em Uberlândia, MG. *Revista Brasileira de Botânica* 28: 295-303.

Hartshorn, G. S. (1980). Neotropical forest dynamics. *Biotropica 12:* 23-30.

Hartshorn, G.S. (1989). Application of gap theory to tropical forest management: natural regeneration on strip clear-cuts in the Peruvian Amazon. *Ecology 70 (3):* 567-569.

Hill, M.O. (1979). *DECORANA: a fortran program for detrended correspondence analysis and reciprocal averaging.* Cornell University.

Kent, M., Coker, P. (1992) (Eds.). *Vegetation description and analysis.* Belhaven Press. 363p.

Kneeshaw, D.D., Bergeron, Y. (1998). Canopy gaps characteristics and tree replacement in the southeastern Boreal Forest, *Ecology 79 (3):* 783-794.

Larpkern, P., Moe, S.R., Totland, O. (2011). Bamboo dominance reduces tree regeneration in a disturbed tropical forest, *Oecologia 165:* 161-168.

Lima, R.A.F. (2005). Estrutura e regeneração de clareiras em Florestas Pluviais Tropicais, *Revista Brasileira de Botânica 28:* 651-670.

Lima, R.A.F., Martini, A.M.Z., Gandolfi, S., Rodrigues, R.R. (2008). Repeated disturbance and canopy disturbance regime in a tropical semideciduous forest, *Journal of Tropical Ecology 24:* 85-93.

Lüttge, U. (1997). *Physiological ecology of tropical plants*, Springer-Verlag 384p..

Martins, S.V., Colletti Jr, R.C., Rodrigues, R.R., Gandolfi, S. (2004). Colonization of gaps produced by death of bamboo clumps in a semideciduous mesophytic forest of south-eastern Brazil, *Plant Ecology 172:* 121-131.

Martins, S.V., Rodrigues, R.R. (2002). Gap-phase regeneration in a semideciduous mesophytic forest, south-eastern Brazil. *Plant Ecology 163:* 51-62.

Martins, S.V., Rodrigues, R.R., Gandolfi, S., Calegari, L. (2009). Sucessão ecológica: Fundamentos e aplicações na restauração de ecossistemas florestais. *In:* Martins, S.V. (org.) *Ecologia de florestas tropicais no Brasil*, Editora UFV, p.19-51.

Mccune, B., Mefford, M.J. (2006). *PC-ORD Multivariate analysis of ecological data*, Version 5.10 MjM Software, Gleneden Beach.

Mitchell, P.L., Whitmore, T.C. (1993). *Use of hemispherical photographs in forest ecology*, Oxford Occasional Papers, Oxford Forest Institute 39p.

Mulkey, S.P., Chazdon, R.L., Smith, A.P. (1996) (Eds.). *Tropical forest plant ecophysiology*, Chapman & Hall 675p.

Nascimento, A.R.T., Felfili, J.M., Fagg, C.W. (2007). Canopy openness and Lai estimates in two seasonally deciduous forest in limestone outcrops in Central Brazil using hemispherical photographs, *Árvore 31 (1):* 167-176.

Nelson, B.W. (2005). Pervasive alteration of tree communities in undisturbed Amazonian forests, *Biotropica 37 (2):* 158-159.

Oldeman, R.A.A. (1990). Dynamics of tropical rain forests. *In:* Holm-nielsen, L.B., Nielsen, I.C., Balslev, H. (eds.) *Tropical forests: Botanical dynamics, speciation and diversity*, Academic Press, p.3-21.

Oliveira, R.E. (1997). *Aspectos da dinâmica de um fragmento florestal em Piracicaba-SP: silvigênese e ciclagem de nutrientes*, São Paulo: ESALQ\USP. Dissertação de Mestrado.

Putz, F.E. & Brokaw, N.V.L. (1989). Sprouting of broken trees on Barro Colorado Island, Panama, *Ecology 70:* 508-512.

Riéra, B. (1995). Rôle des perturbations actuelles et passées dans la dynamique et La mosaïque forestière, *Revie Ecologie (Terre Vie) 50:* 209-223.

Sapotka, I.P., Odén, P.C. (2009). Gap characteristics and their effects on regeneration, dominance and early growth of woody species. *Journal of Plant Ecology (2) 1:* 21-29.

Schnitzer, S. A., Carson, W. P. (2010). Lianas suppress tree regeneration and diversity in treefall gaps, *Ecology Letters 13:* 849-857.

Steege ter, H. (1997) *Winphot 5: a programme to analyze vegetation indices, light and light quality from hemispherical photographs.* Tropenbos Guyana reports 95-2.

Tabanez, A.A.J., Viana, V.M. (2000). Patch structure within Brazilian Atlantic forest fragments and implications for conservation, *Biotropica 32(4):* 925-933.

Tabarelli, M., Mantovani, W., Peres, C.A. (1999). Effects of habitat fragmentation on plant guild structure in the montane Atlantic forest of southeastern Brazil. *Conservation Biology 91:* 119-127.

Tabarelli, M. (2000). Gap-phase regeneration a tropical montane forest: thee effects of gap structure and bamboo species, *Plant Ecology 188:* 149-155.

Ter Braak, C.J.F. 1995. Ordination. In: Jongman, R.H.G., Ter Braak, C.F.J., van Tongeren, O.F.R. (eds.) *Data analysis in community and landscape ecology*, Cambridge University Press, p. 91-173.

van Der Meer, P.J., Bongers, F. (1996). Formation and closure of canopy gaps in the rain forest at Nouragues, French Guiana, *Vegetatio 126:* 167-179.

Vieira, D.M.L., Scariot, A.O. (2006). Principles of natural regeneration of tropical dry forests for restoration, *Restoration Ecology 14:* 11-20.

Whitmore, T.C. (1990). *An introduction to tropical rain forests*, Clarendon Press 226p.

Whitmore, T.C., Brown, N.D., Swaine, M.D., Kennedy, D., Goodwin-Bailey, C.I., Gong, W.-K. (1993). Use of hemispherical photographs in forest ecology: measurement of gap size and radiation total in a Borneo tropical rain forest, *Journal of Tropical Ecology 9:* 131-151.

Yamamoto, S-I. (2000). Forest gap dynamics and tree regeneration, *Journal of Forest Research* 5: 223-229.

Zar, J.H. (1999). (Ed.) *Biostatistical analysis.* Prentice Hall. 662p.

8

The Role of Environmental Heterogeneity in Maintenance of Anuran Amphibian Diversity of the Brazilian Mesophytic Semideciduous Forest

Tiago Gomes dos Santos[1,2],
Tiago da Silveira Vasconcelos[1] and Célio Fernando Baptista Haddad[1]
[1]Universidade Estadual Paulista,
[2]Universidade Federal do Pampa,
Brazil

1. Introduction

Since the 1950s, most ecologists have assumed that animal communities are not simply random assemblages of species (Wells, 2007). Therefore, deterministic factors were pointed out as being responsible for the variation in species diversity and composition along environmental and/or spatio-temporal gradients (Chase & Leibold, 2003). However, according to the recent Hubbell's Neutral Theory, the structure of assemblages (such as diversity and species composition) results from stochastic processes (i.e. ecological drift) that are not influenced by species traits and/or environmental conditions (see details in Tilman, 2004 and Chase, 2007). Therefore the Neutral Theory predicts that species abundance results solely from structured random walks, leaving unexplained the correlations between species traits and their abundances within habitats and/or along environmental gradients (see references in Tilman, 2004).

Some studies highlighted that Neotropical assemblages show little or no structure (Eterovick & Barros, 2003, Afonso & Eterovick, 2007 for anurans and França & Araújo, 2007 for snakes), corroborating the Neutral Theory. On the other hand, partitioning of resources has been historically stressed in terms of both space and time for anuran assemblages (Crump, 1971), where a temporal axis has been considered the first dimension partitioned in the larval phase, and space in adult phase (Toft, 1985). Several studies showed that similar habitats within relatively close distance often have slightly different amphibian assemblages since some species live in a variety of habitats, while others can have more specialized habitat requirements (Snodgrass et al., 2000; Wells, 2007). Therefore, most anuran assemblages appear to be structured since differences in environmental conditions from site to site have explained differences in assemblage features such as richness and abundance of species (Toft, 1982). In addition, other factors such as biological interactions (e.g., predation and competition) and phylogenetic constraints are also considered important to explain structural patterns of assemblages (Zimmerman & Simberlof, 1996; Eterovick & Sazima, 2000; Eason Jr. & Fauth, 2001; Werner et al., 2007).

According to Paton & Crouch (2002) and Bosch & Martínez-Solano (2003), local studies on breeding site preferences of amphibians can provide more accurate information for management purposes than studies on a wider scale, which can be affected by regional variations. In fact, spatial segregation in breeding site occupancy has been recorded for adult and/or larval assemblages of anurans on a local scale (Collins & Wilbur, 1979; Gascon, 1991; Santos et al., 2007; Both et al., 2009) but unfortunately, few studies have applied specific approaches to confirm assemblage structures (see Both et al., 2010; Vasconcelos et al., 2011 for recent examples). Thus, studies on the spatial pattern of assemblages are urgently required to delineate conservation strategies for anurans in ecosystems under strong anthropogenic pressure such as the Mesophytic Semideciduous Forest (MSF), the most fragmented and threatened ecosystem of the Brazilian Atlantic Domain (Viana & Tabanez, 1996). This kind of forest was almost totally devastated because of their soil fertility, smooth relief and high availability of valuable hardwood and because of unscrupulous political interests (Murphy & Lugo, 1986; Prado & Gibbs, 1993; Dean, 1998). Besides, the Semideciduous Forests have been historically neglected as areas for conservation unit creation because of their low level of endemism when compared with the humid forests (Jansen, 1997; Prado, 2000; Pennington et al., 2006).

In the present study, we employed tests of null hypotheses to assess whether patterns of spatial distribution of anuran assemblages differ from a random distribution among aquatic breeding sites monitored at Morro do Diabo State Park (MDSP), one of the four largest remnants of MSF in Brazil (Durigan & Franco, 2006). We also verified the existence of indicator anuran species of environmental heterogeneity on a local scale.

2. Material and methods

2.1 Study area and sampling procedures

We carried out this study in MDSP, a remnant of seasonally dry tropical forest with approximately 33,845 ha in area, located in southeastern Brazil (22°27'--22°40'S, 52°10'--52°22'W), where the altitude ranges from 260 to 599.50 m a.s.l.) (Fig. 1). MDSP is covered by a mosaic of Mesophytic Semideciduous Forest in different stages of regeneration, some small patches of Cerrado *sensu stricto* (savanna-like vegetation), and transitional forests (Durigan & Franco, 2006). The climate is characterized as subtropical with dry winters and wet summers (Cwa type of Köppen's classification) (Leite, 1998), and historical records indicate a mean annual temperature of 22 °C and annual rainfall ranging from 1,100 to 1,300 mm (Faria, 2006).

We monthly monitored (from September 2005 to March 2007) six anuran breeding sites in MDSP: two permanent streams (PS1 and PS2; sections of 500 m in length), two permanent dams (PD1 and PD2), and two temporary ponds (TP1 and TP2; see a complete characterization in Table 1). Physicochemical water measurements were collected and based on a mean of three samples, using a Hach 2100P Turbidimeter (for turbidity) and an YSI 556 Handheld Multiparameter (for the remaining variables).

Permanent dams were located within the borders of forests and presented great depth, low canopy cover, and a bed composed of clay and organic deposits. Dam waters showed low electric conductivity, intermediate oxygenation, and high richness in potential aquatic predators (insects, crustaceans, and fish) (Table 1).

Fig. 1. Geographical localization of the studied area (Morro do Diabo State Park: MDSP), located in the westernmost region of the state of São Paulo, southeastern Brazil.

Permanent streams were located inside the forests and presented a bed mainly composed of sand (PS2) and gravel (PS1), and backwaters with deposits of mud and organic matter. Stream waters were well oxygenated with intermediate conductivity, had a lower temperature than permanent dams and temporary ponds, and high richness of potential aquatic predators (Table 1).

Temporary ponds were shallow and presented a bed covered by leaf litter and other detritus of terrestrial vegetation (such as decaying grasses in TP2) that grows in dry pond basins. Temporary pond water presented high conductivity, low dissolved oxygen, and low richness in potential aquatic predators (Table 1). In addition, temporary waters presented brownish coloration and high salinity, probably due to humic substances and evaporation respectively (Williams, 2006). However, the two temporary ponds differed regarding their canopy cover, since water surface of TP1 was covered by trees whereas TP2 was scarcely shadowed.

We recorded the monthly abundance of anuran species in each breeding site by performing the "surveys at breeding sites" methodology (Scott & Woodward, 1994) during the nocturnal period (from sunset to midnight, when most species had already reduced their calling activities). The search for anurans was made along the perimeters of breeding sites, by recording males engaged in calling activities. Habitat inspections were carried out using artificial light (head lamps), a methodology widely performed and recommend in protocols of inventory and monitoring of amphibians (Heyer et al., 1994), since males of anurans call even if under artificial light. The amount of time spent in each breeding site varied according to its size and complexity (Scott & Woodward, 1994). Additional information regarding MDSP characterization and sampling schedule is available in Santos et al. (2009) and Vasconcelos et al. (2009).

Environmental	Breeding sites					
describers	PD1	PD2	PS1	PS2	TP1	TP2
Geographic coordinates	22°27′03.7″S 52°20′43.3″ W	22°37′00.4″S 52°10′09.5″ W	22°36′16.2″S 52°18′00.8″ W	22°28′30.8″S 52°20′30.9″ W	22°37′10.5″S 52°09′55.8″ W	22°37′07.8″S 52°10′01.9″ W
Altitude (m)	261	264	299	299	263	259
Water features						
Electric conductivity (µS/cm)	18.33	14.67	24.33	21	49.33	33.67
Dissolved O_2 [mg/L (%)]	5.74 (72.3)	3.39 (41.1)	7.88 (90.4)	8.04 (92.8)	1.24 (15.43)	1.25 (15.3)
pH	5.8	4.67	6.61	5.68	5.8	5.52
Salinity (ppt)	0.01	0.01	0.01	0.01	0.02	0.01
Temperature (°C)	27.08	25.12	22.1	22.51	26.52	25.27
Turbidity (ntu)	5.61	26.57	18.03	5.12	10.75	46.74
Movement	io	s	r	r	s	s
Size (m)						
Length	200	25	500	500	70	26
Width	50	5	2.13	2.7	50	27
Depth	>2	>2	0.14	0.26	0.4	0.48
Hydroperiod (months)	18	18	18	18	10	8
Origin	m	m	n	n	n	n
Canopy cover (%)	<5	<5	80	70	90	<5
Edge type	fl, st	st	fl, st	fl, st	fl	fl, st
Bed substrate type	yc, om	yc, om	sa, om, co, gr	sa, om, af	om, yc	om, yc
Vegetation type						
Aquatic	he	he	he	he	he, sh, ar	he, sh
Edges	he, sh, ar	he, sh, ar	he, sh, ar	he, sh, ar	he, sh, ar	he, sh, ar
Matrix	fm, oa	oa, rf, fp	mf	fm	oa, rf	oa, rf
Richness of aquatic predators						
Fish	4	3	4	3	1	1
Insects and crustaceans	10	6	12	12	5	4

Table 1. Localization and environmental characterization of six breeding sites monitored in Morro do Diabo State Park, São Paulo state, southeastern Brazil. Breeding sites: Permanent dams (PD), permanent streams (PS), and temporary ponds (TP); Water movement: running

(r), inlet and outlet flow (io), and standing (s); Hydroperiod: number of months with water from September 2005 to March 2007; Origin: man-made (m), and natural (n); Edge type: flat (fl), and steep (st); Bed substrate type: sand (sa), yellow clay (yc), gravel (gr), arenilitic flagstone (af), accumulation of organic matter and mud (om), and cobble (co); Vegetation type: herbaceous (he), shrubby (sh), and arboreal (ar); Matrix vegetation type: disturbed open area (oa), Forest of Myrtaceae (fm), Forest of *Pinus* (fp), regeneration of Mesophytic Semideciduous Forest (rf), mature Mesophytic Semideciduous Forest (mf); Aquatic predator richness: numbers of families of insects and crustaceans, and number of species of fish collected with dip nets through monthly sampling.

2.2 Statistical analyses

We carried out an environmental representation of the monitored breeding sites in MDSP by calculating the Euclidean Distance index (Krebs, 1999) on abiotic and biotic quantitative measurements (i.e. physicochemical water features, size, canopy cover, and richness in potential aquatic predators). We based Euclidean Distances on transformed (square root) and normalized (by standard deviation) environmental variables, due to deviations of normality of original data and no comparable measurement scales (Clarke & Gorley, 2006). In addition, we tested the existence of spatial patterns in the distribution of anuran assemblages among the six breeding sites monitored by computing similarity analysis (Bray-Curtis index) (Krebs, 1999). We based the similarity matrix on the total abundance of anuran species in each breeding site. Abundance of each species in each breeding site was considered as the highest number of calling males recorded during the monitored period. We adopted this procedure to avoid overestimation of species due to recounting individuals in a serial sampling schedule (Gottsberger & Gruber, 2004; Vasconcelos & Rossa-Feres, 2005; Santos et al., 2007).

We represented dissimilarity and similarity matrices by cluster analysis (UPGMA) (Krebs, 1999), and accessed statistical significance of genuine clusters performing the SIMPROF similarity profile test (Clarke & Gorley, 2006). SIMPROF is a series of permutation tests of the null hypothesis that assumes that the samples are a priori unstructured (i.e. that the breeding sites are unstructured regarding environmental characteristics and/or anuran assemblages). This test is based on an expected profile shape of similarity/dissimilarity obtained by permuting the entries for each variable 1,000 times (i.e. species and/or environmental variables) across that subset of samples; this produces a null condition in which samples have no group structure. The 1,000 permuted values are averaged to produce a mean profile which is statically compared (999 times) with the real similarity profile by absolute distances (Phi) (Clarke & Gorley, 2006). According to Clarke & Gorley (2006), whether environmental variables are responsible for structuring assemblages, it is expected that a plot, based on environmental information, groups the breeding sites in the same way as for a species composition plot. Therefore, we looked for concordance among cluster plots of environmental characteristics and anuran assemblages, in order to explain the spatial patterns of breeding sites used by anurans in MDSP.

We also computed Principal Components Analysis (PCA) (Legendre & Legendre, 1998) to represent together the monthly samples of breeding sites and the species. The purpose of this analysis was to capture as much as possible the variability of the original dataset in a low-dimensional solution, represented by orthogonal axes. Thus, most of the variance is accounted for on the first two or three axes (Clarke & Warwick, 2001; Manly, 2008). We based PCA on a variance/covariance matrix, since we measured all variables in the same unit (i.e. abundance of anurans). Because our data set was composed of temporally serial samples (*sensu* Legendre & Legendre, 1998) and because we were more interested in spatial rather than temporal structures, we used a covariable matrix in PCA, representing time of sampling (i.e. sampling months as dummy variables) to minimize temporal effects in PCA solution.

Finally, we performed the Indicator Species Analysis (ISA) (Dufrêne & Legendre, 1997) to test for the existence of indicator species of environmental heterogeneity. This method is based on a data matrix where there are data groups (a priori established) that can be indicated by some species. According to McCune & Mefford (1999), each species receives an indicator value (IV) of each group, which varies from zero (no indication) to 100 (perfect indication). The null hypothesis in ISA considers that the maximum IV is not greater than that expected by chance. The indicator value is calculated using relative abundance and relative frequency of species across the sample units (considered herein as the monthly records of species abundance in each breeding site). Thus, a good indicator species of a determined group must be frequent and abundant across the samples in this group (Dufrêne & Legendre, 1997). Established groups for ISA were considered based on the clusters indicated by SIMPROF and PCA analyses. Statistical significances of the maximum value indicated for each group were performed using the Monte Carlo permutation test (Manly, 1998) (5,000 times). We based all analyses on log-transformed (log x+1) abundance of anuran species in order to down-weight the contributions of quantitatively dominant species in the similarity analysis, and to linearize relationships in PCA analysis. SIMPROF and ISA analyses were performed using Primer-E 6.1 (Clarke & Gorley, 2006) and PC-ORD 4.0 (McCune & Mefford, 1999) software respectively. PCA was performed using CANOCO 4.0 for Windows software (ter Braak & Smilauer, 1998).

3. Results

We recorded a total of 23 anuran species in the six monitored breeding sites in MDSP and the number of anuran species in the breeding sites ranged from three (PS2) to 17 species (TP1) (Table 2). Anuran abundance ranged from 22 (PS2) to 316 individuals (TP1) (Table 2).

The SIMPROF similarity profile test showed three consistent groups of breeding sites in relation to environmental characteristics: permanent dams, permanent streams, and temporary ponds (Fig. 2A). The first node separated temporary ponds from permanent dams and permanent streams (Fig. 2A). The second node separated permanent streams from permanent dams (Fig. 2A). In addition, the SIMPROF test also showed that anuran species differed from a spatial distribution expected by chance and clustered three consistent groups of breeding sites in MDSP (Fig. 2B). These groups correspond to the same groups previously evidenced by the environmental features (Fig. 2A). The first node separated lotic (permanent streams) from lentic environments (permanent dams and temporary ponds), whilst the second node separated lentic environments in permanent dams and temporary ponds (Fig. 2B). The multivariate structure within the genuine clusters did not differ statistically (Fig. 2).

Species	PD1	PD2	PS1	PS2	TP1	TP2
Chiasmocleis albopunctata	0	0	0	0	7	15
Dendropsophus minutus	0	0	0	0	5	1
Dendropsophus nanus	50	20	0	0	45	25
Elaschistocleis bicolor	0	0	0	0	30	25
Hypsiboas albopunctatus	1	5	5	18	0	0
Hypsiboas lundii	0	0	0	1	0	0
Hypsiboas punctatus	0	12	0	0	0	0
Hypsiboas raniceps	6	0	0	0	20	8
Leptodactylus chaquensis	0	0	0	0	7	1
Leptodactylus fuscus	0	0	0	0	2	4
Leptodactylus labyrinthicus	0	0	0	0	1	0
Leptodactylus mystaceus	0	0	3	0	1	0
Leptodactylus mystacinus	0	0	0	0	20	8
Leptodactylus podicipinus	10	2	6	3	100	30
Physalaemus cuvieri	8	0	7	0	32	15
Pseudis platensis	0	0	0	0	1	0
Rhinella ornata	0	0	13	0	0	0
Rhinella schneideri	3	0	0	0	0	0
Scinax berthae	0	5	0	0	17	11
Scinax fuscomarginatus	60	12	0	0	0	0
Scinax fuscovarius	0	0	0	0	8	3
Scinax similis	0	0	0	0	2	30
Trachycephalus typhonius	0	0	1	0	18	5
Species abundance	138	56	35	22	316	181
Species richness	7	6	6	3	17	14

Table 2. Spatial distribution of anuran species among the six breeding sites, monitored monthly from October 2005 to March 2007 in Morro do Diabo State Park, São Paulo state, southeastern Brazil: permanent dams (PD), permanent streams (PS), and temporary ponds (TP).

PCA analysis showed congruent results with SIMPROF analysis whereas the two-dimensional solution of PCA accounted for 61.4% of the total explained variance (Fig. 3). Third and fourth axes account for few of the total explained variance (only 8.6% and 6%, respectively). PCA 1 (42.3%) showed a tendency to segregate among samples of streams (mainly related to *Hypsiboas albopunctatus*), and samples of permanent dams and temporary ponds (mainly related to *Dendropsophus nanus, Hypsiboas raniceps, Leptodactylus podicipinus, Physalaemus cuvieri*, and *Scinax fuscomarginatus*). On the other hand, PCA 2 (19.1%) showed a tendency to segregate among samples of permanent dams (mainly related to *Dendropsophus nanus* and *Scinax fuscomarginatus*) and samples of temporary ponds (mainly related to *Chiasmocleis albopunctata, Elachistocleis bicolor, Leptodactylus chaquensis, L. mystacinus, L. podicipinus, Physalaemus cuvieri, Scinax similis*, and *Trachycephalus typhonius*).

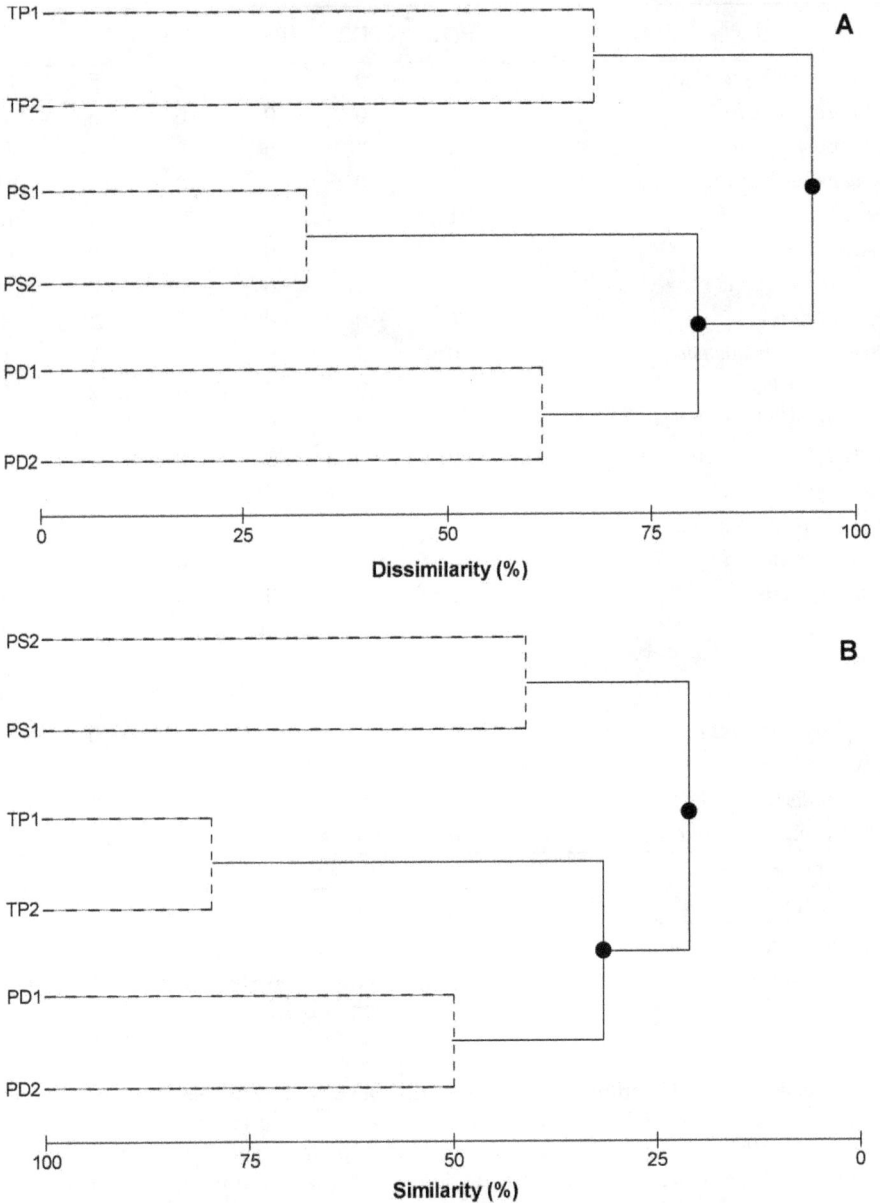

Fig. 2. SIMPROF similarity profile test for environmental characteristics (Euclidian Distance index) (A) and anuran species composition (Bray-Curtis index) (B) recorded in the six breeding sites, monitored monthly from October 2005 to March 2007 in Morro do Diabo State Park, São Paulo state, southeastern Brazil: permanent dams (PD), permanent streams (PS), and temporary ponds (TP). Continuous lines indicate statistically consistent groups (P<0.05), whereas dotted lines indicate no statistical evidence for any structural pattern (P>0.05).

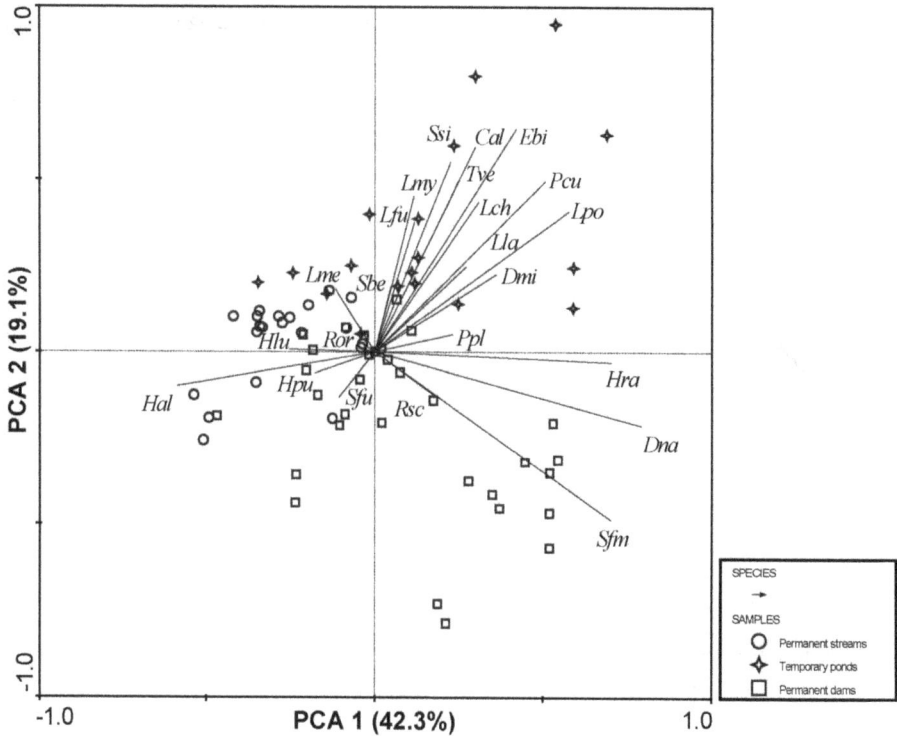

Fig. 3. PCA ordination for 23 anuran species recorded in the six breeding sites, monitored monthly from October 2005 to March 2007 in Morro do Diabo State Park, São Paulo state, southeastern Brazil. Species: *Chiasmocleis albopunctata* (Cal), *Dendropsophus minutus* (Dmi), *D. nanus* (Dna), *Elachistocleis bicolor* (Ebi), *Hypsiboas albopunctatus* (Hal), *H. lundii* (Hlu), *H. punctatus* (Hpu), *H. raniceps* (Hra), *Leptodactylus chaquensis* (Lch), *L. fuscus* (Lfu), *L. labyrinthicus* (Lla), *L. mystaceus* (Lme), *L. mystacinus* (Lmy), *L. podicipinus* (Lpo), *Physalaemus cuvieri* (Pcu), *Pseudis platensis* (Ppl), *Rhinella ornata* (Ror), *R. schneideri* (Rsc), *Scinax berthae* (Sbe), *S. fuscomarginatus* (Sfm), *S. fuscovarius* (Sfu), *S. similis* (Ssi), and *Trachycephalus typhonius* (Tve).

Congruent to the results recorded above, ISA results showed that the frequency of occurrence and abundance of 19 anuran species (about 83% of the total species) is not random when we adopted permanent dams, permanent streams, and temporary ponds as a priori groups in analysis (Table 3, Fig. 4):

Hypsiboas albopunctatus, *H. lundii*, and *Rhinella ornata* were indicator species of permanent streams.

Dendropsophus nanus, *Hypsiboas punctatus*, *Scinax fuscomarginatus*, and *S. fuscovarius* indicated permanent dams.

Chiasmocleis albopunctata, *Dendropsophus minutus*, *Elachistocleis bicolor*, *Hypsiboas raniceps*, *Leptodactylus chaquensis*, *L. fuscus*, *L. mystacinus*, *L. podicipinus*, *Physalaemus cuvieri*, *Scinax berthae*, *S. similis*, and *Trachycephalus typhonius* indicated temporary ponds.

Four anuran species were not indicators of environmental heterogeneity: *Leptodactylus labyrinthicus, L. mystaceus, Pseudis platensis,* and *Rhinella schneideri* (Table 3).

Species	Group	IV	P
Chiasmocleis albopunctata	3	27.8	0.00
Dendropsophus minutus	3	27.8	0.00
Dendropsophus nanus	2	36.2	0.03
Elaschistocleis bicolor	3	44.4	0.00
Hypsiboas albopunctatus	1	52.9	0.00
Hypsiboas lundii	1	13.6	0.04
Hypsiboas punctatus	2	21.9	0.01
Hypsiboas raniceps	3	33.9	0.01
Leptodactylus chaquensis	3	16.7	0.02
Leptodactylus fuscus	3	16.7	0.02
Leptodactylus labyrinthicus	3	5.6	ns
Leptodactylus mystaceus	1	11.3	ns
Leptodactylus mystacinus	3	38.9	0.00
Leptodactylus podicipinus	3	65.1	0.00
Physalaemus cuvieri	3	53.2	0.00
Pseudis platensis	3	5.6	ns
Rhinella ornata	1	13.6	0.04
Rhinella schneideri	2	3.1	ns
Scinax berthae	3	22	0.02
Scinax fuscomarginatus	2	38.7	0.00
Scinax fuscovarius	2	18	0.02
Scinax similis	3	27.8	0.00
Trachycephalus typhonius	3	21.2	0.01

Table 3. Species Indicator Analysis (ISA) for spatial distribution of anuran species in the six breeding sites, monitored monthly from October 2005 to March 2007 in Morro do Diabo State Park, São Paulo state, southeastern Brazil. Groups: 1) permanent streams; 2) permanent dams; and 3) temporary ponds. Indicator values (IV) and statistical significance (P). ns = P>0.05.

Fig. 4. Some anuran species indicators of environmental heterogeneity at Morro do Diabo State Park, São Paulo State, Brazil: *Dendropsophus nanus* (A) and *Scinax fuscomarginatus* (B), typical of permanent dams; *Hypsiboas lundii* (C) and *Rhinella ornata* (D), recorded only in permanent streams; and *Chiasmocleis albopunctata* (E) and *Scinax similis* (F), typical of temporary ponds.

4. Discussion

Our results indicated that anuran assemblages of MDSP were structured and related to the environmental heterogeneity of breeding sites, disagreeing with Hubbell's neutral model of communities (Hubbell, 2001). Werner et al. (2007) also reported that amphibian assemblages of wetlands in Michigan USA, deviated from the neutral model because spatial distribution of species across breeding sites differed from an expected random distribution, and that assemblage structure was also related to the environmental heterogeneity of ponds (i.e. mainly related to gradients of disturbance, productivity, and pond area). In fact, several studies around the world have shown that adult anurans can actively choose breeding sites according to habitat characteristics (Collins & Wilbur, 1979; Eason & Fauth, 2001; Bosch & Martínez-Solano, 2003; Werner et al., 2007).

We expected a spatial structure of the anuran assemblages, since the groups of monitored breeding sites in MDSP (permanent dams, permanent streams, and temporary ponds) differed widely in relation to environmental characteristics. For Neotropical anuran assemblages, studies have also pointed out differential utilization of breeding sites by adults and/or larvae of anurans (Basso, 1990; Hero, 1990; Eterovick & Sazima, 2000; Bertoluci & Rodrigues, 2002; Both et al., 2010; Vasconcelos et al., 2011), although few had effectively used statistical approaches for confirming spatial structure. In the present study, the remarkable degree of concordance between anuran assemblage and environmental plots (i.e. both groups with very similar topology) indicated that the suite of environmental variables has explained spatial patterns of anuran distribution among breeding sites (*sensu* Clarke & Gorley, 2006).

We recorded by SIMPROF and PCA analyses, a primary segregation of stream anuran assemblages in MDSP. In fact, few anuran species (n = 7) occurred in permanent streams (PS1 and PS2) and all of those occurred only in backwaters, except *Rhinella ornata* which was also recorded in riffles and running waters. The low anuran species richness recorded in streams was also reported in other localities of southern and southeastern Brazil (e.g. Bernarde & Anjos, 1999; Bernarde & Machado, 2001; Brasileiro et al., 2005), which may result from four non exclusive hypotheses: i) phylogenetic constraints related to the historic process of colonization in South America (Zimmerman & Simberlof, 1996); ii) pressure of aquatic predators (mainly fish) considered strong in this type of environment (e.g. Gascon, 1991; Magnusson & Hero, 1991); iii) runoff spawns; and iv) morphological limitations in the larval phase (Gascon, 1991). The anuran segregation of streams from the remaining breeding sites (permanent and temporary lentic habitats) differed from the pattern observed in the environmental context, where the external group was composed by temporary ponds, and streams were closely related to the dams. High structural similarity between streams and dams was naturally expected, since dams are originated from dammed streams. However, for anuran assemblages, temporary ponds and dams are more closely related, since both of them are lentic waters, while streams are lotic environments usually unsuitable to colonization.

We recorded through ISA analysis three anuran species as stream indicators: *Hypsiboas albopunctatus*, *H. lundii*, and *Rhinella ornata*. *H. albopunctatus* occurred in backwaters of streams, very similar to swamps (i.e. in wide stream sections with a mud bed, slow-flowing water, and under natural clearings). According to Cei (1980), *H. albopunctatus* lives in open spaces of forests at the edges of standing waters. However, in most studies this species has been reported as being associated with streams (Haddad et al., 1988; Kwet et al., 2002) and swamps (Brasileiro et al., 2005; Santos et al., 2007) with little water flow. *H. albopunctatus* has a generalized reproductive mode (mode 1; Haddad & Prado, 2005) and can be found in human settlements and in disturbed environments (Vasconcelos & Rossa-Feres, 2005; Santos et al., 2007).

Hypsiboas lundii is an anuran species typical from the Brazilian Cerrado (Frost, 2011), totally dependent on gallery forests (Brasileiro et al., 2005). Reproduction of this species is associated with permanent streams of primary and secondary forests (International Union for Conservation of Nature, 2011). Males usually construct basins in banks of streams where eggs are deposited and tadpoles are carried from the basins after flooding caused by heavy rains (mode 4; Haddad & Prado, 2005, Eterovick & Sazima, 2004).

Rhinella ornata is distributed within the Atlantic forest (Frost, 2011). It is a toad that deposits eggs in streams, with exotrophic tadpoles developing in stream backwaters (mode 2; Haddad & Prado, 2005). Our observations indicate that males of this species call in still backwaters and running waters of MDSP streams, highlighting that this toad species is dependent on forest habitats. In fact, a recent study has demonstrated that populations of *R. ornata* undergo genetic erosion due to habitat fragmentation of the Atlantic Coastal Forest (Dixo et al., 2009).

In the current study, we recorded segregation between assemblages of permanent dams and temporary ponds in MDSP. Anuran species richness in permanent dams (PD1 and PD2) was intermediate in relation to permanent streams and temporary ponds (i.e. higher than permanent streams but lower than temporary ponds). In fact, dams have intermediate features between lotic and lentic environments and these are harmful to the reproduction of many species due to changes in physical and chemical features of the water (Esteves, 1998). These statements were corroborated by Both et al. (2008), since they recorded low diversity and high species dominance in the dam environment monitored in southern Brazil. In the present study, dams were dominated by two anuran species (*Dendropsophus nanus* and *Scinax fuscomarginatus*) that have wide distribution in South American biomes and are typically well adapted to anthropogenic disturbances (Duellman, 1999; Frost, 2011; IUCN, 2011). These anuran species have the generalized reproductive mode (mode 1; Haddad & Prado, 2005).

In addition to *Dendropsophus nanus* and *Scinax fuscomarginatus*, two other anuran species were indicators of dams: *Hypsiboas punctatus* and *Scinax fuscovarius*. The former has a wide distribution comprising South and Central America (Duellman, 1999; Vasconcelos et al., 2006), whilst *S. fuscovarius* present more southern distribution in South America (Frost, 2011). Both species have the generalized reproductive mode (mode 1; Haddad & Prado, 2005) and occur in either preserved or disturbed areas (Santos et al., 2007, 2008).

On the other hand, we recorded that temporary ponds (TP1 and TP2) supported high anuran species richness and were spatially structured, disagreeing with the stochastic pattern expected for habitats with low predictability (such as temporary water bodies) (Bonner et al., 1997). The pattern of high anuran species richness that we recorded in temporary ponds can be explained by the high productivity usually reported in temporary environments (Williams & Feltmate, 1992; Brönmark & Hansson, 2005; Williams, 2006), as well as the absence or low abundance of aquatic predators (fish and insects respectively) in comparison with permanent ponds (Heyer et al., 1975; Smith, 1983; Woodward, 1983; Skelly, 1997), and the intermediate-disturbance hypothesis (see Both et al., 2009 for review).

In fact, comparisons of community composition between permanent and temporary waterbodies reveal relatively little overlap regarding the biota (review in Williams, 2006). High species richness has been reported in previous studies on Neotropical anuran assemblages of temporary ponds (e.g. Basso, 1990; Zimmerman & Simberlof, 1996; Santos et al., 2007; Both et al., 2009) and seems to be related to specific features of these environments that make the colonization by several exclusive anuran species advantageous, mainly by those which are more opportunistic. In the present study, we reported many anuran species as indicators of temporary ponds, and our ecological data indicate that most of them (e.g., *Chiasmocleis albopunctata*, *Elachistocleis bicolor*, *Leptodactylus chaquensis*, *Physalaemus cuvieri*, *Scinax similis*, and *Trachycephalus typhonius*) are opportunistic breeders with calling males starting activity after heavy rains (Santos et al., unpubl. data).

Amphibians that use different types of ponds along the permanence gradient tend to have different life-history characteristics (Semlitsch et al., 1996), such as adaptations to deal with drying ponds. Therefore, anuran species that lay eggs in foam nests (e.g. leiuperids and leptodactylids) have an advantage to explore temporary environments due to protection by the foam against desiccation of eggs and/or embryos (Heyer, 1969; Downie, 1988), and probably against wide fluctuations in water levels, typical of temporary breeding sites (Vasconcelos & Rossa-Feres, 2005; Santos et al., 2007). In addition, anuran species with a generalized reproductive mode have high reproductive investments (large clutches), faster larval development, and can also show an evolutive response to explore unpredictable environments such as temporary ponds (Basso, 1990). In our study, 42% of anuran species that were considered as indicators of temporary ponds deposited eggs in foam nests (mode 11: *Physalaemus cuvieri* and *Leptodactylus chaquensis*, mode 13: *L. podicipinus*, and mode 30: *L. fuscus* and *L. mystacinus*), while the remaining 58% presented the generalized reproductive mode (mode 1: *Chiasmocleis albopunctata*, *Elachistocleis bicolor*, *Dendropsophus minutus*, *Hypsiboas raniceps*, *Scinax berthae*, *S. similis*, and *Trachycephalus typhonius*), highlighting that life-history characteristics of these species allow them to explore temporary sites.

The abundance and frequency of occurrence of four anuran species recorded in MDSP did not differ from the randomly expected distribution (*Leptodactylus labyrinthicus*, *L. mystaceus*, *Pseudis platensis*, and *Rhinella schneideri*), and consequently did not indicate association with any breeding site. Three of these were rare at the monitored water bodies (*Leptodactylus labyrinthicus*, *Pseudis platensis*, and *Rhinella schneideri*), which makes interpretations on spatial pattern distribution of these species unfeasible. On the other hand, *Leptodactylus mystaceus* was a ubiquitous anuran species in MDSP. We recorded males of this species calling within mud basins of stream banks and in temporary ponds, as previously reported by Toledo et al. (2003). Reproduction of *L. mystaceus* takes place in sites seasonally flooded by heavy rains (Duellman, 1978) and tadpoles are able to generate foam, probably as an adaptation to survive in places with unpredictable rainfall (Caldwell & Lopez, 1989). Therefore, we believe that the reproductive requirements of *L. mystaceus* seem to be more related to short hydroperiods of breeding sites (such as seasonally flooded banks of permanent streams in MDSP), than to other environmental characteristics, such as lentic or lotic waters.

5. Conclusion

We recorded that anuran assemblages in MDSP were not a random set of species since they were structured according to distinct groups of breeding sites (permanent dams, permanent streams, and temporary ponds). In addition, we also pointed out the existence of species indicators of environmental heterogeneity, i.e. anuran species typical of each group of breeding sites. Regarding indicator species analysis (ISA), we observed lower ISA values associated with anuran indicators of temporary ponds than for those of permanent dams and streams. This result seems to be related to the reproductive patterns of anuran species, since ISA is based on relative abundance and relative frequency of species in the samples (Dufrêne & Legendre, 1997). Therefore, explosive breeding species that occupied temporary ponds for short periods contributed for few samples (i.e. low frequency of occurrence) in ISA analysis, decreasing the final indicator values. On the other hand, anuran species typical of permanent sites remained in streams and dams for longer periods, increasing indicator values. Despite this possible bias in indicator values, we believed in the robustness of ISA results since we recorded a higher number of anuran species as indicators of temporary ponds than of permanent dams and streams. Ecological consistence of ISA can also be supported by other studies showing the same association of anuran species here indicated by ISA with temporary ponds or permanent breeding sites (Brasileiro et al., 2005; Prado et al., 2005; Santos et al., 2007), although direct comparisons are limited due to the lack of similar statistical approaches.

Our results are relevant to conservation proposals since Mesophytic Semideciduous Forests are the most fragmented and threatened ecosystem of the Atlantic Domain and only 2% of this forest type remains in the state of São Paulo as "islands of biodiversity" in an agricultural landscape (Viana & Tabanez, 1996). Therefore, strategies of biodiversity conservation are urgently required for this forest type, and our results have significant implications for anuran conservation actions in remnants of Mesophytic Semideciduous Forest because the importance of heterogeneous breeding sites in the maintenance of high diversity in assemblages of anurans was highlighted. Therefore, the choice of areas for anuran conservation in this forest type must consider the presence of distinct breeding sites such as lotic and lentic environments, as well as sites with different hydroperiods. In this context, the Brazilian Forest Act revision, which hopes to expand agricultural frontiers by clear-cutting the vegetation in riparian zones and legal reserves of private lands, is a recent threat to anuran conservation (Silva et al., 2011). This law proposal, which was approved in the Brazilian Chamber of Deputies, will endanger the water quality and availability, modifying also the hydroperiod of waterbodies used to anuran breeding activity. Besides, the reduction and fragmentation of native vegetation can disconnect areas of shelters and/or foraging from their breeding sites (Becker et al., 2007), as well as promote genetic erosion of anuran species (Dixo et al., 2009).

6. Acknowledgement

This study received financial support from the BIOTA/FAPESP program – The Biodiversity Virtual Institute (www.biota.org.br). We thank Capes and CNPq for the PhD fellowships given to TGS and TSV, respectively. We also thank the Morro do Diabo State Park (MDSP) direction and staff for the logistic support and stimulus and everyone who participated in the fieldwork. We are grateful to Fernando R. Carvalho and Luis G. Gorgatto for fish identification, Luciane Ayres Peres for crustacean identification and Marcia Spies for insect identification and Primer-E use. CFBH thanks FAPESP and CNPq for financial support. All sampling in this study has been conducted with applicable state and federal laws of Brazil (IBAMA permit number 02001.007052/2001).

7. References

Afonso L.G. & Eterovick, P.C. (2007). Microhabitat choice and differential use by anurans in forest streams in southeastern Brazil. *Journal of Natural History*, Vol.41, No.13-16, pp. 937-948, ISSN 0022-2933

Basso, N.G. (1990). Estrategias adaptativas en una comunidad subtropical de anuros. *Cuadernos de Herpetología, Serie Monografías*, Vol.1, pp. 3-70, ISSN 0326-551X

Becker, C.G.; Fonseca, C.R.; Haddad, C.F.B.; Batista, R.F. & Prado, P.I. (2007). Habitat split and the global decline of amphibians. *Science*, Vol.318, pp. 1775-1777, ISSN 1095-9203

Bernarde, P. S. & Anjos, L. (1999). Distribuição espacial e temporal da anurofauna no Parque Estadual Mata dos Godoy, Londrina, Paraná, Brasil (Amphibia: Anura). *Comunicações do Museu de Ciências da PUCRS, Série Zoologia*, Vol.12, pp. 127-140, ISSN 0104-6950

Bernarde, P.S. & Machado, R.A. (2001). Riqueza de espécies, ambientes de reprodução e temporada de vocalização da anurofauna em Três Barras do Paraná, Brasil (Amphibia: Anura). *Cuadernos de Herpetología*, Vol.14, pp. 93-104, ISSN 0326-551X

Bertoluci, J. & Rodrigues, M.T. (2002). Utilização de hábitats reprodutivos e micro-hábitats de vocalização em uma taxocenose de anuros (AMPHIBIA) da Mata Atlântica do Sudeste do Brasil. *Papéis Avulsos de Zoologia*, Vol.42, pp. 287-297, ISSN 0031-1049

Bonner, L.; Diehl, W. & Altig, R. (1997). Physical, chemical and biological dynamics of five temporary dystrophic forest pools in central Mississippi. *Hydrobiologia*, Vol.353, pp. 77-89, ISSN 0018-8158

Bosch, J. & Martínez-Solano, I. (2003). Factors influencing occupancy of breeding ponds in a montane amphibian assemblage. *Journal of Herpetology*, Vol.37, pp. 410-413 ISSN 0022-1511

Both, C.; Kaefer, I.L.; Santos, T.G. & Cechin, S.Z. (2008). An austral anuran assemblage in the Neotropics: seasonal occurrence correlated with photoperiod. *Journal of Natural History*, Vol.42, pp. 205-222, ISSN 0022-2933

Both, C.; Solé, M.; Santos, T.G. & Cechin, S.Z. (2009). The role of spatial and temporal descriptors for Neotropical tadpole communities in southern Brazil. *Hydrobiologia*, Vol.124, pp. 125-138, ISSN 0018-8158

Both, C.; Cechin, S.Z.; Melo, A.S. & Hartz, S.M. (2010). What controls tadpole richness and guild composition in ponds in subtropical grasslands? *Austral Ecology*, Vol.36, pp. 530-536, ISSN 1442-9985

Brasileiro, C.A.; Sawaya, R.J.; Kiefer, M.C. & Martins, M. (2005). Amphibians of an open cerrado fragment in southeastern Brazil. In: *Biota Neotropica*, Vol. 5, 12.07.2011, Available from
http://www.biotaneotropica.org.br/v5n2/pt/abstract?article+BN00405022005
ISSN 1676-0603

Brönmark, C. & Hansson, L.A. (2005). *Biology of Habitats: The Biology of Ponds and Lakes*. Oxford University Press, ISBN 01985-16134, NY

Caldwell, J.P. & Lopez, P.T. (1989). Foam-generating behavior on tadpoles of *Leptodactylus mystaceus. Copeia*, Vol.1989, pp. 498-502, ISSN 0045-8511

The Role of Environmental Heterogeneity in Maintenance of Anuran Amphibian Diversity of the Brazilian Mesophytic Semideciduous Forest

151

Cei, J.M. (1980). Amphibians of Argentina. *Monitore Zoologico Italiano, Monografie,* Vol.2, pp. 1-609

Chase J.M. & Leibold, M.A. (2003). *Ecological Niches: Linking Classical and Contemporary Approaches.* University Chicago Press, ISBN 02261-01800, Chicago

Chase, J.M. (2007). Drought mediates the importance of stochastic community assembly. *Proceedings of National Academy of Sciences of the United States of America,* Vol.104, pp. 17430-17434, ISSN 0027-8424

Clarke, K.R. & Gorley, R.N. (2006). *Software PRIMER v6.* PRIMER-E, Plymouth UK.

Clarke, K.R. & Warwick, R.M. (2001). *Changes in Marine Communities: An Approach to Statistical Analysis and Interpretation.* PRIMER-E, Plymouth, UK

Collins, J.P. & Wilbur, H. (1979). Breeding habits and habitats of the amphibians of the Edwin S. George Reserve, Michigan, with notes on the local distribution of fishes. *Occasional Papers of the Museum of Zoology University of Michigan,* Vol.686, pp. 1-34, ISSN 0076-8413

Crump, M.L. (1971). Quantitative analysis of the ecological distribution of a tropical herpetofauna. *Occasional Papers of the Museum of Natural History,* Vol.3, pp. 1-62, ISSN 0091-7958

Dean, W. (1998). *A ferro e fogo: a história e a devastação da Mata Atlântica brasileira* (1st ed). Companhia das Letras, ISBN 9788571645905, São Paulo, Brazil

Dixo, M.; Metzger, J.; Morgante, J. & Zamudio, K. (2009). Habitat fragmentation reduces genetic diversity and connectivity among toad populations in the Brazilian Atlantic Coastal Forest. *Biological Conservation,* Vol.142, pp. 1560-1569, ISSN 0006-3207

Downie, J.R. (1988). Functions of the foam in the foam-nesting leptodactylid *Physalaemus pustulosus. Herpetological Journal,* Vol.1, pp. 302-307, ISSN 0022-1511

Duellman, W.E. (1978). The biology of an equatorial herpetofauna in Amazon Ecuador. *Miscellaneous Publication, University of Kansas, Museum of Natural History,* Vol.65, pp. 1-352, ISSN 0075-5036

Duellman, W.E. (1999). Distribution Patterns of Amphibians in the South America, In: *Patterns of Distribution of Amphibians: A Global Perspective,* W.E. Duellman (ed.), pp. 255-327, Johns Hopkins University Press, ISBN 08018-61152, Maryland

Dufrêne, M. & Legendre. P. (1997). Species assemblages and indicator species: the need for a flexible asymmetrical approach. *Ecological Monographs,* Vol.67, pp. 345-366, ISSN 0012-9615

Durigan, G. & Franco, G.A.D.C. (2006). Vegetação, In: *Parque Estadual do Morro do Diabo: Plano de Manejo,* H.H. Faria (org.), pp. 111-118, Editora Viena, ISBN 85371-00536, São Paulo, Brazil

Eason Jr., G.W. & Fauth, J.E. (2001). Ecological correlates of anuran species richness in temporary pools: a field study in South Carolina, USA. *Israel Journal of Zoology,* Vol.47, pp. 347-365, ISSN 0021-2210

Esteves, F.A. (1998). *Fundamentos de Limnologia.* Interciência, ISBN 85719-30082, Rio de Janeiro, Brazil

Eterovick, P.C. & Sazima, I. (2000). Structure of an anuran community in a montane meadow in southeastern Brazil: effects of seasonality, habitat, and predation. *Amphibia-Reptilia,* Vol.21, pp. 439-461, ISSN 0173-5373

Eterovick, P.C. & Sazima, I. (2004). *Anfíbios da Serra do Cipó, Minas Gerais, Brasil.* Editora PUC Minas, ISBN 85864-80355, Minas Gerais, Brazil

Eterovick P.C. & Barros, I.S. (2003). Niche occupancy in south-eastern Brazilian tadpole communities in montane meadow streams. *Journal of Tropical Ecology,* Vol.19, pp. 439-448, ISSN 0266-4674

Faria, A.J. (2006). Clima, In: *Parque Estadual do Morro do Diabo: Plano de Manejo*, H.H. Faria (org.), pp. 90-96, Editora Viena, ISBN 85371-00536, São Paulo, Brazil

França, F.G.R. & Araújo, A.F.B. (2007). Are there co-occurrence patterns that structure snake communities in central Brazil? *Brazilian Journal of Biology*, Vol.67, pp. 33-40, ISSN 1519-6984

Frost, D.R. (2011). *Amphibian Species of the World: an Online Reference, Version 5.5*. American Museum of Natural History, New York, 12.07.2011, Available from http://research.amnh.org/herpetology/amphibia/

Gascon, C. (1991). Population and community: level analysis of species occurrences of central Amazonian rain forest tadpoles. *Ecology*, Vol.72, pp. 1731-1746, ISSN 0012-9658

Gottsberger, B. & Gruber, E. (2004). Temporal partitioning of reproductive activity in a Neotropical anuran community. *Journal of Tropical Ecology*, Vol.20, pp. 271-280, ISSN 0266-4674

Haddad, C.F.B.; Andrade, G.V. & Cardoso, A.J. (1988). Anfíbios anuros do Parque Nacional da Serra da Canastra, estado de Minas Gerais. *Brasil Florestal*, Vol.64, pp. 9-20, ISSN 0104-4389

Haddad, C.F.B. & Prado, C.P.A. (2005). Reproductive modes in frogs and their unexpected diversity in the Atlantic Forest of Brazil. *Bioscience*, Vol.55, pp. 207-217, ISSN 1516-3725

Hero, J.M. (1990). An illustrated key to aquatic tadpoles occurring in the Central Amazon rainforest, Manaus, Amazonas, Brasil. *Amazoniana*, Vol.11, pp. 201-62, ISSN 0065-6755

Heyer, W.R. (1969). The adaptive ecology of the species groups of the genus *Leptodactylus* (Amphibia, Leptodactylidae). *Evolution*, Vol.23, pp. 421-428, ISSN 1558-5646

Heyer, W.R.; McDiarmid, R.W. & Weigmann, D.L. (1975). Tadpoles, predation and pond habitats in the tropics. *Biotropica*, Vol.7, pp. 100-111, ISSN 0006-3606

Heyer; M.A. Donnelly; R.W. McDiarmid; L.A.C. Hayek & M.S. Foster (eds.) (1994). *Measuring and Monitoring Biological Diversity: Standard Methods for Amphibians*. Smithsonian Institution Press, ISBN 15609-82845, Washington

Hubbell, S.P. (2001). *The Unified Neutral Theory of Biodiversity and Biogeography*. Monographs in Population Biology. Princeton University Press, ISBN 06910-21287, New Jersey

IUCN, *Red List of Threatened Species*. 2011. Version 2011.1. Electronic database, 12.07.2011, Available from http://www.iucnredlist.org

Jansen, D.H. (1997). Florestas tropicais secas, In: *Biodiversidade*, E.O. Wilson (Ed.), pp. 166-176, Editora Nova Fronteira, ISBN 85209-0792X, Rio de Janeiro

Krebs, C.J. (1999). *Ecological Methodology*. Addison Wesley Educational Publishers, ISBN 03210-21738, California

Kwet, A.; Solé, M.; Miranda, T.; Melchiors, J.; Naya, D.E. & Maneyro, R. (2002). First record of *Hyla albopunctata* Spix, 1824 (Anura: Hylidae) in Uruguay, with comments on the advertisement call. *Boletín de la Asociación Herpetológica Española*, Vol.13, pp. 1-2, ISSN 1130-6939

Legendre, P. & Legendre, L. (1998). *Numerical Ecology*. Elsevier Scientific Publishing Company, ISBN 04448-92494, Amsterdam, The Netherlands

Leite, J.F. (1998). *A Ocupação do Pontal do Paranapanema*. Hucitec, ISBN 85271-04601, São Paulo, Brazil

Magnusson, W.E. & Hero, J.M. (1991). Predation and the evolution of complex oviposition behaviour in Amazon rainforest frogs. *Oecologia*, Vol.86, pp. 310-318, ISSN 0029-8549

Manly, B.J.F. (1998). *Randomization, Bootstrap and Monte Carlo Methods in Biology*. University of Otago, Chapman & Hall, ISBN 04127-21309, New Zealand

Manly, B.J.F. (2008). *Métodos Estatísticos Multivariados: Uma Introdução*. Bookman, ISBN 9788577801855, Rio Grande do Sul, Brazil

McCune, B. & Mefford, M.J. (1999). *PC-ORD Multivariate Analysis of Ecological Data, version 4.2*. MjM Software Design, Oregon

Murphy, P.G. & Lugo, A.E. (1986). Ecology of tropical dry forest. *Annual Review of Ecology and Systematics*, Vol.17, pp. 67–88, ISSN 0066-4162

Paton, P.W.C. & Crouch, W.B. (2002). Using the phenology of pond-breeding amphibians to develop conservation strategies. *Conservation Biology*, Vol.16, pp. 194-204, ISSN 0888-8892

Pennington, R.T.; Ratter, J.A. & Lewis, G.P. (2006). An overview of the plant diversity, biogeography and conservation of Neotropical savannas and seasonally dry forests, In: *Neotropical savannas and seasonally dry forests: plant diversity, biogeography and conservation*, R.T. Pennington, G.P. Lewis & J.A. Ratter (eds), pp. 1-29, CRC Press, ISBN 0849329876, Boca Raton

Prado, D.E. (2000). Seasonally dry forests of tropical South America: from forgotten ecosystems to a new phytogeographic unit. *Edinburgh Journal of Botanic*, Vol.57, No.3, pp. 437–461, ISSN 0960-4286

Prado, D.E. & Gibbs, P.E. (1993). Patterns of species distributions in the dry seasonal forests of South America. *Annals of the Missouri Botanical Garden*, Vol.80, pp. 902–927, ISSN 0026-6493

Prado, C.P.A.; Uetanabaro, M. & Haddad, C.F.B. (2005). Breeding activity patterns, reproductive modes, and habitat use by anurans (Amphibia) in a seasonal environment in the Pantanal, Brasil. *Amphibia-Reptilia*, Vol.26, pp. 211-221, ISSN 0173-5373

Santos, T.G.; Rossa-Feres, D.C. & Casatti, L. (2007). Diversidade e distribuição espaço-temporal de anuros em região com pronunciada estação seca no Sudeste do Brasil. *Iheringia, Série Zoologia*, Vol.97, pp. 37-49, ISSN 0073-4721

Santos, T.G.; Kopp, K.; Spies, M.R.;Trevisan, R. & Cechin, S.Z. (2008). Distribuição temporal e espacial de anuros em área de Campos Sulinos (Santa Maria, RS). *Iheringia, Série Zoologia*, Vol.98, pp. 244-253, ISSN 0073-4721

Santos, T.G.; Vasconcelos, T.S.; Rossa-Feres, D.C. & Haddad, C.F.B. (2009). Anurans of a seasonally dry tropical forest: Morro do Diabo State Park, São Paulo state, Brazil. *Journal of Natural History*, Vol.43, pp. 973-993, ISSN 0022-2933

Scott Jr., N.J. & Woodward, B.D. (1994). Surveys at breeding sites, In: *Measuring and Monitoring Biological Diversity: Standard Methods for Amphibians*, W.R. Heyer; M.A. Donnelly; R.W. McDiarmid; L.A.C. Hayek & M.S. Foster (eds.), pp. 84-92, Smithsonian Institution Press, ISBN 15609-82845, Washington

Semlitsch, R.D.; Scott, D.E.; Pechmann, J.H.K. & Gibbons, J.W. (1996). Structure and dynamics of an amphibian community: evidence from a 16-year study of a natural pond, In: *Long-Term Studies of Vertebrate Communities*, M.L. Cody & J. Smallwood (eds.), pp. 217-248, Academic Press, ISBN 01217-80759, California

Silva, F.R.; Prado, V.H.M. & Rossa-Feres, D. de C. (2011). Value of Small Forest Fragments to Amphibians. *Science*, Vol.332, pp. 1033, ISSN 1095-9203

Skelly, D.K. (1997). Tadpole communities. *American Scientist*, Vol.85, pp. 36-45, ISSN 0003-0996

Smith, D.C. (1983). Factors controlling tadpole populations of the chorus frog (*Pseudacris triseriata*) on Isle Royale, Michigan. *Ecology*, Vol.64, pp. 501-510, ISSN 0012-9658

Snodgrass, J.W.; Komoroski, M.J.; Bryan, A.L. & Burger, J. (2000). Relationships among isolated wetland size, hydroperiod, and amphibian species richness: implications for wetland regulations. *Conservation Biology*, Vol.14, pp. 414-419, ISSN 0888-8892

ter Braak, C.J.F. & Smilauer, P. (1998). *CANOCO Reference Manual and User's Guide to Canoco for Windows: Software for Canonical Community Ordination, version 4*. Microcomputer Power, New York

Tilman, D. (2004). Niche tradeoffs, neutrality, and community structure: a stochastic theory of resource competition, invasion, and community assembly. *Proceedings of National Academy of Sciences of the United States of America*, Vol.101, pp. 10854-10861, ISSN 0027-8424

Toft, C.A. (1982). Community structure of letter anurans in a tropical forest, Makokou, Gabon: a preliminary analysis in the minor dry season. *Revue d'Écologie (La Terre de la Vie)*, Vol.36, pp. 223-232, ISSN 0249-7395

Toft, C.A. (1985). Resource partitioning in amphibians and reptiles. *Copeia*, Vol.1985, pp. 1-21, ISSN 0045-8511

Toledo, L.F.; Zina, J. & Haddad, C.F.B. (2003). Distribuição espacial e temporal de uma comunidade de anfíbios anuros do município de Rio Claro, São Paulo, Brasil. *Holos Environment*, Vol.3, pp. 136-149, ISSN 1519-8634

Vasconcelos, T.S. & Rossa-Feres, D.C. (2005). Diversidade, distribuição espacial e temporal de anfíbios anuros (Amphibia, Anura) na região Noroeste do estado de São Paulo, Brasil. *Biota Neotropica*, Vol.5, 12.07.2011, Available from http://www.biotaneotropica.org.br/v5n2/pt/abstract?article+BN01705022005 ISSN 1676-0603

Vasconcelos, T.S.; Santos, T.G. & Haddad, C.F.B. (2006). Amphibia, Hylidae, *Hypsiboas punctatus*: Extension and filling distribution gaps. *Check List*, Vol.2, pp. 61-62, ISSN 1809-127X

Vasconcelos, T.S.; Santos, T.G.; Rossa-Feres, D.C. & Haddad, C.F.B. (2009). Influence of the environmental heterogeneity of breeding ponds on anuran assemblages from Southeastern Brazil. *Canadian Journal of Zoology*, Vol.87, pp. 699-707, ISSN 0008-4301

Vasconcelos, T.S.; Santos, T.G.; Rossa-Feres, D.C. & Haddad, C.F.B. (2011). Spatial and temporal distribution of tadpole assemblages (Amphibia, Anura) in a seasonal dry tropical forest of southeastern Brazil. *Hydrobiologia*, Vol.673, pp. 93-104, ISSN 0018-8158

Viana, V.M. & Tabanez, A.A.J. (1996). Biology and conservation of forest fragments in the Brazil Atlantic Moist Forest, In: *Forest Patches in Tropical Landscapes*, R. Schella & R. Greenberg (eds.), pp. 151-167, Island Press, ISBN 15596-3426X, Washington

Wells, K.D. (2007). *The Ecology and Behavior of Amphibians*. University of Chicago Press, ISBN 978022-6893341, Illinois

Werner, E.E.; Skelly, D.K.; Relyea, R.A. & Yurewicz, K. (2007). Amphibian species richness across environmental gradients. *Oikos*, Vol.116, pp. 1697-1712, ISSN 0030-1299

Williams, D.D. (2006). *The Biology of Temporary Waters*. Oxford, ISBN 01985-28124, New York

Williams, D.D. & Feltmate, B.W. (1992). *Aquatic Insects*. Oxford University Press, ISBN 08519-87826, UK

Woodward, B.D. (1983). Predator-prey interactions and breeding pond use of temporary-pond species in a desert anuran community. *Ecology*, Vol.64, pp. 1549-1555, ISSN 0012-9658

Zimmerman, B.L. & Simberloff, D. (1996). An historical interpretation of habitat use by frogs in a central Amazonian forest. *Journal of Biogeography*, Vol.23, pp. 27-46, ISSN 0305-0270

Part 4

Impact of Anthropogenic Pressure

9

Seed Dispersal and Tree Spatial Recruitment Patterns in Secondary Tropical Rain Forests

Fred Babweteera
[1]*Department of Forestry, Biodiversity and Tourism,*
Makerere University, Kampala,
[2]*Royal Zoological Society of Scotland, Edinburgh Zoo, Edinburgh,*
[1]*Uganda*
[2]*Scotland*

1. Introduction

Many tropical rain forests are faced with rapid fragmentation and heavy exploitation of flora and fauna (Fa et al., 2005; Laurance, 1998). Studies on effects of forest disturbances, especially logging, have revealed incidental impacts such as damage to the seedlings, saplings and the canopy (e.g. Pereira et al., 2002; White, 1994). However, it is also recognised that the secondary effects of logging may in some cases outweigh the initial damage done by logging. For instance, logging is often accompanied by an increased incidence of hunting, fire and human occupation (Laurance et al., 2006). In addition, these human-induced changes disrupt the ecological processes that are important in maintaining viable populations thus threatening the very survival of forest species. The chain of damaging consequences of these exploitations are believed to lead to the loss of ecological services and loss of timber and non-timber forest products. These result in reduced conservation value of the remnant forests which in turn undermines their sustainability and land productivity.

With the increasing demand for timber and other forest products triggered by growing human populations in the developing countries where these forests are located, it is certain that sustainable management of these remnant forests will be a major challenge (Wright and Muller-Landau, 2006). There is therefore, a need to understand the dynamics of plant and animal populations in secondary tropical forest landscapes. Perhaps of great importance is the understanding of ecological processes that are vital for maintenance of viable tree and animal populations. One of the key ecological processes believed to be affected by forest disturbances and is vital in influencing plant community dynamics is seed dispersal (Barlow and Peres, 2006; Howe and Miriti, 2000). Seed dispersal is crucial for reducing distance- or density-dependent mortality of trees (Hardesty et al., 2006). In addition, within a forest landscape there are sites, such as gaps, that are more favourable for juvenile establishment than others. Consequently, the more widely the seeds of an individual species are dispersed, the greater the chances of the offspring reaching such favourable sites. In tropical rain forests, over 70% of tree species are dispersed by animals (Corlett, 1996; da Silva and Tabarelli, 2000; Gautier-Hion et al., 1985). Seed dispersing animals are believed to influence tree spatial distribution through the seed footprint patterns they create. The seed footprint is determined by the distance over which seeds are dispersed and the density of seeds deposited at any site. Due to the diversity in behavioural ecology among seed dispersing animals, the resulting seed footprints are similarly diverse (Balcomb and Chapman, 2003; Kaplin and Moermond, 1998; Lambert, 2000; McConkey, 2000; Wrangham et al., 1994). Consequently, it is plausible that frugivore diversity in tropical forests may have a strong influence on tree recruitment and spatial distribution (Terborgh et al., 2002). Thus, implying that spatial recruitment of tree species in a forest landscape is altered following loss of some frugivore species.

This study examined seed dispersal and tree spatial recruitment patterns in three tropical forests whose vertebrate populations have been altered differently over the past few decades. The study uses empirical data to test the hypothesis that changes in vertebrate assemblages in tropical rain forests caused by anthropogenic disturbances affect the seed dispersal patterns and subsequent tree spatial recruitment patterns in secondary tropical rain forests. By observing vertebrate assemblages on selected tree species with a range of seed sizes in three tropical rain forests, I sought to address three questions. First, I examine whether there are differences in seed dispersing vertebrate communities in differentially disturbed forests. Second, I determine whether the rate of seed dispersal varies in differentially disturbed forests. Thirdly, I examine whether the observed frugivory patterns are correlated with the tree spatial recruitment patterns. The effects of changes in vertebrate seed disperser community on tree recruitment in secondary forest landscapes are discussed in the wider context of the effectiveness of remnant vertebrate populations in seed dispersal and the possible consequences for tree demography.

2. Methods

2.1 Study sites and tree species

The comparison of seed dispersal and tree recruitment patterns was conducted in three tropical rain forests in Uganda namely: Mabira, Budongo and Kibale Forests (Figure 1). Although these three forests had a similar faunal and floral composition less than a century ago (Hamilton, 1991; Howard, 1991), they now represent a spectrum of disturbance regimes ranging from a highly disturbed and fragmented Mabira Forest, to the moderately disturbed Kibale Forest while Budongo is intermediate. Mabira Forest Reserve is a medium altitude, moist, semi-deciduous forest in Central Uganda (32° 52' – 33° 07' E and 0° 24' – 0° 35' N), covering an area of 306 km². The forest has been subjected to intense anthropogenic disturbances such as logging and hunting which have led to loss of most of its animal populations (Howard, 1991). In addition, vast areas of formerly forested land have been converted to agriculture land. For example, over a period of 15 years (1973 – 1988) it is estimated that 29% of the forest cover was lost and the total forest edge-to-area ratio increased by 29% over the same period (Westman et al., 1989). This resulted in severe forest fragmentation with an estimated fifty thousand people living in the associated enclaves. Budongo Forest Reserve is also a medium altitude, moist, semi-deciduous forest in western Uganda (31° 22' – 31° 46' E and 1° 37' – 2° 03' N), covering an area of 753 km². Although Budongo has been selectively logged since the 1920s, it remains relatively intact with a large population of diurnal primates (Plumptre and Cox, 2006). Mabira and Budongo Forest Reserves are both believed to have had other large vertebrates such as elephants (*Loxodonta africana*) and leopards (*Panthera pardus*) but these were driven to extinction between 1950 and 1980 (Howard, 1991). As forest reserves, logging is still permitted in Mabira and Budongo. On the other hand, the 506 km² Kibale Forest National Park (30⁰ 19' - 30⁰ 32' E and 0⁰ 13' - 0⁰ 41' N) is a moist evergreen forest, transitional between lowland rain forest and montane forest. Kibale is habitat to approximately 280 elephants and has a higher primate biomass than Mabira and Budongo (Plumptre and Cox, 2006). As a national park, Kibale is granted a better protection status than Budongo and Mabira, given that neither logging nor hunting is permitted.

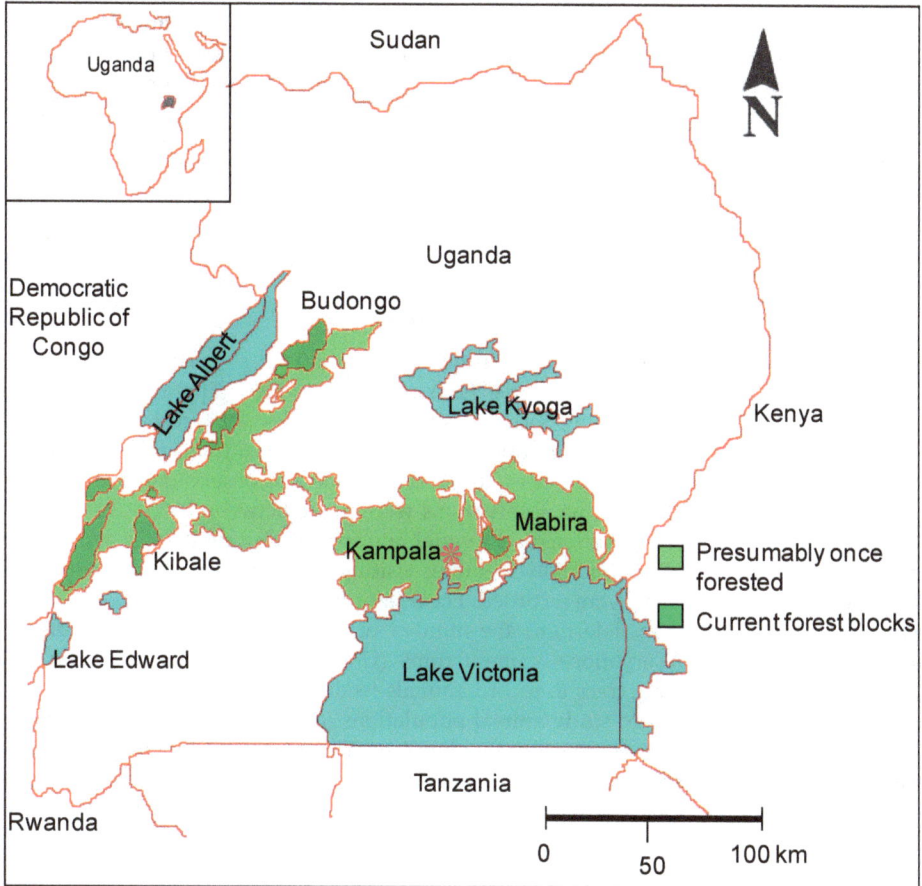

Fig. 1. Map of Uganda showing the forests studied and the areas presumably once forested before 1950s (Hamilton, 1984). Inset is the location of Uganda in Africa

To study seed dispersal and tree spatial recruitment patterns, five tree species were selected on the basis of their fruit/seed size and their occurrence in the three study sites. Fruit/seed size is the major factor limiting vertebrates feeding on fruits and/or seeds of a particular tree (Bollen et al., 2004; Githiru et al., 2002). All five tree species occur in the three forests except for *Ricinodendron heudelotii* that does not grow in Kibale Forest. A brief description of each species is presented in Table 1.

Species	Family	Fruit size (mm)	No of seeds per fruit	Vertebrate dispersers
Balanites wilsoniana Dawe & Sprague	Zygophyllaceae	90	1	Elephants[1]
Chrysophyllum albidum G. Don	Sapotaceae	40	3-4	Large primates and ungulates
Cordia millenii Baker	Boraginaceae	40	1	Primates and ungulates
Ricinodendron heudelotii (Baill.) Pierre ex Pax	Euphorbiaceae	30	2-3	Primates and ungulates[2]
Celtis zenkeri Engl.	Ulmaceae	10	1	Most primates and birds

[1](Babweteera et al., 2007; Chapman et al., 1992);
[2](Feer, 1995; Plumptre et al., 1994).

Table 1. A description of the study tree species

2.2 Vertebrate assemblage and seed dispersal rate

Seed dispersal rate was inferred from the frugivore visitation rates. Vertebrates feeding on the five tree species in each of the three forests were recorded. Three mature fruiting individuals (hereafter referred to as 'focal trees') of each species per forest were identified and observed from time to time for a period of one year. The focal trees of the five species were selected to be at least one kilometre apart and each one was observed at the peak of its fruit ripening for 45 – 75 hours. The observations were made between 0600 – 1200 hrs and 1500 – 1800hrs, recording information on the vertebrate species visiting the tree and the time that each spent feeding. All individual vertebrates visiting the focal trees and observed to be eating the fruits and/or seeds were recorded. Focal sampling was done for each frugivorous species recorded in order to determine the number of fruits consumed per unit time. In addition to the direct observations, camera traps (DSC-P32 Digital Camtrakkers) were mounted beneath the fruiting trees to record animals feeding on fallen fruits. Camera traps have been used successfully to study animal populations (e.g. Carbone et al., 2001; Silveira et al., 2003) and their use is thought to overcome some of the limitations of direct observation such as failure to observe nocturnal feeders or shy frugivores. The camera traps were not mounted to make observations on Celtis trees because of the difficulty in ascertaining whether the photographed animals were feeding on the tiny Celtis fruits. The camera traps were set to make observations during both day and night. The fruiting trees on which they were placed were different from the set used for direct observation. This was done in order to maximize the total observation period for each species, given that the fruiting season for some trees is of short duration. The direct and camera trap observation period for each tree in each forest is summarized in (Table 2).

In addition to the estimate of fruit consumption by arboreal frugivores, an estimate of the rate of fruit removal by vertebrate seed dispersers that feed on fallen fruits were also made. To quantify this, six fruiting trees of each species except for *Celtis*, were selected in each forest. *Celtis* was excluded from the assessment of rate of fruit removal because of the difficulty in ascertaining the fate of the fruit, given its small size. For each individual tree, two fruit piles were randomly placed at 10m (hereafter categorised as NEAR) and 100m (FAR) from the edge of the fruiting tree crown. Fruit piles were placed NEAR and FAR to assess effect of proximity to fruiting tree on rate of fruit removal. Each fruit pile (referred to as 'fruit station') contained 10 ripe fruits. The fruit stations were monitored daily until all the fruits were removed or rotten. An individual fruit was considered 'removed' if the whole fruit was missing or partially eaten with the seed missing.

	Balanites	*Chrysophyllum*	*Cordia*	*Ricinodendron*	*Celtis*
Kibale					
Direct	137	285	216	0	87
Camera traps	1946	1482	1027	0	0
Budongo					
Direct	109	151	221	127	148
Camera traps	1638	1608	1183	1221	0
Mabira					
Direct	146	158	197	121	137
Camera traps	1938	1573	941	1597	0

Table 2. Summary of the direct and camera trap observation hours for frugivory activities on selected tree species in Kibale, Budongo and Mabira Forests. No observations were made on *Ricinodendron* trees in Kibale because they do not exist in this forest

To compare seed dispersal patterns by frugivores in different forests, variations in frugivore body size in the three forests were analysed. The body size is of utmost importance because it is a strong correlate to the quantity of seed dispersal and the distance over which seeds are moved (Lambert, 1998; Lambert, 1999). Limited variation in the body size of frugivores at a particular site causes stereotyped dispersal patterns distinctive of the seed handling and movement patterns of frugivores. In addition, frugivore visitation rates and number of frugivore species visiting each tree species in the three forests were computed as implicit measures of rate of seed dispersal and frugivore preference. The number of individual frugivores visiting each tree species per hour was computed in each forest and ANOVA (SPSS v12) used to test for differences in visitation rate between trees species and forest. The hourly visitation rate data for individual conspecific focal trees in each forest was pooled because there was no significant difference in visitation rates among them for all species. Trees with low visitation rates and narrow ranges of frugivorous species are deemed to be the most vulnerable. To augment the estimated arboreal seed dispersal rate, the rate of ground fruit removal per fruit station was assessed by calculating finite removal rates using the Kaplan-Meier method (Krebs, 1999). The finite removal rate ranges between 0 (0% removal) and 1 (100% removal). To test for differences in fruit removal rates, the calculated finite removal rates were arcsine transformed and used in an ANOVA general linear model procedure. The model included forest type, tree species and distance from the fruiting tree (NEAR or FAR) as the main effects.

2.3 Spatial recruitment of juvenile trees

Spatial juvenile tree recruitment was assessed in square 1-ha plots established around adult conspecific trees for each of the study species. Three plots were established in each of the three forests for each tree species. The selected plots for each species had approximately equal numbers of adult trees of the study species. In each plot, a search for all juveniles (seedlings 0-50 cm in height; sapling 51-400 cm; and poles >400 cm in height but less than 10 cm DBH) of the corresponding tree species was made and the distance to the nearest adult tree was measured. *Balanites* that propagates both sexually and vegetatively, an effort was made to determine whether juveniles originated from root sprouts or seed. Individuals confirmed to be developing from sprouts were omitted from the analysis. To compare the relative dispersion between forests, the cumulative distributions of distances from juveniles to adult trees for each species in each forest were computed (Hamill and Wright, 1986). Pairwise Kolmogorov-Smirnov tests (Dytham, 2003) were then conducted between conspecific plots within each forest to determine whether there were significant differences in the spatial distributions among plots within each forest. Thereafter, the spatial distribution data were pooled for each forest to obtain a single distribution function to enable comparisons between forests using Kolmogorov-Smirnov tests.

3. Results

3.1 Frugivore assemblage and seed dispersal rates

In the three forests a total of 44 frugivore species were recorded, of which 31 were birds, 8 primates, 5 ungulates/omnivores. Five species of rodent seed predators were recorded as well. An overview of the distribution of species and abundance per forest is presented in Appendix. In general, Mabira Forest had the least number of frugivorous species and number of individual frugivores (Figure 2). In addition to fewer frugivore species, there was less variation in the body weight of frugivores in Mabira, whereas the highest variation was in Kibale Forest due to the presence of elephants (Table 3). Mabira also had the highest number of seed predating rodent species (Figure 2).

Frugivore visitation rates were significantly different between the forests ($F = 65$, df = 2, $P < 0.001$). The mean hourly visitation rate was higher in Budongo (2.2 individuals/hr) than in Kibale (1.6 individuals/hr) and Mabira (0.9 individuals/hr). The high visitation rate in Budongo was particularly due to the high frequencies of blue monkeys *Cercopithecus mitis*. The low frugivore visitation rate in Mabira could be an explicit indicator of low vertebrate densities. In addition, small-fruited *Celtis* trees were visited more frequently in all three forests whereas, *Ricinodendron* was the least visited tree (ANOVA; $F = 270$, df = 3, $P < 0.001$; Figure 3). The high visitation rate to *Celtis* compared to the large-fruited trees was mainly due to the large number of frugivorous birds visiting *Celtis* and a preference for large-fruited trees by large frugivores. *Ricinodendron* was the least visited tree and this could be due to the fibrous characteristic of its fruits. Pairwise comparisons of visitation rates to conspecific trees show significant differences between Budongo and Mabira for all tree species whereas visitation rates in Budongo and Kibale were not different except for *Celtis* (Figure 3). *Balanites* was not included in the pairwise comparisons of frugivore visitation rates because, the only observations of frugivores feeding on this species were made by camera traps, for which I could not determine the hourly visitation rate.

A comparison of the estimated quantity of fruit handled by arboreal frugivores (frugivores feeding in the canopy) shows that they handled 20, 36 and 2 fruits per hour in Kibale, Budongo and Mabira respectively. The low quantity of fruit handled in Mabira compared to Budongo and Kibale is due to the low visitation rate and fewer fruits eaten per visiting frugivore. The high quantity handled in Budongo compared to Kibale is attributed to a higher number of chimpanzees and blue monkeys visiting fruiting trees in Budongo (Table 4). Chimpanzees were the most important frugivores in both Kibale and Budongo Forests where they handled over 80% of the fruit. This was largely due to their longer visitation period coupled with a larger quantity of fruit eaten per hour. Chimpanzees and baboons were the only frugivores regularly observed to ingest whole fruits. Other smaller primates ate the pulp and discarded the seed beneath the fruiting trees. However, some of the smaller primates especially the blue monkeys were occasionally seen carrying away a few fruits from the fruiting tree and feeding in the neighbouring trees.

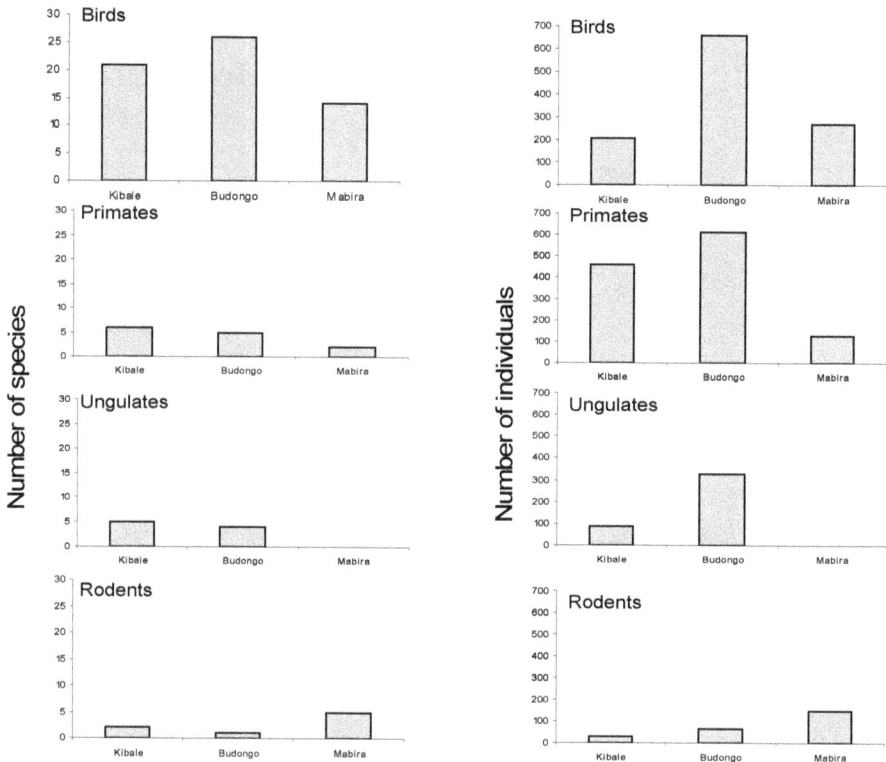

Fig. 2. Species richness and abundance of guilds of vertebrates observed feeding on fruits and seeds of *Celtis, Ricinodendron, Cordia, Chrysophyllum* and *Balanites* in Mabira, Budongo and Kibale Forests.

Forest	Body weight (Kg)			
	25%	75%	Min	Max
Kibale	0.05	6.8	0.02	5000
Budongo	0.03	4.0	0.01	65
Mabira	0.03	0.4	0.02	9

Table 3. Variation in body weight (minimum, maximum and quartile ranges) of frugivores in Kibale, Budongo and Mabira Forests

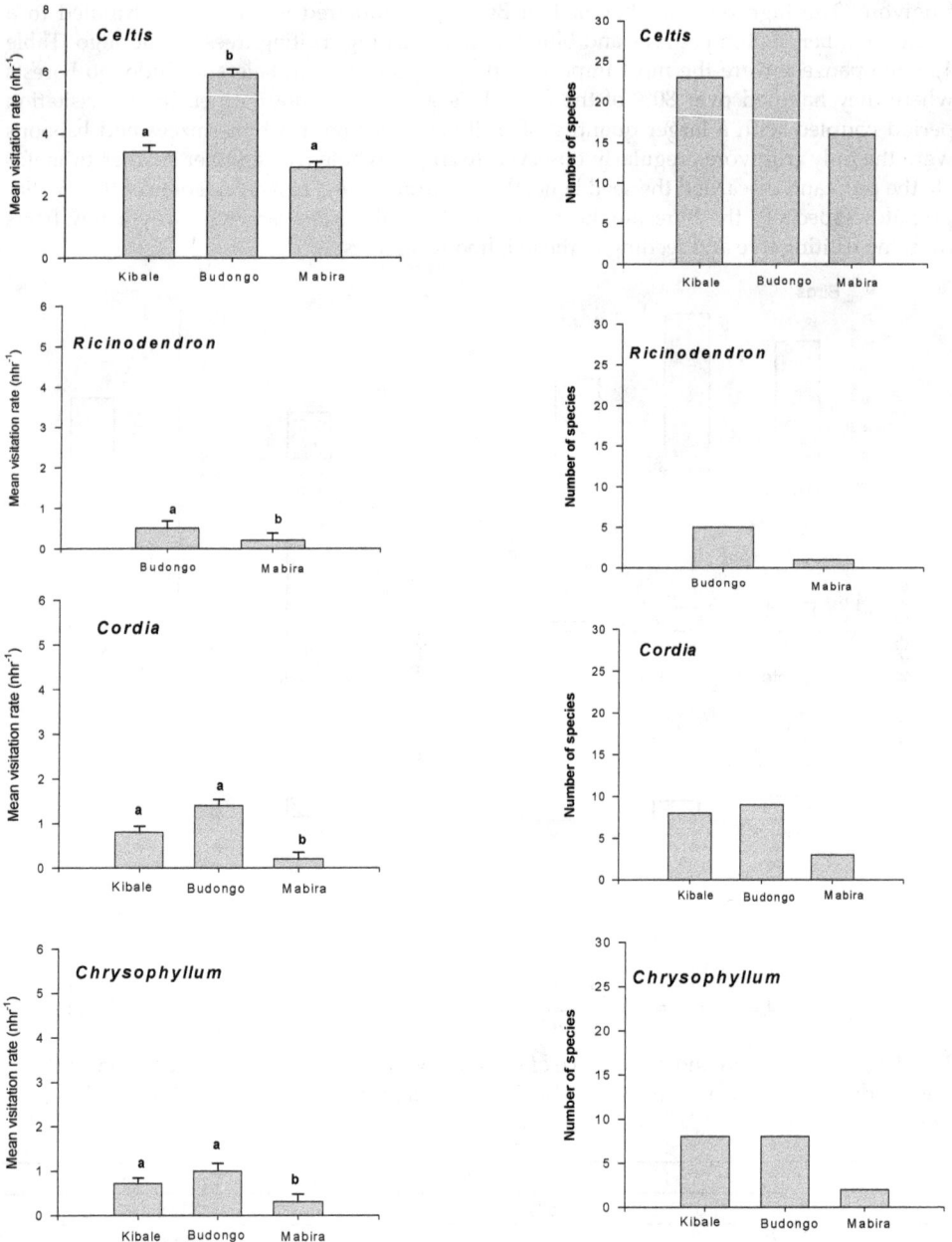

Fig. 3. Frugivore visitation rates (left) and number of frugivorous species visiting (right) different tree species in Kibale, Budongo and Mabira Forests. Bars labelled with different letters represent significantly different mean hourly visitation rates (Tukey HSD) at P < 0.01 (ANOVA). There were no Ricinodendron trees in Kibale.

Species (number of observation hours)	Visits (n)	Duration of visit (hr)			Fruits eaten (n/hr-1)	% of fruits (Total number of fruits) handled
		Mean	25%	75%	Mean	
Kibale (501)						
Chimpanzee*	76	1.33	0.78	1.92	81	80.5 (8187)
Red tailed monkey**	131	0.13	0.05	0.18	52	8.7 (886)
Baboon*	70	0.12	0.07	0.12	78	6.4 (655)
Grey cheeked	27	0.23	0.13	0.42	42	2.6 (260)
mangabey**	24	0.12	0.03	0.15	48	1.4 (138)
Red Colobus**	22	0.08	0.03	0.15	28	0.5 (49)
Black and white colobus**						
Budongo (372)						
Chimpanzee*	138	1.25	0.85	1.6	84	89 (11490)
Blue monkey**	230	0.17	0.07	0.25	29	7 (1134)
Red tailed monkey**	68	0.15	0.05	0.2	51	3.2 (520)
Black and white	26	0.08	0.05	0.13	42	0.5 (87)
Colobus**	2	0.02	0.02	0.02	63	0.3 (3)
Baboon*						
Mabira (355)						
Black mangabey**	46	0.2	0.13	0.32	31	57 (524)
Red tailed monkey**	49	0.1	0.03	0.12	44	43 (216)

* Ingest whole fruit;
** Eat pulp and discard seeds at feeding point

Table 4. Quantity of *Cordia* and *Chrysophyllum* fruits consumed in Kibale, Budongo and Mabira Forests. Duration of visit is represented by mean and quartile ranges.

Estimates of dispersal rates by vertebrates feeding on fruits on the ground at the fruit stations revealed that the rate of removal was significantly affected by the tree species ($F = 61.2$, df = 3, $P < 0.001$) and the forest ($F = 451.6$, df = 2, $P < 0.001$) but not the distance of the fruit station from the fruiting tree. Given that removal rates were not influenced by the proximity of the fruit station to the fruiting tree, the observed ground fruit removal rates at all the fruit stations for each species in each forest were pooled in order to analyse for differences in the rate of fruit removal between forests for each tree species. The pooled data showed that fruit removal rates were higher in both Budongo and Kibale than Mabira (Figure 4). There were no differences in fruit removal rates between Budongo and Kibale for all species except for *Balanites* where hardly any fruits were eaten in Budongo due to the absence of the elephants, the only known frugivores feeding on the fruits. The few fruits of *Balanites* that were removed in Mabira and Budongo were probably eaten by bush pigs or rodents.

Although over 90% of all the fruit was removed in both Kibale and Budongo, the mean duration for 75% of fruits to be removed was significantly lower in Kibale compared to Budongo (6 and 10 days respectively, ANOVA $F = 198$, $P < 0.001$). The faster removal rate in Budongo compared to Kibale may be attributed to a higher density of duikers in Budongo (Appendix). It was further observed that animals consuming the fruit in Budongo and

Kibale ingested the whole fruit since there were no signs of fruit husks or seeds left at the station. In Mabira, 4 out of 96 fruit stations had at least 75% of the fruit removed. Overall 90% of the fruit in Mabira rotted after 3 weeks. It is possible that the few fruits removed in Mabira were consumed by rodent seed predators as I observed fruit husks left at the station. Unlike the arboreal frugivores that preferred *Chrysophyllum* and *Cordia* to *Ricinodendron*, the frugivores feeding on fallen fruits (probably ungulates) did not show any preference for particular fruits.

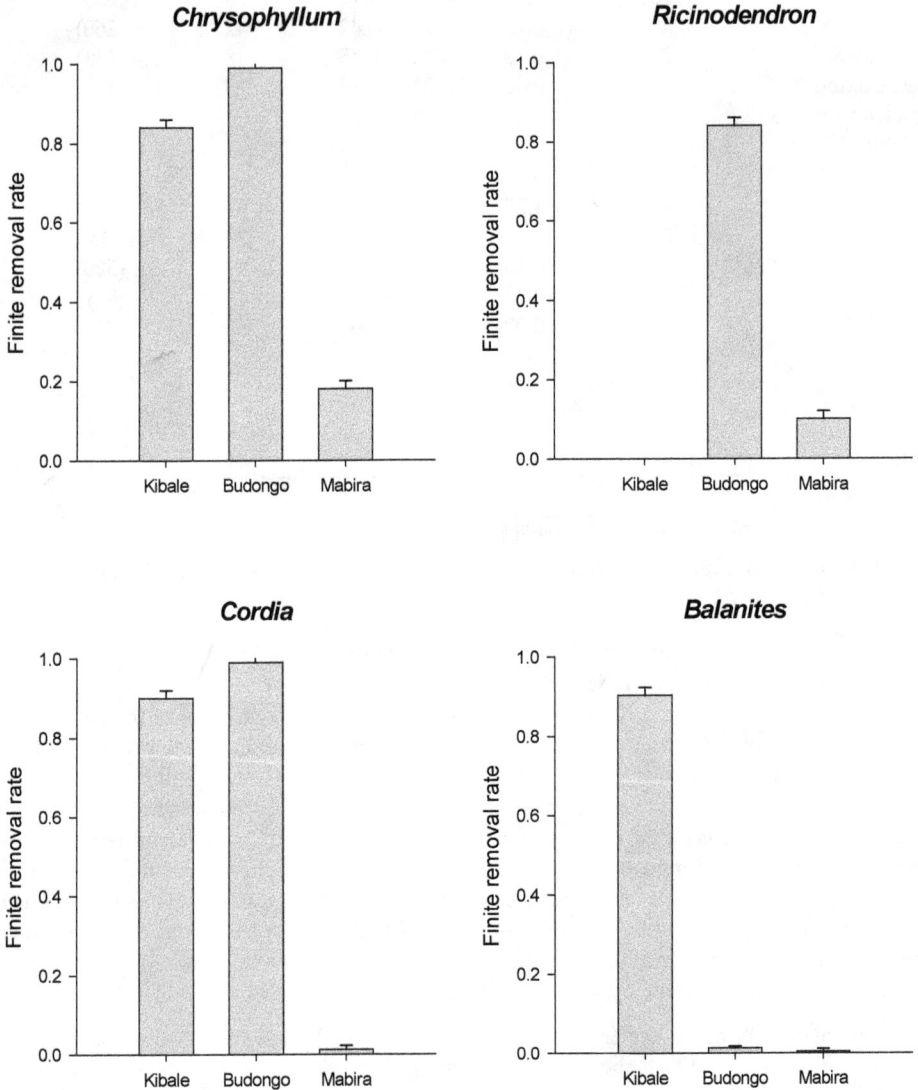

Fig. 4. Fruit removal rates in Kibale, Budongo and Mabira Forests. There were no *Ricinodendron* trees in Kibale.

3.2 Spatial recruitment of juvenile trees

Juveniles of *Cordia* and *Ricinodendron* were not found in any of the plots established in the three forests. The abundance of *Celtis* juveniles was similar between forests whereas juvenile densities of *Chrysophyllum* and *Balanites* varied significantly between forests (Figure 5). *Chrysophyllum* densities were high in both Budongo and Kibale compared to Mabira. In contrast, densities of *Balanites* were highest by a huge margin in Mabira.

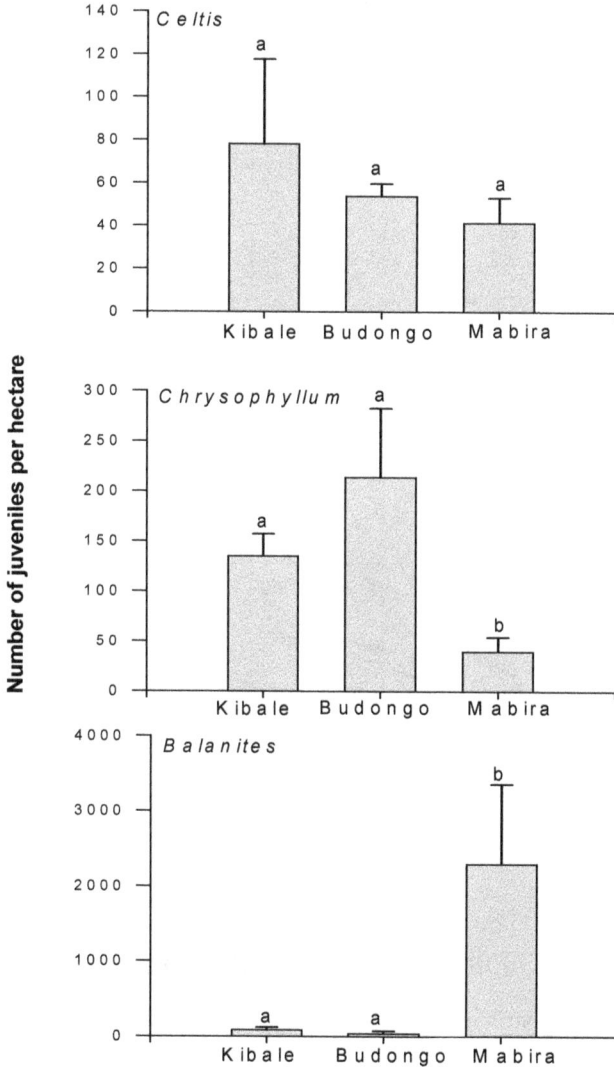

Fig. 5. Juvenile densities (mean + SE) of *Celtis*, *Chrysophyllum* and *Balanites* in Kibale, Budongo and Mabira Forests. Different lower-case letters indicate significant differences (Mann-Whitney U-test; (Dytham, 2003), P < 0.05)

However, an analysis of the juvenile age/size classes shows that juveniles of *Balanites* in Budongo and Mabira were mainly seedlings. In Budongo, none of the seedlings survived to later life stages, whereas 1.4% and 0.3% survived to sapling and pole stages respectively in Mabira. Significant proportions of both *Celtis* and *Chrysophyllum* survived beyond the seedling stage in all three forests.

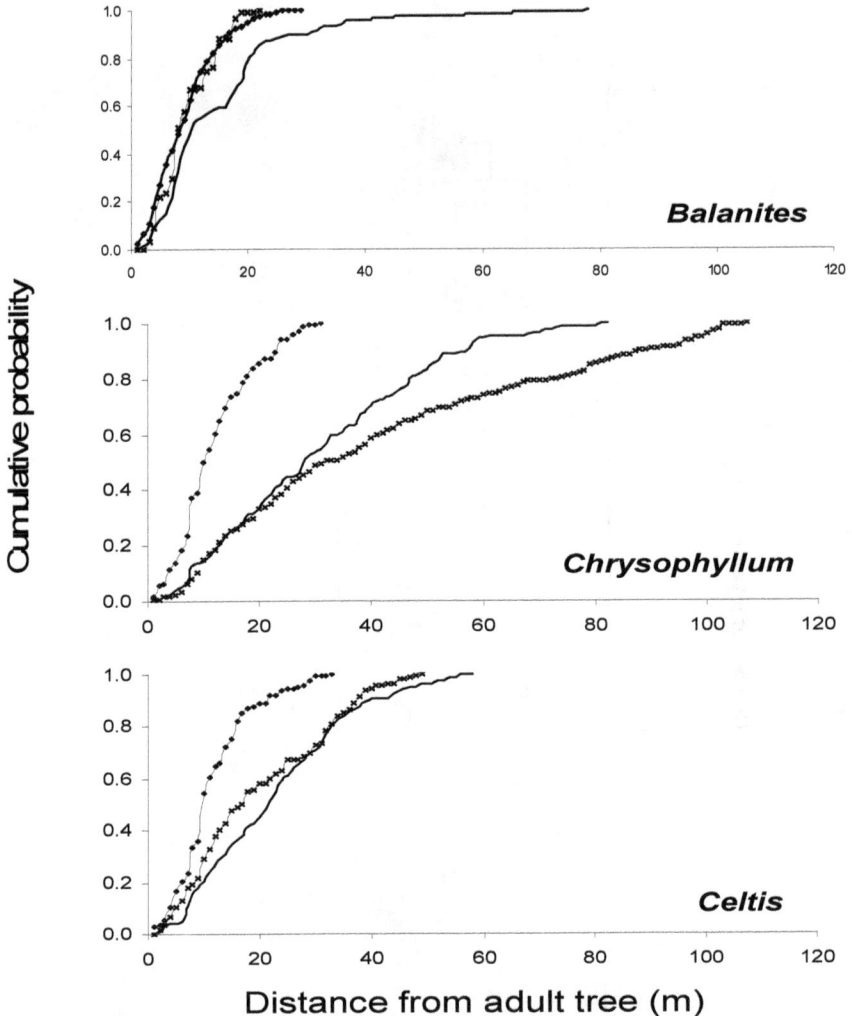

Fig. 6. Observed spatial distribution of juveniles of *Balanites*, *Chrysophyllum* and *Celtis* in Kibale (—), Budongo (xxx) and Mabira (•••) Forests

Pairwise comparison of the pooled data of spatial distributions for conspecific tree plots in each forest shows that the proportions of *Celtis* and *Chrysophyllum* juveniles established beneath adult conspecifics in Kibale and Budongo were not significantly different

(Kolmogorov-Smirnov test $P > 0.05$). However, a larger proportion of juveniles of the two species were established beneath adult conspecifics in Mabira than in Budongo and Kibale ($P < 0.001$). Similarly, the maximum recruitment distance from the mother trees for the two species was lower in Mabira than in Budongo and Kibale (Figure 6). For the large-fruited *Balanites*, the distribution was similar in Mabira and Budongo (Kolmogorov-Smirnov test, $Z = 0.6$ $P > 0.05$) where over 90% of juveniles were established beneath the adult trees (Figure 6). Although the majority of *Balanites* juveniles in Kibale were equally established beneath the adult trees, the spatial distribution was significantly different from that observed in Budongo ($Z = 4.5$, $P < 0.001$) and Mabira ($Z = 4.2$, $P < 0.001$) because of the longer maximum distances over which some juveniles were found (Figure 6).

4. Discussion

4.1 Vertebrate assemblage and seed dispersal rate

Direct and camera trap observations showed a higher species richness and abundance in the less disturbed Kibale and Budongo Forests compared to the heavily disturbed Mabira Forest (Table 2 and Figure 2). Primates are an important frugivorous guild and they contributed most to the frugivore visitation rate and proportion of fruits consumed. The remnant primates in the heavily disturbed Mabira Forest were mainly small-bodied monkeys that often spat seeds beneath the mother trees while feeding compared to the large-bodied primates observed in Kibale and Budongo that ingested the whole fruit. Similarly, ungulates were conspicuously absent in Mabira where they are the favoured bush meat for hunting communities (personal observation).

The frugivore visitation rate and hence rate of seed dispersal was lowest in Mabira and highest in Budongo. The low visitation rate in Mabira is an indicator of low frugivore densities. Low densities of frugivores results in satiation of the disperser community and many mature fruits remain unconsumed (Bas et al., 2006). Although many frugivores were observed in Mabira, almost all were small; 75% of the frugivores weighed less than 0.4kg (Table 3). The loss of large-bodied vertebrates may result in reduced seed dispersal and probably limit the distance over which seeds are moved. Body size is a strong correlate of quantity of seed dispersed and distance over which seeds are moved. The lack of variation in body size implies that frugivore-generated seed footprints in Mabira are likely to be small and homogeneous. A diversity of frugivore-generated seed footprints may be an important means of enhancing the probability of successful tree regeneration through delivery of seed to a variety of safe sites or escaping density dependent mortality. Consequently recruitment of trees in Mabira will not only be impaired by the effects of reduced dispersal rate but also the characteristic short distance dispersal by remnant small-bodied frugivores. In addition to the loss of large vertebrates in Mabira, the forest was characterised by a high frequency of rodent seed predators compared to Kibale and Budongo. This finding is similar to that of Basuta and Kasenene (1987), and Stanford (2000) who found that rodent diversity and abundance increased with logging intensity. Rodent populations are thought to increase in heavily disturbed landscapes due to dense undergrowth in secondary forests that provide safe cover against predators. The increased rodent population in disturbed forests could significantly lower the seed survival probability by increasing seed predation (Kozlowski, 2002). The high density of un-dispersed seeds underneath fruiting trees may exacerbate predator losses. Trees are known to survive seed predation effects through seed predator satiation mechanisms (Fenner and Thompson, 2005). It is possible that the rodents may disperse some seeds in the process of scatter hoarding (Forget, 1990). The significance of seed dispersal by scatter hoarding rodents is not well understood and is an important research subject in heavily disturbed forest landscapes.

Regardless of the vertebrate assemblage differences between forests, the vertebrate assemblage varied between tree species according to the fruit size (Figure 3). This implies that tree species are not equally vulnerable to the loss of vertebrate seed dispersers. The small-fruited Celtis was mainly dispersed by birds, many of which are ubiquitous in all three forests. Similar small-fruited trees may not be adversely affected by forest disturbances. In contrast, large-fruited trees are more vulnerable to disturbance because they depend on large vertebrates that are vulnerable too. Balanites is a notable example of this effect. This species is believed to be dispersed exclusively by elephants (Babweteera et al., 2007; Chapman et al., 1992). In Budongo and Mabira where elephants have become extinct over the past few decades, there were no substitute dispersers of Balanites. There is probably very limited capacity for disperser substitution for large-fruited trees in the disturbed forests.

4.2 Spatial recruitment of juvenile trees

Tree species showed varied recruitment success in different forests. Although, the study did not directly test the factors limiting or enhancing recruitment, implicit inferences indicate that life history, tree fecundity and post dispersal seed and juvenile predation could be the major factors limiting seedling recruitment (Kozlowski, 2002). Establishment of seedlings is a major hurdle for tree regeneration. Early theories suggested coevolution of trees and animal dispersers for which the latter enhance the establishment success of seedlings through gut seed treatment (e.g. Temple, 1977). More recently an experiment on germination of Balanites showed that elephant gut treatment enhance germination by over 50% (Cochrane, 2003). However, significant recruitment of Balanites seedlings in forests where elephants are now extinct was also observed in this study (Figure 5). This provides evidence that germination can also be significant without animal gut treatment. Germination of tree seedlings in tropical forests is influenced by a number of factors including light and moisture regimes, predators, pathogens, forest floor litter and soil disturbance. Experiments that have looked exclusively at the effect of gut passage may have ignored other more important factors influencing seed germination (Robertson et al., 2006).

Recruitment of light demanders (Cordia and Ricinodendron) was limited by unfavourably low light regimes characteristic beneath a closed canopy. Cordia and Ricinodendron are occasionally found in forest gaps in Budongo (personal observation). The two species require high light intensities for establishment and the absence of their juveniles in closed canopy forest underscores the inability of light demanders to recruit outside the forest gaps. Consequently, light demanders require dispersal to enhance their chance of reaching open habitats within a landscape. In Mabira Forest, the seeds of Cordia are dispersed by two small bodied primates; red tail monkey (Cercopithecus ascanius) and black mangabey (Lophocebus aterrimus) while, Ricinodendron was visited by rodent seed predators and no frugivores. The small-bodied vertebrates are likely to disperse the seeds over short distances, thus limiting the probability of seeds reaching open habitats, ultimately leading to lowered recruitment of Cordia and Ricinodendron. On the other hand, low seed production could be the cause of low juvenile densities of *Chrysophyllum* in Mabira. Juveniles of *Chrysophyllum* are shade tolerant and can establish beneath adult conspecifics. They are rarely browsed by ungulates or attacked by insect defoliators. Hence the most likely cause for low recruitment in Mabira compared to Kibale and Budongo is seed limitation which could be a result of low tree fecundity and/or high seed predation rates. Seeds of *Chrysophyllum* are eaten mostly by Gambian rats (*Cricetomys gambianus*) in all three forests. The frequency of visitation by these rodents is highest in Mabira, probably implying a higher seed predation rate (Appendix). Variations in seed dispersal and seed/seedling predation rates of *Balanites* in the three forests accounts for the vast abundance of juveniles in Mabira. *Balanites* trees produce fruit gregariously every 2-3 years (Chapman et al., 1999). Indeed the focal trees observed during this study in all three forests fruited gregariously prior to the commencement of the study. Consequently, the low recruitment in Budongo and Kibale compared to Mabira cannot be attributed to differences in adult tree fecundity. Instead, it could be due to (a) trees in Budongo and Kibale produce less viable fruits, or (b) the absence of predators in Mabira permits the massive recruitment of juveniles. There is no data to support the first hypothesis. The second hypothesis is supported by the fact that *Balanites* seeds are crushed by bush pigs (*Potamochoerus porcus*, (Cochrane, 2003) and the seedlings are browsed mostly by blue duikers (*Cephalophus monticola*, (Babweteera et al., 2007). Consequently, the absence of these potential predators in Mabira (Appendix 1) probably favours the recruitment of *Balanites*. Furthermore, the low density in Kibale could be attributed to dispersal by elephants. A study of seed dispersal by elephants in Kibale showed that elephants visited over 60% of fruiting *Balanites* trees and consumed over 35% of available fruit (Cochrane, 2003). In this study recruitment was assessed in plots around adult conspecifics. Ultimately, it is likely that plots placed at random throughout the forest may have revealed a higher density in Kibale than in Mabira and Budongo where there is no dispersal at all.

With the exception of *Balanites* in Budongo and Mabira, a significant number of seedlings of all study species survive to later life stages. A decrease in number of individuals with increasing age or size is expected for most plant populations (Peet and Christensen, 1987). However, the proportion of *Balanites* juveniles progressing from seedling to pole stage in Mabira (less than 2%) and Budongo (0%) may be insufficient to maintain stable populations in the long-term because, in the event of stochastic mortality, smaller populations are more vulnerable than large populations. In Budongo, although animals capable of dispersing *Balanites* have been lost, the seed and seedling predator populations are intact. This exposes seeds and seedlings to density and/or distance driven mortality factors. Similarly, in Mabira there are no elephants to disperse *Balanites* seeds. However, the survival of a few individuals could be attributed to a lack of seed and seedling predators.

The spatial distribution of juveniles was strongly correlated to the frugivory patterns (Figure 6). The study did not establish the exact parentage of juveniles. Instead it assumed that the observed juveniles were the offspring of the nearest adult tree. There is evidence that seeds can be dispersed hundreds of metres from the mother tree and that germinated seedlings may not be produced by the nearest reproductive adult (Hardesty et al., 2006). However, the noticeable differences between observed juvenile spatial distributions in different forests presented here indicate a strong correlation with the observed frugivory patterns. Consequently, this study provides a meaningful assessment of how forest disturbances affect frugivore activity, which in turn affects the spatial recruitment of trees. In the vertebrate impoverished Mabira Forest, most juveniles were observed recruiting beneath adult conspecifics. This denotes lack of dispersal away from the parent tree. Frugivore species in Mabira were mainly small bodied individuals that often spat seeds beneath or near fruiting trees. Consequently, juveniles of the three tree species were clumped underneath or a few metres from the adult trees in Mabira. Moreover, clumped dispersal footprints are more prominent among the large-fruited trees. For instance *Balanites* is exclusively dispersed by elephants because the fruits and seeds are too large for other frugivores to eat them. The loss of elephants in Budongo and Mabira has obviously left no substitute disperser. Ultimately, the recruitment is restricted to an area immediately beneath adult trees in the two forests. However, even in Kibale where elephants are still present, the spatial distribution of *Balanites* is clumped (Figure 6). This could be due to disperser satiation as a result of mast fruiting and the dependence of the species on a single frugivore (Cochrane, 2003). The observation of clumped distribution patterns in forests with and without animal seed dispersers of *Balanites* probably suggests that studies of seed dispersal should not focus exclusively on the level of juvenile aggregation but instead incorporate a measure of the maximum dispersal distances. In Budongo and Kibale where the large bodied frugivore community is still intact, juveniles of trees that are dispersed further away may have a higher chance for establishment than those dispersed near parent trees or those not dispersed at all.

In conclusion, this study provides evidence of reduced frugivory and seed dispersal activities in heavily disturbed forests due to loss of large vertebrates. However, all tree species are not equally affected by these changes. There is limited capacity for disperser substitution for the large-fruited/seeded trees. Small-fruited/seeded trees dispersed by avian frugivores are unlikely to suffer a major impact on dispersal because many bird species are generalists, resilient to the disturbances. Large-fruited trees should therefore be of particular conservation concern because of the likelihood that they will lose their potential animal dispersers. In addition, this study demonstrates the link between loss of vertebrate seed dispersers and subsequent spatial recruitment patterns of trees. The results underscore the problem about generalising the resilience of tree species to forest disturbances. It is apparent that light demanding species are most vulnerable to the loss of vertebrate seed dispersers given that they are not capable of establishing in closed canopy forest. Consequently, they require a dispersing agent to reach open habitats. Even though open habitats may be common in secondary forests, loss of frugivore species or reduction in their abundance reduces the chance for light demanding tree seeds reaching these sites. In forests where large frugivores are extinct or their populations are reduced, it is plausible that continuous short distance dispersal will lead to spatially clumped tree populations. The long-term population viability of tropical tree species that have clumped distributions resulting from restricted recruitment beneath adult conspecifics is not well understood and could be an important research subject in the future.

5. Appendix

Number of individual vertebrates (direct plus camera trap) and their body weights observed feeding on *Balanites*, *Chrysophyllum*, *Cordia*, *Ricinodendron* and *Celtis* fruits and seeds in Kibale, Budongo and Mabira Forests. Primate, ungulate and rodent body weights after (Kingdon, 1997) and bird body size after Fry et al. (1988; 2000), Fry and Keith (2004), Urban et al. (1986; 1997) and Keith et al. (1992).

Species (common/ *scientific name*)	Body weight (Kg)	Number of individuals		
		Kibale	Budongo	Mabira
Primates				
Chimpanzee *Pan troglodytes*	45	77	181	0
Baboon *Papio anubis*	24	119	7	0
Black and white colobus *Colobus guereza*	13	13	22	0
Grey cheeked mangabey *Cercocebus albigena*	10	55	0	0
Black mangabey *Lophocebus aterrimus*	9	0	0	46
Red Colobus *Procolobus badius*	8	41	0	0
Blue monkey *Cercopithecus mitis*	7	0	308	0
Red tailed monkey *Cercopithecus ascanius*	4	156	94	80
Birds				
Yellow-throated Tinkerbird *Pogoniulus subphulphureus*	0.01	0	62	0
Speckled Tinkerbird *Pogoniulus scolopaceus*	0.02	1	82	25
Little Grey Greenbul *Andropadus gracilis*	0.02	0	17	9
Little Greenbul *Andropadus virens*	0.02	17	26	52
Spotted-flanked Barbet *Tricholaema lachrymose*	0.02	0	7	0
Grey-headed Negrofinch *Nigrita canicapilla*	0.02	0	14	4
Cameroon Sombre Greenbul *Andropadus curvirostris*	0.03	8	50	27
Yellow-whiskered Greenbul *Andropadus latirostris*	0.03	64	59	68
Slender-billed Greenbul *Andropadus gracilirostris*	0.03	13	51	23
Spotted Greenbul *Ixonotus guttatus*	0.04	0	33	0
Common Bulbul *Pycnonotus barbatus*	0.04	13	31	0
Black-billed Barbet *Lybius guifsobalito*	0.04	11	17	0
Green-tailed Bristlebill *Blenda eximia*	0.04	0	0	17
Yellow-spotted Barbet *Buccanodon duchaillui*	0.04	2	2	0
Hairy-breasted Barbet *Lybius hirsutus*	0.05	2	19	7
Violet-backed Starling *Cinnyricinclus leucogaster*	0.05	5	86	0
Grey-throated Barbet *Gymnobucco bonapartei*	0.06	0	1	1
Narina Trogon *Apaloderma narina*	0.06	1	0	0
Red-headed Malimbe *Malimbus rubricollis*	0.06	1	18	0
Purple-headed Glossy Starling *Lamprotornis purpureiceps*	0.07	22	25	19
Yellow-billed Barbet *Trachylaemus purpuratus*	0.09	1		1
Splendid starling *Lamprotornis splendidus*	0.11	12	0	0
Red-eyed dove *Streptopelia semitorquata*	0.2	4	0	0
African Green Pigeon *Treron calva*	0.22	0	6	4
Black-billed Turaco *Tauraco schuetti*	0.24	1	1	0
Crowned Hornbill *Tockus alboterminatus*	0.24	7	2	0
Pied Hornbill *Tockus fasciatus*	0.28	0	9	0

Grey Parrot *Psittacus erithacus*	0.4	0	1	0
Ross's Turaco *Musophaga rossae*	0.4	2	0	0
Great Blue Turaco *Corythaeola cristata*	0.98	16	21	12
Black and white-casqued Hornbill *Ceratogymna subcylindricus*	1.31	2	17	0
Ungulates/omnivores				
Elephant *Loxodonta africana*	5000	62	0	0
Bush pig *Potamochoerus porcus*	65	4	1	0
Weyns duiker *Cephalophus weynsi*	15	2	5	0
Blue duiker *Cephalophus monticola*	5.5	4	307	0
Civet cat *Civetticus civetta*	5	12	14	3
Rodents				
Gambian rat *Cricetomys gambianus*	1.2	1	65	144
Elephant shrew *Rhynchocyon spp*	0.45	0	0	1
Cuvier's tree squirrel *Funiscurius pyrrhopus*	0.25	28	0	1
Long-footed rat *Malacomys longipes*	0.07	0	0	1
Jackson's rat *Praomys jacksoni*	0.04	0	0	1

6. References

Babweteera F, Savill P, Brown N. *Balanites wilsoniana*: Regeneration with and without elephants. Biological conservation (2007) 134:40-47.

Balcomb SR, Chapman CA. Bridging the gap: Influence of seed deposition on seedling recruitment in a primate-tree interaction. Ecological Monographs (2003) 73:625-642.

Barlow J, Peres CA. Effects of single and recurrent wildfires on fruit production and large vertebrate abundance in a central Amazonian forest. Biodiversity and Conservation (2006) 15:985-1012.

Bas JM, Pons P, Gomez C. Exclusive frugivory and seed dispersal of Rhamnus alaternus in the bird breeding season. Plant Ecology (2006) 183:77-89.

Basuta IG, Kasenene JM. Small rodent populations in selectively felled and mature tracts of Kibale Forest, Uganda. Biotropica (1987) 19:260-266.

Bollen A, Van Elsacker L, Ganzhorn JU. Relations between fruits and disperser assemblages in a Malagasy littoral forest: a community-level approach. Journal of Tropical Ecology (2004) 20:599-612.

Carbone C, et al. The use of photographic rates to estimate densities of tigers and other cryptic mammals. Animal Conservation (2001) 4:75-79.

Chapman CA, Wrangham RW, Chapman LJ, Kennard DK, Zanne AE. Fruit and flower phenology at two sites in Kibale National Park, Uganda. Journal of Tropical Ecology (1999) 15:189-211.

Chapman LJ, Chapman CA, Wrangham RW. *Balanites wilsoniana* - Elephant dependent dispersal. Journal of Tropical Ecology (1992) 8:275-283.

Cochrane EP. The need to be eaten: *Balanites wilsoniana* with and without elephant seed-dispersal. Journal of Tropical Ecology (2003) 19:579-589.

Corlett RT. Characteristics of vertebrate-dispersed fruits in Hong Kong. Journal of Tropical Ecology (1996) 12:819-833.

da Silva JMC, Tabarelli M. Tree species impoverishment and the future flora of the Atlantic forest of northeast Brazil. Nature (2000) 404:72-74.

Dytham C. Choosing and Using Statistics: A biologist Guide. (2003) 2nd edn. Oxford: Blackwell.

Fa JE, Ryan SF, Bell DJ. Hunting vulnerability, ecological characteristics and harvest rates of bushmeat species in afrotropical forests. Biological Conservation (2005) 121:167-176.

Feer F. Seed dispersal in African forest ruminants. Journal of Tropical Ecology (1995) 11:683-689.

Fenner M, Thompson K. The ecology of seeds. (2005) Cambridge: Cambridge University Press.

Forget PM. Seed-dispersal of *Vouacapoua americana* (Caesalpiniaceae) by caviomorph rodents in French Guiana. Journal of Tropical Ecology (1990) 6:459-468.

Fry CH, Keith S. The birds of Africa VII. (2004) London: Christopher Helm.

Fry CH, Keith S, Urban EK. The birds of Africa Vol III. (1988) London: Academic Press.

Fry CH, Keith S, Urban EK. The birds of Africa VI. (2000) London: Academic Press.

Gautier-Hion A, et al. Fruit characters as a basis of fruit choice and seed dispersal in a tropical forest vertebrate community. Oecologia (1985) 65:324-337.

Githiru M, Lens L, Bennur LA, Ogol C. Effects of site and fruit size on the composition of avian frugivore assemblages in a fragmented Afrotropical forest. Oikos (2002) 96:320-330.

Hamill DN, Wright SJ. Testing the dispersion of juveniles relative to adults - a new analytic method. Ecology (1986) 67:952-957.

Hamilton A. A field guide to Uganda forest trees. (1991) Kampala: Makerere University Printery.

Hamilton AC. Deforestation in Uganda. (1984) Nairobi: Oxford University Press.

Hardesty BD, Hubbell SP, Bermingham E. Genetic evidence of frequent long-distance recruitment in a vertebrate-dispersed tree. Ecology Letters (2006) 9:516-525.

Howard PC. Nature Conservation in Uganda's Tropical Forest Reserves. (1991) Gland, Switzerland: IUCN.

Howe HF, Miriti MN. No question: seed dispersal matters. Trends in Ecology and Evolution (2000) 15:434-436.

Kaplin BA, Moermond TC. Variation in seed handling by two species of forest monkeys in Rwanda. American Journal of Primatology (1998) 45:83-101.

Keith S, Urban EK, Fry CH. The birds of Africa Vol IV. (1992) London: Academic Press.

Kingdon J. The Kingdon guide to African mammals. (1997) 3rd edn. London: Academic Press.

Kozlowski TT. Physiological ecology of natural regeneration of harvested and disturbed forest stands: implications for forest management. Forest Ecology and Management (2002) 158:195-221.

Krebs JC. Ecological methodology. (1999) 2nd edn. New York: Addison-Welsey Educational Publishers Inc.

Lambert JE. Primate digestion: interactions among anatomy, physiology, and feeding ecology. Evolutionary Anthropology (1998) 7:8-20.

Lambert JE. Seed handling in chimpanzees (*Pan troglodytes*) and redtail monkeys (*Cercopithecus ascanius*): Implications for understanding hominoid and cercopithecine fruit-processing strategies and seed dispersal. American Journal of Physical Anthropology (1999) 109:365-386.

Lambert JE. The fate of seeds dispersed by African apes and cercopithecines. American Journal of Physical Anthropology (2000) 111:204-204.

Laurance WF. Forest fragmentation: another perspective. Trends in Ecology & Evolution (1998) 13:75-75.

Laurance WF, Alonso A, Lee M, Campbell P. Challenges for forest conservation in Gabon, central Africa. Futures (2006) 38:454-470.

McConkey KR. Primary seed shadow generated by Gibbons in the rain forests of Barito Ulu, central Borneo. American Journal of Primatology (2000) 52:13-29.

Peet RK, Christensen NL. Competition and tree death. Bioscience (1987) 37:586-595.

Pereira R, Zweede J, Asner GP, Keller M. Forest canopy damage and recovery in reduced-impact and conventional selective logging in eastern Para, Brazil. Forest Ecology and Management (2002) 168:77-89.

Plumptre A, Cox D. Counting primates for conservation: primate surveys in Uganda. Primates (2006) 47:65-73.

Plumptre A, Reynolds V, Bakuneeta C. The contribution of fruit eating primates to seed dispersal and natural regeneration after selective logging (1994) Oxford: Budongo Forest Project. 101.

Robertson AW, Trass A, Ladley JJ, Kelly D. Assessing the benefits of frugivory for seed germination: the importance of the deinhibition effect. Functional Ecology (2006) 20:58-66.

Silveira L, Jacomo ATA, Diniz-Filho JAF. Camera trap, line transect census and track surveys: a comparative evaluation. Biological Conservation (2003) 114:351-355.

Stanford A. Rodent ecology and seed predation in logged and unlogged forest, Uganda. In: School of Biological Sciences (2000) Bristol: University of Bristol. 171.

Temple SA. Plant-animal mutualism: Coevolution with Dodo leads to near extinction of plant. Science (1977) 197:885-886.

Terborgh J, Pitman N, Silman M, Schichter H, Nunez PV. Maintenance of tree diversity in tropical forests. In: Seed dispersal and frugivory: Ecology, evolution and conservation--Levey DJ, Silva WR, Galetti M, eds. (2002) Wallingford, Oxfordshire: CABI. 1-17.

Urban EK, Fry CH, Keith S. The birds of Africa Vol II. (1986) London: Academic Press.

Urban EK, Fry CH, Keith S. The birds of Africa Vol V. (1997) London: Academic Press.

Westman WE, Strong LL, Wilco BA. Tropical deforestation and species endangerment: the role of remote sensing. Landscape Ecology (1989) 3:97-109.

White LJT. The effects of commercial mechanized selective logging on a transect in lowland rainforest in the Lope-Reserve, Gabon. Journal of Tropical Ecology (1994) 10:313-322.

Wrangham RW, Chapman CA, Chapman LJ. Seed dispersal by forest chimpanzees in Uganda. Journal of Tropical Ecology (1994) 10:355-368.

Wright SJ, Muller-Landau HC. The Future of Tropical Forest Species. Biotropica (2006) 38:287-301.

10

Floristic Composition, Diversity and Status of Threatened Medicinal Plants in Tropical Forests of Malyagiri Hill Ranges, Eastern Ghats, India

S. C. Sahu and N. K. Dhal
Environment and Sustainability Department,
Institute of Minerals and Materials Technology (formerly RRL), CSIR,
Bhubaneswar (Odisha),
India

1. Introduction

Tropical and subtropical forests harbour maximum diversity of plant species found on the earth (WCMC, 1992). These forests are rich in medicinal and economically important plants. Exploitation of these forests has resulted in rapid loss of tropical forests and it is recognized as one of the serious environmental and economic problems all over the world (Hare et al., 1997). A study on floristic composition and species diversity of threatened medicinal plants of tropical forests is ecologically significant besides its usefulness in forest management.

Malyagiri hill ranges belonging to Eastern Ghats of India lies between $21^0 23' 30''$ N latitude and $85^016' 58''$ E longitude, located in the Pallahara Sub-division of Angul District in Odisha. The whole area is endowed with rising and falling hills interspersed with small plains and winding strips of valleys. There are as many as 4 perennial water-falls with dense forests in the northern part known as "Nagira". These provide a congenial niche for the luxuriant growth of marshy plants of various types. "Mankra" a tributary of Brahmani flows in the west of Pallahara which keeps the sub-division aside from National Highway 6. The Lord Siva temple at hill base of Khuluri reserve forest with a perennial water-fall increases the beauty of nature and is a popular tourist place..

The soils of Malyagiri are mainly red and form clays and clay-loams in the valleys (Patra & Choudhury, 1989). The hill ranges experience an extreme climate comprising of 3 distinct seasons; summer, rainy and winter. It enjoys an average annual rainfall of 1421 mm. The highest temperature of 43.9° C is recorded in May and it drops to 14.6 ^{0}c in December.

Malyagiri hill range harbours floristically important tropical deciduous forest of Eastern Ghats, India. The wide range of topographic and climatic conditions favour luxurious growth of vegetation in this hill range (Patra & Choudhury, 1989). Many of the plants have immense medicinal properties. Due to over-exploitation of medicinal plants, fuel wood collection, habitat destruction and grazing, plant diversity of Malyagiri hill range is declining at an alarming rate (Sahu et al., 2010). This may lead to extinction of many valuable species. However, barring a few floristic and ethnobotanical works (Brahmam &

Saxena, 1990; Patra & Choudhury, 1989; Saxena & Dutta, 1975; Saxena et al., 1991), no quantitative study analyzing the vegetation structure of the forest has been undertaken. Therefore, a detailed study was undertaken to analyze the diversity, distribution and population structure of tree species in these forests.

2. Materials and methods

2.1 Field sampling and data analysis

Vegetation analysis was carried out during March 2007 to December 2009 by laying 60 quadrants for each element of vegetation. For all trees ≥15 cm girth at breast height (GBH) were sampled through 20 x 20 m quadrants with sampling intensity of 0.001% based on random sampling methods in tropical dry deciduous forest stand of 2.4 ha area. Individuals with less than 15cm GBH were considered as saplings. The shrubs were sampled through 5 x 5 m and herbs, climbers and saplings were sampled by laying 1 x 1 m quadrants. Herbarium specimens were prepared and the species were identified with the help of regional flora (Gamble & Fischer, 1915-1935; Saxena & Brahmam, 1996). The specimens were deposited in the herbarium (RRL-B) at Institute of Minerals and Materials Technology, Bhubaneswar. The vegetation data were analyzed for 57 tree species. Abundance (A), Frequency (F), Relative frequency (RF), Density (D), Relative density (RD), Basal Area (BA), Relative Basal Area (RBA), Importance Value Index (IVI), Shannon -Wiener index (Shannon and Weaver, 1963) and Simpson's index (Simpson, 1949) were calculated using the quadrant data and following the methods of Misra (1968). IVI of each species was calculated by summing the RF, RD and RBA following the methods of Curtis (1959). Abundance to Frequency (A/F) ratio of each species was calculated to study the dispersion pattern. The range of values for determining dispersion pattern were: regular (< 0.025), random (0.025-0.05) and contiguous (> 0.05) (Curtis and Cottam, 1956). Population structure of tree species was analyzed across the five girth classes. The status and degree of threat to medicinal plants in their natural habitat has been indicated by classifying them according to Red Data Book categories, as defined by the IUCN (Maheswari, 1977; Melville, 1970-71).

Shannon and Weiner's Index (1963) was calculated as follows:

$$H' = - \Sigma \, p_i \log p_i$$

Where, $p_i = n_i / N$

n_i = Importance value for species "i"

N = Total of importance value

Concentration of dominance was calculated following Simpson (1949):

$$Cd = \Sigma \, p_i^2$$

Where p_i is same as the Shannon– Wiener Index.

3. Results and discussion

The dominant forest type of Malyagiri hill range is tropical dry deciduous forest (Champion and Seth, 1968). A total of 1063 trees belonging to 57 species were recorded from 60 sample plots. The dominant tree species in descending order of IVI are *Shorea robusta* Gaertn.f. (44.67), *Terminalia alata* Heyne ex Roth. (31.98), *Madhuca indica* Gmel. (17.3), *Anogeissus latifolia* (Roxb. Ex DC.) Wall.ex Guill. & Perr. (15.64), *Diospyros melanoxylon* Roxb. (13.41) (Table-1).

Name of the species	F (%)	D (ind. ha⁻¹)	BA (m². ha⁻¹)	IVI	Distribution
Aegle marmelos (L.) Corr.	16.7	8.8	0.12	5.17	C
Albizia lebbeck (L.) Benth.	3.3	0.8	0.04	0.93	C
Alstonia scholaris (L.) R.Br.	3.3	0.8	0.01	0.73	C
Anogeissus latifolia (Roxb. ex DC.) Wall.ex Guill. & Perr.	50.0	12.5	0.01	15.64	Ra
Bauhinia purpurea L.	1.7	0.8	0.01	0.45	C
Bombax ceiba L.	3.3	1.3	0.01	0.72	C
Bridelia retusa (L.) Spreng.	1.7	0.3	0.01	0.26	C
Buchanania lanzan Spreng.	38.3	24.5	0.34	8.15	C
Careya arborea Roxb.	5.0	2.5	0.03	1.47	C
Casearia graveolens Dalz.	18.3	10.8	0.11	5.74	C
Cassia fistula L.	11.7	3.3	0.04	2.63	C
Chloroxylon swietiana DC.	15.0	6.5	0.07	4.01	C
Cleistanthus collinus (Roxb.) Benth.ex.Hook.f.	26.7	15.8	0.12	8.13	C
Dalbergia paniculata Roxb.	1.7	0.3	0.02	0.46	C
Dalbergia sisoo Roxb.	1.7	0.3	0.01	0.35	C
Diospyros malabarica (Desr.) Kostel.	20.0	14.0	0.49	6.24	C
Diospyros melanoxylon Roxb.	36.7	19.0	0.56	13.41	Ra
Diospyros montana Roxb.	8.3	4.5	0.16	3.36	C
Erythrina variegata L.	1.7	0.3	0.69	5.32	C
Ficus benghalensis L.	1.7	0.3	0.01	0.26	C
Ficus semicordata Buch.-Ham.ex J.E.Sm.	3.3	0.8	0.02	0.74	C
Ficus mollis. Vahl	16.7	7.0	0.11	4.61	C
Gardenia latifolia Ait.	8.3	5.8	0.13	3.38	C
Glochidion velutinum Wight	3.3	1.3	0.04	1.02	C
Gmelina arborea Roxb.	28.3	14.5	0.21	8.67	C
Haldinia cordifolia (Roxb.) Ridsd.	1.66	0.8	0.05	0.8	C
Ixora pavetta Andr.	10	5.0	0.03	2.67	C
Lagerstroemia parviflora Roxb.	13.3	7.0	0.30	5.58	C
Lannea coromandelica (Houtt.) Merr.	1.7	0.8	0.01	0.39	C
Macaranga peltata (Roxb.) Muell.-Arg.	31.7	17.8	1.23	17.3	C
Madhuca indica Gmel.	3.3	1.3	0.01	17.3	C
Mangifera indica L.	15.0	9.5	1.00	11.54	C

Name of the species	F (%)	D (ind. ha⁻¹)	BA (m². ha⁻¹)	IVI	Distribution
Melastoma malabathricum L.	35.0	18.3	0.26	10.75	Ra
Mitragyna parvifolia (Roxb.) Korth.	13.3	6.3	0.10	3.94	C
Morinda pubescens Sm.	1.7	0.3	0.01	0.27	C
Murraya paniculata (L.) Jack	13.3	6.5	0.06	3.71	C
Nyctanthes arbor-tristis L.	10.0	3.3	0.04	2.39	C
Phyllanthus emblica L.	5.0	2.5	0.04	1.6	C
Polyalthia cerasoides (Roxb.) Bedd.	8.3	4.5	0.04	2.42	C
Protium serratum (Wall. ex Colebr.) Engl.	1.7	0.3	0.03	0.46	C
Pterocarpus marsupium Roxb.	15.0	7.0	0.09	4.25	C
Pterospermum acerifolium (L.) Willd.	13.3	5.3	0.09	3.63	C
Pterospermum xylocarpum (Gaertn.) Sant & Wagh	6.7	3.8	0.04	2	C
Randia malabarica Lam.	1.7	0.3	0.01	0.26	C
Schleichera oleosa (Lour.) Oken	10.0	10.8	0.32	6.17	C
Semecarpus anacardium L.f.	3.3	1.3	0.02	0.86	C
Shorea robusta Gaertn.f.	73.3	59.5	2.88	44.67	Ra
Strychnos potatorum L.f.	11.7	6.5	0.23	4.74	C
Symplocos racemosa Roxb.	11.7	8.3	0.05	3.84	C
Syzygium cumini (L.) Skeels	13.3	6.3	0.21	4.74	C
Terminalia alata Heyne ex Roth.	65.0	49.5	1.62	31.98	Ra
Terminalia arjuna (Roxb.ex DC.) Wight & Arn.	3.3	2.0	0.31	3.13	C
Terminalia bellirica (Gaertn.) Roxb.	5.0	1.5	0.07	1.5	C
Terminalia chebula Retz.	3.3	1.3	0.02	0.87	C
Wendlandia tinctoria (Roxb.) DC.	23.3	12.0	0.16	7.05	C
Xylia xylocarpa (Roxb.) Taub.	10.0	5.8	0.11	3.37	C
Ziziphus xylocarpus (Retz.) Willd.	5.0	32.5	0.01	1	C

Note: F- Frequency; D- Density; BA-Basal Area; IVI-Importance Value Index; Ra: Random; C: Contiguous

Table 1. Phytosociological characteristics of Tree species in Malyagiri hill range, Eastern Ghats

The luxuriant growth of herbs, shrubs and climbers increased the density of the forest vegetation. Among the herbs, *Chromolaena odorata* (L.) R.King & H.Robins., *Andrographis paniculata* (Burm.f.) Wall.ex Nees, *Elephantopus scaber* L., *Curculigo orchioides* Gaertn. were most common species. Important shrub species were *Combretum roxburghii* Spreng., *Holarrhena pubescence* (Buch.-Ham.) Wall.ex G.Don, *Woodfordia fruticosa* (L.) Kurtz, *Lantana camara* L., *Helicteres isora* L. and *Ixora pavetta* Andr. The dominant climbers were *Dioscorea bulbifera* L., *Smilax macrophylla* Roxb., *Ampelocissus latifolia* (Roxb.) Planch. and *Bauhinia vahlii* Wight & Arn.

However, Malyagiri hill range is severely affected by anthropogenic activities. Unsustainable collection of medicinal plants (*Oroxylum indicum* (L.) Vent. and *Cycas cicinalis* L.) for selling purposes by the local people of Pallahara Sub-division (Sahu et al., 2010) is prominent. Most of the local people depend upon the forests for their livelihood for which collection of leaf from *Bauhinia vahlii* Wight & Arn. (Leaf tray), *Phoenix sylvestris* (L.) Roxb. (for brooms, mats etc.), *Diospyros melanoxylon* Roxb. (Bidi) and firewood collection are very common. These are the indications to the anthropogenic activities going on in and around Malyagiri hill ranges.

Out of 57 tree species, five species were randomly distributed and 52 species were contiguously distributed. The study reveals prevalent clumping nature of tree species in the tropical forest of Malyagiri hill ranges. Odum (1971) stated that contiguous distribution is the commonest pattern of plant distribution in nature. Kumar and Bhatt (2006) also reported that most species follow contiguous distribution pattern in foot-hills forests of Garhwal Himalaya and Rao et al. (1990) had similar findings for tree species of a subtropical forest of north-east India. The Shannon-Wiener index (H') was 3.38 and Simpson's index was 1.0. These values indicate that tropical deciduous forests are species diverse systems. The diversity value (H') of 3.38 falls within the range of 0.83-4.1 reported by earlier workers for Sal forest (Rasingam & Parathasarathy, 2009; Shukla, 2009; Singh et al., 1985; Tripathi & Singh, 2009; Visalakshi, 1995).

The mean tree density of the forest was 443 ha^{-1}. The mean stand density of the forest is well within the range of 276-905 stems ha^{-1} reported for trees ≥15 cm GBH in other tropical forests (Bhadra et al., 2010; Nirmal Kumar et al., 2010; Sahu et al., 2007). The value obtained for basal area in the present study is comparable to the Indian tropical forests (Visalakshi, 1995).

Stem density and species richness consistently decreased with increasing girth class of tree species beyond 30-50 cm GBH class (Fig. 1). The highest GBH was in *Ficus benghalensis* (458 cm) followed by *Mangifera indica* (378 cm), *Shorea robusta* (230 cm), *Madhuca indica* (215 cm) and *Schleichera oleosa* (210 cm). Girth class frequency showed reverse J-shaped population structure of trees, which is in conformity with other forest stands in Eastern Ghats such as Shervarayan hills (Kaduvul and Parthasarathy, 1999a) and Kalrayan hills (Kaduvul and Parthasarathy, 1999b).

The mean tree height was 10 m with a height range of 1 to 35 m. Tree distribution by height class intervals shows that 39.1% of individuals were in the height class of 5-10 m, followed by 24.3% in the height class of 10-15 m and 20.4% in the height class of 0-5m (Fig. 2). Only 5.73% of individuals were in the height class of >20 m. The tallest trees were *Shorea robusta* (35 m), *Mangifera indica* (33 m), *Terminalia bellirica* (32 m), *Syzygium cumini* (32 m), *Diospyros malabarica* (27).

The data on species/genus (S/G) ratio helps to compare the rate of species development because high ratio indicates recent diversification. Tropical areas have low species/genus ratio, indicating that the tropical species have emerged over a long period of time (Ricklefs and Miller, 2000). In the present study, all the study sites show lower S/G ratio in the tree layer (1.18), thus showing conformity with the findings of Ricklefs and Miller (2000).

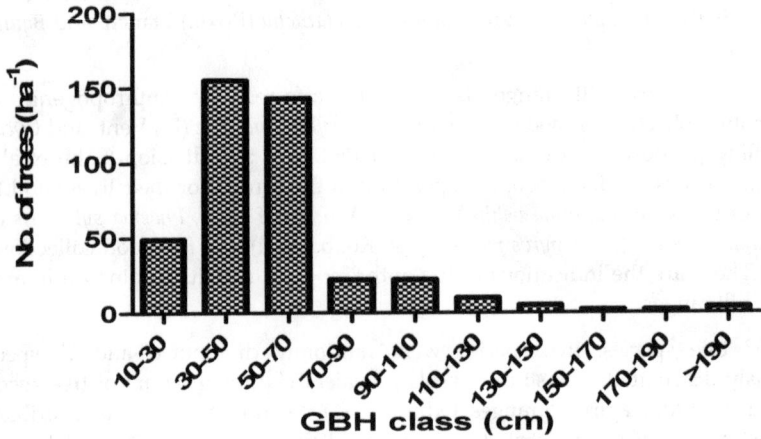

Fig. 1. Distribution of trees in different Girth classes in Malyagiri hill ranges, Eastern Ghats

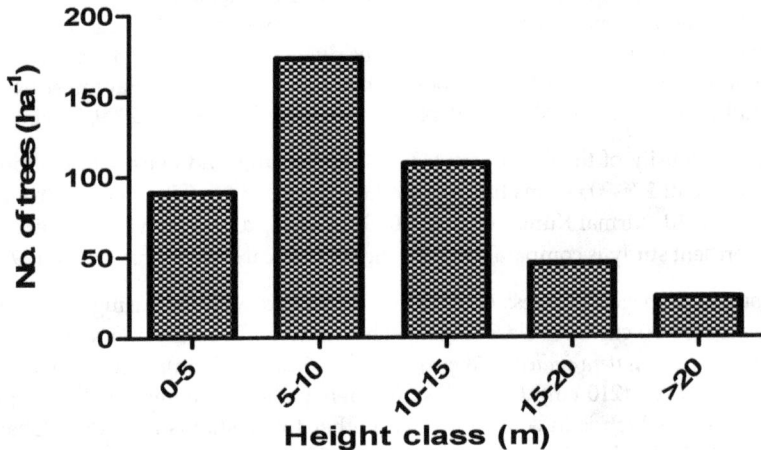

Fig. 2. Distribution of trees in different Height classes in Malyagiri hill ranges, Eastern Ghats

The invasive, exotic species were also found, which can be a serious threat to the forest ecosystem in the future. Important among them are *Ageratum conyzoides* L., *Chromolaena odorata* (L.) R. King & H. Robins., *Crotalaria pallida* Ait., *Hyptis suaveolens* (L.) Poit., *Lantana camara* L., *Mimosa pudica* L., *Parthenium hysterophorus* L. and *Triumfetta rhomboidea* Jacq.

Girth class frequency showed J-shaped population structure of trees exhibited in the study sites are in conformity with many other forest stands in Eastern and Western Ghats such as Shervarayan hills (Kadavul and Parathasarthy, 1999a); Kalrayan hills (Kadavul and Parthasarathy, 1999b); Kakachi (Ganesh et al., 1996); Andaman Islands (Rasingam & Parathasarathy, 2009).

4. Documentation of threatened medicinal plants

Rapid destruction of forests causes severe damage to natural forest of the hill range, thus threatening the very survival of several indigenous medicinal plants. Recent news paper has highlighted the rate of medicinal plants (*Oroxylum indicum* (L.) Vent. and *Cycas cicinalis* L.) collection for selling purposes by the local people of Pallahara Sub-division in Malyagiri hill range (Sahu et al., 2010). Unsustainable collection of medicinal plants has placed them in threatened and vulnerable categories in Conservation Assessment and Management Plan (Ved et al., 2007) of Odisha. Table 2 highlights the 2 critically endangered, 6 endangered and 10 vulnerable species along with their botanical name, voucher specimen number, family, locality, local name, life form, distribution and IUCN status. So it is critical to conserve these medicinal plants locally if not globally. This may be planned through in-situ or ex-situ conservation methods, for preserving the biodiversity of the state of Odisha. In-situ conservation method should be implemented to conserve the medicinal plant resources in their natural habitat (National Park/Wildlife Sanctuary/Biosphere Reserve). Therefore, it is suggested that Malyagiri hills should be declared as a Wildlife Sanstuary in earlier possible. Threatened medicinal plants which will be extinct in near future should be conserved through ex-situ conservation (Botanical gardens, Arboreta or Seed banks).

Sl. No.	Botanical name, Voucher specimen No.	Family, Locality	Local Name	Life form	Distribution	IUCN Status
1	*Caesalpinia digyna* Rottl., 8502	Caesalpiniaceae, Allora	Gilo	Shrub	Peninsular India	Vulnerable
2	*Celastrus paniculata* Willd., 9996	Celastraceae, Kerjeng	Pengu	Climbing Shrub	Myanmar, Thailand	Vulnerable
3	*Garcinia xanthochymus* Hook.f., 8572	Clusiaceae, Panichua	Satyamba	Tree	Eastern Himalayas, Odisha, Myanmar, Thailand	Vulnerable
4	*Gardenia gummifera* L.f., 9759	Rubiaceae, Kerjeng	Gurudu	Shrub	Peninsular India	Vulnerable
5	*Gloriosa superba* L., 11624	Liliaceae, Pallahara	Nanangalia	Climbing herb	Tropical India, S. Africa, Malesia	Vulnerable
6	*Litsea glutinosa* (Lour.) Robins., 8636	Lauraceae, Khuludi	Ledhachhali	Tree	India, Sri Lanka, Malesia	Endangered
7	*Mesua ferrea* L., 12027	Clusiaceae, Khuludi	Nageswar	Tree	Eastern Himalayas, Andaman & Nicobar, Tropical Asia	Endangered
8	*Oroxylum indicum* (L.) Vent., 11811	Bignoniaceae, Khamar	Phanphania	Tree	India, Sri Lanka, Indonesia	Endangered

Sl. No.	Botanical name, Voucher specimen No.	Family, Locality	Local Name	Life form	Distribution	IUCN Status
9	*Paederia foetida* L., 8641	Rubiaceae, Pallahara	Gandhali	Climbing Shrub	North-East India, Andaman & Nicobar, Thailand	Vulnerable
10	*Piper longum* L., 8438	Piperaceae, Pallahara	Pipali	Herb	India, Sri Lanka, Malay	Endangered
11	*Polyalthia cerasoides* (Roxb.) Bedd, 8224	Annonaceae, Allora	Ojhar	Tree	Assam, Odisha,Silhet, Pegu	Vulnerable
12	*Pterocarpus marsupium* Roxb., 10129	Fabaceae, Kerjeng	Bija	Tree	Andhra Pradesh, Bihar, Kerala, Karnataka, Sri Lanka	Endangered
13	*Pueraria tuberosa* (Willd.) DC., 9659	Fabaceae, Khuludi	Bhuin kakharu	Climber	India	Vulnerable
14	*Rauvolfia serpentina* (L.) Benth. Ex Kurtz, 8245	Apocynaceae, Pallahara	Patalagarud	Undershrub	Tropical Himalaya, Deccan Peninsula,Malaya, Sri Lanka	Critically Endangered
15	*Scindapsus officinalis* (Roxb.) Schott, 10139	Araceae, Allora	Kelikadali	Climer	Tropical Himalaya,Sikkim, Andaman,Myanmar	Vulnerable
16	*Strychnos potatorum* L.f., 8820	Strychnaceae, Kerjeng	Nirmali	Tree	West Bengal, Bihar, Sri Lanka, Myanmar	Vulnerable
17	*Symplocos racemosa* Roxb., 12025	Symplocaceae, Khuludi	Lodha	Tree	NE India, Tamil Nadu, Karnataka, Thailand	Endangered
18	*Uraria picta* (Jacq.) Desv.ex DC., 10925	Fabaceae, Khuludi	Ishwarjata	Undershrub	Himalaya, Sri Lanka, SE Asia	Critically Endangered

Table 2. List of Threatened Medicinal plants in Malyagiri hills, Odisha

5. Conclusion

Reverse J-shaped population structure of trees denotes an evolving or expanding population, which needs to be maintained. The unsustainable collection of medicinal plants such as the bark of *Oroxylum indicum* (L.) Vent. and whole plant of *Gloriosa superba* L. and *Uraria picta* (Jacq.) Desv.ex DC., need to be checked to maintain the favourable population structure. Study on floristic composition and diversity will be useful to the conservation researchers and scientists and also to the forest managers for effective management of the forest ecosystem. The present investigation highlights the presence of threatened medicinal plants which need immediate attention for conservation and propagation through in-situ, ex-situ or latest biotechnological approaches (Gene banks, DNA and Pollen storage etc.)

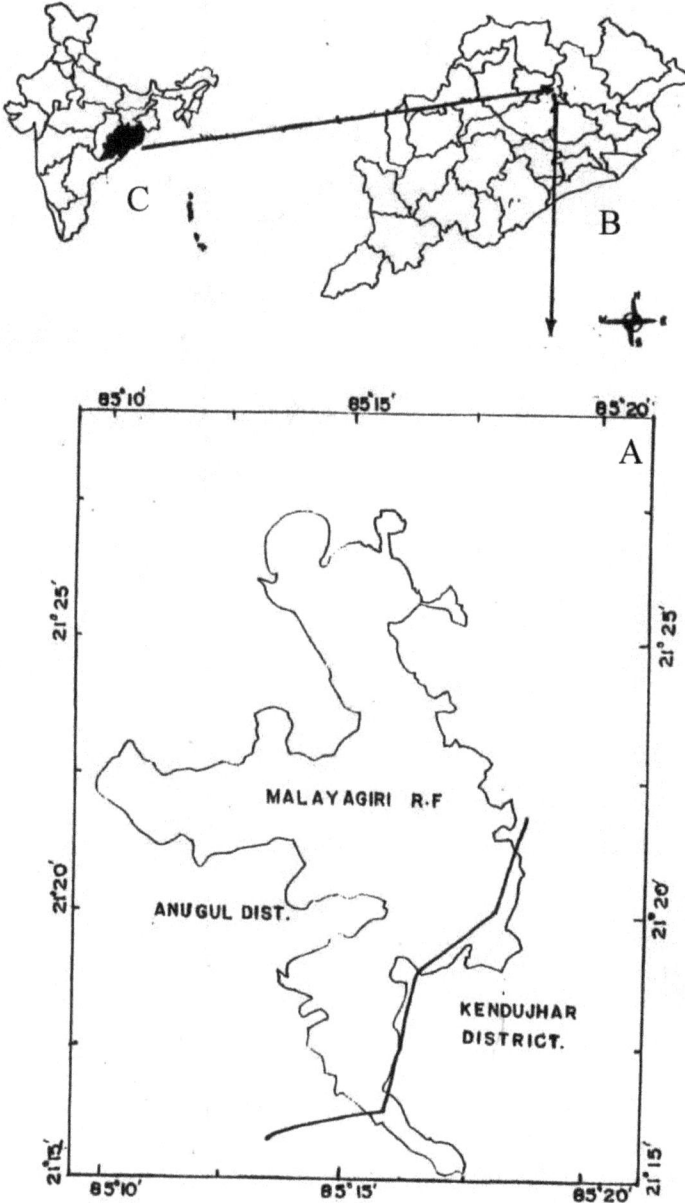

Fig. 3. Study area

Some Threatened Medicinal Plants of Malyagiri hill ranges, Odisha, India

Oroxylum indicum (L.) Vent.

Polyalthia cerasoides (Roxb.) Bedd

Cymbidium aloifolium (L.) Sw.

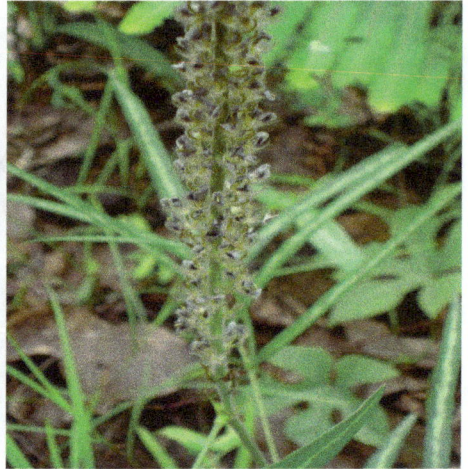

Uraria picta (Jacq.) Desv.ex DC.

6. Acknowledgement

The authors are sincerely thankful to the Director, IMMT, CSIR, Bhubaneswar, for providing necessary laboratory facilities. The help received from the local people involved during the survey is duly acknowledged. The first author is thankful to the Council of Scientific and Industrial Research, New Delhi, for the award of Senior Research Fellowship.

7. References

Bhadra, A.K., Dhal, N.K., Rout, N.C. & Reddy, V.R. (2010). Phytosociology of the tree community of Gandhamardan hill ranges. *The Indian Forester,* 136, 610-620.

Brahmam, M. & Saxena, H.O. (1990). Ethnobotany of Gandhamardan hills. Some noteworthy folk medicinal uses. *Ethnobotany,* 2, 71-79.

Champion, H.G. & Seth, S.K.. (1968). *Revised Survey of Forest Types of India,* New Delhi. Govt. of India.

Curtis, J.T. (1959). *The Vegetation of Wisconsin, An Ordination of Plant Communities.* University Wisconsin Press, Madison, Wisconsin.

Curtis, J.T. & Cotton, G. (1956). *Plant Ecology Workbook, Laboratory Field Manual.* Burgess publishing, Minnesota. pp.193.

Gamble, J.S. & Fischer, C.E.C. (1915-1935). *Flora of Presidency of Madras.* Vols 1-3. Adlard and Son Ltd, London.

Ganesh, T., Ganesan, R., Soubadradevy, M., Davidar, P. & Bawa, K. S. (1996). Assessment of plant biodiversity at a mid-elevation evergreen forest of Kalakad-Mudanthurai Tiger reserve, Western Ghats, India. *Current Science,* 71, 379-92.Hare, M.A., Lantagne, D.O., Murphy, P.G. & Chero, H. (1997). Structure and tree species composition in a subtropical dry forest in the Dominican Republic: Comparision with a dry forest in Puerto Rico. *Tropical Ecology,* 38, 1-17.

Kadavul, K. & Parthasarathy, N. (1999a). Plant biodiversity and conservation of tropical semi-evergreen forest in the Shervarayan hills of Eastern Ghats, India. *Biodiversity Conservation,* 8, 421-439.

Kadavul, K. & Parthasarathy, N. (1999b). Structure and composition of woody species in tropical semi-evergreen forest of Kalayan hills, Eastern Ghats, India. *Tropical Ecology* 40, 247-260.

Kumar, Munesh & V.P. Bhatt. 2006. Plant biodiversity and conservation of forests in foot hills of Garhwal Himalaya. *Journal of Ecology and Application,* 11, 43-59.

Maheswari, J.K. (1977). Conservation of Rare plants- Indian scene vis world scene. *Bulletin of Botanical Survey of India,* 19, 167-173.

Melville, R. (1970-71). *Red Data Book. Vol. 5- Angiospermae.* International Union for Conservation of Nature and Natural resources, Marges.

Mishra, R. (1968). *Ecology Work Book.* Oxford and IBH Publications, Co. New Delhi.

Nirmal Kumar, J.I., Kumar, Rita N., Bhoi, Rohit Kumar., & Sajish, P.R. (2010). Tree species diversity and soil nutrient status in three sites of tropical dry deciduous forest of western India. *Tropical Ecology,* 51, 273-279.

Odum, E.P. (1971). *Fundamentals of Ecology.* 3rd edn. W.B. Saunders Co., Philadelphia. USA.

Patra, B.C. & B.P. Choudhury. (1989). Forest cover of Malyagiri hills in the state of Orissa. *Journal of Economic and Taxonomic Botany,* 13, 315-319.

Rao, P., Barik, S.K., Pandey, H.N., & Tripathi, R.S. (1990). Community composition and tree population structure in a subtropical broad-leaved forest along a disturbance gradient. *Vegetatio,* 88, 151-162.

Rasingam, L. & Parathasarathy, N. (2009). Tree species diversity and population structure across major forest formations and disturbance categories in Little Andaman Island, India. *Tropical Ecology,* 50, 89-102.

Ricklefs, R.E. & Miller, G.L. (2000). Ecology. – W.H. Freeman & Company, New York.

Sahu, S.C., Dhal, N.K., Reddy, C.S., Pattanaik, C., & Brahmam, M. (2007). Phytosociological study of tropical dry deciduous forest of Boudh district, Orissa, India. *Research Journal of Forestry*, 1, 66-72.

Sahu, S.C., Dhal, N.K. & Mohanty, R.C. (2010). Commercialization of a few wild medicinal plants from Deogarh district, Orissa, India. e-planet, 8, 14-16.

Saxena, H.O. & Dutta, P.K. (1975). Studies on ethnobotany of Orissa. *Bulletin of Botanical Survey of India*, 17, 124-131.

Saxena, H.O., Brahmam, M., & Dutta, P.K. (1991). *Ethnobotanical studies in of Orissa*. Pp.123-135. In S.K. Jain (ed.) Contribution to Ethnobotany of India. Scientific Publishers, Jodhpur, India.

Saxena, H.O. & Brahmam, M. (1996). *The Flora of Orissa. Vols I-IV*. Orissa Forest Development Corporation Ltd, Bhuabneswar, India.

Shannon, C.E. & Weaver, W. (1963). *The Mathematical Theory of Communication*. University of Illinois Press, Urbana.

Shukla, R.P. (2009). Patterns of plant species diversity across Terai landscape in north-eastern Uttar Pradesh, India. *Tropical Ecology*, 50, 111-123.

Simpson, E.H. (1949). Measurement of diversity. *Nature*, 163, 688.

Singh, J.S., Singh, S.P., Saxena, A.K., & Rawat, Y.S. (1985). The forest vegetation of Silent Valley, in India. P. 25-52. In: A.C. Chadwick and S.L. Sutton (eds.). *Tropical Rain Forest: The Leeds Symposium*. Leeds Philosophical and Literary Society, Leeds, U.K.

Tripathi, K.P., & Singh, B. (2009). Species diversity and vegetation structure across various strata in natural and plantation forests in Katerniaghat Wildlife Sanctuary, North India. *Tropical Ecology*, 50, 191-200.

Ved D. K., Kinhal G. A., Ravikumar K., Vijayasankar R., Sumathi R., Mahapatra A.K. and Panda P.C. (2007). *CAMP Report*: Conservation assessment and management prioritization for medicinal plants of Orissa, India. Foundation for Revitalisation of Local Health Traditions (FRLHT), Bangalore, India.

Visalakshi, N. (1995). Vegetation analysis of two tropical dry evergreen forests in southern India. *Tropical Ecology*, 36, 117-127.

World Conservation Monitoring Centre. (1992). *Global biodiversity: Status of the Earths Living Resources*. Chapman and Hall. London.

11

Human-Altered Mesoherbivore Densities and Cascading Effects on Plant and Animal Communities in Fragmented Tropical Forests

Nicole L. Michel and Thomas W. Sherry
Department of Ecology and Evolutionary Biology, Tulane University,
USA

1. Introduction

Rainforest loss and fragmentation are proceeding at an alarming rate, and having demonstrable consequences for the relevant plant and animal communities. Between 1984 and 1990 6.5 million hectares of humid tropical rainforest were lost annually worldwide (Hansen & DeFries, 2004). This rate increased to 7.3 million hectares annually between 1990-1997, due largely to accelerating deforestation rates in Asia. Though the pantropical deforestation rate slowed to 5.5 million hectares annually between 2000-2005 due to slowing in much of Asia and Latin America, deforestation continues to increase in Brazil (Hansen et al., 2008). As of 2005, half to two-thirds of humid tropical regions have <50% tree cover, with an additional 20% undergoing selective logging between 2000 and 2005 (Asner et al., 2009; Table 1).

Much of the remaining original forest persists as small, isolated fragments. The highest rates of forest fragmentation are found in North and South America, respectively (Riitters et al., 2000). In the Amazon, the amount of forest in fragments or within 1km of forest edge exceeded the extant forest area by 150% (Skole & Tucker, 1993), and between 1999 and 2002 over 30,000 km of new forest edge was created there annually (Broadbent et al., 2008). Asia and Africa currently have the lowest levels of tropical rainforest fragmentation, although with the rapidly accelerating Southeast Asian deforestation rates this is likely to change (Sodhi et al., 2010a).

Such loss and fragmentation of tropical forest, coupled with other global change phenomena, have profound and diverse effects on the remaining forest fragments. The very complexity of these synergistic effects is only beginning to be documented and comprehended. Recent studies show that tropical forest fragmentation disproportionately affects the largest vertebrates (Henle et al., 2004; Stork et al., 2009; Wilkie et al., 2011). Large predators experience the strongest fragmentation effects, with cascading consequences for the small- and intermediate-sized herbivores regulated by vertebrate predators (Duffy, 2003; Purvis et al., 2000; Redford, 1992). Altered populations of these herbivores, which we term "mesoherbivores" (after Soulé et al.'s (1988) term "mesopredator" referring to small- and intermediate-sized predators), in turn cascade down to affect plant growth and recruitment via altered seed predation and dispersal (Duffy, 2003; Redford, 1992).

	Annual deforestation rate (x10^6 hectares)			Selective logging[3] (%)
	1984-1990[1]	1990-1997[1]	2000-2005[2]	2000s
Latin America	4.31	4.28	3.31	17.2%
Tropical Asia	1.81	2.57	1.88	19.2%
Tropical Africa	0.43	0.37	0.30	27.2%
Pantropical	6.50	7.30	5.50	20.3%

[1] Data from Hansen & DeFries, 2004
[2] Data from Hansen et al., 2008
[3] Percent of regional forest area experiencing selective logging. Data from Asner et al., 2009

Table 1. Annual deforestation rate and selective logging extent in humid tropical forests.

Herein, we review both the direct and indirect effects of fragmentation on tropical forest communities. We review the literature on vertebrate herbivores both because few reviews are available for tropical forests and the pervasiveness and diversity of their ecological effects remain largely undocumented. We begin by reviewing the substantial tropical literature on human overhunting of mesoherbivores and the consequences this has for tropical forest plant communities, in an effort to understand the magnitude of the effects of human-perturbed mesoherbivore populations on tropical forest plant communities.

Ultimately, however, we are particularly interested in the poorly-studied phenomenon wherein loss of vertebrate predators without compensatory human hunting may result in dramatic increases in mesoherbivore populations, which we call "mesoherbivore release." This phenomenon has been observed in multiple temperate systems, as well as tropical savanna and grassland systems (see review in Salo et al., 2010), yet it is poorly studied in tropical forests. This section of the review will synthesize for the first time the sparse and scattered literature involving putative tropical examples of mesoherbivore release, making the case that the cascading consequences of increased mesoherbivore abundance can be as widespread and ecologically destructive as those resulting from mesoherbivore decline. Finally, we address the most pressing conservation and management implications of the research on perturbed mesoherbivore populations, both decline and release. We also identify topics in need of further investigation, in terms of both ecology and conservation.

2. Direct effects of fragmentation

Predictably, rainforest loss and fragmentation are dramatically altering tropical forests, both directly and indirectly. Loss of forest area and, to a lesser degree fragmentation, directly drive population decline and extinction of myriad plant, invertebrate, and vertebrate species pan-tropically (Beier et al., 2002; Fahrig, 1997, 2002, 2003; Laurance, 1999; Laurance et al., 2011; Michalski & Peres, 2007; Newmark, 1991; Sodhi et al., 2008, 2010b; Yaacobi et al., 2007). Pervasive edge effects further alter the microclimate and vegetation structure of remnant fragments, affecting still more species (Fletcher et al., 2007; Laurance, 2000; Laurance et al., 2002, 2011; Lindell et al., 2007; Manu et al., 2007; Norris et al., 2008; Pardini, 2004; Tabarelli et al., 2008). Synergistic interactions between fragmentation and other global change drivers, notably climate change with its concomitant increased drought and fire frequency, and exploitation of wildlife, further exacerbate loss and degradation of remnant tropical rainforest and their plant and animal communities (Brook et al., 2008; Ewers & Didham, 2006; Laurance & Useche, 2009; Laurance & Williamson, 2001; Tabarelli et al., 2008). Roads and other infrastructure also contribute to accelerating declines of mammals, birds, and other organisms in both fragments and intact forests (Benítez-López et al., 2010), e.g., by increasing hunting pressure (Peres, 2001).

Even if a species persists temporarily in fragmented forest, it is likely to experience reduced genetic diversity, inbreeding depression, reduced adaptive potential, and accumulation of deleterious mutations (Keyghobadi, 2007). Inbreeding depression and reduced genetic diversity combine with dispersal limitation, competitive disadvantages, fragment area and isolation effects, and other species- and fragment-level traits to create an "extinction debt," wherein extinctions proceed for decades post-fragmentation (Brooks et al., 1999; Ewers & Didham, 2006; Ferraz et al., 2003; Laurance et al., 2008; Metzger et al., 2009; Tilman et al., 1994), if not at the rate once thought (He & Hubbell, 2011). These high short- and long-term extinction rates, combined with the rapid and accelerating loss and fragmentation of remnant tropical forest, have led many scientists to warn of a tropical biodiversity crisis resulting in a potential mass wave of extinctions rivaling historical extinction events, resulting in an era of novel secondary tropical forests (Bradshaw et al., 2009; Brook et al., 2008; Dirzo & Raven, 2003; Gardner et al., 2007; Laurance, 1999, 2006; Lugo, 2009; but see Wright & Muller-Landau, 2006a,b).

Species declines following rainforest loss and fragmentation are highly non-random. Fragmentation-sensitive species typically share a common suite of traits, including small population size/rarity, large body size, and specializations (see Table 2). Many of these traits (e.g., "slow" life-histories, dietary and habitat specialization) are unique to, or relatively prevalent in tropical species, thus rendering many paradigms (e.g., edge effects) and conservation strategies developed for temperate and boreal forests largely irrelevant when developing conservation plans for tropical forests (Stratford & Robinson, 2005).

Common traits of fragmentation-sensitive species	
Small population size / rarity	High sociality
Large body size	Disturbance- and competition- sensitivity
High trophic position	Specialized diet (e.g., insectivorous and large frugivorous birds)
Fluctuating population dynamics	Specialized habitat use (especially forest interior)
"Slow" life-histories (low fecundity, high annual survival)	Location near edge of altitudinal or biogeographical range
Low dispersal capability	Involved in mutualistic interaction

Table 2. Common traits of fragmentation-sensitive species (from Henle et al., 2004; Kattan et al., 1994; Laurance, 1991; Laurance et al., 2011; Lees & Peres, 2008; Sigel et al., 2006, 2010; Stork et al., 2009)

Tropical forests also tend to harbor diverse, pervasive mutualistic interactions, notably plant-animal interactions (Terborgh & Feeley, 2010). Between 70 and 90% of tropical trees rely on animals for seed dispersal (Muller-Landau & Hardesty, 2005), and >95% of lowland tropical trees are animal-pollinated (Bawa et al., 1985). Thus, habitat loss and fragmentation also alter many important plant-animal interactions, including pollination, seed dispersal, seed predation, and herbivory (Herrera et al., 2011; Laurance, 2005). Some of these altered interactions result from direct effects of habitat loss and fragmentation on plants and animals involved in mutualisms, e.g., extirpations of forest-interior butterflies following loss of their host plants (Brown & Hutchings, 1997; Koh et al., 2004). However, in many cases plant-animal interactions change in response to indirect effects of other animals declining higher in the food chain, in what is known as trophic cascades.

3. Indirect effects of fragmentation: Trophic cascades

Fragmentation effects are particularly strong on large predators, due to their high trophic position, "slow" life-histories that preclude rapid population response to disturbances, large home ranges, and other traits characteristic of fragmentation-sensitive species. Apex predators such as jaguars, pumas, tigers, leopards, large raptors, and large snakes are amongst the first species to disappear from tropical rainforest fragments (Duffy, 2003; Pimm et al., 1988; Purvis et al., 2000; Ray et al., 2005; Redford, 1992; Terborgh, 1992). Moreover, the cascading consequences resulting from apex predator loss can dramatically affect ecosystem structure and function in ways we are just beginning to understand, and thus far exceed the effects of loss of species from lower trophic levels (Dobson et al., 2006; Duffy, 2003; Estes et al., 2011; Schmitz et al., 2010; Terborgh et al., 1999).

Trophic cascades, wherein perturbations of apex predator populations cascade down to affect lower consumer (e.g., herbivore) and producer (i.e., plant) levels (Paine, 1980), were once thought not to occur in tropical forests as a result of their high species diversity and high intraguild predation, leading to functional redundancy and weak links between individual predator and prey species (Polis & Holt, 1992; Polis & Strong, 1996; Shurin et al., 2002; Strong, 1992; Van Bael et al., 2003; Vance-Chalcraft et al., 2007). Terborgh (1992) first presented evidence suggesting a carnivore (jaguar, *Panthera onca*; puma, *Puma concolor*) – large seed predator (peccary, *Pecari* spp.; paca, *Cuniculus* spp.; agouti, *Dasyprocta* spp.) – large-seeded tree (*Dipteryx, Protium*) trophic cascade at Barro Colorado Island, Panama. Three years later, Dial and Roughgarden (1995) published one of the first experimental studies of trophic cascades in tropical rainforest, showing that insectivorous *Anolis* lizards limit herbivorous insects and in turn, herbivory on canopy trees. Shortly thereafter, Letourneau and Dyer (1998) were the first to experimentally show a four-level (beetle – ant – herbivore – plant) trophic cascade in tropical rainforest understory. Within the next few years, a flurry of exclosure experiments demonstrated that insectivorous birds and bats limit arthropod herbivores and herbivory in rain forest canopy and understory (review by Van Bael et al., 2008; Kalka et al., 2008; Michel et al., in review; Williams-Guillén et al., 2008).

Additionally, a number of experimental and observational studies have shown effects of mesopredator release (reviews by Brashares et al., 2010; Prugh et al., 2009). Mesopredator release is a phenomenon wherein small to mid-sized predators increase in abundance following declines of their own predators, resulting in declines in their prey, which are typically small herbivores and insectivores, especially birds and small rodents (Fig. 1; Soulé et al. 1988, Crooks and Soulé 1999). Though first documented in fragmented temperate habitats, Laurance (1994) noted increases in rodents - including one species, *Uromys*, known

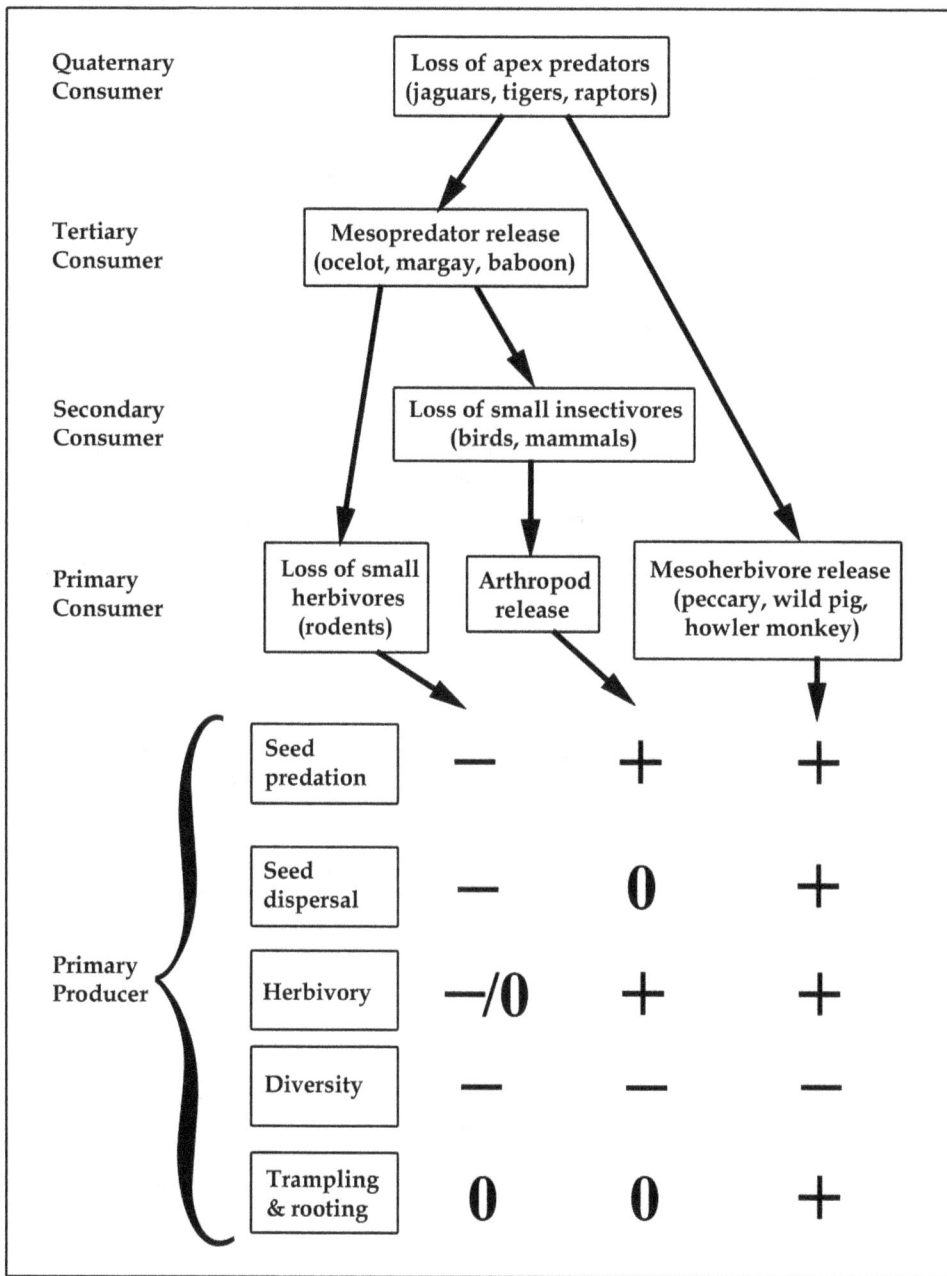

Fig. 1. Diagrammatic representation of the potential cascading consequences of apex predator loss via mesopredator and mesoherbivore release.

to depredate bird nests and consume small vertebrates - in Australian rainforest fragments missing large predators. Increases in other small, generalist predators (e.g., barn owls, red-bellied blacksnakes, and white-tailed rats) were also observed in fragments without large, specialist predators (Laurance, 1997). Elsewhere in Australia, declines and extirpations of dingoes, an apex predator, resulted in continent-wide collapse of native marsupials due to release of invasive house cats and red foxes from predation (Johnson et al., 2007). On Barro Colorado Island, Panama, high avian nest predation has been attributed in part to increased small mammalian nest predators such as white-nosed coatimundi (*Nasua narica*), white-faced capuchin monkeys (*Cebus capucinus*), and opossums following isolation from the mainland, and subsequent loss of large carnivores, when the Panama Canal was built (Loiselle & Hoppes, 1983; Sieving, 1992; Wright et al., 1994). On the Lago Guri islands in Venezuela, former hilltops isolated from nearby mainland by a hydro-electric reservoir, anomalously low bird densities were associated with abundant olive capuchin monkeys (*Cebus olivaceus*), which depredated all artificial nests (Terborgh et al., 1997). Olive baboons (*Papio anubis*) also dramatically increased in Ghana following steep declines in lions (*Pantera leo*) and leopards (*Pantera pardus*), negatively impacting smaller ungulates, primates, rodents, and birds, as well as human livestock, pets, and crops (Brashares et al., 2010).

In summary, strong trophic cascades, resulting in increases in arthropod herbivores and mesopredators, have been repeatedly demonstrated to occur in tropical rainforest despite the functional redundancy and potentially weak links between predators and prey (Terborgh & Feeley, 2010). However, very few experimental studies have focused on the potential cascading consequences of loss of large terrestrial predators on vertebrate herbivores (i.e., of mesoherbivore release), and subsequent plant community effects, despite the fact that large predators have been shown to effectively limit vertebrate herbivores in a variety of habitats worldwide (Estes et al., 2011; Salo et al., 2010). This is likely due, in large part, to logistical, methodological, and financial constraints, including the large scale of impacts and the long generation times of vertebrate herbivores and the trees they frequently impact; responses may not be apparent for years or even decades following perturbations to predator and herbivore populations (Terborgh & Estes, 2010). Furthermore, in many cases, apex predator populations were decimated decades or centuries prior to the development of the theory of trophic cascades (Estes et al., 2011). Moreover, across much of the tropics, the lost ecological services attributable to prey regulation (Dobson et al., 2006, Schmitz et al., 2010) once provided by predators such as jaguars, leopards, and tigers have been compensated - and often overcompensated - for by a new apex predator: human hunters.

4. Humans, herbivores, and tropical tree germination and recruitment

By far the majority of studies documenting cascading effects of altered vertebrate herbivore densities on rainforest plant communities have focused on consequences of extirpated or greatly reduced herbivore populations. Herbivores, especially larger herbivores, often experience declines due to simple area effects of habitat loss and fragmentation (MacArthur & Wilson, 1967; e.g., Dalecky et al., 2002; Laurance et al., 2011), as well as edge effects (Norris et al., 2008; Restrepo et al., 1999), altered vegetation structure, local climate (e.g., droughts), and/or microclimate within fragments (Chiarello, 1999; Fleury & Galetti, 2006; Laurance, 1994; Laurance et al., 2008; Laurance & Williamson, 2001). But tropical forests are rarely impacted by habitat loss and fragmentation alone. Many anthropogenic factors, including logging, fires, introduction of invasive species or disease, and hunting, interact synergistically to exacerbate fragmentation effects on tropical forest residents (Brook et al., 2008; Laurance, 2005, 2008; Laurance & Useche, 2009; Wright 2005, 2010).

Synergistic effects of fragmentation and hunting are particularly strong, as fragmentation - also commonly associated with increased road-building - both a) allows hunters easier access to prey species and to markets, and b) subdivides remnant forest into patches below the minimum area requirement to support sustainable hunting (Peres, 2001). As a result, hunting is having dramatic effects on vertebrates, including herbivores and the plants dependent upon them for pollination and dispersal, resulting in what has come to be known as the "bush meat crisis" (Milner-Gulland et al., 2003; Wright et al., 2007a). Humans have been implicated in extinctions of large vertebrates as far back as 46,000 years ago, when extinctions of Australia's megafauna coincided with human arrival (Roberts et al., 2001). Similarly, humans are partially, if not predominantly, responsible for the North American megafauna extinction, including loss of herbivorous mammoths and mastodons, over 12,000 years ago (Alroy, 2001; Johnson, 2002). The impact of human hunters on large herbivores have only increased as the human population grew exponentially over the last 2,000-3,000 years, and also as it developed modern hunting techniques and weapons, including firearms and traps (Corlett, 2007). Their relative abundance (compared to cats and other apex predators) often makes vertebrate herbivores a prime target for human hunting.

Today human hunting is decimating herbivore populations pan-tropically in intact as well as fragmented forest. In the Congo, 60% of mammals are unsustainably harvested, including 93% of herbivorous ungulates, suggesting catastrophic losses of herbivores and their dependent plants in the near future (Fa et al., 2002). While Fa et al. (2002) estimated that only 20% of Amazonian mammals are unsustainably hunted, Peres and Palacios (2007), using slightly different methods, found that 22 of 30 large mammals, birds, and reptiles significantly declined under high hunting pressure, at rates of up to 74.8%. Large-bodied vertebrates and seed dispersers experienced greater declines than seed predators and browsers. As a result, hunted Amazonian forests will likely experience altered plant communities as large-seeded trees and lianas are no longer able to escape and recruit away from the parent tree (Peres & Palacios, 2007). Indeed, tree species richness is already 55% lower, and density of large-seeded, primate-dispersed trees is 60% lower, in hunted sites near Manu National Park, Peru, where large primates were exterminated and midsize primate populations reduced 80% by hunters (Nuñez-Iturri & Howe, 2007). Another study from the same hunted site found 50-60% lower sapling densities, and declines amongst 72% of tree species, particularly among large-seeded species (Terborgh et al., 2008). Within the Biological Dynamics of Forest Fragments Project (BDFFP) reserves, the large-seeded, mammal-dispersed *Duckeodendron cestroides* experienced a 300% reduction in seed dispersal, 500% reduction in maximum dispersal distance, and 5000% reduction in the number of seeds dispersed 10m beyond the crown in hunted fragments relative to continuous forest (Cramer et al., 2007). Overall, up to one third of Amazonian plants may suffer failed dispersal due to losses of vertebrate seed dispersers (da Silva & Tabarelli, 2000).

Similarly, in the Atlantic coastal forest of Brazil, large-seeded *Astrocaryum aculeatissimum* palms in fragments with low agouti density faced a collapse in seed dispersal, as agoutis disperse more seeds than they predate, and smaller rodents are unable to compensate for the agouti's absence (Donatti et al., 2009). In southern Mexico, fragments <30ha lack most to all large frugivores, and as a result dispersal and recruitment of large-seeded species declined ~50% (Melo et al., 2010). Large frugivorous seed dispersers have also declined in Australian rainforest fragments, affecting 12% of all plant species, particularly large-seeded species (Moran et al., 2009). In Asia, commercial hunting threatens mass extinctions of mammals greater than 1-2 kg, particularly the major seed dispersers – elephants, tapirs, deer, primates, and large birds and bats, all of which face greater hunting pressure than seed predators. As a result, future Asian forests are likely to be dominated by small-fruited and – seeded, fast-growing trees at the expense of large-seeded species (Corlett, 2007).

While all the studies cited above found declines in large-seeded tree species in hunted fragments due to loss of seed dispersers, a number of studies have found the opposite: large-seeded species increase in abundance and diversity due to the loss of hunted vertebrate seed predators. In Panama, hunters favor large-seeded tree species by removing seed predators (e.g., Central American agoutis, *Dasyprocta punctata*, and collared peccaries, *Pecari tajacu*), and thereby increasing seedling survival (Wright et al., 2007b), resulting in 300-500% higher seedling density of two large-seeded palms at heavily hunted sites (Wright et al., 2000). Lianas also benefit from hunting, as most liana seeds are wind-dispersed, then escape competition from small-seeded species that have lost their seed dispersers (Wright et al., 2007b). This effect is particularly pronounced in Los Tuxtlas, Mexico, where thick monospecific carpets of large-seeded seedlings result from heavy rodent predation of small-seeds (Dirzo et al., 2007). In northeastern Costa Rica, large-seeded *Dipteryx panamensis* escaped predation in fragments due to heavy hunting of squirrels (*Sciurus* spp.), agoutis, and peccaries, resulting in higher seedling densities (Hanson & Brunsfeld, 2006). However, even where seed and seedling survival is increased due to the loss of seed predators, the concurrent loss of seed dispersers often results in monospecific seed and seedling carpets clustered underneath conspecific parent trees, eventually resulting in lower tree species diversity (Figure 2; Wright et al., 2000; Wright & Duber, 2001).

Fig. 2. Collared peccary (*Pecari tajacu*; top left), Central American Agouti (*Dasyprocta punctata*; bottom left), and a carpet of undispersed *Attalea butyraceae* fruits at a heavily hunted site in Parque Nacional Camino de Cruces, Panama (all photos by Nicole L. Michel).

Relative seed dispersal and predation rates between hunted fragments and protected forest
tend to be highly plant species-specific (Guariguata et al., 2000, 2002; Wright et al., 2000).
Regardless of which species are impacted or favored by increased hunting pressure on seed
dispersers and predators, it is clear that the loss of large herbivores and frugivores is having
dramatic effects on density and diversity of the remnant plant communities (Wright, 2010;
Wright et al., 2007a). There is some hope, however, that negative density dependence may
allow once declining plant species to rebound as the forest is protected, and once seed
disperser and predator populations to return to their historical levels (Muller-Landau, 2007).

5. Mesoherbivore release

The previous section focused on the consequences of excessive human predation on midsize
and large herbivores and frugivores, the so-called mesoherbivores. However, large non-
human predators also effectively limit vertebrate herbivores in diverse natural systems (Fig.
1; Estes et al. 2011; Salo et al., 2010). What happens when large non-human predators are
removed in the absence of human hunting? Terborgh (1992) first suggested that the loss of
large carnivores (jaguars, pumas) from Barro Colorado Island, Panama could explain the
seemingly higher densities of mesoherbivores such as peccaries, pacas, and agoutis relative
to Cocha Cashu, Peru. Yet, very few studies have since followed up with experimental or
observational studies documenting increased mesoherbivore abundance in tropical forests
following losses of large predators. The few studies that do exist, however, suggest that
mesoherbivore release may have consequences just as catastrophic as their decline.

The best-known and most clear-cut cases of mesoherbivore release occur at temperate
latitudes. In North America, white-tailed (*Odocoileus virginianus*) and mule deer (*O.
hemionus*) populations have exploded following the disappearance of wolves and mountain
lions, with catastrophic effects on plant cover and diversity, nutrient and carbon cycling,
and forest successional rates, with rebounding effects on insects, birds, and other mammals
(see reviews by Côté et al., 2004; McShea et al., 1997). In Yellowstone National Park, in the
western United States, elk (*Cervus canadensis*) populations increased and altered their
foraging behavior when wolves (*Canis lupus*) were extirpated, causing collapses in aspen
and cottonwood (*Populus* spp.) recruitment that did not recover until wolves were
reintroduced (Beschta, 2005; Halofsky & Ripple, 2008a,b; Ripple & Larsen, 2000). Less well-
known in the literature is the eradication of dholes (wild dogs; *Cuon alpinus*) in Bhutan that
allowed wild pig (*Sus scrofa*) densities to increase thirty-fold, resulting in extensive crop
losses and even direct attacks on humans and livestock (Wang et al., 2006; Wangchuk, 2004).

In tropical forests, mesoherbivore release was first suggested at Barro Colorado Island (BCI),
Panama. Terborgh (1992) compared mesoherbivore densities at BCI, with ocelots persisting
but not jaguars and pumas, versus Cocha Cashu (CC), Peru, with all large cats. He reported
mesoherbivore densities 8 to 20 times greater on BCI than CC, and predicted that this could
have a detrimental impact on large-seeded tree species via increased seed predation.
However, this result is controversial, as Wright et al. (1994) pointed out multiple problems
with both the data and its proposed implications. Mesoherbivore densities at BCI may have
been overestimated due to human habituation, and densities at CC may be anomalously low
due to seasonal flooding; indeed other Neotropical sites with large predators have
mesoherbivore densities similar to BCI. Further, predation on and recruitment of the large-
seeded *Dipteryx panamensis*, commonly predated by agoutis, was similar at BCI and CC

despite an order of magnitude difference in agouti abundance between the sites (Terborgh & Wright, 1994). Seed survival of three large-seeded species was also similar at BCI and the nearby mainland after poaching was greatly reduced, and seedling herbivory was lower on BCI (Asquith et al., 1997). Additionally, calculations of total herbivore consumption rates by predators indicates that Manu's jaguars, pumas, and ocelets do not eat sufficient herbivore biomass to effectively limit their populations (Leigh 1999). This led Wright et al. (1994) to conclude that, in comparison to the effects of hunting-driven meso-herbivore loss, "There is, however, no evidence to support the hypothesis that the relatively subtle increases in abundances that may follow the extirpation of large felids are of similar importance."

However, release of tropical mesoherbivores from predation does not always result in "relatively subtle increases in abundance," especially when combined with other factors (e.g., increased food availability) that may amplify mesoherbivore population growth. At Pasoh Forest Reserve, an aseasonal tropical forest in Malaysia have, native wild pig (*S. scrofa*) densities exploded to an estimated $47/km^2$ in 1996, due to a combination of lost natural predators (tigers, leopards) and increased food supply provided by nearby oil palm plantations (Ickes, 2001). These densities are one or two orders of magnitude greater than the observed wild pig densities of $0.1 - 4.2/km^2$ in tropical forest with predators present, and are similar to densities found on Peucang Island, coastal dipterocarp forest with no natural predators ($27-32/km^2$; Ickes, 2001). The hyperabundant pigs had dramatic effects on the natural forest vegetation structure and diversity. Within experimental pig exclosures, the number of seedling recruits increased 3 times, stem density increased by 56%, liana density increased 18%, plant species richness increased by 10%, and height growth increased 52.5% for trees 1-7m tall, though mortality was not affected (Ickes et al., 2001). Female pigs damage an estimated 170,000 saplings/km^2 annually by snapping the stems to build woody nests (Ickes et al., 2003). Pigs build 6 nests/ha annually, causing an estimated 29% of observed tree mortality and 43% of sapling mortality and damage (Ickes et al., 2005). Pigs preferentially select saplings of the dominant and ecologically important Dipterocarpaceae for nest building (Ickes et al., 2005). Yet, tree species have differential ability to regenerate following snapping; the dominant dipterocarps have particularly poor regrowth rates (Ickes et al., 2003). The wild pigs are also implicated in Pasoh's recent invasion by *Clidemia hirta* (Melastomataceae), an invasive shrub colonizing recent treefall gaps and pig-disturbed soil (Peters, 2001). Thus, the ecological release of feral pigs in Pasoh, due to a combination of artificial food inputs (i.e., oil palm) and lost native predators (Ickes, 2001; Ickes & Williamson, 2000), is having and will continue to have dramatic cascading effects on forest structure and tree community composition.

In Gunung Palung National Park in Borneo, recruitment of the dominant dipterocarp canopy trees collapsed during the 1990s, purportedly due, in part, to release of vertebrate seed predators (Curran et al., 1999). Dipterocarps exhibit mast fruiting during El Niño-Southern Oscillation events, and the large extent and synchronous production allow some seeds to escape predation. However, forest fragmentation due to logging and human-induced wildfires has reduced the extent and intensity of the mast fruiting events; and masting events at Pasoh Forest Reserve in Malaysia also have been highly variable in recent years, for unknown reasons (S.J. Wright, pers. comm.). Nonetheless, vertebrate seed predators (including bearded pigs, *Sus barbatus*; primates, rodents, and birds) and seed predation increased dramatically in 1998, particularly in logged and degraded land (Curran et al., 1999; Curran & Leighton, 2000). The combination of high predation, logging, and wildfires contributed to a complete collapse in dipterocarp reproduction: not a single seedling was found following the 1998 masting event, compared to 150,000 seedlings/ha following the previous masting event (Curran et al., 1999).

At Lago Guri in Venezuela, herbivorous howler monkeys (*Alouatta seniculus*), common iguanas (*Iguana iguana*), and leaf-cutter ants (*Acromyrmex* spp. and *Atta* spp.) occur at densities 10-100 times greater than nearby mainland sites (reviewed by Terborgh & Feeley, 2010). As a result, seedling and sapling densities have dramatically decreased due to high mortality and low recruitment, creating "ecological meltdown" (Terborgh et al., 2001, 2006). Mortality and recruitment of both small and large saplings on these islands exceeded comparable rates on the mainland by a factor of 2, primarily due to leaf-cutter ant herbivory (Lopez & Terborgh 2007; Terborgh et al., 2006). Though leaf-cutter ants are clearly not mesoherbivores, this example demonstrates the magnitude of the effects herbivores can have on tropical forest plant communities. Furthermore, hyperabundant mesoherbivores also had species-specific effects on nutrient cycling: iguanas and howler monkeys increased the carbon/nitrogen ratio and, in turn, tree growth, while leaf-cutter ants had the opposite effect (Feeley & Terborgh, 2005). This altered nutrient regime drove a ricocheting bottom-up effect, in which the positive indirect effects of howler monkey density (and concomitant increased nutrient availability) on bird species richness exceeded the negative, direct effects of island area on bird species richness (Feeley & Terborgh, 2008). Indeed, birds maintained territories 3 to 5 times smaller on islands with abundant howler monkeys than on the adjacent mainland (Terborgh et al., 1997). While the herbivore hyperdensities are commonly attributed to the absence of all large and mid-sized predators from the islands (Terborgh et al., 2001, 2006), the very small size of the islands, extensive edge effects, and refuge effect during reservoir inundation also likely contributed (S.J. Wright, pers. comm.).

In each of the previous examples, hyperabundant mesoherbivores were shown to have dramatic, cascading consequences on the rainforest plant communities, yet release from predation may have accounted only in part for the altered herbivore densities. On the contrary, La Selva Biological Station in Costa Rica presents a more clear-cut case of mesoherbivore release. Collared peccaries (*Pecari tajacu*) have increased from a density of ~0/km^2 of trail in the 1970s (T.W. Sherry, pers. obs.) to 14/km^2 in 1993 (Torrealba-Suárez & Rau, 1994), and likely higher today (Romero et al., unpubl. data). While some herds forage in neighboring croplands (Torrealba-Suárez & Rau, 1994), peccaries had similar effect sizes (i.e., no block effects) on vegetation density at 5 paired mammal exclosures located far from any station boundary. This suggests that peccary densities are consistent across the station and thus not dependent on food inputs or experiencing a refuge effect, leaving release from predation a likely driver of the current high peccary densities (Michel et al., unpubl. data).

The question of whether collared peccaries at La Selva are over- or hyperabundant is under debate, and accurate current density estimates are greatly needed. However, the most recent available estimate (14/km^2 in 1993) is itself over double the mean 6.6/km^2 density estimated at 26 unhunted Amazonian sites (Peres & Palacios, 2007), despite some hunting occurring at La Selva (N.L. Michel, pers. obs.). It also exceeds reported densities of 9.6/km^2 at BCI, Panama, despite similar area, hunting pressure, and soil nutrient levels to La Selva (Powers et al., 2005; Wright et al., 2000). Known peccary densities elsewhere in Central America are all lower (3-7.5/km^2; Michel et al., unpubl. data). Furthermore, effect sizes of mammal (primarily collared peccary) exclosure on vegetation metrics at La Selva exceed effect sizes from exclosures at BCI and Gigante Peninsula, Panama by 200-400%, despite sturdier exclosures in Panama that exclude more terrestrial mammals (Michel et al., unpubl. data). Thus all evidence suggests that La Selva's collared peccaries occur at unusually high densities that could be considered overabundant (McShea et al., 1997).

High-density peccaries at La Selva consume 98.6% of *Mucuna holtonii* (liana) seeds on the forest floor (Kuprewicz and García-Robledo, 2010), and reduce seedling densities at La Selva vs. nearby sites with greater hunting pressure (Chazdon et al., in press; Hanson et al., 2006). Peccaries consume stilt roots of *Socratea exorrhiza*, causing high mortality in these once-abundant trees, effects rarely observed at other Central American sites (Lieberman & Lieberman, 1994; N.L. Michel, pers. obs.). Seedling abundances of 30 focal species in exclosures versus controls were significantly different, presumably due to differences in seed predation pressures and herbivory rates (A. Wendt et al., unpubl. data). Peccaries have also reduced woody and herbaceous stem density, canopy cover, and vine and liana density and cover at La Selva (based on mammal exclosures Fig. 3). Moreover, six sites from Costa Rica-Panama show liana tangle frequency varies inversely with peccary density, and liana tangles are 133% more abundant within mammal exclosures than paired controls at La Selva. This research suggests that the hyperabundant collared peccaries at La Selva are having dramatic effects on understory, and even canopy, vegetation. This is particularly troubling, given that vines and lianas provide important foraging and nesting substrate for many organisms, including understory insectivorous birds, a guild that has experienced significant declines at both La Selva and BCI over the past 40 years (Sigel et al., 2006, 2010).

Fig. 3. Understory vegetation density at La Selva Biological Station, Costa Rica, is lower in areas exposed to collared peccaries (left) than within experimental peccary exclosures (right; photos by Nicole L. Michel).

6. Conclusion

6.1 Conservation and management implications

It is clear from this review that fragmentation-induced trophic cascades are having catastrophic impacts on tropical forests worldwide. Regardless of whether mesoherbivores are increasing due to the loss of large predators, over-protection in some ecological reserves (e.g., La Selva), and/or food inputs, or decreasing due to human hunting pressure, it is clear that the cascading effects of perturbations in mesoherbivore populations on tropical forest plant communities and the animals reliant on them are catastrophic, diverse, and pervasive. Indeed, trophic cascades and increased mesoherbivore abundances are considered to be as serious threats to tropical biodiversity as climate change (Terborgh & Feeley 2010).

The solution to the problem is clear: large extents of continuous, or at least connected, forest must be protected and carefully managed to keep populations of both apex predators and mesoherbivores within natural ranges (Soulé, 2010). However clear the solution, its implementation is far from simple for many reasons:

- Large predators have large home ranges and are highly edge-sensitive. In Central America, female jaguars have home range sizes of 10-18.8 km^2, whereas male jaguars roam over 28-40km^2 (Rabinowitz & Nottingham, 1986; Woodruffe & Ginsberg, 1998). As a result, jaguars require a minimum "critical reserve size" of 69 km^2, or 6,900 hectares, to maintain a 50% probability of population persistence (Woodruffe & Ginsberg, 1998). This far exceeds the size of most Central American forest reserves. Tigers, with a female home range size of 16.9 km^2, require even more space: 135 km^2 is the minimum required to maintain a 50% probability of population persistence (Woodruffe & Ginsberg, 1998). Where mesoherbivores are also hunted by humans and prey is thus scarce, predators are likely to need even more space (Cramshaw & Quigley, 1991).
- Human phobias associated with large predators and their impacts on domesticated animal herds contribute to resisting reintroduction of predators to areas from which they have been extirpated, and heavily persecuting predators even beyond reserve borders (Soulé, 2010). Conflict with humans along reserve borders is one of the major causes of predator mortality within protected reserves (Woodruffe & Ginsberg, 1998).
- Many tropical forests are found in countries with limited funds to protect and effectively manage forest reserves (Bruner et al., 2004). The corruption endemic to many national governments in tropical regions further complicates matters (Wright et al., 2007c). Even in Costa Rica and Panama, two of the wealthiest (International Monetary Fund, 2010) and least corrupt (Transparency International, 2010) countries in Central America, poaching is endemic. In Panama, poachers are altering seed dispersal, predation, and recruitment via limitation of mesoherbivores (Wright & Duber, 2001; Wright et al., 2000, 2007b). In Costa Rica, poaching continues to affect vertebrates such as white-lipped peccaries even in the relatively well-protected Corcovado National Park (Carrillo et al., 2000, 2002), and the 260 km^2 Tortuguero National Park was often patrolled by fewer than five forest guards as recently as 2005 (N. Michel, pers. obs.).
- Many people in tropical countries rely upon hunting for either their own subsistence or to trade for food and other necessities (Corlett, 2007). If patrolling forest reserves effectively limits hunting, many neighboring people would need other means of support. This could result in increased slash-and-burn agriculture with its concomitant negative effects on tropical biodiversity (Naughton-Treves et al., 2003).

Many scientists and conservationists argue that simply setting aside land in forest reserves is sufficient to preserve biodiversity, regardless of connectivity or management (e.g., Fahrig, 1997). Yet, others argue that connectivity, i.e., developing biological corridors to connect nearby reserves, is sufficient to preserve biodiversity (e.g., Soulé, 2010). However, if proposed reserves do not already have apex predators, or have degraded habitat unable to support apex predators, or if heavy hunting pressure is limiting mesoherbivore and/or predator populations, the reserve will function in reality as an "Empty Forest" (Redford, 1992; Wilkie et al., 2011). This is a serious problem, as much of the land remaining in tropical regions is either degraded, or in suboptimal habitat, and/or heavily impacted by hunting, and – as discussed above – government funding of reserve protection is often insufficient to control poaching. In order for conservation to succeed, it is imperative that reserves of sufficient size and/or connectivity not only be set aside, but these reserves must also contain sustainable populations of both apex predators (whether extant or reintroduced) and mesoherbivores, i.e., balanced populations to which the natural vegetation and other organisms are adapted. Furthermore, adequate protection needs to be instituted and continued monitoring of predator and herbivore populations, perhaps using a Footprint index (de Thoisy et al., 2010), is crucial in order to prevent destructive trophic cascades and the subsequent loss of tropical biodiversity.

6.2 Future research needs

In terms of future research, we need intensive study of not only what are "normal" and sustainable abundances of mesoherbivores, but also how to maintain these abundances through a combination of vegetation, predator, and/or hunting management. We need better information on all the consequences, both long- and short-term, of human-altered mesoherbivore abundances on native tropical forest communities. There is an urgent need for better information and models on how altered trophic dynamics, especially mesoherbivore abundances, will be affected by other global phenomena, especially climate destabilization. Finally, it is essential to determine the effectiveness of various conservation actions, in order to determine which techniques can best prevent a mass extinction of tropical biodiversity (Brooks et al., 2009).

7. Acknowledgements

The authors thank S. Joseph Wright and Padmini Sudarshana for their helpful reviews of the original manuscript. We thank Walter P. Carson, David and Deborah Clark, Deedra McClearn, Robert Timm, and other La Selva researchers for constructive comments and advice, and thank W.P. Carson as well for access to experimental mammal exclosures on BCI and La Selva. The Explorers Club, National Science Foundation, Organization for Tropical Studies, Smithsonian Tropical Research Institute, and Tulane University all generously supported this research.

8. References

Alroy, J. (2001). A Multispecies Overkill Simulation of the End-Pleistocene Megafaunal Mass Extinction. *Science*, Vol.282, No.5523, (June 2001), pp. 1893-1896, ISSN 0036-8075.

Human-Altered Mesoherbivore Densities and Cascading Effects on Plant and Animal Communities in
Fragmented Tropical Forests

203

Asner, G.P., Rudel T.K., Aide, T.M., Defries, R., & Emerson, R. (2009). A Contemporary Assessment of Change in Humid Tropical Forests. *Conservation Biology*, Vol.23, No.6, (December 2009), pp. 1386-1395, ISSN 0888-8892.

Bawa, K.S., Bullock, S.H., Perry, D.R., Coville, R.E., & Grayum, M.H. (1985). Reproductive Biology of Tropical Lowland Rain Forest Trees. II. Pollination Systems. *American Journal of Botany*, Vol.72, No.3, (March 1985), pp.346-356, ISSN 0002-9122.

Beier, P., Van Drielen, M., & Kankam, B.O. (2002). Avifaunal Collapse in West African Forest Fragments. *Conservation Biology*, Vol.16, No.4, (August 2002), pp. 1097-1111, ISSN 0888-8892.

Benítez-López, A., Alkemade, R., & Verweij, P.A. (2010). The Impacts of Roads and Other Infrastructure on Mammal and Bird Populations: A Meta-Analysis. *Biological Conservation*, Vol.143, No.6, (June 2010), pp. 1307-1316, ISSN 0006-3207.

Beschta, R.L. (2005). Reduced Cottonwood Recruitment Following Extirpation of Wolves in Yellowstone's Northern Range. *Ecology*, Vol.86, No.2 (February 2005), pp. 391–403, ISSN 0012-9658.

Bradshaw, C.J., N.S. Sodhi, & B.W. Brooks. (2009). Tropical Turmoil: A Biodiversity Tragedy in Progress. *Frontiers in Ecology and the Environment*, Vol.7, No.2, (March 2009), pp. 79-87, ISSN 1540-9295.

Brashares, J.S., Prugh, L.R., Stoner, C.J., & Epps, C.W. (2010). Ecological and Conservation Implications of Mesopredator Release, In: *Trophic Cascades: Predators, Prey, and the Changing Dynamics of Nature*, Terborgh, J., & J.A. Estes, (Eds.), pp. 221-240, Island Press, ISBN 978-1597264877, Washington, DC.

Broadbent, E., Asner, G.P., Keller, M., Knapp, D., Oliveira, P., & Silva, J. (2008). Forest Fragmentation and Edge Effects from Deforestation and Selective Logging in the Brazilian Amazon. *Biological Conservation*, Vol.141, No.7, (July 2008), pp. 1745–1757, ISSN 0006-3207.

Brooks, B.W., Sodhi, N.S., & Bradshaw, C.J.A. (2008). Synergisms Among Extinction Drivers Under Global Change. *Trends in Ecology and Evolution*, Vol.23, No.8, (August 2008), pp. 453–460, ISSN 0169-5347.

Brooks, T.M., Pimm, S.L., & Oyugi, J.O. (1999). Time Lag between Deforestation and Bird Extinction in Tropical Forest Fragments. *Conservation Biology*, Vol.13, No.5, (October 1999), pp. 1140-1150, ISSN 0888-8892.

Brooks, T.M., Wright, S.J., & Sheil, D. (2009). Evaluating the Success of Conservation Actions in Safeguarding Tropical Forest Biodiversity. *Conservation Biology*, Vol.23, No.6 (December 2009), pp.1448-1457, ISSN 0888-8892.

Brown, K.S. Jr & Hutchings, R.W. (1997). Disturbance, Fragmentation, and the Dynamics of Diversity in Amazonian Forest Butterflies, In: *Tropical Forest Remnants: Ecology, Management, and Conservation of Fragmented Communities*, Laurance, W.F., & Bierregaard, R.O., (Eds.), pp. 91-110, University of Chicago Press, ISBN 978-0226468990, pp. 91–110, Chicago, IL.

Bruner, A.G., Gullison, R.E., & Balmford, A. (2004). Financial Costs and Shortfalls of Managing and Expanding Protected-Area Systems in Developing Countries. *BioScience*, Vol.54, No.12, (December 2004), pp. 1119-1126, ISSN 0006-3568.

Carrillo, E., Wong, G., & Cuarón, A. (2000). Monitoring Mammal Populations in Costa Rican Protected Areas under Different Hunting Restrictions. *Conservation Biology*, Vol.14, No.6, (December 2000), pp.1580-1591, ISSN 0888-8892.

Carrillo, E., Saenz, J.C., & Fuller, T.K. (2002). Movements and Activities of White-lipped Peccaries in Corcovado National Park, Costa Rica. *Biological Conservation*, Vol.108, No.3, (December 2002), pp. 317-324, ISSN 0006-3207.

Chazdon, R.L., Vilchez Alvarado, B., Letcher, S., Wendt, A., & Sezen, U. (In press). Effects of Human Activities on Successional Pathways: Case Studies from Lowland Set Forests of Northeastern Costa Rica, In: *The Social Life of Forests*, Hecht, S., Morrison, K., & Padoch, C., (Eds.), pp. xxx-xxx, University of Chicago Press, Chicago, IL.

Chiarello, A.G. (1999). Effects of Fragmentation of the Atlantic Forest on Mammal Communities in South-Eastern Brazil. *Biological Conservation*, Vol.89, No.1 (July 1999), pp. 71-82, ISSN 0006-3207.

Corlett, R.T. (2007). The Impact of Hunting on the Mammalian Fauna of Tropical Asian Forests. *Biotropica*, Vol.39, No.3, (May 2007), pp. 292-303, ISSN 0006-3606.

Côté, S.D., Rooney, T.P., Tremblay, J.-P., Dussault, C., & Waller, D.M. (2004). Ecological Impacts of Deer Overabundance. *Annual Review of Ecology, Evolution, and Systematics*, Vol.35, No.1 (December 2004), pp. 113-147, ISSN 0066-4162.

Cramer, J.M., Mesquita, R., Bentos, T., Moser, B., & Williamson, G.B. (2007). Forest fragmentation reduces seed dispersal of *Duckeodendron cestroides*, a Central Amazon endemic. *Biotropica*, Vol.39, No.6, (November 2007), pp. 709–718, ISSN 0006-3606.

Cramshaw, P.G., & Quigley, H.B. (1991). Jaguar Spacing, Activity and Habitat Use in a Seasonally Flooded Environment in Brazil. *Journal of Zoology*, Vol.223, No.3, pp. 357-370, ISSN 0952-8369.

Crooks, K.R., & Soulé, M.E. (1999). Mesopredator Release and Avifaunal Extinctions in a Fragmented System. *Nature*, Vol.400, No.6744 (August 1999), pp. 563–566, ISSN 0028-0836.

Curran, L.M., Caniago, I., Paoli, G.D., Astianti, D., Kusneti, M., Leighton, M., Nirarita, C.E., & Haeruman, H. (1999). Impact of El Niño and Logging on Canopy Tree Recruitment in Borneo. *Science*, Vol.286, No.5447, (December 1999), pp. 2184-2188, ISSN 0036-8075.

Curran, L.M., & Leighton, M. (2000). Vertebrate Responses to Spatiotemporal Variation in Seed Production of Mast-Fruiting Dipterocarpaceae. *Ecological Monographs*, Vol.70, No.1, (February 2000), ISSN 0012-9615.

Dalecky, A., Chauvet, S., Ringuet, S., Claessens, O., Judas, J., Larue, M., & Cosson, J.F. (2002). Large Mammals on Small Islands: Short Term Effects of Forest Fragmentation on the Large Mammal Fauna in French Guiana. *Revue d'Ecologie Terra et la Vie*, Vol.54, No.1 (March 2002), ISSN 0249-7395.

da Silva, J.M.C., & Tabarelli, M. (2000). Tree Species Impoverishment and the Future Flora of the Atlantic Forest of Northeast Brazil. *Nature*, Vol.404, No.6773, (March 2000), pp. 72-74, ISSN 0028-0836.

de Thoisy, B., Richard-Hansen, C., Goguillon, B., Joubert, P., Obstancias, J., Winterton, P., & Brosse, S. (2010). Rapid Evaluation of Threats to Biodiversity: Human Footprint Score and Large Vertebrate Species Responses in French Guiana. *Biodiversity and Conservation*, Vol.19, No.6, (June 2010), pp. 1567-1584, ISSN 0960-3115.

Dial, R., & Roughgarden, J. (1995). Experimental Removal of Insectivores from Rain Forest Canopy: Direct and Indirect Effects. *Ecology*, Vol.76, No. 6, (September 1995), pp. 1821–1834, ISSN 0012-9658.

Dirzo, R., & Raven, P.H. (2003). Global State of Biodiversity and Loss. *Annual Review of Environment and Resources*, Vol.28, No.1, (November 2003), pp. 137-167, ISSN 1543-5938.

Dirzo, R., Mendoza, E., & Ortíz, P. (2007). Size-Related Differential Seed Predation in a Heavily Defaunated Neotropical Rain Forest. *Biotropica*, Vol.39, No.3, (May 2007), pp. 355-362, ISSN 0006-3606

Dobson, A., Lodge, D., Alder, J., Cumming, G.S., Keymer, J., McGlade, J., Mooney, H., Rusak, J.A., Sala, O., Wolters, V., Wall, D., Winfree, R., & Xenopoulos, M.A. (2006). Hatitat Loss, Trophic Collapse, and the Decline of Ecosystem Services. *Ecology*, Vol.87, No.8, (August 2006), pp. 1915-1924, ISSN 0012-9658.

Donatti, C.I., Guimarães, P.R. Jr., & Galetti, M. (2009). Seed Dispersal and Predation in the Endemic Atlantic Rainforest Palm *Astrocaryum aculeatissimum* across a Gradient of Seed Disperser Abundance. *Ecological Research*, Vol.24, No.6, (November 2009), pp. 1187-1195, ISSN 0912-3814.

Duffy, J.E. (2003). Biodiversity Loss, Trophic Skew, and Ecosystem Functioning. *Ecology Letters*, Vol.6, No.8, (August 2003), pp. 680-687, ISSN 1461-0248.

Estes, J.A., Terborgh, J., Brashares, J.S., Power, M.E., Berger, J., Bond, W.J., Carpenter, S.C., Essington, T.E., Holt, R.D., Jackson, J.B.C., Marquis, R.J., Oksanen, L., Oksanen, T., Paine, R.T., Pikitch, E.K., Ripple, W.J., Sandin, S.A., Scheffer, M., Shcoener, T.W., Shurin, J.B., Sinclair, A.R.E., Soulé, M.E., Virtanen, R., & Wardle, D.A. (2011). Trophic Downgrading of Planet Earth. *Science*, Vol.333, No. 6040, (July 2011), pp. 301-306, ISSN 0036-8075.

Ewers, R.M., & Didham, R.K. (2006). Confounding Factors in the Detection of Species Responses to Habitat Fragmentation. *Biological Reviews*, Vol.81, No.2 , (May 2006), pp. 117-142, ISSN 1464-7931.

Fa, J.E., Peres, C.A., & Meeuwig, J. (2002). Bushmeat Exploitation in Tropical Forests: An Intercontinental Comparison. *Conservation Biology*, Vol.16, No.1, (February 2002), pp. 232–237, ISSN 0888-8892.

Fahrig, L. (1997). Relative Effects of Habitat Loss and Fragmentation on Population Extinction. *The Journal of Wildlife Management*, Vol.61, No.3, (July 1997), pp. 603-610, ISSN 1937-2817.

Fahrig, L. (2002). Effect of Habitat Fragmentation on the Extinction Threshold: A Synthesis. *Ecological Applications*, Vol.12, No.2, (April 2002), pp. 346-353, ISSN 1051-0761.

Fahrig, L. (2003). Effects of Habitat Fragmentation on Biodiversity. *Annual Review of Ecology, Evolution, and Systematics*, Vol.34, No. 1, (December 2003), pp. 487-515, ISSN 0066-4162.

Feeley, K.J., & Terborgh, J.W. (2005). The Effects of Herbivore Density on Soil Nutrients and Tree Growth in Tropical Forest Fragments. *Ecology*, Vol.86, No.1, (January 2005), pp. 116–124, ISSN 0012-9658.

Feeley, K.J., & Terborgh, J.W. (2008). Direct Versus Indirect Effects of Habitat Reduction on the Loss of Avian Species from Tropical Forest Fragments. *Animal Conservation*, Vol.11, No.5, (October 2008), pp. 353-360, ISSN 1469-1795.

Ferraz, G., Russell, G.J., Stouffer, P.C., Bierregaard, R.O., Pimm, S.L., & Lovejoy, T.E. (2003). Rates of Species Loss from Amazonian Forest Fragments. *Proceedings of the National Academy of the Sciences of the USA*, Vol.100, No.24, (November 2003), pp. 14069–14073, ISSN 1091-6490.

Fletcher, R.J. Jr., Ries, L., Battin, J., & Chalfoun, A.D. (2007). The Role of Habitat Area and Edge in Fragmented Landscapes: Definitely Distinct or Inevitably Intertwined? *Canadian Journal of Zoology*, Vol.85, No.10, (October 2007), pp. 1017-1030, ISSN 0008-4301.

Fleury, M., & Galetti, M. (2006). Forest Fragment Size and Microhabitat Effects on Palm Seed Predation. *Biological Conservation*, Vol.131, No.1, (July 2006), pp. 1-13, ISSN 0006-3207.

Gardner, T.A., Barlow, J., Parry, L.W., & Peres, C.A. (2007). Predicting the Uncertain Future of Tropical Forest Species in a Data Vacuum. *Biotropica*, Vol.39, No.1, (January 2007), pp. 25-30, ISSN 0006-3606.

Guariguata, M.R., Arias-Le Claire, H. & Jones, G. (2002) Tree Seed Fate in a Logged and Fragmented Forest Landscape, Northeastern Costa Rica. *Biotropica*, Vol.34, No.3, (September 2002), pp. 405–415, ISSN 0006-3606.

Guariguata, M.R., Rosales Adame, J.J., & Finegan, B. (2000). Seed Removal and Fate in Two Selectively Logged Lowland Forests with Contrasting Protection Levels. *Conservation Biology*, Vol.14, No.4, (August 2000), pp. 1046-1054, ISSN 0888-8892.

Halofsky, J.S., & Ripple, W.J. (2008a). Fine-Scale Predation Risk on Elk After Wolf Reintroduction in Yellowstone National Park, USA. *Oecologia*, Vol.155, No.4, (April 2008), pp. 869–877, ISSN 0029-8549.

Halofsky, J.S., & Ripple, W.J. (2008b). Linkages Between Wolf Presence and Aspen Recruitment in the Gallatin Elk Winter Range of Southwestern Montana, USA. *Forestry*, Vol.81, No.2, (April 2008), pp. 195–207, ISSN 0015-752X.

Hansen, M.C., & DeFries, R.S. (2004). Detecting Long-Term Global Forest Change Using Continuous Fields of Tree-Cover Maps From 8-km Advanced Very High Resolution Radiometer (AVHRR) Data for the Years 1982-99. *Ecosystems*, Vol.7, No.7, (November 2004), pp. 695-716, ISSN 1432-9840.

Hansen, M.C., Stehman, S.V., Potapov, P.V., Loveland, T.R., Townshend, J.R.G., DeFries, R.S., Pittman, K.W., Arunarwati, B., Stolle, F., Steininger, M.K., Carroll, M., & DiMiceli, C. (2008). Humid Tropical Forest Clearing from 2000 to 2005 Quantified by Using Multitemporal and Multiresolution Remotely Sensed Data. *Proceedings of the National Academy of the Sciences of the USA*, Vol.105, No.27, (July 2008), pp. 9439-9444, ISSN 1091-6490.

Hanson, T., Brunsfeld, S. & Finegan, B. (2006). Variation in Seedling Density and Seed Predation Indicators for the Emergent Tree *Dipteryx panamensis* in Continuous and Fragmented Rain Forest. *Biotropica*, Vol.38, No.6, (November 2006), pp. 770-774, ISSN 0006-3606.

He, F., & Hubbell, S.P. (2011). Species-Area Relationships Always Overestimate Extinction Rates from Habitat Loss. *Nature*, Vol.473, No.7347, (May 2011), pp. 368-371, ISSN 0028-0836.

Henle, K., Davies, K.F., Kleyer, M., Margules, C., & Settele, J. (2004). Predictors of Species Sensitivity to Fragmentation. *Biodiversity and Conservation*, Vol.13, No.1, (January 2004), pp. 207-251, ISSN 0960-3115.

Herrera, J.M., García, D., Martínez, D., & Valdés, A. (2011). Regional vs Local Effects of Habitat Loss and Fragmentation on Two Plant-Animal Interactions. *Ecography*, Vol.34, No.4, (August 2011), pp. 606-615, ISSN 0906-7590.

Ickes, K. (2001). Hyper-Abundance of Native Wild Pigs (*Sus scrofa*) in a Lowland Dipterocarp Rain Forest of Peninsular Malaysia. *Biotropica*, Vol.33, No.4, (December 2001), pp. 682-690, ISSN 0006-3606.

Ickes, K., Dewalt, S.J., & Appanah, S. (2001). Effects of Native Pigs (*Sus scrofa*) on Woody Understorey Vegetation in a Malaysian Lowland Rain Forest. *Journal of Tropical Ecology*, Vol.17, No.2, (March 2001), pp. 191–206, ISSN 0266-4674.

Ickes, K., DeWalt, S.J., & Thomas, S.C. (2003). Resprouting of Woody Saplings Following Stem Snap by Wild Pigs in a Malaysian Rain Forest. *Journal of Ecology*, Vol.91, No.2, (April 2003), pp. 222-233, ISSN 1365-2745.

Ickes, K., Paciorek, C.J., & Thomas, S.C. (2005). Impacts of Nest Construction by Native Pigs (*Sus scrofa*) on Lowland Malaysian Rain Forest Saplings. *Ecology*, Vol.86, No.6, (June 2005), pp. 1540-1547, ISSN 0012-9658.

Ickes, K., & Williamson, G.B. (2000). Edge Effects and Ecological Processes – Are they on the Same Scale? *Trends in Ecology and Evolution*, Vol.15, No.9, (September 2000), p. 373, ISSN 0169-5347.

International Monetary Fund. (2010). World Economic Outlook Database, In: *World Economic and Financial Surveys*, 01.08.2011, Available from: <http://www.imf.org/external/pubs/ft/weo/2011/01/weodata/index.aspx>.

Johnson, C.N. (2002). Determinants of Loss of Mammal Species During the Late Quaternary 'Megafauna' Extinctions: Life History and Ecology, but Not Body Size. *Proceedings of the Royal Society B: Biological Sciences*, Vol.269, No.1506 (November 2002), pp. 2221-2227, ISSN 1471-2954.

Johnson, C.N., Isaac, J.L., & Fisher, D.O. (2007). Rarity of a Top Predator Triggers Continent-Wide Collapse of Mammal Prey: Dingoes and Marsupials in Australia. *Proceedings of the Royal Society B: Biological Sciences*, Vol.274, No.1608, (February 2007), pp. 341-346, ISSN 1471-2954.

Kalka, M.B., Smith, A.R., & Kalko, E.K.V. (2008). Bats Limit Arthropods and Herbivory in a Tropical Forest. *Science*, Vol.320, No.5872, (April 2008), p. 71, ISSN 0036-8075.

Kattan, G.H., Alvarez-Lopez, H., & Giraldo, M. (1994). Forest Fragmentation and Bird Extinctions: San Antonio Eighty Years Later. *Conservation Biology*, Vol.8, No.1 (March 1994), pp. 138-146, ISSN 0888-8892.

Keyghobadi, N. (2007). The Genetic Implications of Habitat Fragmentation for Animals. *Canadian Journal of Zoology*, Vol.85, No.10, (October 2007), pp. 1049-1064, ISSN 0008-4301.

Koh, L.P., Dunn, R.R., Sodhi, N.S., Colwell, R.K., Proctor, H.C., & Smith, V.S. (2004). Species Coextinctions and the Biodiversity Crisis. *Science*, Vol.305, No.5690, (September 2004), pp. 1632–1634, ISSN 0036-8075.

Kuprewicz, E.K., & García-Robledo, C. (2010). Mammal and Insect Predation of Chemically and Structurally Defended *Mucuna holtonii* (Fabaceae) Seeds in a Costa Rican Rain Forest. *Journal of Tropical Ecology*, Vol.26, No.3, (May 2010), pp. 263-269, ISSN 0266-4674.

Laurance, W.F. (1991). Ecological Correlates of Extinction Proneness in Australian Tropical Rain Forest Mammals. *Conservation Biology*, Vol.5, No.1, (March 1991), pp. 79–89, ISSN 0888-8892.

Laurance, W.F. (1994). Rainforest Fragmentation and the Structure of Small Mammal Communities in Tropical Queensland. *Biological Conservation*, Vol.69, No.1, (January 1994), pp. 23–32, ISSN 0006-3207.

Laurance, W.F. (1997). Responses of Mammals to Rainforest Fragmentation in Tropical Queensland: A Review and Synthesis. *Wildlife Research*, Vol.24, No.5, (October 1997), pp. 603–612, ISSN 1035-3712.

Laurance, W.F. (1999). Introduction and Synthesis. *Biological Conservation*, Vol.91, No.2-3, (December 1999), pp. 101-107, ISSN 0006-3207.

Laurance, W.F. (2000). Do Edge Effects Occur Over Large Spatial Scales? *Trends in Ecology and Evolution*, Vol.15, No.4, (April 2000), pp. 134–135, ISSN 0169-5347.

Laurance, W.F. (2005). The Alteration of Biotic Interactions in Fragmented Tropical Forests, In: *Biotic Interactions in the Tropics: Their Role in the Maintenance of Species Diversity*, Burslem, D.F.R.P., Pinard, M.A., & S.E. Hartley, (Eds.), pp. 441-458, Cambridge University Press, ISBN 978-0521609852, Cambridge, UK.

Laurance, W.F. (2006). Have we Overstated the Tropical Biodiversity Crisis? *Trends in Ecology and Evolution*, Vol.22, No.2 (February 2006), pp. 65-70, ISSN 0169-5347.

Laurance, W.F. (2008). Theory Meets Reality: How Habitat Fragmentation Research has Transcended Island Biogeographic Theory. *Biological Conservation*, Vol.141, No.7, (July 2008), pp. 1731-1744, ISSN 0006-3207.

Laurance, W.F., & Useche, D.C. (2009). Environmental Synergisms and Extinctions of Tropical Species. *Conservation Biology*, Vol.23, No.6, (December 2009), pp. 1427-1437, ISSN 0888-8892.

Laurance, W.F., & Williamson, G.B. (2001). Positive Feedbacks among Forest Fragmentation, Drought, and Climate Change in the Amazon. *Conservation Biology*, Vol.15, No.6, (December 2001), pp. 1529-1535, ISSN 0888-88892.

Laurance, W.F., Camargo, J.L.C., Luizão, R.C.C., Laurance, S.G., Pimm, S.L., Bruna, E.M., Stouffer, P.C., Williamson, G.B., Benítez-Malvido, J., Vasconcelos, H.L., Van Houtan, K.S., Zartman, C.E., Boyle, S.A., Didham, R.K., Andrade, A., & Lovejoy, T.E. (2011). The Fate of Amazonian Forest Fragments: A 32-Year Investigation. *Biological Conservation*, Vol.144, No.1, (January 2011), pp. 56-67, ISSN 0006-3207.

Laurance, W.F., Lovejoy, T.E., Vasconcelos, H.L., Bruna, E.M., Didham, R.K., Stouffer, P.C., Gascon, C., Bierregaard, R.O., Laurance, S.G., & Sampaio, E. (2002). Ecosystem Decay of Amazonian Forest Fragments: a 22-Year Investigation. *Conservation Biology*, Vol.16, No.3, (June 2002), pp. 605–618, ISSN 0888-8892.

Laurance, W.F., Laurance, S.G., & Hilbert, D.W. (2008). Long-Term Dynamics of a Fragmented Rainforest Mammal Assemblage. *Conservation Biology*, Vol.22, No.5, (October 2008), pp. 1154-1164, ISSN 0888-8892.

Lees, A.C., & Peres, C.A. (2008). Conservation Value of Remnant Riparian Forest Corridors of Varying Quality for Amazonian Birds and Mammals. *Conservation Biology*, Vol.22, No.2, (April 2008), pp. 439-449, ISSN 0888-8892.

Leigh, E.G., Jr. (1999). *Tropical Forest Ecology: A View From Barro Colorado Island* (1st edition), Oxford University Press, ISBN 978-0195096033, New York, NY, USA.

Letourneau, D.K., & Dyer, L.A. (1998). Experimental Tests in Lowland Tropical Forest Shows Top-Down Effects Through Four Trophic Levels. *Ecology*, Vol.79, No.5, (July 1998), pp. 1678–1687, ISSN 0012-9658.

Human Altered Mesoherbivore Densities and Cascading Effects on Plant and Animal Communities in
Fragmented Tropical Forests

209

Lieberman, M., & Lieberman, D. (1994). Patterns of Density and Dispersion of Forest Trees, In: *La Selva: Ecology and Natural History of a Neotropical Rainforest*, McDade, L.A., Bawa, K.S., Hespenheide, H.A., & Hartshorn, G.S., (Eds.), pp. 106-119, University of Chicago Press, ISBN 978-0226039503, Chicago, IL.

Lindell, C.A., Riffell, S.K., Kaiser, S.A., Battin, A.L., Smith, M.L., & Sisk, T.D. (2007). Edge Responses of Tropical and Temperate Birds. *Wilson Journal of Ornithology*, Vol.119, No.2, (June 2007), pp. 205-220, ISSN 1559-4491.

Loiselle, B.A., & Hoppes, W.G. (1983). Nest Predation in Insular and Mainland Lowland Rainforest in Panama. *Condor*, Vol.85, No.1, (February 1983), pp. 93-95, ISSN 0010-5422.

Lopez, L., & Terborgh, J. (2007). Seed Predation and Seedling Herbivory as Factors in Tree Recruitment Failure on Predator-Free Forested Islands. *Journal of Tropical Ecology*, Vol.23, No.2, (March 2007), pp. 129–137, ISSN 0266-4674.

Lugo, A.E. (2009). The Emerging Era of Novel Tropical Forests. *Biotropica*, Vol.41, No.5, (September 2009), pp. 589-591, ISSN 0006-3606.

MacArthur, R. H., & Wilson, E.O. (1967). *The Theory of Island Biogeography*, Princeton University Press, ISBN 978-0691088365, Princeton, NJ.

Manu, S., Peach, W., & Cresswell, W. (2007). The Effects of Edge, Fragment Size and Degree of Isolation on Avian Species Richness in Highly Fragmented Forest in West Africa. *Ibis*, Vol.149, No.2, (April 2007), pp. 287–297, ISSN 1474-919X.

McShea, W.J., Underwood, H.B., & Rappole, J.H. (Eds.). (1997). *The Science of Overabundance: Deer Ecology and Population Management*. Smithsonian Books, ISBN 978-1560986812, Washington, D.C.

Melo, F.P.L., Martínez-Salas, E., Benítez-Malvido, J., & Ceballos, G. (2010). Forest Fragmentation Reduces Recruitment of Large-Seeded Tree Species in a Semi-Deciduous Tropical Forest of Southern Mexico. *Journal of Tropical Ecology*, Vol.26, No.1, (January 2010), pp. 35-43, ISSN 0266-4674.

Metzger, J.P., Martensen, A.C., Dixo, M., Bernacci, L.C., Ribeiro, M.C., Godoy Texeira, A.M., & Pardini, R. (2009). Time-Lag in Biological Responses to Landscape Changes in a Highly Dynamic Atlantic Forest Region. *Biological Conservation*, Vol.142, No.6, (June 2009), pp. 1166-1177, ISSN 0006-3207.

Michalski, F., & Peres, C.A. (2007). Disturbance-Mediated Mammal Persistence and Abundance-Area Relationships in Amazonian Forest Fragments. *Conservation Biology*, Vol.21, No.6, (December 2007), pp.1626-1640, ISSN 0888-8892.

Michel, N.L., Sherry, T.W., & Carson, W.P. (In Review, Oecologia). Non-Trophic Effects of Large Vertebrates Reverse Trophic Cascades in Wet-Tropical Forest Understory.

Milner-Gulland, E.J., Bennett, E.L. & Society for Conservation Biology 2002 Annual Meeting Wild Meat Group. (2003). Wild Meat: the Bigger Picture. *Trends in Ecology and Evolution*, Vol.18, No.7, (July 2003), pp. 351–357, ISSN 0169-5347.

Moran, C., Catterall, C.P., Kanowski, J. (2009). Reduced Dispersal of Native Plant Species as a Consequence of the Reduced Abundance of Frugivore Species in Fragmented Rainforest. *Biological Conservation*, Vol.142, No.3, (March 2009), pp. 541-552, ISSN 0006-3207.

Muller-Landau, H.C. (2007). Predicting the Long-Term Effects of Hunting on Plant Species Composition and Diversity in Tropical Forests. *Biotropica*, Vol.39, No.3, (May 2007), pp. 372–384, ISSN 0006-3606.

Muller-Landau, H.C., & Hardesty, B.D. (2005). Seed Dispersal of Woody Plants in Tropical Forests: Concepts, Examples, and Future Directions, In: *Biotic Interactions in the Tropics: Their Role in the Maintenance of Species Diversity*, Burslem, D.F.R.P., Pinard, M.A., & S.E. Hartley, (Eds.), pp. 267-309, Cambridge University Press, ISBN 978-0521609852, Cambridge, UK.

Naughton-Treves, L., Mena, J.L., Treves, A., Alvarez, N., & Radeloff, V.C. (2003). Wildlife Survival Beyond Park Boundaries: the Impact of Slash-and-Burn Agriculture and Hunting on Mammals in Tambopatu, Peru. *Conservation Biology*, Vol.17, No.4, (August 2003), pp. 1106-1117, ISSN 0888-8892.

Newmark, W.D. (1991). Tropical Forest Fragmentation and the Local Extinction of Understory Birds in the Eastern Usambara Mountains, Tanzania. *Conservation Biology*, Vol.5, No.1, (March 1991), pp. 67–78, ISSN 0888-8892.

Norris, D., Peres, C.A., Michalski, F., & Hinchsliffe, K. (2008). Terrestrial Mammal Responses to Edges in Amazonian Forest Patches: A Study based on Track Stations. *Mammalia*, Vol.72, No.1, (March 2008), pp. 15-23, ISSN 0025-1461.

Nuñez-Iturri, G., & Howe, H.F. (2007). Bushmeat and the Fate of Trees with Seeds Dispersed by Large Primates in a Lowland Rain Forest in Western Amazonia. *Biotropica*, Vol.39, No.3, (May 2007), pp. 348-357, ISSN 0006-3606.

Paine, R.T. (1980). Food Webs: Linkage, Interaction Strength, and Community Infrastructure. *Journal of Animal Ecology*, Vol.49, pp. 667–685, ISSN 1365-2656.

Pardini, R. (2004). Effects of Forest Fragmentation on Small Mammals in an Atlantic Forest Landscape. *Biodiversity and Conservation*, Vol.13, No.13, (December 2004), pp. 2567-2586, ISSN 0960-3115.

Peres, C.A. (2001) Synergistic Effects of Subsistence Hunting and Habitat Fragmentation on Amazonian Forest Vertebrates. *Conservation Biology*, Vol.15, No.6, (December 2001), pp. 1490–1505, ISSN 0888-8892.

Peres, C.A., & Palacios, E. (2007). Basin-Wide Effects of Game Harvest on Vertebrate Population Densities in Amazonian Forests: Implications for Animal-Mediated Seed Dispersal. *Biotropica*, Vol.39, No.3, (May 2007), pp. 304–315, ISSN 0006-3606.

Peters, H.A. (2001). *Clidemia hirta* Invasion at the Pasoh Forest Reserve: An Unexpected Plant Invasion in an Undisturbed Tropical Forest. *Biotropica*, Vol.33, No.1, (March 2001), pp. 60-68, ISSN 0006-3606.

Pimm, S.L., Jones, H.L. & Diamond, J. (1988). On the Risk Of Extinction. *American Naturalist*, Vol.132, No.6, (December 1988), pp. 757–785, ISSN 0003-0147.

Polis, G.A., & Holt, R.D. (1992). Intraguild Predation: the Dynamics of Complex Trophic Interactions. *Trends in Ecology & Evolution*, Vol.7, No.3, (March 1992), pp. 151-154, ISSN 0169-5347.

Polis, G.A., & Strong, D.R. (1996). Food Web Complexity and Community Dynamics. *American Naturalist*, Vol.147, No.5, (May 1996), pp. 813–846, ISSN 0003-0147.

Powers, J.S., Treseder, K.K., & Lerdau, M.T. (2005). Fine Roots, Arbuscular Mycorrhizal Hyphae and Soil Nutrients in Four Neotropical Rain Forests: Patterns Across Large Geographic Distances. *New Phytologist*, Vol.165, No.3 (March 2005), pp. 913-921, ISSN 1469-8137.

Prugh, L.R., Stoner, C.J., Epps, C.W., Bean, W.T., Ripple, W.J., Laliberte, A.S., & Brashares, J.S. (2009). The Rise of the Mesopredator. *BioScience*, Vol.59, No.9, (October 2009), pp. 779-791, ISSN 0006-3568.

Human Altered Mesoherbivore Densities and Cascading Effects on Plant and Animal Communities in
Fragmented Tropical Forests

211

Purvis, A., Gittleman, J.L., Cowlishaw, G., & Mace, G.M. (2000). Predicting Extinction Risk in Declining Species. *Proceedings of the Royal Society B: Biological Sciences*, Vol.267, No.1451, (July 2000), pp. 1947-1952, ISSN 1471-2954.

Rabinowitz, A.R., & Nottingham, B.G. Jr. (1986). Ecology and Behaviour of the Jaguar (*Panthera onca*) in Belize, Central America. *Journal of Zoology*, Vol.210, No.1, (September 1986), pp. 149-159, ISSN 0952-8369.

Ray, J., Redford, K.H., Steneck, R., Berger, J. (Eds.). (2005). *Large Carnivores and the Conservation of Biodiversity*. Island Press, ISBN 978-1559630801, Washington, DC.

Redford, K. (1992). The Empty Forest. *BioScience*, Vol.42, No.6, (June 1992), pp. 412–422, ISSN 0006-3568.

Restrepo, C., Gomez, N. & Heredia, S. (1999). Anthropogenic Edges, Treefall Gaps, and Fruit–Frugivore Interactions in a Neotropical Montane Forest. *Ecology*, Vol.80, No.2, (March 1999), pp. 668–685, ISSN 0012-9658.

Riitters, K., Wickham, J., O'Neill, R., Jones, B., & Smith, E. (2000). Global-Scale Patterns of Forest Fragmentation. *Conservation Ecology*, Vol.4, No.2, (September 2000), art. 3. Available from: <http://www.consecol.org/vol4/iss2/art3>.

Ripple, W.J., & Larsen, E.J. (2000). Historic Aspen Recruitment, Elk, and Wolves in Northern Yellowstone National Park, USA. *Biological Conservation*, Vol.95, No.3, (October 2000), pp. 361–370, ISSN 0006-3207.

Roberts, E.G., Flannery, T.F., Ayliffe, L.K., Yoshida, G.H., Olley, J.M., Prideaux, G.J., Laslett, G.M., Baynes, A., Smith, M.A., Jones, R., & Smith, B.L. (2001). New Ages for the Last Australian Megafauna: Continent-Wide Extinction About 46,000 Years Ago. *Science*, Vol.292, No.5523, (June 2001), pp. 1888–1892, ISSN 0036-8075.

Salo, P., Banks, P.B., Dickman, C.R., & Korpimäki, E. (2010). Predator Manipulation Experiments: Impacts on Populations of Terrestrial Vertebrate Prey. *Ecological Monographs*, Vol.80, No.4, (November 2010), pp. 531-546, ISSN 0012-9615.

Schmitz, O.J., Hawlena, D., & Trussell, G.C. (2010). Predator Control of Ecosystem Nutrient Dynamics. *Ecology Letters*, Vol.13, No.10, (October 2010), pp. 1199-1209, ISSN 1461-0248.

Shurin, J.B., Borer, E.T., Seabloom, E.W., Anderson, K., Blanchette, C.A., Broitman, B., Cooper, S.D., & Halpern, B.S. (2002). A Cross-Ecosystem Comparison of the Strength of Trophic Cascades. *Ecology Letters*, Vol.5, No.6, (November 2002), pp. 785–791, ISSN 1461-0248.

Sieving, K.E. (1992). Nest Predation and Differential Insular Extinction Among Selected Forest Birds of Panama. *Ecology*, Vol.73, No.6, (December 1992), pp. 2310-2328, ISSN 0012-9658.

Sigel, B.J., Robinson, W.D., & Sherry, T.W. (2010). Comparing Bird Community Responses to Forest Fragmentation in Two Lowland Central American Reserves. *Biological Conservation*, Vol.143, No.2, (February 2010), pp. 340-350, ISSN 0006-3207.

Sigel, B.J., Sherry, T.W., & Young, B.E. (2006). Avian Community Response to Lowland Tropical Rainforest Isolation: 40 Years of Change at La Selva Biological Station, Costa Rica. *Biotropica*, Vol.20, No.1, (January 2006), pp. 111-121, ISSN 0006-3606.

Skole, D.S., & Tucker, C.J. (1993). Tropical Deforestation and Habitat Fragmentation in the Amazon: Satellite Data from 1978 to 1988. *Science*, Vol.260, No.5116, (June 1993), pp. 1905–1910, ISSN 0036-8075.

Sodhi, N.S., Posa, M.R.C, Lee, T.M., & Warkentin, I.G. (2008). Effects of Disturbance or Loss of Tropical Rainforest on Birds. *Auk*, Vol.125, No.3, (July 2008), pp. 511-519, ISSN 0004-8038.

Sodhi, N.S., Koh, L.P., Clements, R., Wanger, T.C., Hill, J.K., Hamer, K.C., Clough, Y., Tscharntke, T., Posa, M.R.C., & Lee, T.M. (2010b). Conserving Southeast Asian Forest Biodiversity in Human-Modified Landscapes. *Biological Conservation*, Vol.143, No.10, (October 2010), pp. 2375-2384, ISSN 0006-3207.

Sodhi, N.S., Posa, M.R.C., Lee, T.M., Bickford, D., Koh, L.P., & Brook, B.W. (2010a). The State and Conservation of Southeast Asian Biodiversity. *Biodiversity and Conservation*, Vol.19, No.2, (February 2010), pp. 317-328, ISSN 0960-3115.

Soulé, M.E. (2010). Conservation Relevance of Ecological Cascades, In: *Trophic Cascades: Predators, Prey, and the Changing Dynamics of Nature*, Terborgh, J., & J.A. Estes, (Eds.), pp. 337-352, Island Press, ISBN 978-1597264877, Washington, DC.

Soulé, M.E., Bolger, D.T., Alberts, A.C., Wright, J., Sorice, S., & Hill, S. (1988). Reconstructing Dynamics of Rapid Extinctions of Chaparral-Requiring Birds in Urban Habitat Islands. *Conservation Biology*, Vol.2, No.1, (March 1988), pp. 75–92, ISSN 0888-8892.

Stork, N.E., Coddington, J.A., Colwell, R.K., Chazdon, R.L., Dick, C.W., Peres, C.A., Sloan, S., & Wills, K. (2009). Vulnerability and Resilience of Tropical Forest Species to Land-Use Change. *Conservation Biology*, Vol.23, No.6, (December 2009), pp. 1438-1447, ISSN 0888-8892.

Stratford, J.A., & Robinson, W.D. (2005). Gulliver Travels to the Fragmented Tropics: Geographic Variation in Mechanisms of Avian Extinction. *Frontiers in Ecology and the Environment*, Vol.3, No.2, (March 2005), pp. 91-98, ISSN 1540-9295.

Strong, D.R. (1992). Are Trophic Cascades All Wet? Differentiation and Donor-Control in Speciose Ecosystems. *Ecology*, Vol.73, No.3, (June 1992), pp. 747–754, ISSN 0012-9658.

Tabarelli, M., Lopes, A.V., & Peres, C.A. (2008). Edge-Effects Drive Tropical Forest Fragments Towards an Early-Successional System. *Biotropica*, Vol.40, No.6 (November 2008), pp. 657-661, ISSN 0006-3606.

Terborgh, J. (1992). Maintenance of diversity in tropical forests. *Biotropica*, Vol.24, No.2 (June 1992), pp. 283–292, ISSN 0006-3606.

Terborgh, J., & Estes, J.A. (2010). Preface, In: *Trophic Cascades: Predators, Prey, and the Changing Dynamics of Nature*, Terborgh, J., & J.A. Estes, (Eds.), pp. xiii-xx, Island Press, ISBN 978-1597264877, Washington, DC.

Terborgh, J., Estes, J.A., Paquet, P., Ralls, K., Boyd-Heigher, D., Miller, B.J., & Noss, R.F. (1999). The Role of Top Carnivores in Regulating Terrestrial Ecosystems, In: *Continental Conservation: Scientific Foundations of Regional Reserve Networks*, Soulé, M., & J. Terborgh, (Eds.), pp. 39–54, Island Press, ISBN 978-1559636988, Washington, DC.

Terborgh, J. & Feeley, K. (2010). Propagation of Trophic Cascades via Multiple Pathways in Tropical Forests, In: *Trophic Cascades: Predators, Prey, and the Changing Dynamics of Nature*, Terborgh, J., & J.A. Estes, (Eds.), pp. 125-140, Island Press, ISBN 978-1597264877, Washington, DC.

Terborgh, J., Feeley, K. Silman, M., Nuñez, P., & Balukjan, B. (2006). Vegetation Dynamics on Predator-Free Land-Bridge Islands. *Journal of Ecology*, Vol.94, No.2, (March 2006), pp. 253–263, ISSN 1365-2745.

Terborgh, J., Lopez, L., Nuñez, P., Rao, M., Shahabuddin, B., Orihuela, G., Riveros, M., Ascanio, R., Adler, G.H., Lambert, T.D., & Balbas, L. (2001). Ecological Meltdown in Predator-Free Forest Fragments. *Science*, Vol.294, No.5548, (November 2001), pp. 1923-1926, ISSN 0036-8075.

Terborgh, J., Lopez, L., & Tello, J. (1997). Bird Communities in Transition: The Lago Guri Islands. *Ecology*, Vol.78, No.5, (July 1997), pp. 1494–1501, ISSN 0012-9658.

Terborgh, J., Nuñez-Ituri, G., Pitman, N., Cornejo, F., Alvarez, P., Pringle, B., Swamy, V., & Paine, T. (2008). Tree Recruitment in an "Empty" Forest. *Ecology*, Vol.89, No.6, (June 2008), pp. 1757–1768, ISSN 0012-9658.

Terborgh, J., & Wright, S.J. (1994). Effects of Mammalian Herbivores on Plant Recruitment in Two Neotropical Forests. *Ecology*, Vol.75, No.6, (September 1994), pp. 1829-1833, ISSN 0012-9658.

Tilman, D., May, R.M., Lehman, C.L., & Nowak, M.A. (1994). Habitat Destruction and the Extinction Debt. *Nature*, Vol.371, No.1, (September 1994), pp. 65-66, ISSN 0028-0836.

Torrealba-Suárez, I.M., & Rau, J.M. (1994). Biology of collared peccary groups *Tayassu tajacu* and their damage provocated in crops neighboring La Selva Biological Station, Costa Rica: Final report 1993. (Unpubl.) Universidad Nacional de Costa Rica, Programa Regional Manejo de Vida Silvestre.

Transparency International. (2010). Corruption Perceptions Index 2010, 01.08.2011, Available from:<http://www.transparency.org/policy_research/surveys_indices/cpi/2010>

Van Bael, S.A., Brawn, J.D., & Robinson, S.K. (2003). Birds Defend Trees from Herbivores in a Neotropical Forest Canopy. *Proceedings of the National Academy of the Sciences of the USA*, Vol.100, No.14, (July 2008), pp. 8304-8307, ISSN 1091-6490.

Van Bael, S.A., Philpott, S.M., Greenberg, R., Bichier, P., Barber, N.A., Mooney, K.A., & Gruner, D.S. (2008). Birds as Predators in Tropical Agroforestry Systems. *Ecology*, Vol.89, No.4, (April 2008), pp. 928-934, ISSN 0012-9658.

Vance-Chalcraft, H.D., Rosenheim, J.A., Vonesh, J.R., Osenberg, C.W., & Sih, A. (2007). The Influence of Intraguild Predation on Prey Suppression and Prey Release: A Meta-Analysis. *Ecology*, Vol.88, No.11, (November 2007), pp. 2689-2696, ISSN 0012-9658.

Wang, S., Curtis, P.D., & Lassoie, J.P. (2006). Farmer Perceptions of Crop Damage by Wildlife in Jigme Singye Wangchuck National Park, Bhutan. *Wildlife Society Bulletin*, Vol.34, No.2, (June 2006), pp. 369-365, ISSN 0091-7648.

Wangchuk, T. (2004). Predator–Prey Dynamics: The Role of Predators in the Control of Problem Species. *Journal of Bhutan Studies*, Vol.10, No.1, (Summer 2004), pp. 68–89, ISSN 1608-411X.

Wilkie, D.S., Bennett, E.L., Peres, C.A., & Cunningham, A.A. (2011). The Empty Forest Revisited. *Annals of the New York Academy of Sciences, The Year in Ecology and Conservation Biology*, Vol.1223, No.1, (March 2011), pp. 120-128, ISSN 0077-8923.

Williams-Guillén, K., Perfecto, I., & Vandermeer, J. (2008). Bats Limit Insects in a Neotropical Agroforestry System. *Science*, Vol.320, No.5872, (April 2008), p. 70, ISSN 0036-8075.

Woodroffe, R., & Ginsberg, J.R. (1998). Edge Effects and Extinction of Populations Inside Protected Areas. *Science*, Vol.280, No.5372, (June 1998), pp. 2126–2128, ISSN 0036-8075.

Wright, S.J. (2005). Tropical Forests in a Changing Environment. *Trends in Ecology and Evolution*, Vol.20, No.10, (October 2005), pp. 553-560, ISSN 0169-5347.

Wright, S.J. (2010). The Future of Tropical Forests. *Annals of the New York Academy of Sciences, The Year in Ecology and Conservation Biology*, Vol.1195, No.1, (March 2010), pp. 1-27, ISSN 0077-8923.

Wright, S.J., & Duber, H.C. (2001). Poachers and Forest Fragmentation Alter Seed Dispersal, Seed Survival, and Seedling Recruitment in the Palm *Attalea butyraceae*, with Implications for Tropical Tree Diversity. *Biotropica*, Vol.33, No.4, (December 2001), pp. 583–595, ISSN 0006-3606.

Wright, S.J., Gompper, M.E., & DeLeon, B. (1994). Are Large Predators Keystone Species in Neotropical Forests? The Evidence from Barro Colorado Island. *Oikos*, Vol.71, No.2, (November 1994), pp. 279–294, ISSN 0030-1299.

Wright, S.J., Hernandez, A., & Condit, R. (2007b). The Bush Meat Harvest Alters Seedling Banks by Favoring Lianas, Large Seeds, and Seeds Dispersed by Bats, Birds, and Wind. *Biotropica*, Vol.39, No.3, (May 2007), pp. 363–371, ISSN 0006-3606.

Wright, S.J., & Muller-Landau, H.C. (2006a). The Future of Tropical Forest Species. *Biotropica*, Vol.38, No.3, (May 2006), pp. 287–301, ISSN 0006-3606.

Wright, S.J., & Muller-Landau, H.C. (2006b). The Uncertain Future of Tropical Forest Species. *Biotropica*, Vol.38, No.4, (July 2006), pp. 443-445, ISSN 0006-3606.

Wright, S.J., Sanchez-Azofeifa, G.A., Portillo-Quintero, C., Davies, D. (2007c). Poverty and Corruption Compromise Tropical Forest Reserves. *Ecological Applications*, Vol.17, No.5, pp. 1259-1266, ISSN 1051-0761.

Wright, S.J., Stoner, K.E., Beckman, N., Corlett, R.T., Dirzo, R., Muller-Landau, H.C., Nuñez-Iturri, G., Peres, C.A., & Wang, B.C. (2007a). The Plight of Large Animals in Tropical Forests and the Consequences for Plant Regeneration. *Biotropica*, Vol.39, No.3, (May 2007), pp. 289-291, ISSN 0006-3606.

Wright, S.J., Zeballos, H., Domínguez, I., Gallardo, M.M., Moreno, M.C., & Ibáñez, R. (2000). Poachers Alter Mammal Abundance, Seed Dispersal and Seed Predation in a Neotropical Forest. *Conservation Biology*, Vol.14, No.1, (February 2000), pp. 227–239, ISSN 0888-8892.

Yaacobi, G., Ziv, Y., & Rosenzweig, M.L. (2007). Habitat Fragmentation May Not Matter to Diversity. *Proceedings of the Royal Society B: Biological Sciences*, Vol.274, No.1624 (October 2007), pp. 2409-2412, ISSN 1471-2954.

12

A Review of Above Ground Necromass in Tropical Forests

Michael Palace[1,*], Michael Keller[1,2], George Hurtt[3] and Steve Frolking[1]

[1]*Complex Systems Research Center, Morse Hall,*
University of New Hampshire, Durham, NH,
[2]*International Institute of Tropical Forestry, USDA Forest Service, Rio Piedras, PR,*
[3]*Department of Geography, University of Maryland, College Park, MD,*
USA

1. Introduction

Tropical forests are marked by high biological diversity and complex vegetation dynamics that result in a spatially diverse array of forest stand structures (Richards 1952, Denslow 1987, Salati and Vose 1984, Terborgh 1992, Ozanne et al. 2003). Knowledge of the forest structure is vital for estimation of carbon stocks and fluxes (Houghton et al. 2000, 2001), habitat and faunal distributions (Schwarzkopt and Rylands 1989), and interactions between the biosphere and atmosphere (Keller et al. 2004a). With deforestation and land use change occurring throughout the tropics, improved understanding of these dynamic and complex forests are vital for the development of regional and global carbon budgets (Nobre et al. 1991, Werth and Avissar 2002, Houghton et al. 2001, Davidson and Artaxo 2004).

The death and subsequent decomposition of trees is an important component in the forest ecosystem carbon cycling and directly tied to the forest structure (Denslow, 1987; Harmon and Franklin, 1989). The dead portions of trees and branches, termed coarse woody debris (CWD) or above ground coarse necromass (here after termed necromass for this chapter), are an important component in the carbon cycle of forests accounting for 20-40% of carbon storage and 12% of the total above ground respiration (Harmon and Sexton 1996, Brown 1997, Palace et al. 2007). Necromass is also important in nutrient cycling and provides habitat for many organisms (MacNally et al. 2001, Norden and Paltto, 2001).

The dynamics of necromass production and loss through disturbance and decay are poorly understood and quantified in tropical forests (Martius and Banderia 1998, Eaton and Lawrence 2006). The slow process of decomposition is dependent upon the chemical and structural complexity of wood as well as the influence of a multitude of organisms involved with decomposition. Decomposition rates depend upon physical climate properties that vary over time. The production of necromass through the death of whole trees or portions of trees is episodic ranging greatly over temporal and spatial scales (Wessman 1992). This range in scale makes necromass measurement difficult, requiring large plots or long

---

* Corresponding Author

transects to catch rare large tree falls, as well as long periods of study to estimate both necromass production and decomposition (Harmon et al. 1986, Palace et al. 2008).

In this paper, we review necromass studies conducted in tropical forested ecosystems. We describe and define important terms and components in necromass research. In conjunction with this discussion, we examined various methodologies designed to measure these components and current literature involved with field based estimates of necromass. A simple model was developed to examine pool and decay estimates throughout these forested regions where literature estimates were unavailable. General relationships between necromass components were explored such as proportion of necromass to biomass and fallen to standing dead necromass.

2. Methods

We reviewed literature that dealt with field measurements of above ground coarse woody necromass stocks, production of dead wood, and decomposition of necromass. We gathered sources on necromass, with a focus on tropical forests through library searches, references cited in seminal ecological articles, peer suggestion, Web of Science© (http:/apps.isiknowledge.com) and an Yahoo© newsgroup focusing on dead wood (http:// groups.yahoo.com/group/dead_wood). We avoided the abundant studies focused on fine litter dynamics or soil respiration, although these aspects of carbon cycling would be important for comprehensive review and site comparison of carbon budgets. This study did not examine remote sensing or modeling literature with regard to necromass, although these two approaches may provide fruitful means for estimation and understanding of necromass production and cycling (Frolking et al. 2009).

More than 100 papers were examined for necromass stock, production and decay information and field estimated values. Data relevant to tropical forests is presented in Table 1 and 2. Data in Table 1 presents measured necromass components and information about the site location. Table 2 includes site information along with measured and estimated values of production and decomposition rates. Methodology is presented in a discussion about each necromass component. We recorded stocks of necromass, production of dead wood, and decomposition rates when available. We also attempted to gather biomass estimates from other papers at a site when biomass was not presented originally in the necromass literature and in some cases we were able to contact authors directly for biomass information (personal communications, Simon Grove and Michael Liddell, Keller et al. 2001, Asner et al. 2002, Baker et al. 2004). In each of the following three sections, necromass stocks, production of necromass, and decomposition of necromass, we review the component, present methodologies, and review the literature pertaining to the tropical forests.

2.1 Coarse woody debris stocks

Necromass is defined as the mass of all dead material and usually in reference to dead plant material. Necromass stocks aboveground include fine litter and coarse woody debris (CWD) where CWD has generally been defined as necromass with a diameter greater than 2 cm (Harmon et al. 1986). Necromass, is often divided into two categories: (1) fallen or downed necromass, and (2) standing dead wood (snags) (Harmon et al., 1986). For this review we will use the term necromass to refer to coarse woody debris. Similar stocks for coarse and fine material are found below ground. Below ground necromass is not treated in this review, but can be a significant portion of total dead material in a forested and savanna ecotone. Necromass stocks or pools have further been divided into groups dependent upon whether the material has fallen to the ground or whether it remains standing. The diameter of necromass and the degree of decomposition (decay class) have been used to further refine necromass categories (Harmon et al. 1986).

Measurement of fallen necromass is done primarily by one of two methods, line intercept or plot sampling (Harmon et al. 1986). Another method, relascope sampling (Gove et al. 2002) has not been used in tropical field studies and is not discussed here. Line intercept sampling (also termed planar intercept sampling) uses a straight line where all pieces of necromass that intersect the line are measured (Image 1).

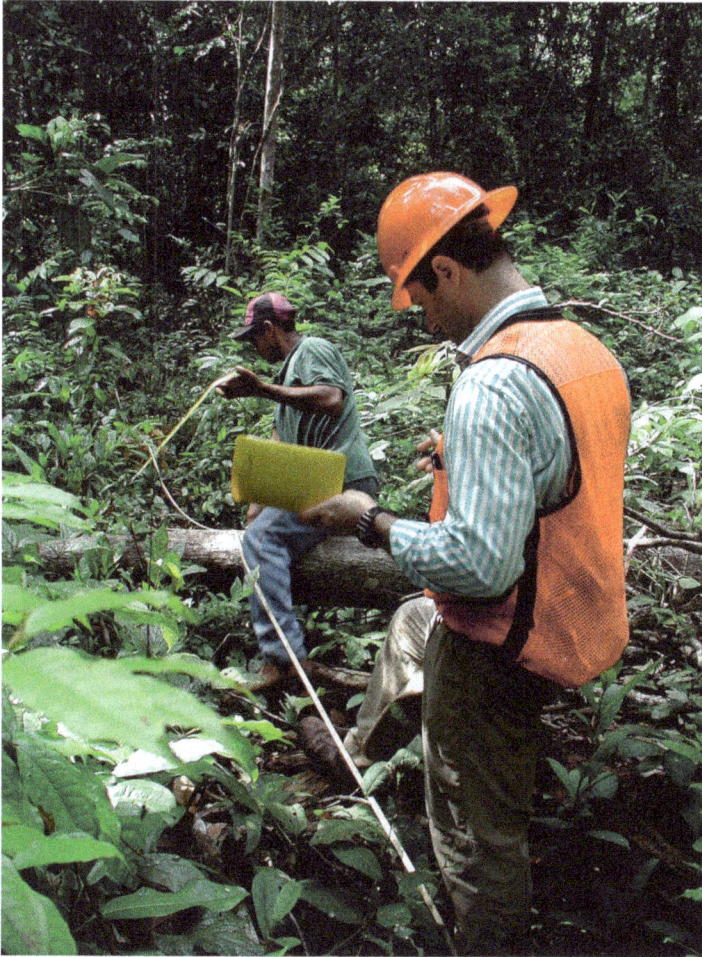

Image 1. Line intercept sampling in a selectively logged forest in the Tapajos, National Forest, Brazil.

Volume (V) (m^3 ha^{-1}) of necromass for an individual transect is calculated using the following equation:

$$V = \frac{\pi^2 \Sigma \, (d_n)^2}{(8 * L)}$$

where d_n is the diameter of a piece of necromass at the line intercept and L is the length of the transect used in sampling (De Vries, 1986).

In plot based sampling a fixed area is determined and all pieces of necromass are measured in that area. Plot measurements of necromass require more work, but retain spatial information that can be compared with other biometric or other environmental variables. Plot estimates of fallen and standing necromass use a variety of methods to estimate the volume. These include the assumption that a piece of necromass is a cylinder, a frustum, or the use of multiple measurements along the length of the log to calculate the volume. Taper functions have been used to calculate the volume of fallen and standing dead (Rice et al. 2004; Palace et al. 2007, 2008).

Fallen necromass has been divided into diameter size classes. Depending on the sampling methodology, diameter can have different meanings. In line intercept sampling, the fallen necromass diameter is only measured at the point in which the two-dimensional plane is intersected by the piece of necromass (Brown 1974). For plot level sampling, diameter often refers to the average diameter of the entire log, along which multiple diameters have been measured (Harmon et al. 1986). Small diameter fallen necromass is often grouped for to tallies to save labor (Brown 1974). Many studies have used a diameter of 2 cm as a low-end cutoff for sampling although there are a few exceptions.

Plot and line intercept sampling provide measurements of volume. An exception to this is when all pieces of necromass are weighed in a plot. Five studies in our review, published prior to 1980, weighed all pieces of necromass (Table 1). In order to quantify necromass from volume estimates, measurements of the densities of necromass pieces are required. More highly decayed logs theoretically should have less mass (Harmon et al. 1995). A common approach to quantification of mass is the stratification of necromass into decay classes and the application of decay-class-wide densities to the volume quantified by decay class. Other approaches to the estimation of necromass density include application of average density of live trees (Gerwing 2002, Nascimento and Laurence 2002), application of guesses (Gerwing 2002), and use of values from other sites (Rice et al. 2004). One study did not mention how mass was derived from volume estimates (Uhl and Kauffman 1990). Another used measured values for classes of necromass, such as trunks, prop roots, branches, and twigs (Robertson and Daniel 1989).

Decay classes are easily determined by the data collector and allow for a stratification of necromass sampling. Densities are measured for a sample of all coarse dead wood. Measurement approaches for density include weighing entire pieces of coarse dead wood, disks cut out of a log, and smaller plugs or samples across a cut disk (Harmon et al. 1986, Chambers et al. 2000, Keller et al. 2004b) (Image 2 and 3). In all cases, samples must be dried to a constant weight. Keller et al. (2004b) used a constant 60 degrees C. Large void spaces, created by organisms like termites or beetles, are often not considered in necromass density estimates. Larger samples used in density estimate include these void spaces, but smaller samples need to account for this. Keller et al. (2004b) used digitized images of disks cut through large pieces of fallen necromass in order to measure void spaces adjust density estimates accordingly (Image 4). This methods has been adopted by other studies (Palace et al. 2007, Baker et al. 2007, Chao et al. 2008)

Reference	Location	Type of Site	Forest Type	Total Necromass (Mg ha-1)	AG Biomass (Mg ha-1)	Fallen CWD (Mg ha-1)	Standing Dead (Mg ha-1)	Necro Biomass Ratio
Baker et al. 2007, Baker et al. 2004	Southern Peru	UF	TRF	17.7	234.1	13.5	4.2	
Bernhard-Reversat et al. 1978	Yapo, Ivory Coast	LD	TRF	3.6	429.4			0.01
Bernhard-Reversat et al. 1978	Banco, Ivory Coast	UF	TRF	3.8	538.3			0.01
Brown et al.,1992	Acre, Brazil	2nd	TMF	13.0	95.0			0.14
Brown et al.,1992	Acre, Brazil	UF	TMF	43.0	320.0			0.13
Brown et al. 1995	Rondonia, Brazil	UF	TMF	40.0	285.0			0.14
Buxton 1981	Tsavo National Park, Kenya	2nd	TMF	3.2				
Carey et al. 1994	Venezuela	UF	TMF		326.0			0.00
Carey et al. 1994	Venezuela	UF	LMMF		397.0			0.00
Chambers et al. 2000, 2001	Manaus, Brazil	UF	LEF	21.0				
Chao et al. 2008	Northern Peru	UF-flooded	TRF	10.3	214.8			
Chao et al. 2008	Northern Peru	UF-white sand	TRF	45.8	226.1			
Chao et al. 2008	Northern Peru	UF-clay	TRF	30.9	261.4			
Chao et al. 2009a	NE Amazonia	UF	TMF	74.5	434.5	58.7	15.8	
Chao et al. 2009a	NE Amazonia	UF	TMF	62.8	229.7	54.1	8.7	
Chao et al. 2009a	NE Amazonia	UF	TMF	38.4	347.8	20.5	17.9	
Chao et al. 2009a	NW Amazonia	UF	TRF	31.4	254.7	27.1	4.3	
Chao et al. 2009a	NW Amazonia	UF	TRF	41.1	235.3	24.0	17.1	
Chao et al. 2009a	NW Amazonia	UF	TRF	21.5	251.0	17.9	3.6	
Chao et al. 2009a	NW Amazonia	UF	TRF	27.4	251.1	24.9	2.5	
Chao et al. 2009a	NW Amazonia	UF	TRF	25.5	260.0	17.8	7.7	
Chao et al. 2009a	NW Amazonia	UF	TRF	37.9	253.3	28.1	9.8	
Chao et al. 2009a	NW Amazonia	UF	TRF	15.4	263.0	12.4	3.0	
Chao et al. 2009a	NW Amazonia	UF	TRF	18.6	260.4	12.2	6.4	
Clark et al., 2002	La Selva, Costa Rica	UF	TRF	52.8	155.8	46.3	6.5	0.34
Cochrane et al. 1999	Tailandia, Para, Brazil	DF-first burn	TMF	50.0	220.0			0.23
Cochrane et al. 1999	Tailandia, Para, Brazil	DF-second burn	TMF	71.0	129.0			0.55
Cochrane et al. 1999	Tailandia, Para, Brazil	DF-third burn	TMF	116.0	47.0			2.47
Cochrane et al. 1999	Tailandia, Para, Brazil	UF	TMF	53.0	242.0			0.22

Reference	Location	Type of Site	Forest Type	Total Necromass (Mg ha^{-1})	AG Biomass (Mg ha^{-1})	Fallen CWD (Mg ha^{-1})	Standing Dead (Mg ha^{-1})	Necro Biomass Ratio
Collins 1981	Nigeria	DF-fires	S	2.5		1.4	1.1	
Cummings et al. 2002	Rondônia, Brazil	OF	DF	32.4	270.3			
Cummings et al. 2002	Rondônia, Brazil	UDF	TMF	30.5	337.8			
Cummings et al. 2002	Rondônia, Brazil	SE	SE	20.8	319.9			
Delaney et al. 1998	Venezuela	UF	TTW	2.4	13.1	1.0	1.4	0.18
Delaney et al. 1998	Venezuela	UF	TDF	4.8	141.2	1.9	2.9	0.03
Delaney et al. 1998	Venezuela	UF	TMDF	6.6	330.0	3.8	2.8	0.02
Delaney et al. 1998	Venezuela	UF	TMF	33.3	346.9	18.5	14.8	0.10
Delaney et al. 1998	Venezuela	UF	LMMF	42.3	341.1	21.0	21.3	0.12
Delaney et al. 1998	Venezuela	UF	MWF	34.5	325.5	8.2	26.3	0.11
Eaton & Lawrence 2006	Southern Mexico	cleared	DTF	51.6		51.6		
Eaton & Lawrence 2006	Southern Mexico	2nd-1-5 y	DTF	19.9		19.9		
Eaton & Lawrence 2006	Southern Mexico	2nd-6-12 y	DTF	11.4		11.4		
Eaton & Lawrence 2006	Southern Mexico	2nd-12-16 y	DTF	15.0		15.0		
Eaton & Lawrence 2006	Southern Mexico	Montana	DTF	37.5	147.0	37.5		0.25
Edwards & Grubb 1977	New Guinea	UF	LMMF	10.9	490.0			0.02
Gerwing 2002	Paragominas, Brazil	MD-L	TMF	76.0	245.0	68.0	8.0	0.31
Gerwing 2002	Paragominas, Brazil	HD-L	TMF	149.0	168.0	140.0	9.0	0.89
Gerwing 2002	Paragominas, Brazil	LD-fire	TMF	101.0	177.0	70.0	31.0	0.57
Gerwing 2002	Paragominas, Brazil	HD-fire	TMF	152.0	50.0	128.0	24.0	3.04
Gerwing 2002	Paragominas, Brazil	UF	TMF	55.0	309.0	33.0	22.0	0.18
Golley et al. 1973, 1969	Darien, Panama	UF	TMF	17.6	370.5			0.05
Golley et al. 1973, 1969	Darien, Panama	UF	TMF	6.2	263.5			0.02
Golley et al. 1973, 1969	Darien, Panama	UF	TMF	4.8	271.4			0.02
Golley et al. 1973, 1969	Darien, Panama	UF	TMF	19.1	176.6			0.11
Golley et al. 1973, 1969	Darien, Panama	UF	TMF	102.1	279.2			0.37
Golley et al. 1975	Darien, Panama	UF	TMF	14.6	377.8			0.04
Golley et al. 1975	Darien, Panama	UF	RF	4.9	284.1			0.02
Greenland & Kowal 1960	Kade, Ghana	2nd- 50 y	MTF	71.8	213.7			0.34

Reference	Location	Type of Site	Forest Type	Total Necromass (Mg ha⁻¹)	AG Biomass (Mg ha⁻¹)	Fallen CWD (Mg ha⁻¹)	Standing Dead (Mg ha⁻¹)	Necro Biomass Ratio
Greenland & Kowal 1960	Yangambi, Belgian Congo	2nd- 18 y	MTF	17.4	128.7			0.14
Grove 2001, biomass Liddell personal	Australia	Logged	TRF	7.3				
Grove 2001, biomass Liddell personal	Australia	2nd	TRF	5.1				
Grove 2001, biomass Liddell personal	Australia	UF	TRF	9.3	270.0			0.03
Harmon et al. 1995	Quintana Roo, Mexico	MD-H	TDF	48.5	112.0	45.7	2.8	0.43
Harmon et al. 1995	Quintana Roo, Mexico	MD-F	TDF	128.2	112.0	72.4	55.8	1.14
Harmon et al. 1995	Quintana Roo, Mexico	MD-H	TDF	60.8	133.0	47.3	13.5	0.46
Harmon et al. 1995	Quintana Roo, Mexico	MD-F	TDF	122.5	133.0	1.6	120.9	0.92
Harmon et al. 1995	Quintana Roo, Mexico	HD-H	TDF	28.0	94.0	21.2	6.8	0.30
Harmon et al. 1995	Quintana Roo, Mexico	HD-fire	TDF	118.6	94.0	25.8	92.8	1.26
Harmon et al. 1995	Quintana Roo, Mexico	UF	TDF	38.0	209.0	24.8	13.2	0.18
Hughes et al. 2000	Tuxtlas, Mexico	UF	TEF	14.0	382.5	14.0		0.04
Kauffman et al. 1988	Rio Negro, Venezuela	DF	TRF	48.5	47.5			1.02
Kauffman et al. 1988	Rio Negro, Venezuela	DF	TRF	26.9	80.1			0.34
Kauffman et al. 1988	Rio Negro, Venezuela	UF	TRF	1.6	11.4			0.14
Kauffman et al. 1988	Rio Negro, Venezuela	UF	TRF	5.3	38.7			0.14
Kauffman et al. 1988	Rio Negro, Venezuela	UF	TRF	42.4	20.6			2.06
Kauffman et al. 1988	Rio Negro, Venezuela	UF	TRF	12.9	51.1			0.25
Kauffman et al. 1988	Rio Negro, Venezuela	UF	TRF	14.4	238.6			0.06
Kauffman et al. 1988	Pernambuco, Brazil	2nd pre-fire	TDF		72.0	63.1		
Kauffman et al. 1988	Pernambuco, Brazil	2nd-post fire	TDF		72.0	11.9		
Keller et al. 2004b, Asner et al. 2002	Cauaxi, Brazil	RIL	TMF	89.6	203.0	74.7		0.44
Keller et al. 2004b, Asner et al. 2002	Cauaxi, Brazil	CL	TMF	129.4	203.0	107.8		0.64
Keller et al. 2004b, Asner et al. 2002	Cauaxi, Brazil	UF	TMF	66.2	203.0	55.2		0.33
Keller et al. 2004b, Keller et al. 2001	Tapajos, Brazil	RIL	TMF	91.4	282.0	76.2		0.32

Reference	Location	Type of Site	Forest Type	Total Necromass (Mg ha-1)	AG Biomass (Mg ha-1)	Fallen CWD (Mg ha-1)	Standing Dead (Mg ha-1)	Necro Biomass Ratio
Keller et al. 2004b, Keller et al. 2001	Tapajos, Brazil	UF	TMF	60.8	282.0	50.7		0.22
Kira 1978	West Malaysia	UF	TRF	50.9	421.3			0.12
Kira 1978	West Malaysia	UF	TRF		374.2			
Kira 1978	West Malaysia	UF	TRF		379.8			
Klinge 1973	Central Amazon, Brazil	UF	TRF	18.0	731.7	18.0		0.02
Lang & Knight 1979	Barro Colorado, Panama	UF	TRF					
Lang & Knight 1983	Panama	2nd-60 yr	TMF					
Martius 1997	Central Amazon, Brazil	UF	RF	11.4				
Martius & Bandeira 1998	Manaus, Brazil	UF	TMF	9.5		9.5		
Nascimento & Laurence 2002	Mauaus, Brazil	UF	TMF	31.0	356.2	24.8	6.14	0.09
Odum 1970	El Verde, Puerto Rica	UF	TRF					
Palace et al. 2007	Mato Grosso, Brazil	RIL	TMF	67.0	263.0	67.0	8.8	0.29
Palace et al. 2007	Mato Grosso, Brazil	UF	TMF	44.9	281.0	44.9	5.3	0.18
Palace et al. 2008, Keller et al. 2001, 2004b	Tapajos, Brazil	UF	TMF	50.7	282.0	50.7	7.7	0.21
Palace et al. 2008, Keller et al. 2001, 2004b	Tapajos, Brazil	RIL	TMF	72.6	282.0	72.6	12.9	0.30
Rice et al. 2004	Tapajos, Brazil	UF	TMF	70.0	298.1			0.23
Rice et al. 2004	Tapajos, Brazil	UF	TRF	99.6	298.1			0.33
Roberston & Daniel 1989	Australia	UF	Yman	1.8	387.0	1.0	0.8	0.00
Roberston & Daniel 1989	Australia	UF	Oman	14.3	465.0	4.8	9.4	0.03
Saldarriaga et al. 1988	Columbia and Venezeula	2nd-14 yr	TRF	15.2	60.8			0.25
Saldarriaga et al. 1988	Columbia and Venezeula	2nd- 20 yr	TRF	0.7	100.0			0.01
Saldarriaga et al. 1988	Columbia and Venezeula	2nd- 20 yr	TRF	1.6	73.4			0.02
Saldarriaga et al. 1988	Columbia and Venezeula	2nd- 20 yr	TRF	0.8	79.8			0.01
Saldarriaga et al. 1988	Columbia and Venezeula	2nd- 20 yr	TRF	1.0	116.0			0.01
Saldarriaga et al. 1988	Columbia and Venezeula	2nd- 20 yr	TRF	9.0	61.9			0.15
Saldarriaga et al. 1988	Columbia and Venezeula	2nd- 30 yr	TRF	8.5	130.0			0.07

Reference	Location	Type of Site	Forest Type	Total Necromass (Mg ha⁻¹)	AG Biomass (Mg ha⁻¹)	Fallen CWD (Mg ha⁻¹)	Standing Dead (Mg ha⁻¹)	Necro Biomass Ratio
Saldarriaga et al. 1988	Columbia and Venezeula	2nd- 35 yr	TRF	1.0	129.0			0.01
Saldarriaga et al. 1988	Columbia and Venezeula	2nd- 40 yr	TRF	5.7	186.0			0.03
Saldarriaga et al. 1988	Columbia and Venezeula	2nd- 60 yr	TRF	32.4	133.0			0.24
Saldarriaga et al. 1988	Columbia and Venezeula	2nd- 60 yr	TRF	2.7	244.0			0.01
Saldarriaga et al. 1988	Columbia and Venezeula	2nd- 60 yr	TRF	34.2	169.0			0.20
Saldarriaga et al. 1988	Columbia and Venezeula	2nd- 80 yr	TRF	8.2	160.0			0.05
Saldarriaga et al. 1988	Columbia and Venezeula	2nd- 80 yr	TRF	7.6	213.0			0.04
Saldarriaga et al. 1988	Columbia and Venezeula	2nd- 80 yr	TRF	40.4	173.0			0.23
Saldarriaga et al. 1988	Columbia and Venezeula	2nd- 80 yr	TRF	10.0	170.0			0.06
Saldarriaga et al. 1988	Columbia and Venezeula	UF	TRF	22.3	268.0			0.08
Saldarriaga et al. 1988	Columbia and Venezeula	UF	TRF	53.0	322.0			0.16
Saldarriaga et al. 1988	Columbia and Venezeula	UF	TRF	15.1	326.0			0.05
Saldarriaga et al. 1988	Columbia and Venezeula	UF	TRF	14.6	335.0			0.04
Scott et al. 1992	Roraima, Brazil	UF	LEF	5.1		4.1	1.0	
Summers 1998	Central Amazon, Brazil	DF	TMF	88.8				
Tanner 1980	Jamaica	UF	DTF	7.6	231.0			0.03
Tanner 1980	Jamaica	UF	DTF	12.0	338.0			0.04
Uhl & Kauffman 1990, Kauffman & Uhl 1990	Paragominas, Brazil	P	TMF	40.0		51.5		
Uhl & Kauffman 1990, Kauffman & Uhl 1990	Paragominas, Brazil	LF	TMF	178.8		178.8		
Uhl & Kauffman 1990, Kauffman & Uhl 1990	Paragominas, Brazil	2nd	TMF	23.0	28.0	27.7		0.82
Uhl & Kauffman 1990, Kauffman & Uhl 1990	Paragominas, Brazil	UF	TMF	51.0	250.0	55.6		0.20
Uhl et al. 1988	Paragominas, Brazil	LD-young	TMF	78.3	29.4			2.66
Uhl et al., 1988	Paragominas, Brazil	LD-old	TMF	87.8	86.1			1.02
Uhl et al. 1988	Paragominas, Brazil	LD-old	TMF	49.9	88.9			0.56

Reference	Location	Type of Site	Forest Type	Total Necromass (Mg ha⁻¹)	AG Biomass (Mg ha⁻¹)	Fallen CWD (Mg ha⁻¹)	Standing Dead (Mg ha⁻¹)	Necro Biomass Ratio
Uhl et al. 1988	Paragominas, Brazil	MD-young	TMF	16.3	5.8			2.81
Uhl et al. 1988	Paragominas, Brazil	MD-young	TMF	21.6	8.3			2.60
Uhl et al. 1988	Paragominas, Brazil	MD-young	TMF	47.4	16.8			2.82
Uhl et al. 1988	Paragominas, Brazil	MD-young	TMF	11.2	17.0			0.66
Uhl et al. 1988	Paragominas, Brazil	MD-old	TMF	20.3	37.0			0.55
Uhl et al. 1988	Paragominas, Brazil	MD-old	TMF	20.4	32.8			0.62
Uhl et al. 1988	Paragominas, Brazil	HD-young	TMF	10.4	7.6			1.37
Uhl et al. 1988	Paragominas, Brazil	HD-young	TMF	8.6	15.5			0.55
Uhl et al. 1988	Paragominas, Brazil	HD-old	TMF	5.7	4.7			1.21
Uhl et al. 1988	Paragominas, Brazil	UF	TMF	41.9	300.0			0.14
Uhl et al. 1988	Paragominas, Brazil	UF	TMF	35.6	327.6			0.11
Uhl et al. 1988	Paragominas, Brazil	UF	TMF	48.1	284.7			0.17
Wilcke et al. 2005	Andes, Ecuador	UF	MWF	9.1		7.6	1.5	
Yoda & Kira 1982	Pasoh, Malaysia	UF	TRF	46.7				
Yoneda et al. 1977	Pasoh, Malaysia	UF	TRF					
Yoneda et al. 1990	Sumatra, Indonesia	UF	TMF	55.0	408.0	39.0	16.0	0.13

Forest types are: TRF – tropical rain forest, TMF – tropical moist forest, LMMF – lower montane moist forest, S – Savanna, SE – Savanna Edge, TTW –tropical thorn woodland, TDF – tropical dry forest, TMDF – tropical moist dry forest, MWF – tropical montane wet forest, RF – riverine forest, YMan – young mangrove, Oman – old mangrove.
Type of site definitions: UF – undisturbed, LD – low disturbance, 2nd – secondary forest (type or year of regrowth), OF – open forest, UDF – undisturbed dense forest, SE – savanna edge, Montana – cleared, MD-L(H, F) – moderate disturbance from logging (hurricane, fire), HD-L(H) heavy disturbance from logging (hurricane, fire), CL – conventional selective logging, RIL – reduced impact logging, LD – light disturbance.
Blank spaces indicate no data available from source. AG – above ground biomass, CWD – coarse woody debris.

Table 1. Reviewed literature for tropical forest necromass stocks.

Decay classes are usually described in two or more categories (Harmon et al. 1986, Chambers et al. 2000). These decay classes range from newly fallen necromass to highly decayed material that can be broken apart by hand (Harmon et al., 1995; Keller et al., 2004b) (Image 5). For a good description of a five decay class description, please refer to Keller et al. (2004). Chao et al. (2008) compared a three and five decay class system and found comparable patterns between the two systems and concluded that a three class system is preferable because of the higher sample sizes obtained by grouping classes is statistically advantageous.

Image 2. Density measurements using a plug and tenon extractor attached to a hand-held power drill. Section of necromass sampled along a line intercept with section of fallen necromass being chain sawn into section for extraction.

Image 3. Multiple samples extracted from a section of fallen necromass using the tenon extractor.

Image 4. Example of void space in a section of fallen necromass.

Image 5. Example of the variation in decay classes, a sound piece of necromass to the left and a highly decayed and friable piece of necromass to the right.

Reference	Location	Type of Site	Forest Type	Production Measured (Mg ha-1 y-1)	Decay (k) Measured (y-1)	Production Est. (biomass *0.02) (Mg ha-1 y-1)	Estimated Decay (k) (y-1)	Production Est. (necromass *0.15) Mg ha-1 y-1
Baker et al. 2007, Baker et al. 2004	Southern Peru	UF	TRF	3.8	0.21	4.68	0.26	2.66
Bernhard-Reversat et al. 1978	Yapo, Ivory Coast	LD	TRF			8.59	2.39	0.54
Bernhard-Reversat et al. 1978	Banco, Ivory Coast	UF	TRF			10.77	2.83	0.57
Brown et al.,1992	Acre, Brazil	2nd	TMF			1.9	0.15	1.95
Brown et al.,1992	Acre, Brazil	UF	TMF			6.4	0.15	6.45
Brown et al. 1995	Rondonia, Brazil	UF	TMF			5.7	0.14	6
Buxton 1981	Tsavo National Park, Kenya	2nd	TMF	0.31	0.07			0.48
Carey et al. 1994	Venezuela	UF	TMF	8.8		6.52		
Carey et al. 1994	Venezuela	UF	LMMF	9.53		7.94		
Chambers et al. 2000, 2001	Manaus, Brazil	UF	LEF	4.2	0.13, 0.17			3.15
Chao et al. 2008	Northern Peru	UF-flooded	TRF			4.3	0.42	1.55
Chao et al. 2008	Northern Peru	UF-white sand	TRF			4.52	0.1	6.87
Chao et al. 2008	Northern Peru	UF-clay	TRF			5.23	0.17	4.64
Chao et al. 2009	NE Amazonia	UF	TMF	6.7		8.69	0.12	11.18
Chao et al. 2009	NE Amazonia	UF	TMF	8.6		4.59	0.07	9.42
Chao et al. 2009	NE Amazonia	UF	TMF	6.3		6.96	0.18	5.76
Chao et al. 2009	NW Amazonia	UF	TRF	2.5		5.09	0.16	4.71
Chao et al. 2009	NW Amazonia	UF	TRF	7.9		4.71	0.11	6.17
Chao et al. 2009	NW Amazonia	UF	TRF	5.9		5.02	0.23	3.23
Chao et al. 2009	NW Amazonia	UF	TRF	4.4		5.02	0.18	4.11
Chao et al. 2009	NW Amazonia	UF	TRF	5.5		5.2	0.2	3.83
Chao et al. 2009	NW Amazonia	UF	TRF	4.8		5.07	0.13	5.69
Chao et al. 2009	NW Amazonia	UF	TRF	4.8		5.26	0.34	2.31
Chao et al. 2009	NW Amazonia	UF	TRF	4.1		5.21	0.28	2.79
Clark et al., 2002	La Selva, Costa Rica	UF	TRF	4.9	0.09	3.12	0.06	7.92
Cochrane et al. 1999	Tailandia, Para, Brazil	DF-first burn	TMF			4.4	0.09	7.5

Reference	Location	Type of Site	Forest Type	Production Measured (Mg ha^{-1} y^{-1})	Decay (k) Measured (y^{-1})	Production Est. (biomass 0.02) (Mg ha^{-1} y^{-1})	Estimated *Decay (k) (y^{-1})	Production Est. (necromass * 0.15) Mg ha^{-1} y^{-1}
Cochrane et al. 1999	Tailandia, Para, Brazil	DF-second burn	TMF			2.58	0.04	10.65
Cochrane et al. 1999	Tailandia, Para, Brazil	DF-third burn	TMF			0.94	0.01	17.4
Cochrane et al. 1999	Tailandia, Para, Brazil	UF	TMF			4.84	0.09	7.95
Collins 1981	Nigeria	DF-fires	S	1.27	0.51			0.37
Cummings et al. 2002	Rondônia, Brazil	OF	DF			5.41	0.17	4.86
Cummings et al. 2002	Rondônia, Brazil	UDF	TMF			6.76	0.22	4.58
Cummings et al. 2002	Rondônia, Brazil	SE	SE			6.4	0.31	3.12
Delaney et al. 1998	Venezuela	UF	TTW		0.06	0.26	0.11	0.36
Delaney et al. 1998	Venezuela	UF	TDF		0.2	2.82	0.59	0.72
Delaney et al. 1998	Venezuela	UF	TMDF		0.52	6.6	1	0.99
Delaney et al. 1998	Venezuela	UF	TMF		0.03	6.94	0.21	5
Delaney et al. 1998	Venezuela	UF	LMMF		0.13	6.82	0.16	6.35
Delaney et al. 1998	Venezuela	UF	MWF		0.11	6.51	0.19	5.18
Eaton & Lawrence 2006	Southern Mexico	cleared	DTF	0.11	0.38			7.74
Eaton & Lawrence 2006	Southern Mexico	2nd-1-5 y	DTF	0.11	0.38			2.98
Eaton & Lawrence 2006	Southern Mexico	2nd-6-12 y	DTF	0.11	0.38			1.71
Eaton & Lawrence 2006	Southern Mexico	2nd-12-16 y	DTF	0.91	0.38			2.25
Eaton & Lawrence 2006	Southern Mexico	Montana	DTF		0.38	2.94	0.08	5.62
Edwards & Grubb 1977	New Guinea	UF	LMMF			9.8	0.9	1.64
Gerwing 2002	Paragominas, Brazil	MD-L	TMF			4.9	0.06	11.4
Gerwing 2002	Paragominas, Brazil	HD-L	TMF			3.36	0.02	22.35
Gerwing 2002	Paragominas, Brazil	LD-fire	TMF			3.54	0.04	15.15
Gerwing 2002	Paragominas, Brazil	HD-fire	TMF			1	0.01	22.8
Gerwing 2002	Paragominas, Brazil	UF	TMF			6.18	0.11	8.25
Golley et al. 1973, 1969	Darien, Panama	UF	TMF			7.41	0.42	2.64
Golley et al. 1973, 1969	Darien, Panama	UF	TMF			5.27	0.85	0.93
Golley et al. 1973, 1969	Darien, Panama	UF	TMF			5.43	1.13	0.72
Golley et al. 1973, 1969	Darien, Panama	UF	TMF			3.53	0.18	2.87
Golley et al. 1973, 1969	Darien, Panama	UF	TMF			5.58	0.05	15.32
Golley et al. 1975	Darien, Panama	UF	TMF			7.56	0.52	2.19
Golley et al. 1975	Darien, Panama	UF	RF			5.68	1.16	0.74
Greenland & Kowal 1960	Kade, Ghana	2nd- 50 y	MTF			4.27	0.06	10.77
Greenland & Kowal 1960	Yangambi, Belgian Congo	2nd- 18 y	MTF			2.57	0.15	2.61
Grove 2001, biomass Liddell personal								1.09
Grove 2001, biomass Liddell personal								0.76
Grove 2001, biomass Liddell personal						5.4	0.58	1.4
Harmon et al. 1995	Quintana Roo, Mexico	MD-H	TDF			2.24	0.05	7.28
Harmon et al. 1995	Quintana Roo, Mexico	MD-F	TDF			2.24	0.02	19.23
Harmon et al. 1995	Quintana Roo, Mexico	MD-H	TDF			2.66	0.04	9.12
Harmon et al. 1995	Quintana Roo, Mexico	MD-F	TDF			2.66	0.02	18.38
Harmon et al. 1995	Quintana Roo, Mexico	HD-H	TDF			1.88	0.07	4.2
Harmon et al. 1995	Quintana Roo, Mexico	HD-fire	TDF			1.88	0.02	17.79

Reference	Location	Type of Site	Forest Type	Production Measured (Mg ha⁻¹ y⁻¹)	Decay (k) Measured (y-1)	Production Est. (biomass *0.02) (Mg ha⁻¹ y⁻¹)	Estimated Decay (k) (y-1)	Production Est. (necromass * 0.15) Mg ha⁻¹ y⁻¹
Harmon et al. 1995	Quintana Roo, Mexico	UF	TDF			4.18	0.11	5.7
Hughes et al. 2000	Tuxtlas, Mexico	UF	TEF			7.65	0.55	2.1
Kauffman et al. 1988	Rio Negro, Venezuela	DF	TRF			0.95	0.02	7.28
Kauffman et al. 1988	Rio Negro, Venezuela	DF	TRF			1.6	0.06	4.04
Kauffman et al. 1988	Rio Negro, Venezuela	UF	TRF			0.23	0.14	0.24
Kauffman et al. 1988	Rio Negro, Venezuela	UF	TRF			0.77	0.15	0.8
Kauffman et al. 1988	Rio Negro, Venezuela	UF	TRF			0.41	0.01	6.36
Kauffman et al. 1988	Rio Negro, Venezuela	UF	TRF			1.02	0.08	1.94
Kauffman et al. 1988	Rio Negro, Venezuela	UF	TRF			4.77	0.33	2.16
Kauffman et al. 1988	Pernambuco, Brazil	2nd pre-fire	TDF			1.44		
Kauffman et al. 1988	Pernambuco, Brazil	2nd-post fire	TDF			1.44		
Keller et al. 2004b, Asner et al. 2002						4.06	0.05	13.45
Keller et al. 2004b, Asner et al. 2002						4.06	0.03	19.4
Keller et al. 2004b, Asner et al. 2002						4.06	0.06	9.94
Keller et al. 2004b, Keller et al. 2001						5.64	0.06	13.72
Keller et al. 2004b, Keller et al. 2001						5.64	0.09	9.13
Kira 1978	West Malaysia	UF	TRF	3.3	0.3	8.43	0.17	7.64
Kira 1978	West Malaysia	UF	TRF			7.48		
Kira 1978	West Malaysia	UF	TRF			7.6		
Klinge 1973	Central Amazon, Brazil	UF	TRF			14.63	0.81	2.7
Lang & Knight 1979	Barro Colorado, Panama	UF	TRF		0.46			
Lang & Knight 1983	Panama	2nd-60 yr	TMF					
Martius 1997	Central Amazon, Brazil	UF	RF	6	0.33			1.71
Martius & Bandeira 1998	Manaus, Brazil	UF	TMF					1.43
Nascimento & Laurence 2002	Mauaus, Brazil	UF	TMF			7.124	0.23	4.64
Odum 1970	El Verde, Puerto Rica	UF	TRF		0.11			
Palace et al. 2007	Mato Grosso, Brazil	RIL	TMF			5.26	0.07	10.05
Palace et al. 2007	Mato Grosso, Brazil	UF	TMF			5.62	0.11	6.74
Palace et al. 2008, Keller et al. 2001, 2004						5.64	0.1	7.61
Palace et al. 2008, Keller et al. 2001, 2004						5.64	0.07	10.89
Rice et al. 2004	Tapajos, Brazil	UF	TMF	5	0.17	5.96	0.09	10.5
Rice et al. 2004	Tapajos, Brazil	UF	TRF	5	0.12	5.96	0.06	14.94

Reference	Location	Type of Site	Forest Type	Production Measured (Mg ha⁻¹ y⁻¹)	Decay (k) Measured (y-1)	Production Est. (biomass *0.02) (Mg ha⁻¹ y⁻¹)	Estimated *Decay (k) (y-1)	Production Est. (necromass * 0.15) Mg ha⁻¹ y⁻¹
Roberston & Daniel 1989	Australia	UF	Yman	0.1	0.06	7.74	4.27	0.27
Roberston & Daniel 1989	Australia	UF	Oman	0.97	2	9.3	0.65	2.14
Saldarriaga et al. 1988	Columbia and Venezeula	2nd-14 yr	TRF			1.22	0.08	2.28
Saldarriaga et al. 1988	Columbia and Venezeula	2nd- 20 yr	TRF			2	2.86	0.11
Saldarriaga et al. 1988	Columbia and Venezeula	2nd- 20 yr	TRF			1.47	0.92	0.24
Saldarriaga et al. 1988	Columbia and Venezeula	2nd- 20 yr	TRF			1.6	2	0.12
Saldarriaga et al. 1988	Columbia and Venezeula	2nd- 20 yr	TRF			2.32	2.32	0.15
Saldarriaga et al. 1988	Columbia and Venezeula	2nd- 20 yr	TRF			1.24	0.14	0.15
Saldarriaga et al. 1988	Columbia and Venezeula	2nd- 30 yr	TRF			2.6	0.31	1.35
Saldarriaga et al. 1988	Columbia and Venezeula	2nd- 35 yr	TRF			2.58	2.58	1.28
Saldarriaga et al. 1988	Columbia and Venezeula	2nd- 40 yr	TRF			3.72	0.65	0.15
Saldarriaga et al. 1988	Columbia and Venezeula	2nd- 60 yr	TRF			2.66	0.08	0.86
Saldarriaga et al. 1988	Columbia and Venezeula	2nd- 60 yr	TRF			4.88	1.81	4.86
Saldarriaga et al. 1988	Columbia and Venezeula	2nd- 60 yr	TRF			3.38	0.1	0.41
Saldarriaga et al. 1988	Columbia and Venezeula	2nd- 80 yr	TRF			3.2	0.39	5.13
Saldarriaga et al. 1988	Columbia and Venezeula	2nd- 80 yr	TRF			4.26	0.56	1.23
Saldarriaga et al. 1988	Columbia and Venezeula	2nd- 80 yr	TRF			3.46	0.09	1.14
Saldarriaga et al. 1988	Columbia and Venezeula	2nd- 80 yr	TRF			3.4	0.34	6.06
Saldarriaga et al. 1988	Columbia and Venezeula	UF	TRF			5.36	0.24	1.5
Saldarriaga et al. 1988	Columbia and Venezeula	UF	TRF			6.44	0.12	3.35
Saldarriaga et al. 1988	Columbia and Venezeula	UF	TRF			6.52	0.43	7.95
Saldarriaga et al. 1988	Columbia and Venezeula	UF	TRF			6.7	0.46	2.27
Scott et al. 1992	Roraima, Brazil	UF	LEF					0.76
Summers 1998	Central Amazon, Brazil	DF	TMF					13.32
Tanner 1980	Jamaica	UF	DTF	2	0.26	4.62	0.61	1.14
Tanner 1980	Jamaica	UF	DTF	2	0.17	6.76	0.56	1.8
Uhl & Kauffman 1990, Kauffman & Uhl 1990								6
Uhl & Kauffman 1990, Kauffman & Uhl 1990								25.95
Uhl & Kauffman 1990, Kauffman & Uhl 1990						0.56	0.02	3.45

Reference	Location	Type of Site	Forest Type	Production Measured (Mg ha⁻¹ y⁻¹)	Decay (k) Measured (y-1)	Production Est. (biomass *0.02) (Mg ha⁻¹ y⁻¹)	Estimated Decay (k) (y-1)	Production Est. (necromass * 0.15) Mg ha⁻¹ y⁻¹
Uhl & Kauffman 1990, Kauffman & Uhl 1990						5	0.1	7.65
Uhl et al. 1988	Paragominas, Brazil	LD-young	TMF			0.59	0.01	11.75
Uhl et al., 1988	Paragominas, Brazil	LD-old	TMF			1.72	0.02	13.17
Uhl et al. 1988	Paragominas, Brazil	LD-old	TMF			1.78	0.04	7.49
Uhl et al. 1988	Paragominas, Brazil	MD-young	TMF			0.12	0.01	2.45
Uhl et al. 1988	Paragominas, Brazil	MD-young	TMF			0.17	0.01	3.24
Uhl et al. 1988	Paragominas, Brazil	MD-young	TMF			0.34	0.01	7.11
Uhl et al. 1988	Paragominas, Brazil	MD-young	TMF			0.34	0.03	1.68
Uhl et al. 1988	Paragominas, Brazil	MD-old	TMF			0.74	0.04	3.05
Uhl et al. 1988	Paragominas, Brazil	MD-old	TMF			0.66	0.03	3.06
Uhl et al. 1988	Paragominas, Brazil	HD-young	TMF			0.15	0.01	1.56
Uhl et al. 1988	Paragominas, Brazil	HD-young	TMF			0.31	0.04	1.29
Uhl et al. 1988	Paragominas, Brazil	HD-old	TMF			0.09	0.02	0.86
Uhl et al. 1988	Paragominas, Brazil	UF	TMF			6	0.14	6.28
Uhl et al. 1988	Paragominas, Brazil	UF	TMF			6.55	0.18	5.34
Uhl et al. 1988	Paragominas, Brazil	UF	TMF			5.69	0.12	7.22
Wilcke et al. 2005	Andes, Ecuador	UF	MWF	0.82	0.09			1.37
Yoda & Kira 1982	Pasoh, Malaysia	UF	TRF					7.01
Yoneda et al. 1977	Pasoh, Malaysia	UF	TRF		0.32			
Yoneda et al. 1990	Sumatra, Indonesia	UF	TMF	3.8	0.11	5.69	0.12	8.25

Forest types are: TRF – tropical rain forest, TMF – tropical moist forest, LMMF – lower montane moist forest, S – Savanna, SE – Savanna Edge, TTW –tropical thorn woodland, TDF – tropical dry forest, TMDF – tropical moist dry forest, MWF – tropical montane wet forest, RF – riverine forest, YMan – young mangrove, Oman – old mangrove.

Type of site definitions: UF – undisturbed, LD – low disturbance, 2nd – secondary forest (type or year of regrowth), OF – open forest, UDF – undisturbed dense forest, SE – savanna edge, Montana – cleared, MD-L(H, F) – moderate disturbance from logging (hurricane, fire), HD-L(H) heavy disturbance from logging (hurricane, fire), CL – conventional selective logging, RIL – reduced impact logging, LD – light disturbance.

Blank spaces indicate no data available from source

Table 2. Reviewed literature for tropical necromass production and decomposition rates compared with estimated production and decomposition rates.

Decay class estimates of density were used in 23 studies, five studies weighed all material, and 11 did not use decay class density estimates, but density site averages or were unclear as to their methodology. Fifteen studies reported their decay class density estimates in detail (Table 3). Decay classes and density measurements for such decay classes were similar across many studies (Table 3). Harmon et al. (1995), Eaton and Lawrence (2006), Keller et al. (2004b), and Palace et al. (2007), all used five decay classes in their studies. Although Palace et al. (2007) and Keller et al. (2004b) conducted field work in the same biome, moist tropical forest, Eaton and Lawrence (2006) worked in a dry tropical forest. Eaton and Lawrence (2006) found consistent decay class density estimates between biomes suggesting that the apparently arbitrary classification is robust. Clearly, site specific density measurements should be most accurate approach. However, Palace et al. (2007) suggested, that coarse dead wood density measurements may be applied across broad areas provided that the decay classes are defined uniformly across the sites.

Location and Reference	Forest Type	Group	Decay Groups or Classes						
Southern Peru									
Baker et al. 2007	TRF	Density Adjusted for Void	DC1 0.54	DC2 0.53	DC3 0.40	DC4 0.37	DC5 0.38		
Northern Peru									
Chao et al. 2008	TRF	Three Classes	DC1	DC2	DC3				
Chao et al. 2008	TRF	Plug Density	0.57	0.42	0.30				
Chao et al. 2008	TRF	Void Density	0.03	0.03	0.03				
Chao et al. 2008	TRF	Density Adjusted for Void	0.55	0.41	0.23				
Northern Peru									
Chao et al. 2008	TRF	Five Classes	DC1	DC2	DC3	DC4	DC5		
Chao et al. 2008	TRF	Plug Density	-	0.55	0.42	0.30	0.21		
Chao et al. 2008	TRF	Void Density	-	0.03	0.04	0.04	-		
Chao et al. 2008	TRF	Density Adjusted for Void	-	0.53	0.41	0.28	0.21		
Manaus, Brazil									
Chambers et al., 2000	LEF		Heavy Wood 0.85	Light Wood 0.45					
La Selva, Costa Rica									
Clark et al., 2002	TRF		Sound 0.45	Partial Decomp. 0.35	Full Decomp. 0.25				
Venezuela			Sound	Intermed.	Rotten	small 2.5-4.99	med 5-9.99		
Delaney et al., 1998	TTW		NW	NW	NW	0.29	NW		
Delaney et al., 1998	TDF		NW	0.59	0.40	0.29	0.70		
Delaney et al., 1998	TMDF		0.42	0.37	0.25	0.24	0.22		
Delaney et al., 1998	TMF		0.58	0.59	0.50	0.29	0.29		
Delaney et al., 1998	TLMF		0.52	0.39	0.31	0.20	0.12		
Delaney et al., 1998	TMF		NW	0.48	0.32	0.19	0.13		
Southern Mexico									
Eaton and Laurence 2006	DTF		DC1 None	DC2 0.74	DC3 0.78	DC4 0.62	DC5 0.36		
Australia									
Grove 2001	TRF	Yellow fibrous	DC1 0.29	DC2 0.23	DC3 0.18	DC4 0.11	DC5 None		
Grove 2001	TRF	Hard	0.35	0.36	0.38	0.35	None		
Grove 2001	TRF	Brown crumbly	0.29	0.23	0.20	0.16	None		
Grove 2001	TRF	Brown fibrous	0.29	0.23	0.20	0.23	None		
Grove 2001	TRF	Red Block	None	0.23	0.19	0.17	None		
Quintana Roo, Mexico									
Harmon et al., 1995	TDF	By Species	DC1	DC2	DC3	DC4	DC5		
		Used Range	0.25 - 0.81	0.19 - 0.84	0.06 - 0.81	0.49 - 0.64	0.22		
Australia			Mod. Decayed	Very Decayed	Extremely Decayed				
Roberston and Daniel 1989	Mangrove	Trunk	0.34	0.34	0.23				
Roberston and Daniel 1989		Prop Roots	0.51	0.28	0.39				
Roberston and Daniel 1989		Branches	0.60	0.43	0.28				
Roberston and Daniel 1989		Twigs	0.63	0.41	0.35				
Reserva Ducke, Manaus, Brazil			Fine Wood	Coarse Wood					
Martius and Bandeira 1998	TMF	Fresh to Dry Mass	0.55	0.46					
Ecuadorian Andes, Ecuador			Intact	Rotten	Full Decomp.	Bark			
Wilcke et al. 2005	TMonF		0.38	0.22	0.25	0.37			
Tapajos, Brazil			DC1	DC2	DC3	DC4	DC5	Small 2-5 cm	Med. 5-10
Keller et al., 2004	TMF	Plug Density	0.61	0.71	0.63	0.58	0.46	0.36	0.45
Keller et al., 2004	TMF	Void Density	0.02	0.02	0.08	0.21	0.26	NA	NA
Keller et al., 2004	TMF	Density Adjusted for Void	0.60	0.70	0.58	0.45	0.34	0.36	0.45
Juruena, Brazil			DC1	DC2	DC3	DC4	DC5	Small 2-5 cm	Med. 5-10
Palace et al. 2007	TMF	Plug Density	0.72	0.70	0.66	0.67	0.44	0.52	0.50
Palace et al. 2007	TMF	Void Density	0.01	0.02	0.08	0.12	0.20	NA	NA
Palace et al. 2007	TMF	Density Adjusted for Void	0.71	0.69	0.60	0.59	0.33	0.52	0.50
Sumatra, Indonesia									
Yoneda et al. 1990	TMF	Density	DC0 0.69	DC1 0.58	DC2 0.56	DC3 0.41	DC4 0.28	DC5 0.15	

Table 3. Density and decay class estimates for tropical forest. Blank spaces indicate no data available from source.

In our review of the literature, we found a total of 49 papers on tropical necromass, with 24 papers reporting stock. All but five of the 24 papers used a volume sampling method, either plots (20 studies) or line intercepts (seven studies) or both methods (six studies). Fallen necromass was measured using line intercept sampling and standing dead was measured using plots in a few studies (Nascimento and Laurence 2002, Palace et al. 2007). One study used plots except for one area in which dense understory prohibited movement and line intercept sampling was used (Grove 2001). Baker et al. (2007) compared line intercept with plot sampling and found similar values for necromass volume and estimated the fallen mass. In our review, reported values of necromass stock were evenly distributed between disturbed and undisturbed sites. Many studies included both undisturbed and disturbed plots. Standing dead and fallen necromass were both measured in 21 articles, with the ratio of standing dead to total necromass ranging from 6% in a disturbed forest and 98% at a heavily disturbed site (Gerwing 2002, Harmon et al., 1995). In undisturbed forests, standing dead to total fallen necromass stock measurements ranged from 11% to 76% (Palace et al. 2007, Delaney et al., 1998). We do not present averages of necromass stock or other components because this would be misleading; the literature examined do not represent a statistical sample of the necromass or forest types found in the tropics.

Size class criteria differed slightly among studies. Of the 23 studies that reported stock estimates, the majority explained their size class methodology. A cutoff of less than 2 cm was used in seven of 23 studies that reported size class methodology. Six studies used a cutoff of 2.5 cm. Many used a cutoff greater than 10 cm. Six studies used a 10 cm cutoff to define the difference between small and large diameter necromass. Approximately half of all studies used a cutoff of 10 cm for standing dead measurement. We suggest standardization, with the use of three size classes, small diameter (2-5 cm), medium diameter (5-10 cm), and large diameter (greater than 10 cm) for fallen CWD.

Standing dead trees or snags include whole dead trees and portions of dead trees that remain upright (Harmon et al. 1986). In tropical forests, standing dead was measured 65% less frequently than fallen CWD. Many studies use a percentage of total fallen necromass to estimate standing dead necromass (Keller et al. 2004b). The size of standing dead included in tallies differs between studies. Palace et al. (2007) and many others have used a cutoff of 10 cm dbh, while others have measured standing dead down to 2 cm dbh (Edwards and Grubb 1997). The methodology of height measurement also varies among studies. Visual estimates or average heights (Rice et al. 2004) were used when standing dead heights are not measured. For more precise studies, measuring tapes and clinometers or laser range finders have been used (Palace et al. 2007). No studies in tropical forest that we examined included in their methodology specific mention of stumps or standing dead less than 1.3 m in height in estimates of standing dead other than Palace et al. (2007).

The stock of coarse woody debris contributes a large percentage to the total carbon pool in any forest. In tropical forests, fallen necromass was found to range from 1.0 to 178.8 Mg ha^{-1} (Table 1). In dry tropical forests, fallen necromass amounts tended to be lower than moist tropical forests, with dry forests ranging from 2.5 (Collins 1981) to 118.6 Mg ha^{-1} (Harmon et al., 1995) in a heavy logged area. In moist tropical forests necromass ranged from 2.4 (Delaney et al., 1998) to 178.8 Mg ha^{-1} (Uhl and Kauffman 1990, Kauffman and Uhl 1990) (Table 1). In tropical forest areas outside of the Brazilian Amazon researchers found necromass ranging from 3.8 to 6.0 Mg C ha^{-1} in montane forest in Jamaica (Tanner 1980), 22.4 Mg C ha^{-1} in wet forest in Costa Rica (Clark et al 2002) and 22.5 Mg C ha^{-1} in dipterocarp forests in Malaysia (Yoda and Kira 1982). In the Brazilian Amazon, where much recent work on necromass is concentrated, estimates of fallen necromass in undisturbed moist forests in terra firma included 42.8 Mg C ha^{-1} (Summers 1998) and 48.0 Mg C ha^{-1} (Rice et al. 2004 on the high end and 27.6 Mg C ha^{-1} (Keller et al. 2004), 15 Mg C ha^{-1} (Brown et al 1995), and 16.5 Mg C ha^{-1} (Gerwing 2002) on the low end. Other studies examined necromass in secondary forests and the effects of logging on necromass (Gerwing 2002, Uhl et al 1988, Keller et al. 2004b). The proportion of necromass to total above ground mass can be surprisingly high, 18 to 25% (Keller et al. 2004b; Rice et al. (2004) even in unmanaged forests. These values are for the Tapajos National Forest near Santarem, Brazil where Saleska et al. (2003) hypothesized that the 1997-1998 El Niño drought led to substantial mortality prior to the necromass measurements cited above.

2.2 Production of necromass

Death of whole trees or portions of trees creates necromass. Mechanisms that lead to tree death include forest disturbances at various scales. The spatial scale of disturbances ranges from branch-falls and small gaps to landscape level blowdowns due to microbursts that can cover thousands of hectares (Nelson et al. 1994). Tree mortality in tropical forest plots ranges from 0.001 to 0.07 per year (Carey et al. 1994, Phillips and Gentry 1994). Disturbance in tropical forests include individual tree processes, landscape level processes, and regional and climate influences. These processes and influences function on different temporal and spatial scales and are variable in the impact they have on tropical forests (Chambers et al. 2007, Frolking et al. 2009).

Tree mortality in tropical forests is driven on the individual tree level by competition, primarily for nutrients and light (Prance and Lovejoy 1985, Martinez-Ramos et al. 1988, Lieberman et al. 1989). As a tree dies and falls to the forest floor, a gap in the canopy is created (Denslow 1987). These gaps are important in an ecological sense because they are involved with tree regeneration dynamics and species diversity and distribution (Schemske and Browkaw 1981, Denslow 1987, Vitousek and Denslow 1986). Gaps increase light levels in understory, release nutrients, and create structural habitat for some species of flora, fauna, and fungi (Schemske and Browkaw 1981, Denslow 1987, Vitousek and Denslow 1986, Dickinson et al. 2000, Svenning 2000). Blackburn and Milton (1996) examined gap production and progressive enlargement of gaps as natural disturbances instead of catastrophic events. Young and Hubbell (1991) also found that trees were more likely to fall into gaps and suggested that gaps may be more persistent in tropical forests then previously thought. The persistence of gaps also predicts the locations where necromass is likely to collect. This spatial coincidence has not been tested.

Mortality of trees in the tropics is also influenced by fungi, insects and other animals, and the trees themselves (Denslow 1987). Branch fall as a source of necromass has rarely been quantified although it has been recognized as is a major disturbance for seedlings growing in the understory (Lang and Knight 1983, Aide 1987, Clark and Clark 1991, van der Meer and Bongers 1996). The diversity of trees in the mosaic that is a tropical forest landscape makes it rare for a single insect infestation to create denuded canopies and cause the death of many trees (Janzen 1987). Vines entangling adjacent crowns may cause the death of a single tree to result in tree falls that involve several neighboring trees (personal observation). Some species of *Ficus*, strangler figs, have constricting vines that eventually kill the host tree (Windsor et al. 1988). Epiphytic vegetation load has also been tied to tree mortality (Prance 1985). Trees can also die as a result of genetic programming as is the case for monocarpic trees such as *Tachigalia versicolor* (Kitajima and Augspurger, 1989).

On non-degraded moist and wet forests, fires are rare events that do not propagate easily (Prance 1985). However, this belief is being challenged with studies of forest drying during El Niño events (Nepstad et al. 2002). Fire has also been shown to be an influential disturbance on white-sand forests in the Amazon (Anderson 1981). Fire in the Amazon is strongly influenced by people (Cardoso et al. 2003). Lightning may also cause fires and localized mortality in tropical forests (Magnusson et al. 1996).

Disturbances are also influenced by weather and topography (Bellingham and Tanner 2000). Topography was found to be influential on disturbances and thus was reflected in the local species distributions (Gale 2000). Tropical trees tend to have shallow root mass for nutrient exploitation and buttresses for structural support and have been shown to easily topple (Prance 1985). Disturbances also include larger scale processes such as microbursts, blowdowns, volcanoes, and landslides in the tropics (Nelson et al. 1994, Sanford et al. 1986, Lawton and Putz 1988, Garwood et al. 1979). Hurricanes have been shown to have an influence on the tropical forests in the Caribbean and elsewhere (Lugo and Scatena 1996, Walker et al. 1996). Spatial patterns and recent trends in tree mortality have also been attributed to ENSO events (Condit et al. 1995, Malhi et al. 2004).

Approximately half of the studies we reviewed compared undisturbed forests with forests experiencing disturbance due to anthropogenic factors, such as selective logging. Selective logging is a practice that fells a few trees per hectare (Peireira et al. 2002). This type of logging has been shown to affect substantial areas in the Brazilian Amazon and in tropical Asia (Asner et al. 2005, Curran et al. 2004). Other human influenced disturbances in the literature of tropical necromass included fire, agriculture, conversion to pasture through deforestation, and repeated disturbances due to a combination of fire and agricultural practices (Table 1). The number of sites in our literature review was evenly distributed among the undisturbed and disturbed forests. We excluded a study by Feldpausch et al. (2005) because that study measured the amount of necromass generated by selective logging, but did not measure total necromass stocks in either logged or undisturbed forests.

Few studies have measured the production of necromass in tropical forests. Approaches to the estimation of necromass production include allocating a portion of Net Primary Productivity (NPP), a portion of existing biomass (Palace et al. 2007), or mortality estimates of trees (Rice et al. 2004, Baker et al. 2007). Flaws associated with these methods include the lack of variation over time, lack of spatial influence, lack of size class estimates, and a lack of knowledge of the proportion of necromass that remains standing and an assumption that the system is in a steady state.

Necromass production has been directly measured using repeated surveys on the same plots by marking necromass or removing it at each survey period (Harmon et al. 1986, Clark et al. 2002). Few necromass production studies conducted repeated surveys (Palace et al. 2008, Chao et al. 2008). This lack of repeated sampling limits the understanding of longer term influences of weather patterns such as El Niño or ability to assess if the forest is at a steady state or if it is recovering from a larger scale disturbance (e.g. Saleska et al. 2003).

Rare events such as blowdowns or even large tree falls complicate sampling design and interpretation of necromass data. For example, a tree fall of diameter 150 cm DBH drastically altered the measured flux of necromass in one sampling interval from a study at the Tapajos National Forest in Brazil (Palace et al. 2008). Trees of this size occur with a frequency of only about 0.079 per ha (Keller et al. 2001) at Tapajos. Assuming adequate sampling of 100 ha blocks, there are only 7.9 trees of this size class per block. Assuming an annual mortality rate of 1.7 (Rice et al. 2004) then the chance of seeing a fall of this size is $(1-0.983^{7.9})$ or 12.7% per year. In the study of Palace et al. (2008), a much larger sample area would have been required to record a large tree death annually. Larger but less frequent disturbances such as blowdowns (Nelson et al. 1994) require even more extensive sampling designs.

A compilation of studies that directly measured necromass production is present in Table 2. Of the 48 papers reviewed here only 30% made measurement or estimates of necromass production. Eaton and Lawrence (2006) measured production in several disturbed sites and in one undisturbed site in dry tropical forest in southern Mexico. Their estimate removed and measured new necromass four times over a two year period for an undisturbed forest was 0.91 Mg ha^{-1} yr^{-1}. Tanner (1980) estimated necromass production in a Jamaican forest to be 2.0 Mg ha^{-1} yr^{-1} using repeated samples over four years. Other estimates in dry tropical studies include 0.1 and 0.97 Mg ha^{-1} yr^{-1} conducted by Buxton (1981) and Collins (1981) respectively. Kira (1978) directly measured necromass production of 3.3 Mg ha^{-1} yr^{-1} in Pasoh Forest in western Malaysia. Clark et al. (2002) measured influx of necromass to be 4.8 Mg ha^{-1} yr^{-1} using a repeated survey in Costa Rica. In a 4.5 year study in the Brazilian Amazon, Palace et al. (2008), measured necromass production to be 6.7 Mg ha^{-1} yr^{-1}. Large size class necromass (>10 cm DBH) production was 4.7 Mg ha^{-1} yr^{-1}. The production of small size class necromass (< 2cm DBH) was 0.8 Mg ha^{-1} yr^{-1} and medium size class necromass ($\geq$2cm and $\leq$10 cm) was 1.2 Mg ha^{-1} yr^{-1}. Interestingly, Rice et al. (2004) estimated necromass production based on mortality of trees > 10 cm DBH at 4.8 Mg ha^{-1} yr^{-1} for a nearby forest area. This suggests that mortality based approaches will underestimate necromass production.

2.3 Decomposition of necromass

Decomposition of wood is generally a slow process that involves biological, chemical, and physical processes. The sequence that these processes act on dead wood varies over time due to changes in physical climate and the chemical and physical makeup of the wood over its decay life. Each piece of dead wood has a unique chemical and physical makeup (Kaarik 1974). The difference in chemical and physical composition starts with differences in live trees. Differences among trees depend on tree species (wood characteristics), nutrient composition of soil, climate, tree health (including infections by insects, microbes, and fungus), and how the tree died (Harmon et al. 1986, Martius 1997). Differences within trees may also be important due to internal variation in wood density (Noguiera et al. 2005).

Wood decomposition involves unique fauna (Dickinson and Pugh 1974). Decay organism can be grouped into three categories, bacteria, fungi, and macroorganisms (Dickinson and Pugh 1974). The presence of fauna and their own growth efficiencies, nutrient requirements, and temperature and moisture requirements dictate the overall decomposition process. Each of these categories of organisms acts on wood differently and are important at different time in the temporal sequence of wood decay (Kaarik 1974). In the tropics, wood fragmentation is primarily caused by termites (Buxton 1981). This fragmentation occurs on highly decayed logs or parts of logs. In addition, termites remove the wood to other places (Collins 1981).

The placement of the wood on the ground can dictate the rate of its decay. Logs on hills tend to accumulate more soil on the uphill side, creating a wetter microclimate beneficial for many decomposing soil organisms (Harmon et al. 1986). In the Brazilian *varzea* forest, (a flooded forest type), the season that the wood falls is influential on its immediate and longer term decomposition rate (Martius 1997). Decomposition of smaller litter occurs rapidly, often less than one year, while larger CWD can have a turnover time close to a century (Mackensen et al. 2003).

The estimation of necromass decomposition rates uses two major approaches, chronosequences and time series (Harmon and Sexton 1996). In a time series, individual pieces of wood are followed over time (Harmon and Sexton 1996). In chronosequence studies, varying ages of coarse dead wood are examined at a single point in time (Harmon et al. 1986). Dates of necromass production have been made using disturbance records, living stumps, seedlings, dendrochronology, fall scars, and bent trees (Harmon et al. 1999). Some researchers have conducted a combination of chronosequences and time series (Harmon and Sexton 1996, Chambers et al. 2000).

Within sample chronosequences or time series, decomposition may be studied by mass loss, density change, uniform substrate decomposition, radioisotopes, respiration rates, mineralization, enzyme activity, and selective inhibition experiments (Swift et al. 1979, Harmon and Sexton 1996, Harmon et al. 1999). The majority of studies in the tropics have used mass loss, density change, or chamber systems to measure the respiration.

Measurement of decomposition through mass loss requires multiple measurements of necromass over time (Buxton 1981, Harmon et al. 1995, Chambers et al. 2000). This can only be done accurately if moisture content can be measured accurately and non-destructively. Alternatively, changes in density can be used as a surrogate for mass loss. It is important for density measurements to account for void spaces. Void spaces in logs must be accounted for in density measurements either by using large pieces of necromass (e.g., Chambers et al., 2000, Clark et al., 2002) or by separately quantifying void space (Keller et al., 2004a, Palace et al. 2007). Direct measurements have also been made in the laboratory with CWD removed from the field (Richards 1952, Chambers 2001).

Respiration studies have been conducted on necromass in a number of ways. Essentially all methods depend upon isolating, sections, full pieces, or extracted samples of necromass in a chamber. Chambers may be attached to the surface of necromass or necromass pieces may be inserted into chambers. The chambers are sealed and CO_2 concentration is measured directly by infra-red detection (Chambers et al. 2001) or, in older studies, CO_2 emitted is absorbed in alkali (Swift et al. 1979; Marra and Edmonds 1994). Respiration will underestimate necromass loss because it does not account for dissolution and fragmentation. However, there are indications that for tropical moist forests, respiration is the major pathway for CO_2 loss. Chambers et al. (2004) estimated that 80% of mass loss in necromass resulted for respiration. This was done by using the ratio from a respiration study and a mass loss study (Chambers et al. 2000, 2001)

Substrate decomposition studies have also been conducted (Harmon et al. 1995). In this method, uniform substrates such as Popsicle sticks or wooden dowels are placed in the field and measured over time. These studies provide information on the temporal variability of decomposition and also provide a standard for comparison of decomposition rates across the sites.

Radioisotopes have been used as tagging agents for materials to estimate leaching and soil organic matter formation (Wedin et al. 1995, Carvalho et al. 2003). Studies have been done by injecting isotopes into litter, but for necromass this is difficult (Harmon et al. 1999).

A compilation of field measured decay rates and estimated decay rates based on a mortality estimate of 0.02 yr[-1] are presented in Table 2. Of the 46 papers we reviewed only 35% made measurement of necromass decomposition. Estimates of necromass decomposition rates in the Brazilian Amazon are rare. Chambers et al. (2000) used two different methods (closed chamber using an infra-red analyzer and measured mass loss) for estimates of 0.13 y[-1] and 0.17 y[-1] for each method. Palace et al. (2008) estimated decomposition rates using a steady state model. Their estimate of decay for all pieces of wood is 0.17 y[-1] for large (>10 cm diameter), 0.21 y[-1] for medium (5-10 cm diameter), and 0.47 y[-1] for small size (2-5 cm diameter) class necromass. No other study that we know of has data for these smaller size classes and their production decomposition rates for tropical forests. Other tropical forest necromass decomposition rates range from 0.03 y[-1] (Delaney et al. 1998) to 0.51 yr[-1] (Collins 1981). An extremely high decay rate of 2.0 y[-1] was estimated in a tropical mangrove forest by Robertson and Daniel (1989). We found only two studies that estimated decomposition rates for standing dead with estimates being 0.461 y[-1] (Lang and Knight 1979) and 0.115 y[-1] (Odum 1970). Palace et al. (2008) estimated the movement of standing dead through the pool to be 0.24 y[-1].

2.4 A simple model to expand and compare literature results

When production or decomposition rates were lacking in the literature that we reviewed, estimates were generated using the following methods and rationale. A production estimate of necromass was generated using the biomass value and a mortality rate of 0.02 y[-1]. We used a mortality rate at the upper end of the range for old-growth tropical forests (Philips and Gentry, 1994) but feel that this is a reasonable estimate, since most biomass studies do not include smaller diameter trees and lianas. In addition, mortality rates we used to estimate necromass production often underestimate necromass production because, branch fall is not included (Palace et al. 2007).

Using the production estimate divided by the stock measured values, we calculated decomposition rates. If no biomass estimate was available, we estimated necromass production to be 0.15 y[-1] of the total necromass stock. Though these estimates of decay and production amounts are prone to error and hypothetical in nature, they allow us to attempt to compare sites and biomes in tropical forests. Comparison of field data and model estimates also allow us to evaluate the assumption of steady state for a variety of sites.

3. Discussion

3.1 Methodology

Methodology was comparable among sites with similarity in decay classes and size classes used. Although there are some discrepancies among papers, we believe that stock estimates are broadly comparable. Biomass estimates were not done at all sites and we suggest that biomass be measured whenever necromass is examined. Production and decomposition measurements were both lacking at many sites and the existing measurements lacked consistency. We suggest longer temporal studies, allowing for a better understanding of the dynamics of these fluxes and their relation to meteorological parameters.

Different methods for fallen necromass quantification may be used depending upon the question being asked by the researcher, such as fuel load amount (Uhl and Kauffman 1990) or biometry and respiration estimates (Chambers et al. 2000, Keller et al. 2004b). In concert with previous evaluations for other regions, we find that line intercept sampling is generally easier to adapt to field conditions where sufficient area is available for sampling. For example, Grove (2001) switched from plot based work to line intercept sampling when confronted with dense understory. Baker et al. (2007) compared line intercept sampling with plot based sampling measuring similar amounts of fallen necromass. Palace et al. (2007) conducted line intercept sampling at the same site as Rice et al. (2004) and had similar estimates with similar uncertainties. Palace et al. (2007) suggest that line intercept sampling was six times as efficient and took about one third the amount of time and with half the field crew. Plot estimates require more movement than line intercept sampling and become especially difficult in logged sites or in sites with dense under-story.

Fallen necromass stock was almost 1.5 times more frequently measured than standing dead. This is likely due to difficulty measuring the height of standing dead in a complex and dense forest canopies common in many tropical sites. We believe that stock estimates tend to be the easiest and most accurate necromass component to measure when compared to decomposition and production rates, which require multiple samples over time.

Methodology for decay classes was similar among studies. Much of the recent literature cites Harmon et al. (1995) in regard to decay classification. It is likely that this paper has set the standard for decay classes used. Implementation of the decay classification may vary across the sites in necromass studies. We do not know of tests for field classification differences across the sites in tropical forests. A number of studies we examined had similar decay class definitions (Harmon et al. 1995, Eaton and Lawrence 2006, Keller et al. 2004b, Palace et al. 2007). In addition, some studies had similar decay class density measurements (Table 3). Chao et al. (2008) compared a three and five decay classification and suggest using a three class system because of less interpretation issues and higher sample numbers in each category. A wider range of decaying logs in a class may increase the variability within a class, though Chao et al. (2008) did not find this to be true.

Decomposition and production estimates of dead wood both have unique difficulties. Decomposition is a complex process; however estimates for decomposition rates based on a variety of methods often give similar results (Palace et al. 2007, Chambers et al. 2001). Production estimates need to cover a large enough area to capture the rare episodic tree fall events. While trading space for time is helpful for quantifying necromass productions, long term studies that could link necromass to weather changes and other aspects of interannual variability would help us better understand variability in carbon dynamics.

Tropical forests contain a large number of tree species and this creates difficulty when measuring decomposition rates (Chambers et al. 2000). Decomposition rate measurements maybe be misleading when only a few species of trees or a few trees are only examined for a short period of time. Chambers et al. (2000) developed a regression for decay that incorporates temperature, moisture, and necromass diameter. An exponential relationship has been shown between microbial activity and temperature, until temperature is so high that proteins are damaged and enzymes denature (Mackensen et al. 2003).

Many of the studies (38%) only examined one component of necromass dynamics. A combination of methods and components measured is preferable, allowing for the comparison of production and decomposition rates with stock estimates at the beginning and end of the study. Comparison to other measurable ecological parameters, such as NPP, woody increment, and mortality rates proves helpful in better understanding necromass dynamics. Necromass and biomass estimates should be done in conjunction at research sites. Finally, studies using the same methodology are beneficial to necromass research (Palace et al. 2007).

3.2 Comparison among sites

Necromass studies in tropical forests are few in number and concentrated in Central American dry forests and areas of the Eastern Amazon, especially in the State of Pará, Brazil. Many of the sites were highly disturbed due to logging activity, agriculture, fire, and in African, one case elephants (Buxton 1981, Uhl et al. 1988, Eaton and Lawrence 2006). We estimated production and decay estimates for these areas, but admit that our steady state approach is ill-suited to these sites.

The proportion of necromass to biomass is highly variable among among sites (Figure 1a) ranging from 0.01 in an undisturbed site in the Ivory Coast (Bernhard-Reversat et al., 1978) to 3.04 (Gerwing, 2002) in a heavy logged and burned site in Paragominas, Brazil. In undisturbed forests there appears to be a peak in the necromass with middle values of the biomass distribution (Figure 1b). Beyond that peak as biomass increases the proportion of necromass decreases. The highest biomass sites may have been undisturbed for long periods resulting in low necromass. We drew a hypothetical limit to illustrate such a relationship. High biomass and low necromass sites were often from studies that used small plots that do not reflect the landscape spatial distribution of biomass and necromass. Small plots may be chosen with the "majestic forest bias" that tends toward high biomass and little recent disturbance (Keller et al. 2001, Chave et al. 2001). Disturbed sites filled in are of the lower portion of biomass and higher necromass in Figure 1b. Chao et al. (2009a) found a significant but weak relation between biomass and necromass in a study across Amazonia ($r^2=0.12$). There work showed a stronger relation between living wood density and necromass, indicating that denser woods decay more slowly, with an additional insight as to differences in turnover time between Eastern and Western Amazonia. Use of living wood density is probably the best indicator of necromass at sites.

Standing dead and fallen necromass have been found to be proportionally related, even at disturbed sites (Palace et al. 2007). A regression examining just undisturbed sites was found to be significant ($r^2 = 0.22$, p=0.003; Figure 2). No such relation was found in our examination of disturbed sites. Nonetheless, standing necromass accounts for a large proportion of the total necromass stock, up to 66% in an undisturbed forest and 98% at a heavily disturbed site, and should be included in future field measurements (Palace et al. 2007, Harmon et al. 1995).

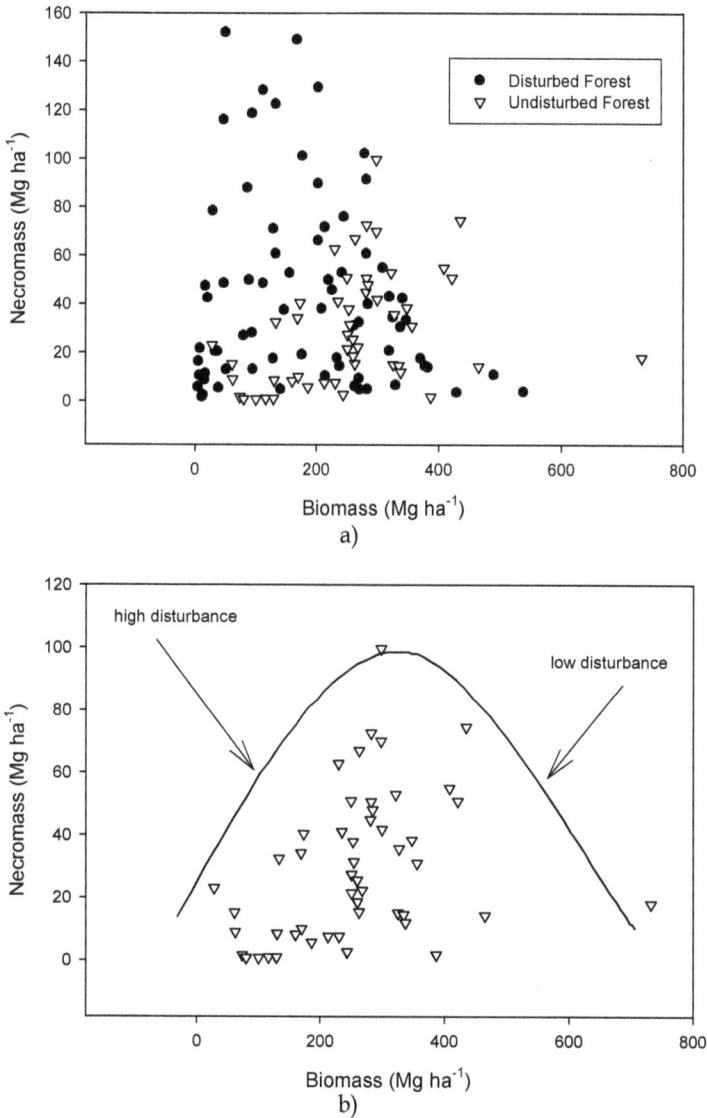

Fig. 1. a) Biomass and Necromass field measured values in undisturbed and disturbed tropical forests. b) Biomass and Necromass field measured values in undisturbed tropical forests showing areas of high and low disturbance.

We did not make comparisons of size classes among sites because few studies separated data by size class. In addition, comparisons among studies for necromass size classes are difficult because of differences in the limits for size classes themselves (Table 3). Still some studies indicate that smaller size classes (generally less than 10 cm diameter) contribute up to 21% of the total CWD stock (Uhl and Kauffman 1990, Palace et al. 2007) and we suggest that smaller size classes be included in field measurements. Smaller size classes decay more quickly and may contribute more to the overall site respiration budget (Harmon et al. 1986, Palace et al. 2008). Chambers et al. (2000), showed a relation with decomposition rates and necromass diameter. Palace et al. (2007, 2008) using production and stock estimates grouped by size classes were able to estimate decomposition rates for the size classes using a simple model.

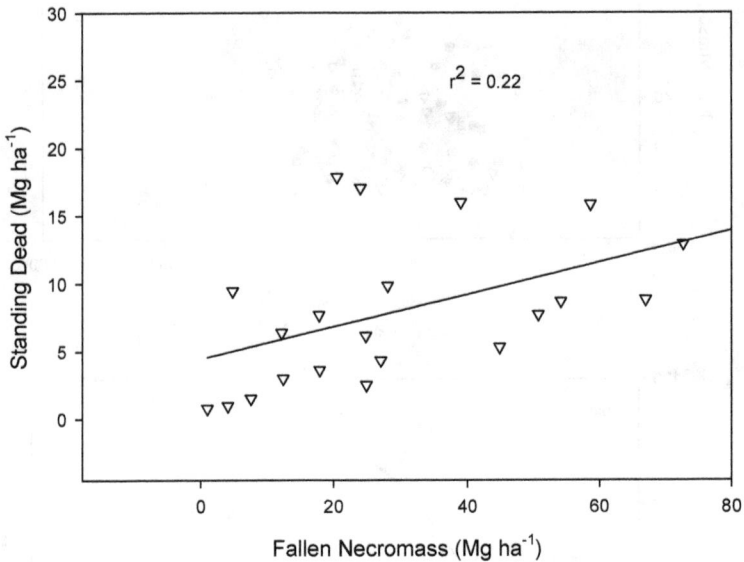

Fig. 2. Fallen necromass and standing dead field measured values in undisturbed tropical forests.

Trees lose branches through several processes that do not lead to whole tree mortality. For example, shaded lower branches may be shed and physical damage may result from crown interactions or animal activity. Mortality estimates used to determine necromass may underestimate production due to branch fall that is not associated with the death of a tree. Determination of the source of necromass would aid in quantifying branchfall. These small and medium classes are likely to include a substantial component from branchfall. Chambers et al. (2001) estimated branch-fall to be 0.9 Mg ha^{-1} y^{-1} based upon a comparison of field measured allometries and an optimized model tree structure based on the hydraulic constraints to tree architecture. Palace et al. (2008) examined the source of necromass by field classification of necromass as either branch, trunk, or unidentifiable. Necromass derived from tree trunks dominated the large size class in both necromass production and in pools. The other size classes were more evenly distributed among sources. Palace et al. (2008) found significant differences between logged and undisturbed forest treatments for the proportions of trunk, branch, and unidentified material within both production and pool necromass. Proportions between groups (production and pool estimates) and within a treatment were also found to be significantly different according to Palace et al. (2008). Chao et al. (2009b) found that trees die differently across Amazonia, with Northwest Amazonian trees dying more often in blowdowns with multiple deaths, while Northeastern Amazonian trees die as single tree events and die standing.

In our review, we used measured necromass stock and either an estimated production (biomass * 0.02) or decomposition rate (necromass * 0.15) to generate production and decay rates when they were missing from the literature. Using these rates and stocks, we examined if sites were at steady state. No sites were found to be at steady state. Either these sites were not at steady state or the generalized assumptions of production and decomposition rates may not accurately reflect real world values. It is not reasonable to expect all sites to be at steady state. Plots were often too small to represent landscape necromass dynamics.

We compared total necromass with field measured necromass production and found a significant relation, though the r^2 value was low ($r^2 = 0.28$, p=0.014; Figure 3). Chao et al. (2009a) found a similar relation with mortality mass input and necromass ($r^2 = 0.28$, p=0.003). Though both of these relations are significant, we are cautious about drawing conclusion with regression with low r^2 values.

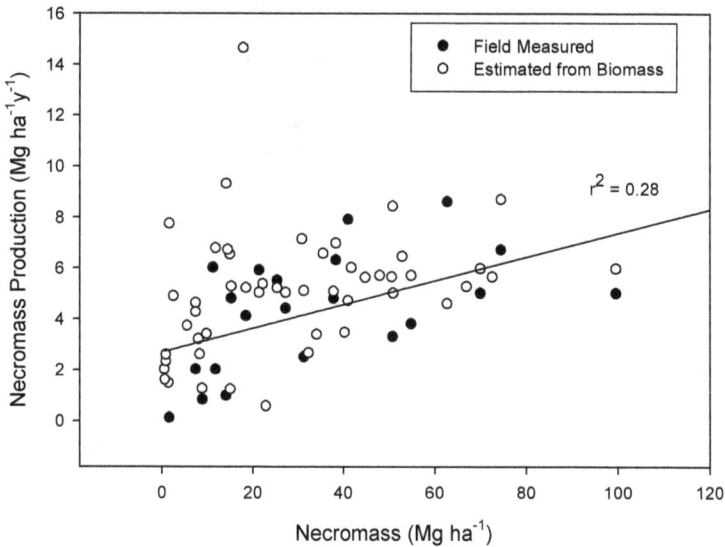

Fig. 3. A comparison of undisturbed forests for field measured necromass production and estimated necromass production from biomass with field measured necromass stocks. Only field measured necromass production is used in our regression.

We found that measured decomposition rates and those estimated by our simple model were similar. Higher decomposition rates associated with lower necromass stocks suggests that decomposition rates are an important control. We caution that this conclusion depends upon our model estimates using necromass production equal to a fixed proportion of biomass (0.02 y^{-1}). Higher decomposition rates may be associated with forests that experience higher disturbance rates as hypothesized by Baker et al. (2004) and Malhi et al. (2004) based upon a comparison between western and eastern Amazon forests. Baker et al. (2004) discussed the variation in wood density and how this determines the biomass in

Amazonian forests. Wood density variation is attributed to nutrient cycling and this influences species assemblages. The syndrome suggested by these two studies is that high-turnover forests have low density wood that in turn decomposes faster.

4. Conclusion

We compiled data from existing studies and compared pools and fluxes of necromass among tropical forest sites. General relationships among necromass components were explored such as necromass to biomass proportions and fallen to standing dead necromass. Methodology was comparable across the literature for necromass production and stock estimates. Fallen stock was 1.5 times more frequently measured than standing dead. We calculated production and decomposition rate estimates through the use of a simple model when these values were not available. General relations and proportions between necromass components were explored and were found to vary greatly. In undisturbed forests, we found weak but significant relations between fallen necromass and standing dead, as well as total necromass and measured necromass production. In undisturbed forests there appears to be a peak in the necromass with middle values of the biomass distribution. Beyond that peak as biomass increases the proportion of necromass decreases. The ratio of necromass to biomass ranged from 0.4 % in an undisturbed forest to 304% in a disturbed forest. Standing dead necromass accounts for a large proportion of the total CWD stock, up to 66% in an undisturbed forest and 98% at a heavily disturbed site, and should be included in further field estimates. We found that localized variability is high and complicates or hinders the development of general relationships of necromass components across the tropics. Many of the studies (37%) only examined only one component of necromass dynamics. We stress the importance of measuring multiple necromass components and ideally conducting these measurements over years or even decades in order to improve our knowledge of necromass dynamics in tropical forests.

5. Acknowledgements

We thank Christina Czarnecki for help with finding and organizing references. We also thank H.H. Shugart and John Aber for comments on this chapter. This work was supported by the NASA Terrestrial Ecology Program (NNX08AL29G) and the NASA New Investigator Program (NNX10AQ82G)

6. References

Aide, TM., 1987. Limbfalls: A major cause of sapling mortality for tropical forest plants. Biotropica 19(3), 284-285.

Anderson, A. 1981. White-sand vegetation of Brazilian Amazonia. Biotropica 13(3), 199-210.

Asner, G. P., Knapp, D.E., Broadbent, E.N., Oliveira, P.J.C., Keller, M., Silva. J.N.,2005. Selective logging in the Brazilian Amazon. Science 310, 480-482.

Asner, G., M. Palace, M. Keller, M., Pereira , J. Silva, J. Zweede, 2002. Estimating canopy structure in an Amazon forest from laser rangefinder and IKONOS satellite observations. Biotropica 34(4), 483-492.

Baker, T.R., Coronado, E.N.H., Phillips, O.L., Martin, J., van der Heijden, G.M.F., Garcia, M., Espejo, J.S., 2007. Low stocks of coarse woody debris in a southwest Amazonian forest. Oecologia, 152(3), 495-504.

Baker, T.R., Phillips, O.L., Malhi, Y., Almeida, S., Arroyo, L., Di Fiore, A., Erwin, T., Killeen, T.J., Laurance, S.G., Laurance, W.F., Lewis, S.L., Lloyd, J., Monteagudo, A., Neill, D.A., Patino, S., Pitman, N.C.A., Silva, N.M., Martinez, R.V., 2004. Variation in wood density determines spatial patterns in Amazonian forest biomass. Global Change Biology 10, 545-562.

Bellingham, P.J. and Tanner, E.V.J., 2000. The influence of topography on tree growth, mortality, and recruitment in a tropical montane forest. Biotropica, 32(3), 378-384.

Bernhard-Reversat, F., C. Huttel, and G. Lemée. 1978. Structure and functioning of evergreen rain forest ecosystems of the Ivory Coast. Pages 557-574 in Tropical Forest Ecosystems: A State-of-the-Knowledge Report. UNESCO, Paris.

Blackburn, G. A., Milton, E.J. 1996. Filling the Gaps: Remote Sensing Meets Woodland Ecology. Global Ecology and Biogeography Letters, 5, No. 4/5, Remote Sensing and GIS in the Service of Ecology and Biogeography: A Series of Case Studies. pp. 175-191.

Brown, I.F., Martinelli, L. A., Thomas, W. W., Moreira, M.Z., Ferreira, C.A.C., Victoria, R.A., 1995. Uncertainty in the biomass of Amazonian forests: an example from Rondonia, Brazil. Forest Ecology and Management 75, 175-189.

Brown, I.F., Nepstad, D.C., Pires, O., Luz, L.M. and Alechandre, A.S., 1992. Carbon storage and land-use in extractive reserves, Acre, Brazil. Environ. Conservation, 19, 307-315.

Brown, J.K., 1974. Handbook for inventorying downed woody material. USDA Forest Service, Ogden, Utah, 1-24.

Brown, S., 1997. Estimating biomass and biomass change of tropical forests: A primer. FAO Forestry Paper 134. Food and Agriculture Organization of the United Nations (FAO), Rome, Italy, 1-55.

Buxton, R. D. 1981. Termites and the turnover of dead wood in an arid environment. Oecologia 51, 379-384.

Cardoso, M. F., Hurtt, G. C., Moore, B., Nobre, C. A. and Prins, E. M. (2003), Projecting future fire activity in Amazonia. Global Change Biology, 9: 656–669. doi: 10.1046/j.1365-2486.2003.00607.x

Carey, E.V., Brown, S., Gillespie, A.J.R., Lugo, A.E., 1994. Tree Mortality in Mature Lowland Tropical Moist and Tropical Lower Montane Moist Forests of Venezuela. Biotropica, 26(3), 225-265.

Chambers, J.Q., Higuchi, N., Schimel, J.P., Ferreira L.V., Melack, J.M., 2000. Decomposition and carbon cycling of dead trees in tropical forests of the central Amazon. Oecologia 122, 380-388.

Chambers, J. Q., Tribuzy, E.S., Toledo, L.C., Crispim, B.F., Higuchi, N., Dos Santos, J., Araújo, A.C., Kruijt, B., Nobre, A.D., Trumbore, S.E., 2004. Tropical Forest Ecosystem Respiration. Ecological Applications 14(4), s72-s88.

Chambers, J.Q., G.P. Asner, D.C. Morton, L.O. Anderson, S.S. Saatchi, F.D.B Espírito-Santo, M. Palace, C. Souza, (2007). Regional ecosystem structure and function: ecological insights from remote sensing of tropical forests. Trends in Ecology and Evolution, 22(8), 414–423.

Chao, K.-J., Phillips, O.L., Baker, T.R., 2008. Wood density and stocks of coarse woody debris in a northwestern Amazonian landscape. Can. J. For. Res. 38, 795-805.

Chao, K.-J., Phillips, O.L., Baker, T.R., Peacock, J., Lopez-Gonzalez, G., Vásquez Martínez, R., Monteagudo, A., and Torres-Lezama, A., 2009a. After trees die: quantities and determinants of necromass across Amazonia. Biogeosciences Discuss., 6, 1979-2006.

Chao, K.-J., Phillips, O.L., Monteagudo, A., Torres-Lezama, A., and Vásquez Martínez, R., 2009b. How do trees die? Mode of death in northern Amazonia. Journal of Vegetation Science, 20, 260-268.

Chave, J., Riera, B., Dubois, M. A., 2001. Estimation of biomass in a neotropical forest of French Guiana: spatial and temporal variability. Journal of Tropical Ecology. 17: 79-96.

Clark, D.B., Clark, D.A.,1991. The Impact of Physical Damage on Canopy Tree Regeneration in Tropical Rain Forest. The Journal of Ecology, 79(2) 447-457.

Clark, D.B., Clark, D.A., Brown, S., Oberbauer, S.F., Veldkamp, E., 2002. Stocks and flows of coarse woody debris across a tropical rain forest nutrient and topography gradient. Forest Ecology and Management 164, 237-248.

Cochrane, M. A., Alencar, A., Schulze, M. D., Souza, C. M., Nepstad, D. C., Lefebvre, P. Davidson, E., 1999. Positive feedbacks in the fire dynamics of closed canopy tropical forests, Science 284, 1832-1835

Collins, N.M., 1981. The role of termites in the decomposition of wood and leaf litter in the southern Guinea savanna of Nigeria. Oecologia 51, 389-399.

Condit, R., Hubbell, S. P., Foster, R. B., 1995, Mortality Rates of 205 Neotropical Tree Species and the Responses to a Severe Drought, Ecol. Monogr. 65, 419–439.

Cummings, D.L., Kauffman, J.B., Perry, D.A., Hughes, R.F., 2002. Aboveground biomass and structure of rainforests in the southwestern Brazilian Amazon. Forest Ecology and Management 163, 293-307.

Curran, L. M., Trigg, S.N., McDonald, A.K., Astiani, D., Hardiono, Y.M. Siregar, P., Caniago, I., Kasischke, E., 2004, Lowland Forest Loss in Protected Areas of Indonesian Borneo. Science 303, 1000-1002.

Davidson, EA., Artaxo, P., 2004. Globally significant changes in biological processes of the Amazon Basin: results of the Large-scale Biospehere-Atmosphere Experiment. Global Change Biology 10(5), 519

Delaney, M., Brown, S., Lugo, A.E. , Torres-Lezama, A., Quintero, N.B., 1998. The quantity and turnover of dead wood in permanent forest plots is six life zones of Venezuela. Biotropica 30(1), 2-11.

Denslow, J.S., 1987. Tropical rainforest gaps and tree species diversity. Annual. Review Ecology and Systematics 18, 431-451.

De Vries, P.G., 1986. Sampling theory for forest inventory. A teach-yourself course. Springer-Verlag Berlin Heidelberg, Wageningen, 399 pp.

Dickinson, C.H., Pugh, G.J.F., 1974. Biology of Plant Litter Decomposition. Academic Press, NYC. Vol 1 and 2.

Eaton, J. M., Lawrence, D., 2006. Woody debris stocks and fluxes during succession in a dry tropical forest. Forest Ecology and Management 232, 46-55.

Edwards, P.J., Grubb, P.J., 1977. Studies of mineral cycling in mountain rainforest in New Guinea, I. The distribution of organic matter in the vegetation and soil. Journal of Ecology, 65, 943-969.

Feldpausch, T.R., Jirka, S., Passos, C.A.M., Jasper, F., Riha, S.J., 2005. When big trees fall: damage and carbon export by reduced impact logging in southern Amazonia. Forest Ecology and Management 219, 199-215.

Frolking, S., M. Palace, D.B. Clark, J.Q. Chambers, H.H. Shugart, G.C. Hurtt, (2009). Forest disturbance and recovery - a general review in the context of space-borne remote sensing of impacts on aboveground biomass and canopy structure. J. Geophys. Res., 114, G00E02, doi:10.1029/2008JG000911.

Gale, N., 2000. The relationship between canopy gaps and topography in a western Ecuadorian rain forest. Biotropica 32(4a):653-661.

Garwood, N., Janos, D.P., Brokaw, N., 1979. Earthquake-caused landslides: a major disturbance to tropical forests. Science 205, 997-999.

Gerwing, J.J., 2002. Degradation of forests through logging and fire in the eastern Brazilian Amazon. Forest Ecology and Management 157, 131-141.

Golley, F. B., J. T. McGinnis, R. G. Clements, G. I. Child, and M. I. Duever. 1975 . Mineral cycling in a tropical moist forest ecosystem. University of Georgia Press, Athens. 272 pp.

Golley, F.B., 1973. Nutrient Cycling and Nutrient Conservation, in F. B. Golley, editor, Tropical Rain Forest Ecosystems: 137-56, Elsevier Scientific Publishing Company, New York.

Golley, F.B., 1969. Caloric value of wet tropical forest vegetation. Ecology, 50, 517-519.

Gove, J.H., Ducey, M.J., Valentine, H.T., 2002. Multistage point relascope and randomized branch sampling for downed coarse woody debris estimation. Forest Ecology and Management 155, 153–162.

Greenland, D.J., Kowal, J.M.L., 1960. Nutrient content of the moist tropical forest of Ghana. Plant and Soil, 12, 154-174.

Grove, S.J., 2001. Extent and composition of dead wood in Australian lowland tropical rainforest with different management histories. Forest Ecology and Management 154, 35-53.

Harmon, M.E., Franklin, J.F., Swanson, F.J., Sollins, P., Gregory, S.V., Lattin, J.D., Anderson, N.H., Cline, S.P., Aumen, N.G., Sedell, J.R., Lienkaemper, G.W., Cromack, K., Cummins, K.W., 1986. Ecology of coarse woody debris in temperate ecosystems. Advances in Ecological Research 15, 133-302.

Harmon, M. E., Franklin, J. F., 1989. Tree seedlings on logs in Picea-Tsuga forests of Oregon and Washington. Ecology 70, 48-59.

Harmon, M.E., Whigham, D.F., Sexton, J., Olmsted, I., 1995. Decomposition and mass of woody detritus in the dry tropical forests of the northeastern Yucatan peninsula, Mexico. Biotropica 27(3), 305-316.

Harmon, M.E., Sexton, J.. 1996. Guidelines for Measurements of Woody Debris in Forest Ecosystems. Publication No. 20. U.S. LTER Network Office:University of Washington, Seattle, WA, USA. 73 pp.

Harmon ME, Franklin JF, Swanson FJ, Sollins P (1999) Measuring decomposition, nutrient turnover, and stores in plant litter. In: Robertson GP, Coleman DC, Bledsoe CS, Sollins P (eds) Standard soil methods for long-term ecological research. Oxford University Press, New York, pp 202–240

Houghton, R.A., Lawrence, K.T., Hackler, J.L. Brown, S., 2001. The spatial distribution of forest biomass in the Brazilian Amazon: a comparison of estimates. Global Change Biology 7, 731-746.

Houghton, R.A., Skole, D.L., Nobre, C.A., Hackler, J.L., Lawrence, K.T., Chomentowski, W.H., 2000. Annual fluxes of carbon from the deforestation and regrowth in the Brazilian Amazon. Nature 403, 301-304.

Hughes, R. F., J. B. Kauffman, Jaramillo, V.J., 2000.. Ecosystem-Scale Impacts of Deforestation and Land Use in a Humid Tropical Region of Mexico. Ecological Applications, 10(2) 515-527.

Janzen, D.H., 1987. Insect diversity of a Costa Rican dry forest: why keep it, and how? Biological Journal of the Linnean Society, 30(4), 343-356

Kaarik, AA. 1974. Decomposition of wood. In Dickinson, C.H., Pugh, G.J.F., 1974. Biology of Plant Litter Decomposition. Academic Press, NYC, pp. 129–174.

Kauffman, J. B., Uhl, C., 1990. Interactions of anthropogenic activities, fire, and rain forests in the Amazon Basin. Pages 117–134 in J. G. Goldammer, editor. Fire in the tropical biota: ecosystem processes and global challenges. Ecological Studies, Volume 84. Springer-Verlag, Berlin, Germany.

Kauffman, J.B., Uhl, C., Cummings, D.L., 1988. Fire in the Venzuelan Amazon. 1. Fuel biomass and fire chemistry in the evergreen rainforest of Venezuela. Oikos, 53, 167–175.

Keller, M., Palace, M., Hurtt, G., 2001. Biomass estimation in the Tapajos National Forest, Brazil: examination of sampling and allometric uncertainities. Forest Ecology and Management 154, 371-382.

Keller, M., Alencar. A., Asner, G.P., Braswell, B., Bustamante, M., Davidson, E., Feldpausch, T., Fernandes, E., Goulden, M., Kabat, P., Kruijt, B., Luizão, F., Miller, S., Markewitz, D., Nobre. A.D., Nobre, C.A., Filho, N.P., Rocha, H., Dias, P.S., von Randow, C., Vourlitis, G.L., 2004a. Ecological Research in the Large Scale Biosphere Atmosphere Experiment in Amazônia (LBA): Early Results. Ecological Applications 14(4) Supplement., S3-S16.

Keller, M., Palace, M., Asner, G.P., Pereira, R., Silva, J.N.M., 2004b. Coarse woody debris in undisturbed and logged forests in the eastern Brazilian Amazon. Global Change Biology 10(5), 784-795.

Kira, T. 1978. Community architecture and organic matter dynamics in tropical lowland rain forests of Southeast Asia with special reference to Pasoh Forest, West Malaysia. Pages 561-590 in Tropical Trees as Living Systems. P.B. Tomlinson and M.H. Zimmerman, editors. Cambridge: Cambridge University Press.

Kitajima, K. and Augspurger, C.K., 1989. Seed and seedling ecology of a monocarpic tropical tree, Tachigalia versicolor. Ecology, 70, 1102-1114.

Klinge, H., 1973. Biomassa y materia orgánica del suelo in el ecosistema de la pluviselva centro-amazonica. Acta Cientifica Venezolana. 24, 174-181.

Lang, G. E., and D. H. Knight. 1979. Decay rates for boles of tropical trees in Panama. Biotropica 11:316–317.

Lang, G.E., Knight, D.H., 1983. Tree growth, mortality, recruitment, and canopy gap formation during a 10-year period in a tropical moist forest. Ecology 64, 1075-1080.

Lawton, RO., Putz, FE., 1988. Natural disturbance and gap-phase regeneration in a wind-exposed tropical cloud forest. Ecology, 69(3), 764-777.

Lieberman, M., Lieberman, D., Peralta, R.. 1989. Forests are not just swiss cheese: canopy stereogeometry of non-gaps in tropical forests. Ecology 70(3), 550-552.

Lugo, AE., Scatena, FN., 1996. Background and catastrophic tree mortality in tropical moist, wet, and rain forests. Biotropica 28(4a), 585-599.

Mackensen, J., Bauhus, J., Webber, E., 2003. Decomposition rates of coarse woody debris – A review with particular emphasis on Australian tree species. Australian Journal of Botany 51, 27-37.

MacNally, R., Parkinson, A., Horrocks, G., Conole, L., Tzaros, C., 2001. Relationships between terrestrial vertebrate diversity, abundance and avaiability of coarse woody debris on south-eastern Australian floodplains. Biological Conservation 99, 191-205.

Magnusson, W.E., Lima, A.P., Lima, O.P., 1996. Group lightning mortality of trees in a newtropical forest. Journal of Tropical Ecology 12, 899-903.

Malhi, Y., et al. 2004. The above-ground coarse woody productivity of 104 neotropical forest plots. Global Change Biology 10(5), 563.

Marra, J.L., Edmonds, R.L., 1994. Coarse woody debris and forest floor respiration in an old-growth coniferous forest on the Olympic Peninsula, Washington, USA. Can. J. For. Res. 24, 1811-1817.

Martinez-Ramos, M., Alvarez-Buylla, E., Sarukhan, J., Pinero, D., 1988. Treefall age determination and gap dynamics in a tropical forest. Journal of Ecology 76, 700-716.

Martius, C., 1997. Decomposition of wood. In Junk, Wolfgang J., 1997. Ecological Studies 126. The Central Amazon Floodplain. Ecology of a Pulsing System. Springer, NYC, 267-276.

Martius, C., Bandeira, A.G., 1998. Wood litter stocks in tropical moist forest in central Amazonia. Ecotropica 4, 115-118.

Nascimento, H.E.M., Laurance, W.F., 2002. Total aboveground biomass in central Amazonia rainforests: a landscape-scale study. Forest Ecology and Management, 168, 311-321.

Nelson, B.W., Kapos, V., Adams, J.B., Oliveria, W.J., Braun, O.P.G., Doamaral, I.L., 1994. Forest disturbance by large blowdowns in the Brazilian Amazon. Ecology 75(3), 853-858.

Nepstad, D.C., Decarvalho, C.R., Davidson, E.A., Jipp, P.H., Lefebvre, P.A., Negreiros, G.H, Da Silva, E.D., Stone, T.A., Trumbore, S.E., Vieira, S. 2002. The role of deep roots in the hydrological and carbon cycles of Amazonian forests and pastures. Nature, 372, 666 – 669.

Nobre, C., Sellers, P., Shukla, J., 1991. Amazonian deforestation and regional climate change. Journal of Climate 4, 957-988.

Nogueira, E.M., Nelson, B.W., Fearnside, P.M., 2005. Wood density in dense forest in central Amazonia, Brazil. Forest Ecology and Management 208, 261–286.

Nordén, B., Paltto, H., 2001. Wood-decay fungi in hazel wood: species richness correlated to stand age and dead wood features. Biological Conservation 101, 1-8.

Odum, H.T., 1970. Summary: an emerging view of the ecological system at El Verde. In: Odum, H.T., Pigeon, R.F. (eds.). A tropical rain forest. A study of irradiation and ecology at El Verde, Puerto Rico. U.S. Atomic Energy Commission, Oak Ridge, I191-I289.

Ozanne, C.M.P., Anhuf, D., Boulter, S. L., Keller, M., Kitching, R.L., Korner, C., Meinzer, F. C., Mitchell, A.W., Nakashizuma, T., Dias, P.L.S., Stork, N.E., Wright, S.J., Yoshimura, M.. 2003. Biodiversity meets the atmosphere: A global view of forests canopies. Science 301, 183-186.

Palace, M., M. Keller, G.P. Asner, J.N.M. Silva, C. Passos, (2007). Necromass in undisturbed and logged forests in the Brazilian Amazon. Forest Ecology and Management, 238, 309-318

Palace, M., M. Keller, H. Silva, (2008). Necromass production: studies in undisturbed and logged Amazon forests. Ecological Applications: 18, 873–884.

Pereira, R., Zweede, J.C., Asner, G.P., Keller, M.M., 2002. Forest canopy damage and recovery in reduced impact and conventional logging in eastern Para, Brazil. Forest Ecology and Management 168, 77-89.

Phillips, O.L., Gentry, A.H., 1994. Increasing turnover through time in tropical forests. Science 263, 954-958.

Prance, G.T., Lovejoy, T.E., 1985. Key Environments Amazonia. Pergamon Press, NYC, 422 pp.

Rice, A.H., Pyle, E.H., Saleska, S.R., Hutyra, L., Camargo, P.B., Portilho, K., Marques, D.F., Palace, M., Keller, M., Wofsy, S.C., 2004. Carbon balance and vegetation dynamics in an old-growth Amazonian forest. Ecological Applications 14(4), s55-s71.

Richards, P.W., 1952. The Tropical Rain Forest: An Ecological Study. Cambridge University Press, Cambridge, UK. 450 pp.

Robertson, A.I., Daniel, P.A., 1989. Decomposition and the annual flux of detritus from fallen timber in tropical mangrove forests. Limnol. Oceanogr. 34(3), 640-646.

Salati, E., Vose, PB., 1984. Amazon basin: a system in equilibrium. Science 225, 129-138.

Saldarriaga, J.G., West, D.C., Tharp, M.L., Uhl, C., 1988. Long-term chronosequence of forest succession in the upper Rio Negro of Colombia and Venezuela. Journal of Ecology, 76, 938-958.

Saleska, S.R., Miller, S.D., Matross, D.M., Goulden, M.L., Wofsy, S.C., da Rocha, H.R., de Camargo, P.B., Crill, P., Daube, B.C., de Freitas, H.C., Hutyra, L., Keller, M., Kirchhoff, V., Menton, M., Munger, J.W., Pyle, E.H., Rice, A.H., Silva, H. 2003.

Carbon in Amazon forests: Unexpected seasonal fluxes and disturbance-induced losses. Science 302 (5650), 1554-1557.

Sanford, RL., Braker, H.E., Hartshorn, G.S., 1986. Canopy openings in a primary neotropical lowland forest. Journal of Tropical Ecology 2, 277-282.

Schemske. D.W., Brokaw, N., 1981. Treefalls and the Distribution of Understory Birds in a Tropical Forest. Ecology 62(4), 938-945.

Schwarzkopt, L., Rylands, A.B., 1989. Primate species richness in relation to habitat structure in Amazonian rainforests fragments. Biological Conservation 48, 1-12.

Scott, D.A., Proctor, J., Thompson, J., 1992. Ecological studies on a lowland evergreen rain forest on Maracá Island, Roraima, Brazil. II. Litter and nutrient cycling. Journal of Ecology 80, 705-717.

Summers, P. M., 1998. Estoque, decomposicao, e nutrientes da liteira grossa em floresta de terra firme, na Amazonia Central. Pages 118. Ciencias de Florestas Tropicais. Instituto Nacional de Pesquisas da Amazonia, Manaus, Brazil.

Svenning, J.C., 2000. Small canopy gaps influence plant distributions in the rain forest understory. Biotropica 322, 252-261.

Swift, M.J., Heal, O.W., Anderson, J.M., 1979. Decomposition in Terrestrial Ecosystems. University of California Press, Berkeley, 372 pp.

Tanner, E.V.J., 1980. Studies on the biomass and productivity in a series of Montane rain forests in Jamaica. The Journal of Ecology 68(2), 573-588.

Terborgh, J., 1992. Maintenance of diversity in tropical forests. Biotropica 24, 283-292.

Uhl, C., Kauffman, J.B., 1990. Deforestation, Fire Susceptibility, and Potential Tree Responses to Fire in the Eastern Amazon. Ecology, 71(2), 437-449.

Uhl, C., K. Clark, N. Dezzeo, Maquirrino, P., 1988. Vegetation dynamics in Amazonian treefall gaps. Ecology 69, 751–763.

van der Meer, P.J., Bongers, F., 1996. Patterns of Tree-Fall and Branch-Fall in a Tropical Rain Forest in French Guiana. The Journal of Ecology, 84(1) 19-29.

Vitousek. P.M., Denslow, J.S., 1986. Nitrogen and Phosphorus Availability in Treefall Gaps of a Lowland Tropical Rainforest. The Journal of Ecology, 74(4) 1167-1178.

Walker, L.R., Zarin, D.J., Fetcher, N., Myster, R.W., Johnson, A.H., 1996. Ecosystem development and plant succession on landslides in the Caribbean. Biotropica 28, 566-576.

Wedin, D.A., Tieszen, L.L., Dewey, B., Pastor, J., 1995. Carbon Isotope Dynamics During Grass Decomposition and Soil Organic Matter Formation. Ecology 76(5), 1383-1392.

Werth, D., Avissar, R., 2002. The local and global effects of Amazon deforestation, J. Geophys. Res. 107, doi:10.1029/2001JD000717.

Wessman, C.A., 1992. Spatial scales and global change: Bridging the gap from plots to GCM grid cells. Ann. Rev. Ecol. Syst. 23, 175–200.

Wilcke, W., Hess, T., Bengel, C., Homeier, J., Valarezo, C., Zech W., 2005. Coarse woody debris in a montane forest in Ecuador: mass, C and nutrient stock, and turnover. Forest Ecology and Management 205, 139–147.

Windsor, DM., Morrsion , D.W., Estribi, M.A., de Leon, B., 1988. Phenology of fruit and leaf production by "strangler" figs on Barro Colorado Island, Panama. Experimentia. 45, 647-653.

Yoda, K., Kira, T., 1982. Accumulation of organic matter, carbon, nitrogen, and other nutrient elements in the soils of a lowland rainforest at Pasoh, Peninsular Malaysia. Japan Journal of Ecology 32, 275-291.

Yoneda, T. Yoda, K., Kira, T., 1977. Accumulation and decomposition of big wood litter in Pasoh Forest, West Malaysia. Jpn. J. Ecol, 27, 53-60.

Yoneda, T., Tamin, R., Ogino, K., 1990. Dynamics of aboveground big woody organs in a foothill dipterocarp forest, West Sumatra, Indonesia. Ecol Res, 5, 111-130.

Young, T.P., Hubbell, S.P., 1991. Crown Asymmetry, Treefalls, and Repeat Disturbance of Broad-Leaved Forest Gaps. Ecology, 72(4) 1464-1471.

Part 5

Geographic Information System and Remote Sensing

13

Seasonal Pattern of Vegetative Cover from NDVI Time-Series

Dyah R. Panuju and Bambang H. Trisasongko

Department of Soil Sciences and Land Resource, Bogor Agricultural University, Bogor, Indonesia

1. Introduction

Indonesia manages various forested land utilizations, for instance natural forest and plantations. In the past, natural forest had been exploited throughout the country, mainly in the islands of Sumatera and Borneo (Nawir & Rumboko, 2007). Greater criticisms on forest exploitation lead to a moratorium which needs to be monitored frequently. Most of forest concessions were converted into plantations (Kartodihardjo & Supriono, 2000). Popular plantations developed in the country have been rubber and oil palm. Successful management of the plantation as well as natural forest has been under careful examination. Assessment of woody vegetation could be taken using field surveys or remote sensing. Although having possibilities to capture detailed datasets, field survey is lacking in terms of implementation: lengthy data capture and least favorable to remote areas. In these situations, remotely sensed data play a crucial role.

Current practice of the vegetative assessment has been single temporal analysis assisted by remote sensing data; some with extension to multi-temporal analysis. Mono-temporal analysis was shown useful for biomass estimation, using optical (Foody *et al.*, 2001) or Synthetic Aperture Radar (SAR) data (Raimadoya *et al.*, 2005). Nonetheless, changing climate reveals difficulties to determine exact condition of the vegetative land cover. Season is probably the most intriguing factor in this case, especially in long-term land cover changes (Lambin, 1999). The study suggested the importance of seasonality in remote sensing data analysis to account fluctuating forest state under climate change.

In this study, we employed seasonal adjustment of time-series statistical method to understand phenology and detect disturbance on some woody vegetation. Seasonal adjustment methods have been developed particularly to forecast and vastly applied on financial studies. Nonetheless, this has been rarely applied to environmental studies, except in meteorology and hydrology. X12-ARIMA is one of the latest versions of X method that dealing with seasonal adjustment time-series. It decomposes single value of time-series data into three components, namely trend, seasonal and irregular. Seasonal components can be used to test seasonality or impact of season on the vegetation index. Test on global trend of the index and irregular aspect can also be inferred from the method. This research employed NDVI time-series of SPOT VEGETATION. NDVI has been a popular index in vegetation related studies. Foody et al. (2001) found that NDVI has some weaknesses, for instance: (1) relationship between NDVI and biomass was usually asymptotic that limits its ability to represent large amount of biomass accurately, and (2) sensitivity of NDVI to represent vegetation greenness varied upon environmental circumstances. Despite its weaknesses, the advantages of using NDVI to understand vegetative cover phenomena is due to its long-term availability, easy in calculation, and its relationship with some vegetative parameters such as greenness, and biomass (Hess et al., 1996).

2. Remote sensing of tropical forest cover

Remote sensing application on tropical forest was first demonstrated by Tucker et al. (1984). Using Advanced Very High Resolution Radiometer (AVHRR), the study indicated large clearing which was associated with large-scale dwelling. Since then, tropical forest gains particular attentions, employing multispectral, hyperspectral and radar datasets. It appears that Landsat was one of popular multispectral sensor employed for tropical forest assessment and monitoring. Sader et al. (1989) pioneered Landsat applications in tropical forest biomass estimation. Although not highly successful, the research informed that fairly mature successional forest were undetectable by means of Normalized Difference Vegetation Index (NDVI), which has led to many attempts to improve the achievement.

It is well understood that tropical regions have severe cloud cover. Although there have been numerous attempts to reduce or eliminate atmospheric disturbance (for instance a paper by Richter, 1996), the issue persists. Longer wavelength sensor such as Synthetic Aperture Radar has been exploited to minimize atmospheric disturbance (Reddy, 2008). At first, single polarimetric SAR datasets were employed, mainly on VV (vertical transmit and receive) or HH (horizontal transmit and receive). C-band and L-band SAR systems were largely reported, however, higher wavelength (L- or P-band) is preferable (Trisasongko, 2009). Exploiting the advance of processing technology, many SAR sensors have been mounted into spaceborne platform. Currently, users have options to obtain dual polarimetric SAR; some SAR systems allow data capture in fully polarimetric mode. Despite limited sensors capable to acquire in fully polarimetric mode, several applications of fully polarimetric SAR were successfully demonstrated for tropical forest monitoring (for instance Trisasongko et al., 2007; Trisasongko, 2010).

Forests, including the ones in tropical regions, reveal complexity in monitoring. Seasonality has been one of major trends in forest applications of remote sensing, especially in conjunction with climate studies. Studies on seasonal pattern of vegetative cover have been published in attempt to understand impacts of climate change to vegetation growth. Brando et al. (2010), for instance, studied climate variations seasonally and interannually to understand drought in Amazon using time-series data of MODIS Enhanced Vegetation Index. In the same region, Koltunov et al. (2009) presented a study confirming association between logging activity and forest phenology in dry seasons of 2000-2002. Using the time-series of NDVI (Normalized Difference Vegetation Index) of AVHRR or MODIS some scholars showed that vegetation phenology has been affected by climate dynamic particularly precipitation and/or temperature (Wang et al., 2008; Meng et al., 2011).

3. Time-series analysis

Time-series analysis of remotely sensed data, as shown earlier, has gained special attention supported by availability of wide-coverage, high temporal satellite data. Earlier works exploited descriptive analysis, and focusing on visual analysis on graphical representation of remote sensing data of selected sites. Goetz et al. (2006), for instance, presented useful graphical representations of NDVI time-series data to assess post-fire forest regrowth in Canada. Present advance using X12-ARIMA demonstrated that time-series data were indispensable to characterize fire spot in tropical region (Panuju et al., 2010).

Analysis of time-series data has been employed for following aims: (1) to describe characteristics of the data; (2) to explain variance of any variable associated to surrogate variables; (3) to predict a variable beyond the time span; or (4) to control processes (Chatfield, 1984). Underlying assumptions of time-series analysis are variance homogeneity and stationary of the data. Lu et al. (2001) successfully employed decomposition technique to visually separate herbaceous from woody vegetation of temperate region.

Basic formula of X12-ARIMA is ARIMA and the X technique. ARIMA is a standard time series introduced by Box and Jenkins (1970), while the X technique is a seasonal adjustment procedure started from X0 to the recent form the-X12 (Shiskin et al., 1965). Since the background of developing procedure is financial application, most of terminology to process the decomposition represents business and financial related activities. The application of the ARIMA model on environmental study was widely reported by hydrological studies for instance by Hipel and McLeod (1994). The ARIMA model is a modification of ARMA process which is a mixed model of autoregressive and moving average (Wei, 2006):

$$\phi_p(B)Z_t = \theta_q(B)a_t \tag{1}$$

where $\Phi_p(B)=1-\Phi_pB-....-\Phi_pB^p$, and $\theta_q(B)=1-\theta_qB-....-\theta_qB^q$. Meanwhile the ARIMA is denoted as follows:

$$\phi_p(B)(1-B)^d Z_t = \theta_0 + \theta_q(B)a_t \tag{2}$$

where the difference between both equation are on the differencing process represents by $1-B)^dZ_t$.

4. Methodology

4.1 Test site

The study was located in Jambi Province, Indonesia (Fig. 1), between 101.15°E – 104.43°E and 0.76°S – 2.77°S. The province is famous for two key national parks: The Berbak National Park, largest swamp forest in South-East Asia which is also a Ramsar Convention on Wetlands site (www.ramsar.org), and The Kerinci-Seblat National Park, a natural habitat of gigantic flowers such as *Amorphophallus* and *Rafflesia*. Jambi also hosts Bukit Tigapuluh NP where aboriginal tribes of Talang Mamak and Orang Rimba reside. The province is also an important region for Indonesia which has contributed in oil and gas (BPS, 2010) and oil palm estate (http://aplikasi.deptan.go.id/bdsp/newlok.asp). The important estates include rubber and oil palm, which are managed by locals, government and private companies. Rubber and oil palm estates are found throughout the province.

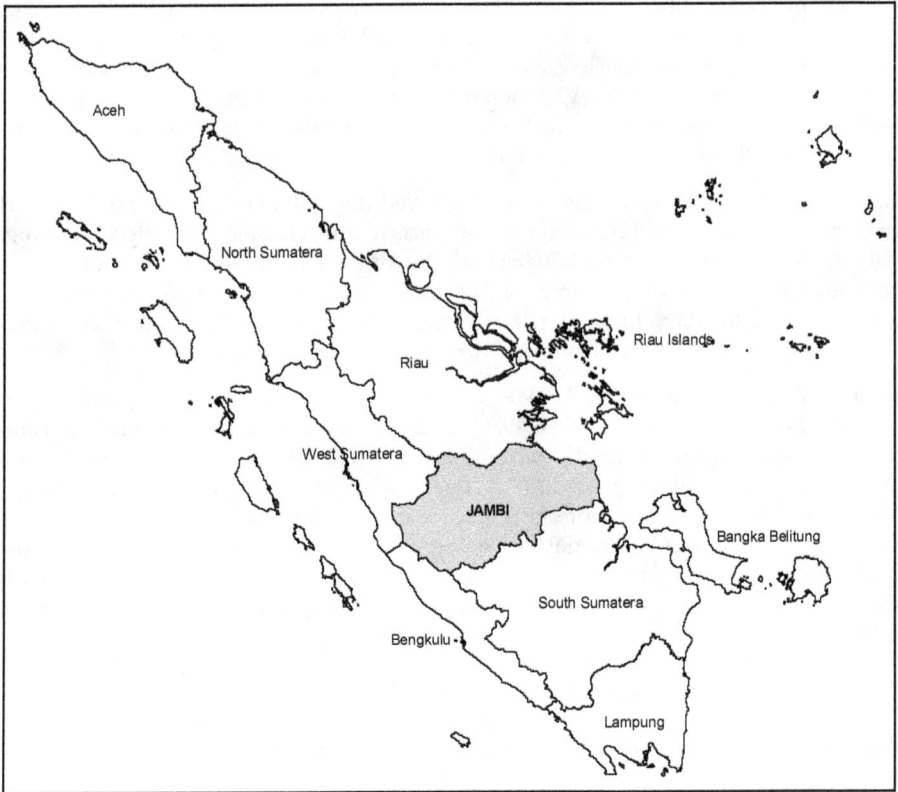

Fig. 1. Site location

4.2 Remote sensing and field data

In this research, the main remote sensing datasets were time-series NDVI SPOT VEGETATION (SPOT-VGT) and high resolution ALOS AVNIR-2 imageries. SPOT-VGT NDVI data were collected from the Flemish Institute for Technological Research (VITO), Belgium. VITO website (http://free.vgt.vito.be) provided 10-day NDVI composite images, which were then constructed into single database from April 1998 to December 2010. To assist time-series analysis on selected land cover types, a set of ALOS AVNIR-2 imageries provided by Indonesian Ministry of Agriculture were used. The imageries were georeferenced according to Indonesian Base Map specification (WGS 1984, geographic coordinate system). Google Earth (http://earth.google.com) was also utilized to assure locations for samples.

Field surveys were designated to obtain current land use in selected regions. Selection was made considering vastness of the province. First field visit was conducted in October 2010, which covered following regencies (*kabupatens*): Batanghari, Bungo, Tebo, Muaro Jambi,

Sarolangun, Tanjung Jabung Barat and Merangin (Fig. 2). Another field visit was arranged to focus on specific areas of Bungo and Kerinci regencies in July 2011.

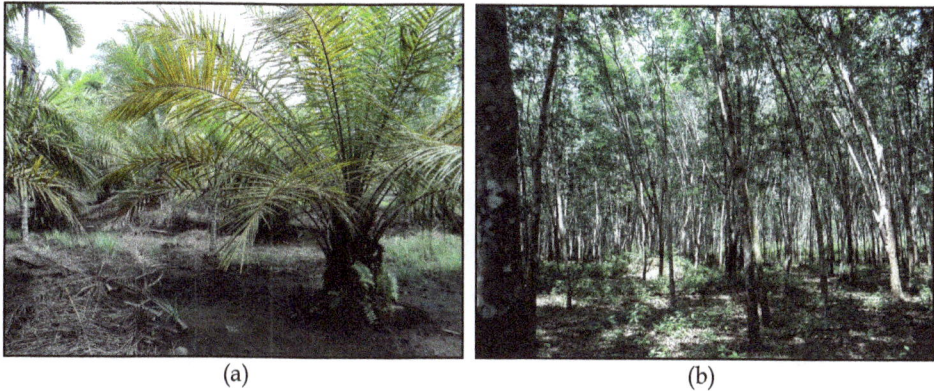

<table>
<tr><td>(a)</td><td>(b)</td></tr>
</table>

Fig. 2. Field survey which successfully collected data in most of regencies of Jambi Province. (a) Oil palm plantation (Tanjung Jabung Barat), and; (b) Rubber plantation (Merangin).

4.3 Data analysis

First step on this research was samples identification for forest, rubber and oil palm area. We selected two regions of interest (ROI) for each land cover to develop time series NDVI of SPOT-VGT. Each land cover was delineated for about 25 pixels of SPOT-VGT or 25 kilometers. All those samples were named Forest1, Forest2, Rubber1, Oilpalm1, Rubber2 and Oilpalm2 (Fig. 3). Forest1 was a wetland forest (Berbak National Park) while forest2 was taken from mountainous region (Kerinci Seblat National Park). Rubber1 was selected from Tebo Regency (101.82°E-102.82°E; 0.88°S-1.91°S) while rubber2 was from Merangin Regency (101.55°E-102.57°E; 1.63°S-2.77°S). Then, the oilpalm1 and oilpalm2 were from Tanjung Jabung Barat (102.64°E-103.66°E; 0.75°S-1.50°S) and Merangin Regency respectively.

NDVI from each 25 samples were then averaged. Since the NDVI of SPOT-VGT was the 10-days synthesis product (S10) and procedure of X12-ARIMA only accommodated monthly data, we computed maximum value of the S10 NDVI. Some researchers including Jia et al. (2002) noted that maximum of NDVI will produce best representation of monthly data and anticipate effect of cloud cover which reducing NDVI value.

Time series analysis was then processed using X12-ARIMA procedure. It was utilized to determine seasonal model of the NDVI series of each sample. The model usually was represented as (p, d, q) (P, D, Q), where p symbolizes autoregressive order, d denotes differencing and q stands for moving average order. According to Wei (2006), seasonal ARIMA model is denoted as follow:

$$\Phi_P(B^s)\phi_p(B)(1-B)^d(1-B^s)^D Z_t = \theta_q(B)\Theta_Q(B^s)a_t \qquad (3)$$

Where Φ is autoregressive function and θ expresses the moving average process. Further discussion on seasonal model refers to Findley et al. (1998). The model was analyzed using

ARIMA auto-modeling software developed by Eurostat which freely accessible in 2002. Based on the auto-modeling, the model will be selected from seasonal or non seasonal model with some alternatives order of autoregressive, differencing and moving average order. Since the X12-ARIMA decomposing series data into three components, we used the decomposition process to compare differences among land covers. Even though there are multiplicative and log alternatives, all series were only performed the additive model option. The model was tested based on residual analysis using BIC (*Bayesian Information Criterion*) or SBC (*Schwartz's Bayesian Criterion*). The parameter was calculated using this equation:

$$SBC(M) = N \ln \hat{\sigma}_a^2 + M \ln N \qquad (4)$$

where M denotes number of parameters and N refers to number of series data. The best model will produce the smallest BIC or SBC. Furthermore, to test randomness of error Ljung Box suggested probability calculation of χ^2. If χ^2 is less than 10%, then ARIMA automodel will reject the option. We employed TRAMO (*Time series Regression with ARIMA noise, Missing value and Outliers*) option to understand history of series. The TRAMO generated possible option to explore any noise such as missing and extreme value due to any disturbances into the series. Indicative option will lead conclusion into the certain time of evidences perturbing NDVI.

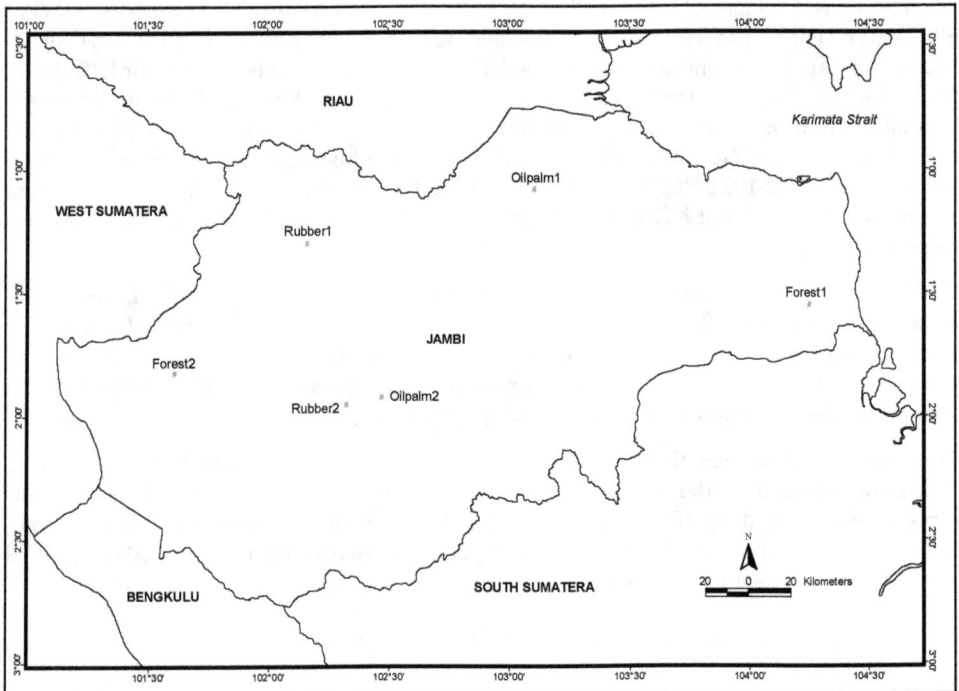

Fig. 3. Sampling locations.

5. Results and discussion

5.1 Comparison of forest phenologies

Study on phenology of vegetative cover is important to understand plant and animal behavior. Seasonality of vegetation behavior is apparently a resultant of micro- and macro-climatic aspects as well as other living species activities. Studies on phenology by biologists mostly focus on physiological aspects of plants and phenophases, including leafing, flowering and fruiting (van Schaik *et al.*, 1993). In this study, we discuss a different perspective to investigate phenology. Physiological aspect combined with biotic and abiotic factors as well as sensors condition and disturbance were assumed simultaneously epitomized by vegetation index (NDVI). Decomposition of NDVI series was expected to explain plant characteristics and activities occurring in the tropical forest. Since the spatial resolution of SPOT Vegetation is considered coarse, decomposition process was intended to deliver some details of the series. Lu *et al.* (2001) showed that decomposition of NDVI time series of AVHRR, particularly trend and seasonal components, was able to separate woody and herbaceous vegetation by using STL (seasonal trend decomposition based on loess). This research extended Lu *et al.* (2001) results to decompose NDVI series of forested and plantation areas.

Natural forest cover in Jambi is usually managed as a National Park. Under this scheme, forest cover is expected to be undisturbed and preserved to play its ecological function. Nonetheless, threads remain, mostly for plantations or other anthropogenic activities such as shifting cultivation. Some reports note that local inhabitants surrounding national parks usually take advantage of forest for their daily life. This phenomenon should be understood as a function of forest to support people nearby. In Jambi, involvement of people on forest utilization apparently depends on ecosystem itself and access into the forest. Therefore, it was necessary to obtain at least two different types of forest, wetland and montane forests. In these areas, we attempted to explore decomposition of time series NDVI to understand evidences on both ecosystems.

The general parameters of ARIMA model is displayed in Table 1. Apparently, both forests produced similar model with some parameters were a slight better on Kerinci-Seblat National Park (KSNP). The (p, d, q) indicates that both wetland and montane forests fit into ARIMA model. Seasonal model is shown fairly obvious on both forest types, indicated by (P, D, Q) units. This suggests that seasonality was quite apparent on both forests, hence monitoring solely based on single date should be avoided. The results also advise that different seasons should be taken into account for selecting imageries in multitemporal analysis, for instance by means of higher optical or SAR sensors. Due to lacking of available datasets, imageries of different season in multitemporal analysis such as land use detection and change have been utilized. In this case, as this research suggests, detailed assessment are needed to avoid bias in detection. The standard error (SE) of residual and BIC of KSNP series were smaller than from Berbak National Park (BNP). The model is modestly better fitted on KSNP.

Furthermore, the decomposition of NDVI in two Jambi forests is presented in Figure 4. It is shown that original series as well as their decomposition of both forest ecosystems were quite different. The original series of BNP wetland forest which is located nearby the Eastern coastline was more stable rather than samples obtained from the mountainous forest of KSNP. According to van Schaik *et al.* (1993), the phenology of tropical plants has been affected mostly by biotic and abiotic factors, such as water, light, and nutrients availability.

Parameters	BNP forest	KSNP forest
(p, d, q) (P, D, Q)$_{12}$	(0, 0, 0) (0, 1, 1)	(0, 0, 0) (0, 1, 1)
SE(res)	0.083	0.077
Q-val	12.431	12.489
BIC	-254.704	-271.326

Table 1. Parameters of seasonal time series model for forests

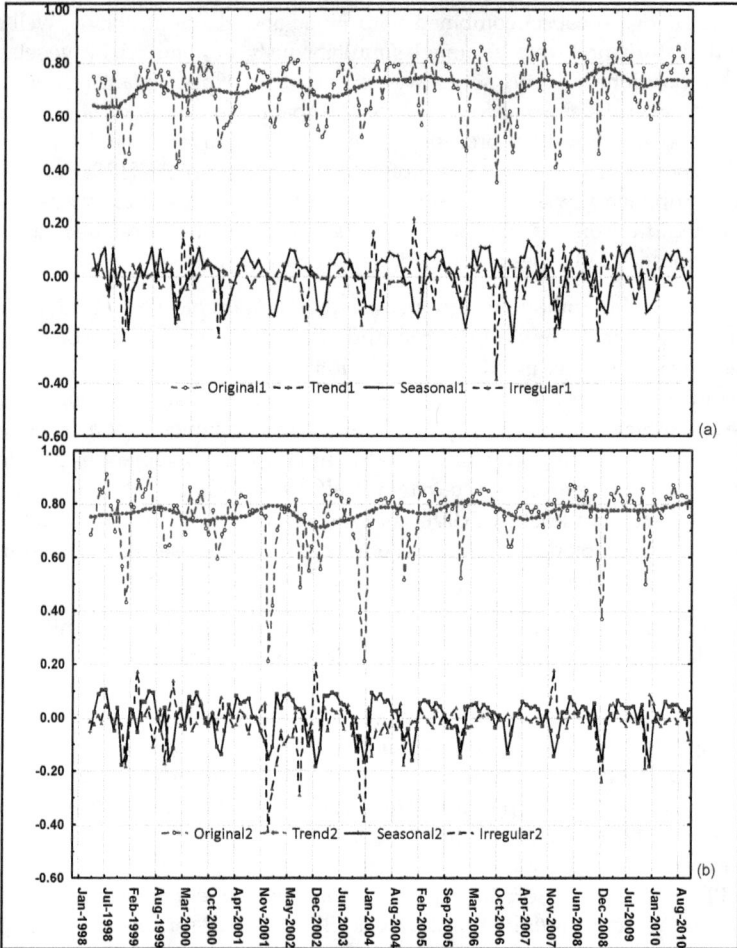

Fig. 4. Original and decomposition series of forest taken from Berbak National Park (a) and Kerinci Seblat National Park (b)

Continuous water availability in wetland area was suspected to be a primary cause on the stability. Original NDVI amplitude of BNP is about 0.5 with minimum value of 0.4 and the largest at about 0.9. Meanwhile, the NDVI amplitude of KSNP is around 0.7 with minimum value 0.2 and the maximum at 0.9. It suggests that disturbances were severe at the montane forest. Second field visit supports this argument. During the visit, several shifting cultivations in the KSNP were observed. Local inhabitants have planted several economic commodities such as cinnamons, duku (*Lansium domesticum* Corr.) and pinang (*Areca catechu* L.). It appears that there has been a mutual agreement between the government and local inhabitants to access some of KSNP areas as agroforestry to support local economies. In turn, local inhabitants are asked to assist KSNP protection program.

Trend series of both types of forests were similar. However, it is discernible that seasonal factor of BNP illustrates larger amplitude than KSNP. This might be related to the fact that BNP is located in wetland areas (altitude of 0-20 meters), where tidal dynamics (about 2-2.5 meters, according to Wetlands International, Indonesia Office) play a role at different season. Another interesting finding was that the irregular components of KSNP fluctuated more than BNP. This would be the subject of future investigation.

5.2 Comparison of plantation phenologies

Jambi has been one of important provinces who manage an immense coverage of plantations. Oil palm and rubber have been the popular estate commodities in Jambi, along with cinnamon, tea and Aquilaria (local name: gaharu). However, only oil palm and rubber are discussed in this chapter. Results of seasonal time series modeling are presented in following table.

Parameters	Oilpalm1	Oilpalm2	Rubber1	Rubber2
$(p, d, q)\ (P, D, Q)_{12}$	(0, 0, 0) (0, 1, 1)	(0, 1, 1) (0, 1, 1)	(0, 0, 0) (0, 1, 1)	(0, 0, 0) (0, 1, 1)
SE(res)	0.073	0.084	0.101	0.085
Q-val	18.722	19.684	21.851	23.272
BIC	-296.633	-240.155	-202.475	-253.079

Table 2. Parameters of seasonal time series model for plantations

It is clear that seasonality was important to both oil palm and rubber at all cases. This also suggests that remote sensing data analysis involving more than two acquisitions should be carefully taken to avoid preconceived notions. At some points, oil palm produced a better certainty than rubber. Some parameters such as standard error of residual and BIC of oil palm were smaller than of the rubber (Table 2).

To save space, we presented only Merangin region for the assessment of oil palm and rubber plantations (Fig. 5). The results show that the original series of NDVI of rubber generated smaller amplitude than the oil palm. NDVI of rubber spans from 0.6 to 0.9 while oil palm slightly varies, from 0.5 to 0.9. This is due to gaps often found in the oil palm blocks due to various reasons, including diseases. The moderate to large oil palm plantations (about tens of thousand hectares) are usually set up factories on-site, which would severely reduce NDVI values. The trend components of both plants were fairly similar. It tends to increase by time depending on age. Seasonal trend shows different pattern during the periods. Irregular components indicate noise and any disturbances on specific times, which cannot be describe further at this time.

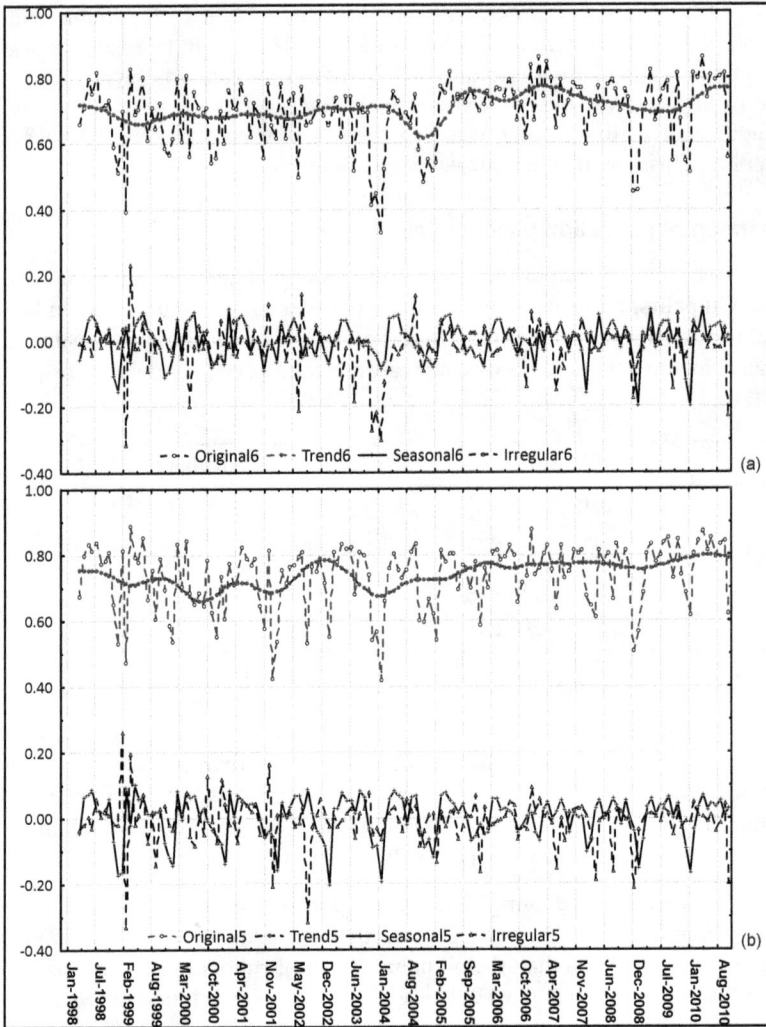

Fig. 5. Original and decomposition series of oil palm (a) and rubber (b) plantations.

5.3 Disturbances on forest and plantations

On previous subsections, disturbances could be indicated by time series datasets. To observe more on this disturbance, the TRAMO was employed in identification of any unusual observation of the series. The unusual observation could be due to biotic or abiotic causes, which was not determined at the moment. Instead, our focus was in detection ability of TRAMO. Result of TRAMO detection is shown in Table 3.

Outlier type	Forest1	Forest2	Oilpalm1	Oilpalm2	Rubber1	Rubber2
AO	(Oct-2006)		-	-	-	-
TC	-	(Jan-2002); (Dec-2003)	-	-	-	(Dec-2003)
LS	-	-	-	-	-	-
IO	-	-	-	-	-	-

Table 3. Detection of disturbance by TRAMO

Comparison with higher remote sensing data on Forest1 (BNP) indicated false-alarm produced by TRAMO. Using Landsat ETM+ images dated June 3, 2007 (LE71250612007154SGS01) and July 5, 2007 (LE71250612007186PFS00), the forested area has been shown intact with minimum disturbance by human. Detection on KSNP (Forest2) could be considered successful as indicated by second field trip. The survey observed that agroforestry estates (cinnamons, duku and pinang) were at mature stage (various ages). Merangin site which hosts Rubber2 is considered dynamic regency where several estate conversions (changing commodities) were reported.

6. Conclusion

NDVI has been proven useful to study the dynamic of vegetative covers using time-series approach. In this study, seasonality in NDVI time-series data was explored to understand various responses provided by woody vegetation. Using SPOT-VGT data, we showed that seasonality should be taken into account in multitemporal or time-series analyses. We found that NDVI time series of wetland and montane forests could be modeled by seasonal ARIMA, although level of confidence was found higher on montane forest (KSNP). Seasonal models of both types of forest were similar as well. Exploration on seasonal ARIMA model also found that NDVI series of wetland forest (BNP) tend to be more stable, which indicated a stable environment with least deviations on water, light and/or nutrients. Higher NDVI span on KSNP signified variable response from vegetative cover. Field surveys indicated some contributions from agroforestry (cinnamons, duku and pinang) in KSNP which might be contributing factors to more fluctuating NDVI values. NDVI ranges were found quite similar between oil palm and rubber plantations. Slightly bigger range on oil palm might be related to some gaps in plantation; one of them is Crude Palm Oil (CPO) on-site production facilities. This research also discovered some disturbances from time-series data. Although a false-alarm fault was observed, TRAMO was shown useful to identify disruptions in KSNP which was related to agroforestry.

7. Acknowledgment

We are indebted to VITO who provided SPOT-VGT datasets, Ms. Gusmaini for her assistance in downloading the datasets and Ms. Lili Suryani in field surveys. Several Landsat ETM+ images were downloaded from Earth Explorer provided by the USGS.

8. References

Badan Pusat Statistik (BPS, Indonesian Statistical Agency). (2010). *Trends of the Selected Socio-Economic Indicators of Indonesia, August 2010*. ISSN 2085.5664

Box, G.E.P. & Jenkins, G.M. (1970). *Time Series Analysis: Forecasting and Control*. Holden-Day, ISBN 0816210942, San Francisco

Brando, P.M.; Goetz, S.J.; Baccini, A.; Nepstad, D.C.; Beck, P.S.A. & Christman, M.C. (2010). Seasonal and interannual variability of climate and vegetation indices across the Amazon. *PNAS*, Vol.107, No.33, (August 2010), pp. 14685-14690, ISSN 0027-8424

Chatfield, C. (1984). *The Analysis of Time Series: An Introduction, 3rd Edition*, Chapman and Hall, ISBN 0412260301, London, UK

Findley, D.F.; Monsell, B.C.; Bell, W.R.; Otto, M.C. and Chen, B.C. (1998). New capabilities and methods of the X-12-ARIMA Seasonal Adjustment Program. *Journal of Business and Economic Statistics*, Vol. 16, No. 2, (April 1998), pp. 127-157, ISSN 1537-2707

Foody, G.M.; Cutler, M.E.; McMorrow, J.; Pelz, D.; Tangki, H.; Boyd, D.S. & Douglas, I. (2001). Mapping the biomass of Bornean tropical rain forest from remotely sensed data. *Global Ecology and Biogeography*, Vol.10, No.4, (July 2001), pp. 379-387, ISSN 1466-8238

Goetz, S.J.; Fiske, G.J. & Bunn, A.G. (2006). Using satellite time series data sets to analyze fire disturbance and forest recovery across Canada. *Remote Sensing of Environment*, Vol.101, No.3, (April 2006), pp. 352-365, ISSN 0034-4257

Hess, T.; Stephens, W. & Thomas, G. (1996). Modelling NDVI from decadal rainfall data in the Nort East Arid Zone of Nigeria. *Journal of Environmental Management*, Vol. 48, No. 3, (November 1996), pp. 249-261, ISSN 0301-4797

Jia, G.J.; Epstein, H.E. & Walker, D.A. (2002). Spatial characteristics of AVHRR-NDVI along latitudinal transects in Northern Alaska. *Journal of Vegetation Science*, Vol.13, No.3, (June 2002), pp. 315-326, ISSN 1654-1103

Kartodihardjo, H. & Supriono, A. (2000). The impact of sectoral development on natural forest conversion and degradation: The case of timber and tree crop plantations in Indonesia. *CIFOR Occasional Paper* No. 26(E). ISSN 0854-9818

Koltunov, A.; Ustin, S.L.; Asner, G.P & Fung, I. (2009). Selective logging changes forest phenology in the Brazilian Amazon: Evidence from MODIS image time series analysis. *Remote Sensing of Environment*, Vol.113, No.11, (November 2009), pp. 2431-2440, ISSN 0034-4257

Lambin, E.F. (1999). Monitoring forest degradation in tropical regions by remote sensing: Some methodological issues. *Global Ecology and Biogeography*, Vol.8, No.3-4, (May 1999), pp. 191-198, ISSN 1466-8238

Lu, H.; Raupach, M.R. & McVicar, T.R. (2001). Decomposition of vegetation cover into woody and herbaceous components using AVHRR NDVI time series. *CSIRO Land and Water Technical Report* 35/01, CSIRO, Canberra, Australia.

Meng, M.; Ni, J, & Zong, M. (2011). Impacts of changes in climate variability on regional vegetation in China: NDVI based analysis from 1982 to 2000. *Ecological Research*, Vol.26, No.2, (March 2011), pp. 421-428, ISSN 0912-3814

Nawir, A.A. & Rumboko, L. (2007). History and state of deforestation and land degradation. In: *Forest Rehabilitation in Indonesia: Where to after More Than Three Decades?* Nawir, A.A.; Murniati & Rumboko, L., pp. 11-32, Center for International Forestry Research, ISBN 978-979-14-1205-6, Bogor, Indonesia

Panuju, D.R.; Trisasongko, B.H.; Susetyo, B.; Raimadoya, M.A. & Lees, B.G. (2010). Historical fire detection of tropical forest from NDVI time-series data: Case study on Jambi, Indonesia. *ITB Journal of Science*, Vol.42A, No.1, (March 2010), pp. 49-66, ISSN 1978-3043

Raimadoya, M.A.; Dobson, M.C.; van Rensburg, R. & Trisasongko, B.H. (2005). Envisat-Indonesia Radar Biomass Experiment (EIRBEX), *Proceedings of the 2004 Envisat & ERS Symposium*, ESA SP-572, ISBN 92-9092-883-2, Salzburg, Austria, 6-10 September 2004

Reddy, M.A. (2008). *Remote Sensing and Geographical Information Systems*. Third Edition. BS Publications, ISBN 81-7800-135-7, Hyderabad, India

Richter, R. (1996). Atmospheric correction of satellite data with haze removal including a haze/clear transition region. *Computers & Geosciences*, Vol.22, No.6, (July 1996), pp. 675-681. ISSN 0098-3004

Shiskin, J.; Young, A.H. & Musgrave, J.C. (1965). Summary of the X-11 variant of the census method II Seasonal Adjustment Program. Available at: http://fraser.stlouisfed.org/publications/BusCycD/page/1985/download/3075/1985_1965-1969.pdf

Trisasongko, B.; Lees, B. & Paull, D. (2007). Discrimination of scatterer responses on tailings deposition zone using radar polarimetry. *Sensing and Imaging: An International Journal*, Vol.8, No.3-4, pp. 111-120, DOI: 10.1007/s11220-007-0037-8. ISSN 1557-2064

Trisasongko, B.H. (2009). Tropical mangrove mapping using fully-polarimetric radar data. *ITB Journal of Science*, Vol.41A, No.2, (September 2009), pp. 98-109. ISSN 1978-3043

Trisasongko, B.H. (2010). The use of polarimetric SAR data for forest disturbance monitoring. *Sensing and Imaging: An International Journal*, Vol.11, No.1, pp. 1-13. DOI: 10.1007/s11220-010-0048-8. ISSN 1557-2064

Tucker, C.J.; Holben, B.N. & Goff, T.E. (1984). Intensive forest clearing in Rondonia, Brazil, as detected by satellite remote sensing. *Remote Sensing of Environment*, Vol.15, No.3, (June 1984), pp. 255-261. ISSN 0034-4257

Van Schaik, C.P.; Terborgh, J.W. & Wright, S.J. (1993). The phenology of tropical forests: Adaptive significance and consequences for primary consumers. *Annual Reviews of Ecology and Systematics*, Vol. 24, pp. 353-377. ISSN 0066-4162

Wang, J.; Meng, J.J. & Cai, Y.L. (2008). Assessing vegetation dynamics impacted by climate change in the Southwestern Karst region of China with AVHRR NDVI and AVHRR time series. *Environmental Geology*, Vol.54, No.6, (May 2008), pp. 1185-1195, ISSN 0943-0105

Wei, W.W.S. (2006). *Time Series Analysis: Univariate and Multivariate Methods*. Second edition. Pearson Addison Wesley, ISBN 0-321-32216-9, Boston, USA.

14

Natural Forest Change in Hainan, China, 1991-2008 and Conservation Suggestions

Mingxia Zhang and Jianguo Zhu
Kunming Institute of Zoology,
The Chinese Academy of Sciences, Kunming, Yunnan,
China

1. Introduction

Hainan Island, a tropical island located in southern China, is one of the conservation hotspots of the world (Myers et al, 2000), and represents a large proportion of China's tropical area. In the past hundred years, the natural forest has decreased dramatically in this island. The first distinct and fast decrease was during the World War II, when the Island was occupied by Japanese, a large area of forest was logged. The natural forest has decreased from 169 200 km^2 to 120 000 km^2 from the year 1933 to 1950. The second big loss was from 1950 to 1987, when log was treated as an important natural resource by the government. During this period, many large log companies were established and the forest area decreased from 120 000 km^2 to 39 120 km^2 (Lin & Zhang, 2001). From 1952 to 1990, 3950 km^2 of rubber plantation was established on the island, mainly distributed at the elevation of 0-800 m asl (li, 1995), where most of the original tropical rain forest stood. As the forest decreased, 11 of the log companies transformed from log to plantation industry starting from 1983. After 1984, the local government began to reduce the log quota and reverted to protect the forest. In 1994, the cutting of natural forest was totally banned by the government of Hainan Province (Lin & Zhang, 2001; Zhang et al., 2010). After that, eucalyptus was planted in some barren or logged areas, usually between 800-1300 m. The deforestation took place from coastal plain and mesa to inland hill and basin, and finally to mountainous area in the middle of the island. The main factors affecting the tropical forest were not the same in different phases. From Han Dynasty to 1933, it was due to aboriginal cultivation; from 1933 to 1950, it was due to plundering cutting and destroying (Lin & Zhang, 2001; Zhang et al., 2010); and then, the ultimate cause was due to fast increasing local population and changes in policy, such as crops-economy.

Although it was under the continual pressure of being logged and encroachment of plantation, the remaining large patches of forests still keep unique ecosystem in the central mountainous area of the Island. It harbors many endemic species such as Hainan gibbon (*Nomascus hainanus*), Hainan partridge (*Arborophila ardens*) and Hainan peacock pheasant (*Polyplectron katsumatae*). It is critical to identify these forest patches and the changes that have occurred for understanding the status of conservation priorities in the future.

Geographic Information System (GIS) and Remote Sensing (RS) images have proved to be effective in land cover mapping, habitat evaluation and environmental risk assessment (Osborne, 2001; Mumby and Edwards, 2002; Moufaddal, 2005). This study aims to analyze the changes in natural forest and plantations on Hainan Island between 1991-2008 by using GIS and RS and trying to explore the driving factors of changes based on local policies, and give suggestions for future conservation plan.

2. Method

2.1 Study area

Hainan Island is located between 108°36'–111°04' E and 18°09'–20°11' N, with an area of 33 920 km^2. The topography range in elevation is 0-1884 m asl. The island is more mountainous in the middle, and flattened in northern and coastal parts (Zhang et al., 2010; Meng et al., 2011). Annual rainfall is generally high, between 900 and 2500 mm. The rainy season is from May-October and dry season is from November-April. The vegetation shows high diversity. The vertical zonation of vegetation in the mountainous areas with high rainfall encompasses lowland rainforest below 600 m, montane and ravine rainforest from 600 to 1200 m, and evergreen broadleaf forest above 1200 m. In the dry and rigid area, the vegetation developed into seasonal forest or tropical conifer forest while on the ridge and the top of the hill, the vegetation became, evergreen or dwarf forest. There is also a small area of mangroves located by the seashore (Wang & Zhang, 2002; Zhang et al., 2010). The plantations on the island include eucalyptus (*Eucalyptus* spp.), rubber (*Hevea brasiliensis*), horsetail (*Casuarina* spp.), and fruit orchards.

2.2 Data collection

To cover the whole Hainan Island and to detect forest change, nine Landsat Thematic Mapper and Enhanced Thematic Mapper-plus (TM/ETM+) images of two time period were needed; the path/row numbers of the images were 123-124/46-47. Landsat TM images of the years 1988 and 1991 were obtained from the website of Globe Land Cover Facility (GLCF, http://glcfapp.umiacs.umd.edu:8080/esdi/index.jsp), and ETM+ images of 2004, 2007 and 2008 were obtained from the website of the U.S. Geological Survey (USGS) (http://glovis.usgs.gov/). The resolution of the images were 30 m, they were geo-referenced to a Gauss Kruger/Krasovsky coordinates with a Root Mean Square (RMS) <1 pixel. As 80% of the island was covered by 12447 images which were taken in 1991 and 2008, so we defined our study period as 1991-2008.

In order to collect ground truth data for mapping and validation, we took samples from field by stratified method called Gradsect sampling (Austin and Heyligers, 1989). We partitioned the whole island into 23 cells of 30 minutes in longitude and 20 minutes in latitude. In each cell, the sample size was determined in proportion to the number of pixels in each environmental class: if the class occupied more than 9999 pixels, we took at least 30 samples; if it occupied 1000-9999 pixels, we took at least 10 samples; if it occupied 100-999 pixels, we took at least two samples; if 40-99 pixels, we took at least one sample; if <40 pixels, no sample was collected (Fig. 1). Field surveys were implemented in 2005, one from April to May for the rainy season, and another from October to November for the dry season. The field survey spent 47 days in total. A total of 1225 ground truth samples were collected from the field surveys.

For the images taken in 2008, we randomly selected half of our ground truth samples (613) obtained from the field survey as a reference for classification, and the other half (612) were used for evaluation of the results. For the images taken in 1991, the ground truth data was acquired from 1:100,000 topographic maps produced by the State Bureau of Surveying and Mapping, Beijing, China in 1981. For example, if one area is covered by rubber on the topographic map, and it was still rubber during the field survey of 2005, the place will be defined as rubber. We also consulted local nature reserve staffs for vegetation information. A total of 1315 samples were collected for the interpretation of image in 2008, 658 of them was randomly selected for use in the classification and the rest were used to evaluate the accuracy of the resultant map.

2.3 Images classification

The images were interpreted by using the software Erdas 9.0, with a method of combing the supervised/unsupervised classification or namely guided clustering (Bauer et al., 1994; Reese et al., 2002). The procedure followed is as below:

1. Cut the cloud out of image by hand;
2. Randomly divide the classification samples into two halves;
3. Classified the remote sense images by using supervised classification (with half of the sample from result of step 2);
4. Evaluated the result of step 3 with another half sample from result of step 2, and clipped away those classes with an accuracy higher than 75%;
5. Clipped the images with different classes of samples, ran unsupervised classification on the samples, and then Class A was divided as class A1, A2, A3... An, and these sub-classes were saved as the template for step 6;
6. Used the result of step 5 to classify the images by using supervised classification, and then recombined the classes to Class A, Class B...
7. Used another half of the samples obtained from result of step 2 to evaluate the result of step 6, and then clipped away those classes with an accuracy higher than 75%;
8. Repeated steps 5-7, till the highest accuracy was obtained;
9. Mosaic the results together;
10. Used the evaluation samples to evaluate the overall accuracy.

2.4 Analysis

The change in the area of natural forest and plantations were compared by different elevations of 0-380 m, 380-760 m, 760-1140 m, 1140-1520 m and above 1520 m. The reason of change was analyzed based on local policy change history. The map of natural forest was overlapped with the nature reserves boundaries, using the software ArcGis 9.0, to detect possible conservation gaps, if any.

The resultant map included villages (including bare land and crop), urban areas, water bodies, natural forest and plantations. Here, we defined the forest as areas with a minimum of 40% canopy closure, at 2 m high or above, within 100 * 100 m^2 (0.01 km^2) squares. Those areas planted with artificial mono species were defined as plantations.

3. Result

The remote sensing images were classified to get the land cover map of 1991 and 2008; the overall accuracy of the resultant map was 81% for 1991, and 78% for 2008 (Fig. 1).

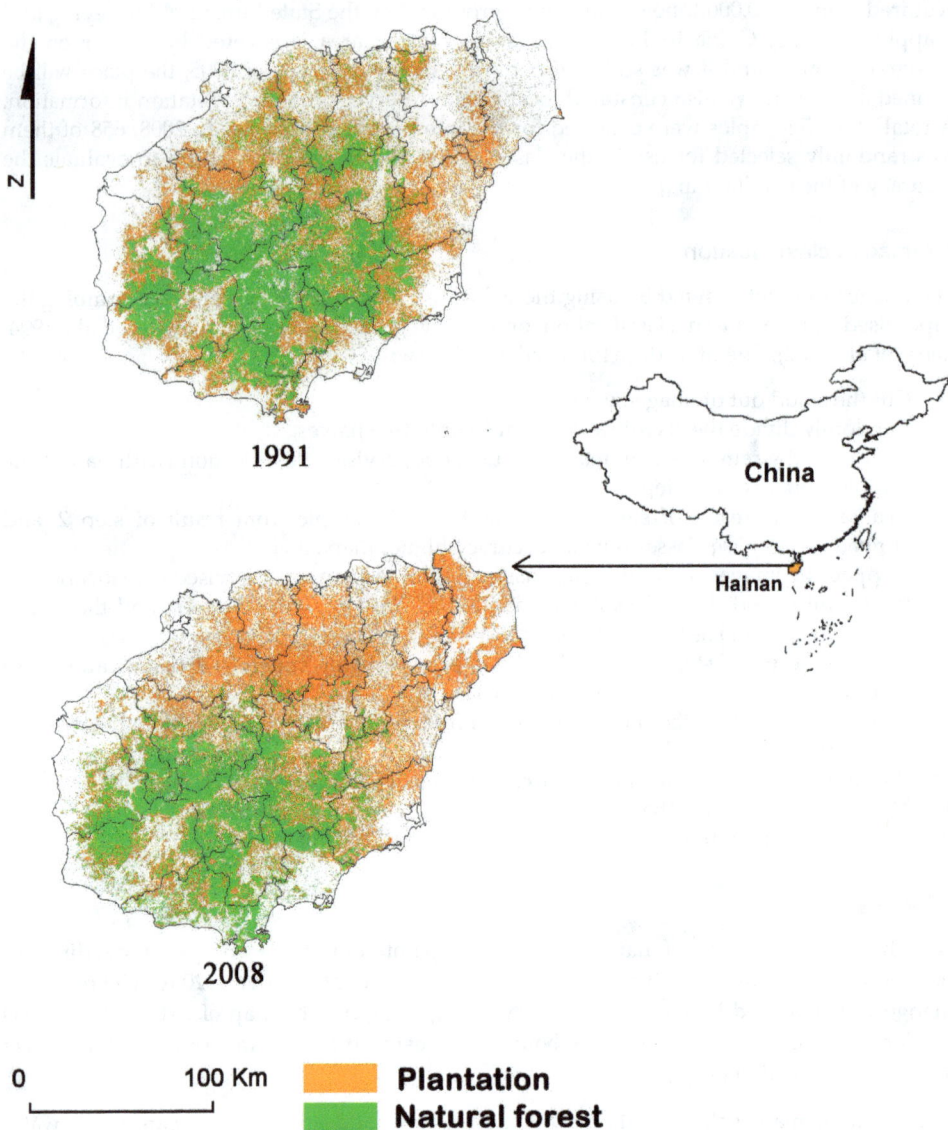

Fig. 1. Landcover map of Hainan Island, China in 1991 and 2008, interpreted from Landsat images.

Land cover type	Area in 1991 (km²)	Area in 2008 (km²)	Area change (km²)	Change percentage (%)
Natural forest	7314	5852	-1462	-20
Plantation	7807	10736	2929	38
Total	**15121**	**16588**	**567**	**10**

Table 1. The area change of natural forest and plantation from 1991 to 2008 on Hainan Island, China

The area changes of forest and plantations are shown in Table 1. During the 18 years, the natural forest decreased by 20%, while plantation increased by 38%. In different elevations, the 18 years' change of land cover type showed different trends. The natural forest decreased by 1557 km² below elevation of 760 m, while the plantation increased by 2896 km². The change above 760 m was minor in comparison with the change in the lower elevation; the natural forest increased by 95 km², and the plantation also increased by 32 km² (Fig. 2).

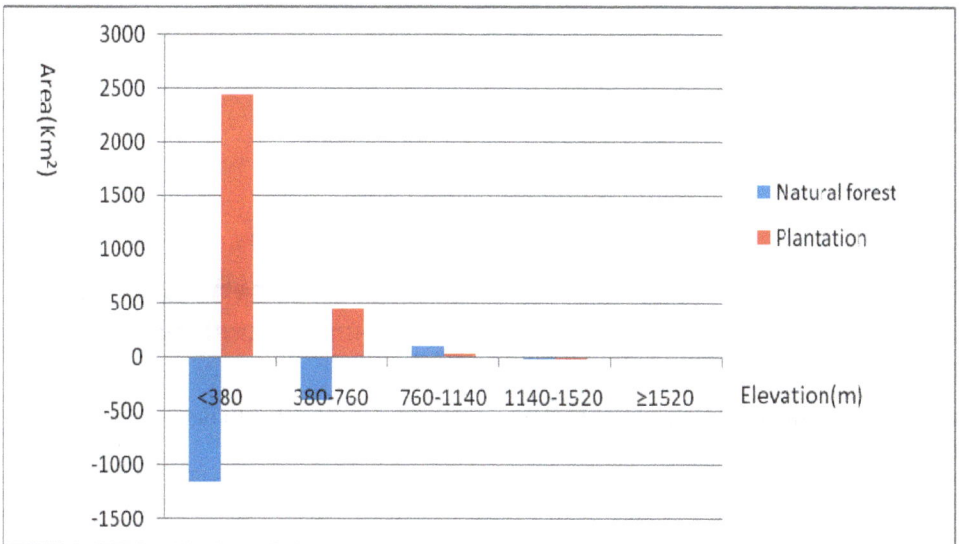

Fig. 2. Area changes of natural forest and plantation in different elevations in Hainan Island, China, 1991-2008.

The remained natural forests of the whole island were mainly distributed in the central and moutainous area, the west part of these forests were well protected due to the establishiment of several large nature reserves. But in the lower east part, the nature reserves covered only a small proportion of the forest (Fig. 3).

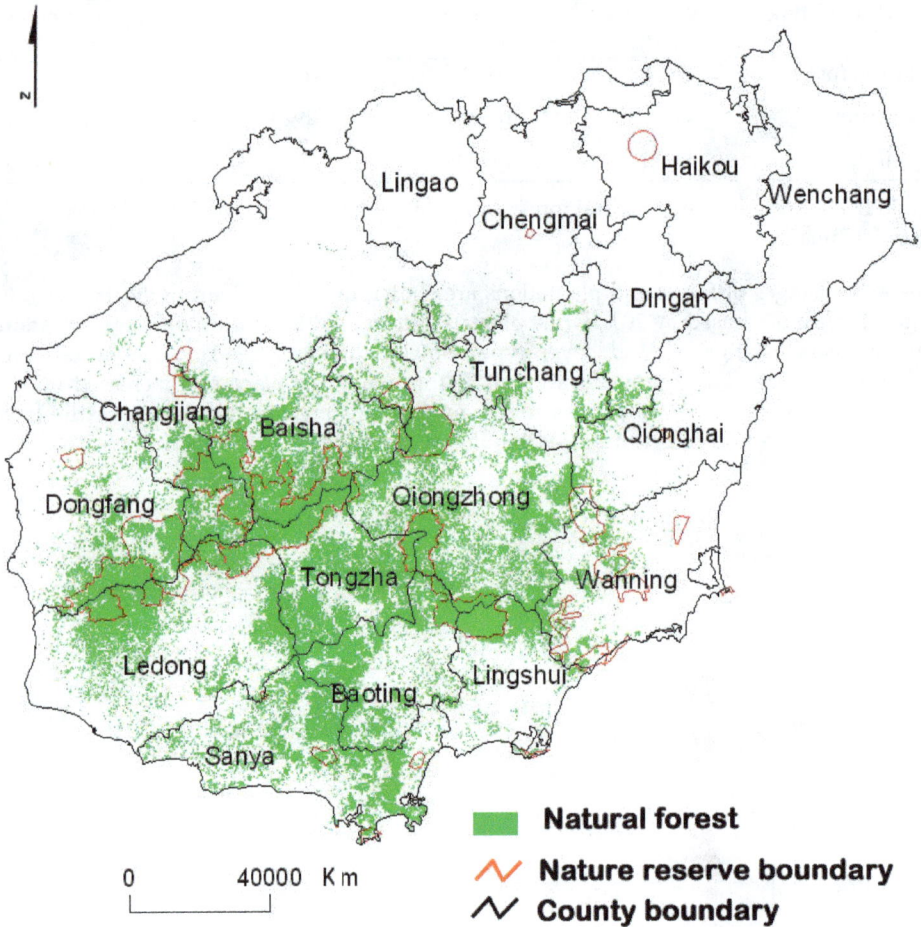

Fig. 3. The natural forest of Hainan, China in 2008 and overplayed with nature reserve boundaries.

4. Discussion

Although the natural forest of Hainan kept the trend of decreasing in the past two decades, the "green" area of the Island on the remote sense image had increased when plantations were included. The forest cover rate was used as an environment-friendly criteria in the state government's annual statistics report, in which the plantation was also defined as forest (Xue, 2011). But the eco-function of mono species plantation has been widely debated, especially if the plantation was built on the degraded land area rather than by replacing the natural forest, and if the native species were planted rather than the exotic species, they were more likely to contribute to the biodiversity (Bremer & Farley, 2010; Xue, 2011). In Hainan Island, the plantations were mainly exotic species (such as eucalyptus, rubber, oil palm and some of the fruit crops) and were cultivated after the forest was cleared. The conversion of natural forests into crop plantations, not only cause severe loss of local biodiversity, but also release considerable amounts of carbon dioxide into the atmosphere (Li et al., 2007; Cotter et al., 2009). In this study, we seperated plantations from the natural forest, and we do think the results were more appropriate for the evaluation of the ecosystem. Thus, we suggest that the plantations should be considered seperately from forest in the future goverment statistics reports, to release more accurate and useful information on the ecosystem changes for policy making.

The land cover change was largely dependant on the local policy and people's livelihood (Rao & Pant, 2001; Zhao et al., 2006). On the topographic map of Hainan Island, more than 95% of townships were distributed below the elevation of 800 m. The growth of human population led to the expansion of agricultural land. The rubber plantings was encouraged during 1970s-1980s. The rubber plantation and other crops land together had wiped out the natural forest between 0-760 m. The natural forest cover above 760 m expanded slowly (Fig. 2) because of lower human density and inappropriate condition for rubber and crops plantations. Even after the logging-ban policy of the local goverment began from 1994 (Lin & Zhang, 2001; Zhang et al., 2010), the collection of fuel or husbandary grazing in the forest could still disturb the forest. In the future, we suggest to freeze the expansion of additional plantations in the low elevation area, and begin to take procedures for low land forest recovery.

About 8.4% of Hainan Island was protected by nature reserves, the coverage was geographically biased toward its central and west mountainous areas with higher elevation, rugged terrain, and fertile soils. Nature reserve coverage was not enough to capture biodiversity features in lowlands, north and northeast plains (Wu et al., 2011). To improve the conservation system of Hainan, more nature reserves should be established in the north and northeast plains, and also the west mountainous region, such as forest patches not covered by the current nature reserve system in Sanya, Baoting, Tongzha and Ledong (Fig. 3).

5. References

Austin, M. P and P. C Heyligers. 1989. Vegetation survey design for conservation: gradsect sampling of forests in northeast New South Wales. Biological Conservation, 50: 13-32.

Bauer, M. E., T. E. Burk, A. R. Ek, P. R. Coppin, S. D. Lime, T. A. Walsh, D. K. Walters, W. Befort, and D. F. Heinzen. 1994. Satellite inventory of Minnesota forest resources. Photogrammetric Engineering and Remote Sensing 60: 287-298.

Bremer, L. L., and K. A. Farley. 2010. Does plantation forestry restore biodiversity or create green deserts? A synthesis of the effects of land-use transitions on plant species richness. Biodiversity and Conservation: 1-23.

Cotter, M., Martin, K., Sauerborn, J. 2009. How do "renewable products" impact biodiversity and ecosystem services - The example of natural rubber in China. Journal of Agriculture and Rural Development in the Tropics and Subtropics 110(1): 9-22.

Li, H., T. M. Aide, Y. Ma, W. Liu, and M. Cao. 2007. Demand for rubber is causing the loss of high diversity rain forest in SW China. Biodiversity and Conservation 16: 1731-1745.

Li, Y. D. 1995. The variance of tropical forest in Hainan Island and strategies for biodiversity conservation. Forest Research 8(4): 455-461. (in Chinese)

Lin, M., and Y. Zhang. 2001. Dynamic Change of Forest in Hainan Island. Geographical Research 20: 706-710. (in Chinese with English abstract)

Meng, J., Y. Lu, X. Lei, and G. Liu. 2011. Structure and floristics of tropical forests and their implications for restoration of degraded forests of China's Hainan Island. Tropical Ecology 52: 177-191.

Moufaddal, W. M. 2005. Use of satellite imagery as environmental impact assessment tool: a case study from the new egpytial red sea coastal zone. Environmental Monitoring and Assessment 107: 427-452.

Mumby, P. J., and A. J. Edwards. 2002. Mapping marine environments with IKONOS imagery: enhanced spatial resolution can deliver greater thematic accuracy. Remote Sensing of Environment 82: 248-257.

Myers, N., R. A. Mittermeier, C. G. Mittermeier, G. A. B. da Fonseca, and J. Kent. 2000. Biodiversity hotspots for conservation priorities. Nature 403: 853-858.

Osborne, P. E., J. C. Alonso, and R. G. Bryant. 2001. Modelling landscape-scale habitat use using GIS and remote sensing: a case study with great bustards. Journal of Applied Ecology 38: 458-471.

Rao, K. S., and R. Pant. 2001. Land use dynamics and landscape change pattern in a typical micro watershed in the mid elevation zone of central Himalaya, India. Agriculture, Ecosystems and Environment 86: 113-124.

Reese, H. M., T. M. Lillesand, D. E. Nagel, J. S. Stewart, and R. Goldmann. 2002. Statewide land cover derived from multiseasonal Landsat TM data- A retrospective of the WISCLAND project. Remote Sensing of Environment 82: 224-237.

Wang, B. S., Zhang W.Y., 2002. The groups and features of tropical forest vegetation of Hainan Island. Guihaia 22(2): 107-115. (in Chinese with English abstract)

Wu, R. D., Ma, G. Z., Long, Y. C., Yu, J. H., Li, S. N., Jiang, H. S. 2011. The performance of nature reserves in capturing the biological diversity on Hainan Island, China. Environmental Science and Pollution Research 18: 800-810.

Xue, J. C. 2011. China's new forests aren't as green as they seem. Nature 477: 371.

Zhang, M., J. R. Fellowes, X. Jiang, W. Wang, B. P. L. Chan, G. Ren, and J. Zhu. 2010. Degradation of tropical forest in Hainan, China, 1991-2008: Conservation implications for Hainan gibbon (*Nomascus hainanus*). Biological Conservation 143: 1397-1404.

Zhao, S., C. Peng, H. Jiang, D. Tian, X. Lei, and X. Zhou. 2006. Land use change in Asia and the ecological consequences. Ecological Research 21: 890-896.

15

Measuring Tropical Deforestation with Error Margins: A Method for REDD Monitoring in South-Eastern Mexico

Stéphane Couturier[1], Juan Manuel Núñez[1] and Melanie Kolb[3]
[1]*Laboratorio de Análisis Geo-Espacial, Instituto de Geografía, UNAM,*
[2]*Centro de Investigación en Geografía y Geomática 'Ing. Jorge L. Tamayo' (CentroGeo),*
[3]*Comisión Nacional para el Uso y Conservación de la Biodiversidad (CONABIO),*
Mexico

1. Introduction

In the second half of the twentieth century, high rates of land use and land cover (LULC) change with severe deforestation trends have caused ecosystem degradation and biodiversity loss all throughout the tropical and sub-tropical belts (Lambin et al. 2003). Estimating the rate of change in tropical forest cover has become a crucial component of global change monitoring. For example, the viability of worldwide schemes such as the reduction of emissions from deforestation and degradation (REDD) depends on an accurate change estimate. Much research has covered the subject of tropical deforestation and degradation (Achard et al., 2010), however, there is so far very little information on the accuracy of quantitative estimates, leaving much room for uncertainty at regional and global scales. In Mexico, for example, the national projections for the rate of deforestation in the past three decades have ranged from 260,000 to 1,600,000 ha/year according to the record of academic studies and official reports (Velázquez et al., 2002). The estimate depended on the total area under study, on remote sensing materials and ground measurements involved in the computation of change rates, but above all none of the studies did contemplate a sampling scheme that would permit the computation of error margins for the rate of change. As a consequence, the alleged recent reduction in deforestation is subject to much political controversy in Mexico. Although, recent advances in Geographic Information Science (GIS) have been made for the accuracy assessment of maps, a standard method for assessing land cover change has not yet been established.

This chapter presents a methodological framework for the measurement of tropical deforestation in Southeast Mexico, based on the experience of accuracy assessment of regional land cover maps and on-site measurements of tropical forest cover in Mexico. In this chapter, we first describe the status of the accuracy assessment of forest cover change maps, an emerging branch of research in GIS. We review the studies that relate to the measurement of deforestation in Mexico and focus on studies where the method for measuring forest cover change is explicitly described. Another section is dedicated to the challenges related to forest canopy change definitions for the assessment and to a sampling design that would encompass the extent of both change and non-change classes. We discuss the need for systematic data as one of the technical limitations to achieve robust estimates. The next sections focus on the framework that is being developed as well as the planned application of the framework in the case of forest cover maps in Southeastern Mexico. As a conclusion, special emphasis lies on the distinctive features which make this case a pioneering experience for deforestation assessments as well as a possibly valuable benchmark for cartographic agencies dealing with forest cover mapping in other sub-tropical regions of the world. Recommendations are drawn for the design of future REDD norms and regulations in Mexico.

2. A review of forest cover change studies and reliability issues

2.1 Deforestation globally and the emergence of REDD

According to the Global Forest Assessment of the year 2010, tropical deforestation is estimated at   16 million ha per year in the period 1990-2000, and a 13 million ha per year in the period 2000-2010 (FAO, 2010). This assessment is a report from the Food and Agriculture Organization (FAO), based on a global database of national estimates of forest area change for the period 1990-2010. These figures reflect in fact a significant institutional effort, at national level, of many sub-tropical countries, for tropical forest mapping since the 1980s. It is thought that the estimated reduction of net forest loss between the 1990s and 2000s is largely due to afforestation, natural forest regrowth, reforestation and forest plantations (Achard et al., 2010). However, the gross deforestation rate is still unacceptably high by the standards of global change processes that have trespassed several internationally recognized planetary boundaries (Rockström et al., 2009), especially biodiversity loss and climate change.

Carbon emissions and fluxes from fossil fuels, cement production and various non-tropical land use changes, mainly as a result of our modern urban consumption habits worldwide, contribute for an estimated 85% of the anthropogenic emissions of greenhouse gases, a major driver of climate change (van der Werf, 2009). The remaining 15% is contributed by deforestation, as well as peat and forest degradation in the tropics, principally through the release of carbon dioxide. This latter emission, estimated at 1.5 +- 0.4 GtC yr-1 is considered significant in the global carbon budget. As a consequence, international discussions were initiated at the United Nations Framework Convention on Climate Change (UNFCCC) 11th Conference Of Parties (COP 11, 2005) on the issue of REDD in sub-tropical countries. The need to provide incentives for REDD was, however, not mentioned until COP-15 (Copenhagen Accord, 2009) in the final declaration of the Heads of State and governments. This declaration encourages the 'immediate establishment of a mechanism including REDD-plus to enable the mobilization of financial resources from the developed countries'. Decision 4/CP.15 deals with the establishment of 'robust and transparent national forest monitoring systems and, if appropriate, sub-national systems'. Indeed, the largest uncertainties of the global carbon budget are on the side of the land-use change balance (IPCC fourth Assessment report: Solomon et al., 2007). Sub-tropical countries are thus expected to demonstrate that they are fulfilling requirements in the framework of the REDD mechanism.

2.2 Forest cover change studies in Mexico

In the United States of Mexico (USM, hereafter 'Mexico'), according to official information in 2007, the extent of forest ecosystems (tropical and temperate forest) was estimated at 65.3

million hectares, which means a significant loss compared to an estimated 69.2 million hectares in 1993. However, the official document "Mexico's REDD+ vision" (CONAFOR, 2010), states that the country went from losing on average 354,035 hectares of forest extent each year for the period 1993-2002 to 155,152 ha for the period 2002 -2007. This means a decrease in deforestation rates of over 50%, highlighting the fact that for the last 5 years 99.9% of the deforestation has occurred in the tropical forests.

This information on deforestation is built on the basis of visual interpretation of medium resolution satellite imagery at the national level (1:250 000 scale) (Mas et al., 2002). As a consequence the interpretation of the deforestation phenomenon is limited to this scale. Additionally, an explicit estimate of error margins for this calculation is not provided. In fact, numerous studies in the country have focused on measuring deforestation, but the diverging results have contributed to a perception of high uncertainty and none has been able to offer error margins to the calculations. This situation of high discrepancies about the forest cover loss is illustrated in table 1, where academic studies and official sources are separated and compared.

Academic sources		Official sources	
Source	(ha/year)	Source	(ha/year)
Grainger, 1984	1,600,000	FAO, 1988	615,000
Repetto, 1988	460,000	SARH, 1992	365,000
Castillo et al., 1989	746,000	SARH, 1994	370,000
Myers, 1989	700,000	FAO, 1995	678,000
Toledo, 1989	1,500,000	FAO, 1997	508,000
Masera et al., 1997	668,000	CONAFOR, 2004	260,000
Velázquez et al., 2002	550,000	FAO (Torres, 2004)	775,800
Sánchez-Colón et al., 2008	484,000	SEMARNAT, 2006	365,000
Sum	6,708,000	Sum	3,936,800
Average	838,500	Average	492,100
Stdev	±451,417	Stdev	±181,851

Table 1. Deforestation rates estimated in Mexico for the last three decades.

Variations in the inputs, projections, scale and timing, have probably contributed to the high variability of the results, and many difficulties have hampered efforts for a unified methodology that might have permitted statistical information on its reliability. Estimates of rates of deforestation seem contradictory and a consequence is a low credibility of the sources and an institutional weakness at designing regulation policies.

Since 2001, the National Commission of Forests (CONAFOR), dependent of the National Environmental Agency in Mexico (SEMARNAT), is in charge of updating the vegetation cover and its change in Mexico, in parallel with the regional LULC cartography produced by the National Institute of Statistics, Geography and Informatics (INEGI: 2002 'Serie III' map, and 2007 'Serie IV' map). None of this cartography to date has been generated with an international standard accuracy assessment scheme as described in this chapter. Since 2004, CONAFOR has established a periodical forest inventory every 5 years ('Inventario Nacional Forestal y de Suelos', INFyS: CONAFOR, 2008); The Mexican territory is monitored, based on a systematic grid of ground plots over the entire vegetation cover of Mexico.

2.3 Accuracy assessment of maps

Global reports on deforestation and forest degradation stem from the FAO and are based on a global database of national estimates of forest area change. These national estimates are obtained through governmental agencies of sub-tropical countries, using Land Use/ Land Cover (LULC) maps at a regional scale, intermediate between local (> 1:50,000) and continental (1:5,000,000). However, the quality of these LULC maps are usually unquestioned, taken for granted, just as if each spatial unit on the map perfectly matched the key on the map, which in turn perfectly matched ground reality. The minimum mapping unit, which defines the scale of the map, is commonly the only information available about the spatial accuracy of these maps and no statistically grounded reliability study is applied as a plain step of the cartographic production process.

Since the 1990s, the classification of satellite imagery has become the standard for LULC mapping programs at the regional scale. However, the classification process is affected by different types of error (Green and Hartley, 2000; Couturier et al., 2009) related in part to the limited discrimination capacity of the spaceborne remote sensor. Indeed, the difficult distinction, on the satellite imagery, between categories (or 'thematic classes') of a cartographic legend (e.g. a density grade of a forest cover) can cause a high percentage of errors on the map, especially maps that were generated by coarse resolution, global satellite sensors. This is why a forest management policy whose strategy is simply 'process map information and rely on the quality of the map' is highly questionable.

In Mexico, as discussed earlier, none of the regional cartography is evaluated using a statistically grounded assessment. This is most unfortunate since the statements of the CONAFOR governmental agency on recent deforestation rates is based on these maps (online geoportal: CONAFOR, 2008). These official statements and figures are then passed unquestioned on to the FAO database. Moreover, the absence of such estimate indicates that these figures stand without error margins, and as such, without statistical validity, so that the deforestation rate may remain the focus of controversial academic and public discussions nationwide. It is worth stating that the online availability of the satellite imagery – a feature advertized by this governmental agency - does not increase the reliability of a parameter derived from the imagery. The extraction of the parameter based on colour tones of the satellite imagery available online is far from trivial and it is simply impossible for a user to quantitatively derive the global reliability of the cartography from internet access to the imagery.

An error bar is sometimes present aside the legend of National Institute of Statistics and Geography (INEGI) maps and indicates an estimate of positional errors in the process of map production. However, the procedure leading to this estimate is usually undisclosed, and any objective interpretation of this estimate by the user is thus discouraged (Foody, 2002). Moreover, such error bar indicates a very reduced piece of information with respect to the thematic accuracy of the map.

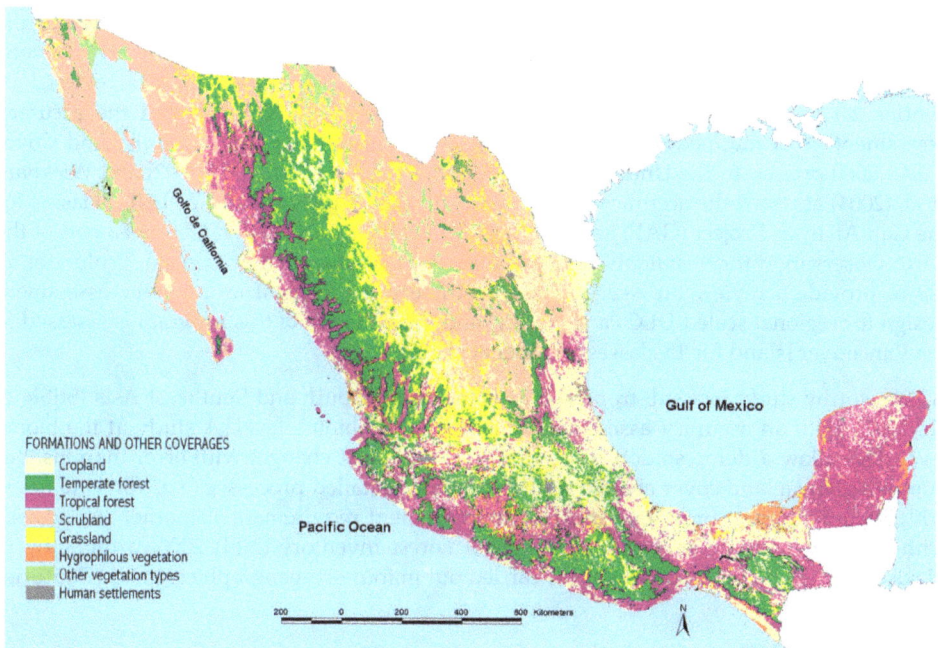

Fig. 1. National Forest Inventory map of Mexico in the year 2000 (taken from Mas et al., 2002)

Instead, the accuracy of a cartographic product is a statistically grounded quantity, which gives the user a robust estimate of the agreement of the cartography with respect to reality. Such estimate is essential when indices derived from cartography – i.e. spatial extent statistics, deforestation rates, land use change analysis - are released to the public or to intergovernmental environmental panels. The accuracy of a map also serves as a measurement of the risk undertaken by a decision maker using the map. On the other hand, this information also allows error propagation modeling through a GIS (Burrough, 1994) in a multi-date forest monitoring task. The construction of the statistically grounded accuracy estimate is generally named 'accuracy assessment'.

Assessing the accuracy of LULC maps is a common procedure in geo-science disciplines, as a means, for example, of validating automatic classification methods on a satellite image. Generally, map accuracy is measured by means of reference sites and a classification process more reliable than the one used to generate the map itself. The classified reference sites are then confronted with the map, assuming that the reference site is "the truth". Stehman and Czaplewski (1998) have proposed a standard structure for accuracy assessment designs, which are divided into three phases:

1. Representative selection of reference sites (sampling design),
2. Definition, processing and classification of the selected reference sites (verification design),
3. Comparison of the map label with the reference label (synthesis of the evaluation).

Agreement or disagreement is recorded in error matrices, or confusion matrices (Card, 1982), on the basis of which various reliability (accuracy) indices may be derived. For regional scale LULC maps, because of budget constraints and the large extension of the map, the complexity of accuracy assessments is increased. Only relatively recently, comprehensive accuracy assessments have been built and applied to the regional or continental LULC maps. In Europe, Büttner and Maucha (2006) reported the accuracy assessment of 44 mapped classes (including 3 forest classes) of the CORINE Land Cover (CLC) 2000 project. In the United States of America (USA), Laba et al. (2002) and Wickham et al. (2004) assessed the accuracy of 1992 maps of, respectively, 29 and 21 LULC classes for the Gap Analysis Project (GAP) and the National Land Cover Data (NLCD). As a part of the Earth Observation for Sustainable Development (EOSD) program of Canada, Wulder et al. (2006) provide a review on issues related to these three steps of an accuracy assessment design for regional scale LULC cartography, and the accuracy of this program is assessed in the Vancouver Island for 18 classes (Wulder et al., 2007).

A noteworthy study in a sub-tropical area is the one in South and Southeast Asia (Stibig et al., 2007), with an accuracy assessment obtained at the biome level. A study at the biome level does allow a deforestation study (forest – non forest change) with error margins, but does not allow a land cover change study with more detailed processes (e.g. 'forest to forest with alteration'), also important in REDD management requirements. Another study deals with the accuracy assessment of the National Forest Inventory (NFI) 2000 cartography in Mexico (figure 1). This assessment was carried out in four eco-geographical areas (Couturier et al., 2010).

2.4 Accuracy assessment of forest cover change maps

Operational forest mapping at the national level using satellite imagery is now a regular task for most of the sub-tropical countries. However, reducing the uncertainty in the national and global carbon budget for REDD mechanisms requires the capability to estimate changes of forest extents in a reliable manner. Technical capabilities and statistical tools have advanced since the early 1990s. Methods have been implemented for forest cover change at national level(e.g. Velázquez et al., 2002), based on either coarse (e.g. Advanced Very High Resolution Radiometer (AVHRR), Moderate Resolution Imaging Spectroradiometer (MODIS), SPOT-VEGETATION) or medium (e.g. Landsat, SPOT) resolution sensors or a combination of both:

- Identification of areas of rapid forest cover change from coarse resolution imagery
- Analysis of wall-to-wall coverage from coarse resolution imagery to identify locations of large deforestation fronts for further analysis with a sample of medium resolution imagery
- Analysis of wall-to-wall coverage from medium resolution imagery from visible or radar sensors

Several studies state that coarse resolution imagery alone should not be used to map changes in forested areas, owing to uncertainty levels (e.g. Achard et al., 2010, Couturier, 2010), which are higher than levels of area changes (Fritz et al., 2009). Land cover maps obtained through coarse resolution imagery can serve as a prior stratification against which future change can be assessed. The use of medium resolution imagery for historical assessment of deforestation has been boosted by the recent free availability of the Landsat Global Land Survey Database(www.glovis.usgs.gov). However, in all cases no accuracy assessment (with a sampling design and higher resolution imagery) of forest cover change has been achieved.

A method has been developed at South Dakota State University (SDSU), as part of the NASA Land Cover and Land Use Change program, to improve the measurement of deforestation at pan-tropical level (Hansen et al., 2008). The method is based on a prior stratification of tropical forests according to forest cover change probabilities derived from time series of coarse resolution imagery. An analysis of medium resolution imagery on the stratified layer permits a rectification or refinement of the first step stratification. This method allowed a targeted sampling of medium resolution imagery, which saves costs because of the synergy coarse – medium resolution data. However, the method did not provide a rigorous protocol for error estimation. For example, one of the challenging features of an assessment design is related to sampling intensity (ratio of sampled surface over total studied surface) for the most extended non-change classes. The strategy does not precisely address this sampling challenge and the results are possibly affected by a strong bias in areas where coarse resolution imagery indicates no-change, because change may have been missed due to the limitation of coarse resolution imagery.

Other academic efforts have focused on making operational the analysis of medium – resolution imagery for comprehensive forest change estimation. For example, the Forest Resources Assessment 2010 programme (FAO, 2010) prepares a Remote Sensing Survey of 20 km x 20 km plots placed on an extensive systematic grid (around 0.9% of the land surface in sub-tropical areas). This approach is expected to deliver globally to regionally accurate estimates of forest cover change in periods 1990-2000 and 2000-2005 for those countries or regions where sampling intensity is sufficient (e.g. Brazil: Broich et al., 2009; the entire Congo River basin: Duveiller et al., 2008). In some regions, this approach has been assessed against wall-to-wall cartography based on medium resolution imagery (e.g. for Brazil: Eva et al., 2010).

Whether through wall-to-wall or sample-based approaches, information derived from fine spatial resolution imagery is the most appropriate data to rigorously assess the accuracy of land-cover change estimation (Achard et al., 2010). For this purpose, the European Space Agency (ESA) is launching an action with the Joint Research Council (JRC) to build a database of high resolution satellite imagery susceptible to produce better estimates of forest cover change in Latin America and South East Asia up to the year 2010.

Díaz-Gallegos et al. (2010) have proposed and applied an accuracy assessment scheme to regional land cover change for the first time in Mexico. The assessed LULC maps are official national level INEGI Serie I (year 1978: INEGI, 1980) and National Forest Inventory (year 2000) maps over several states of Southeast Mexico. The assessment is based on a systematic aerial photograph coverage, and is well adapted to available reference material in Mexico. However, the sampling intensity (43 pairs of photograph) was probably not sufficient to ensure a statistical representation over change and non-change classes. Additionally, some features in the sampling design (e.g. stratification per center of aerial photograph) impeded the calculation of error margins from the accuracy indexes obtained in the study.

Finally, on all the above-cited studies (including the popular FRA study), estimates of deforestation were considered with a minimum mapping unit above 5 hectares. This means only an extensive component of deforestation is measured, and in particular, these estimates do not correspond to the forest definition emitted by FAO as an international standard, as will be seen in the next section.

3. Challenges for the reliable measurement of deforestation

3.1 Forest definitions and forest cover change definitions

As adopted by the United Nations Framework Convention on Climate Change (UNFCCC) at the 7th Conference of the Parties (COP-7, 2001) under the 'Marrakesh Accords', 'For LULC and forestry activities under Article 3, paragraphs 3 and 4 of the Kyoto Protocol (http://unfccc.int/kyoto_protocol/items/2830.php), the following definitions shall apply: (a) 'Forest' is a minimum area of land of 0.05-1.0ha with tree crown cover (or equivalent stocking level) of more than 10-30 percent with trees with the potential to reach a minimum height of 2-5m at maturity in situ. A forest may consist either of closed forest formations where trees of various stories and undergrowth cover a high proportion of the ground, or open forest....' COP-7 further noted that parties recognize that there should be some flexibility. To date, most countries are defining forests with a minimum crown cover of 30%.

As any definition choice would, this official definition (the FAO definition of forest) leads to a number of challenges for consistent forest monitoring worldwide. For instance, a minimum area of 0.05 – 1.0 ha implies that deforestation (understood officially as the 'direct human-induced conversion of forested land to non forested land', UNFCCC, Marrakech Accords, 2001) can certainly not be derived from cartography at 1:250,000 scale (whether generated by medium or coarse resolution imagery); Clearings due to the establishment of large scale mechanized agriculture may be detectable on the coarse scale map but not the removal of forest patches of 0.05 – 1.0 ha. Therefore, coarse scale cartography may detect the amount of a *specific type* of deforestation, which is not deforestation under FAO definition. However, this specific type of deforestation (large area deforestation) is the one reported in FAO worldwide reports and not FAO defined deforestation. It seems though that the FAO definitions of forest and deforestation agreed under the UNFCCC will also serve as a reference for the future REDD mechanism (The Marrakech Accords).

A difficulty in any definition of forest cover change is to handle a sufficiently small minimum area of forest (e.g. 0.5 ha in the FAO definition) and a compatible scale of the available cartography from governmental agencies. Another potential difficulty is related to the variety of vegetation types in a diverse environment, some within the FAO definition of forests and some outside this definition but within the 'Other Forested Land' (OFL) definition. At national level it may be desirable to count the removal of such vegetation cover as deforestation because of its ecological function. Yet the inclusion of many vegetation types within the deforestation count may cause greater levels of uncertainty in deforestation figures. Additionally, there is no official definition of forest *degradation*, but in a REDD-plus context, it is directly related to a loss of carbon stocks in forests due to human activities.

3.2 The sampling design challenge

Apart from the forest definition issue, the *a posteriori* (posterior to mapping efforts) spatial detection of estimated change, and the general dominance of non-change on the map, both pose a challenge for their validation with reference sites. The selection of reference sites is a statistical sampling issue (Cochran, 1977), where strategies have varied according to the application and complexity of the spatial distribution. Stehman (2001) defines the probability sampling, where each piece of mapped surface is guaranteed a non-null probability of inclusion in the sample, as being a basic condition for statistical validity. In most local scale applications, reference sites are selected through simple random sampling. Two-stage (or double) random sampling has been preferred in many studies in the case of regional cartography; in a first step, a set of clusters is selected through, for example, simple random sampling. This technique permits much more control over the spatial dispersion of the sample, which means much reduction of costs (Zhu et al., 2000), and was adopted in the first regional accuracy assessments in the USA, for LULC maps of 1992 (Laba et al., 2002; Stehman et al., 2003).

A random, stratified by class sampling strategy means that reference sites are sampled separately for each mapped class (Congalton, 1988). This strategy is useful if some classes are sparsely represented on the map and, therefore, difficult to sample with simple random sampling. This strategy was adopted by Stehman et al. (2003) and Wickham et al. (2004) at the second stage of a double sampling design and might be useful for the assessment of change classes.

Finally, systematic sampling refers to the sampling of a partial portion of the mapped territory, where the portion has been designed as sufficiently representative of the total territory. This strategy, adopted as a first stratification step, is attractive for small scale datasets and reference material of difficult access. Wulder et al. (2006) define a systematic stratum for the future (and first) national scale accuracy assessment of the forest cover map in Canada.

4. A framework for deforestation measurement with error margins

This research proposes a framework for reliable deforestation measurement in Mexico, with key features based on forest definitions and sampling design. This framework is aimed at contributing to technical specifications for REDD monitoring in Mexico.

4.1 Remote sensor discrimination capacity

As discussed earlier, the deforestation according to UNFCCC talks (FAO definition) cannot be derived through coarse scale (e.g. 1:250,000) data because the minimum area of 0.05 – 1.0 ha is not resolved by coarse scale data. Clearings for large agriculture (usually) mechanized projects or for a massive newly settled migrant population (e.g. relocation programs in Indonesia, colonization of land in the Amazon), may be detected with coarse resolution imagery based on digital analysis (see PRODES (2010) in Brazil). On the contrary, small agricultural clearings or clearings for peripheral settlements require higher resolution data (< 50m x 50m, achieved by medium resolution imagery). Even smaller clearings or degradation of the forest canopy require high resolution (10m x 10m or smaller) imagery and a greater visual control on the interpretation of the imagery. With the experience from earlier studies, we propose to handle three levels of sensor discrimination capacity for the assessment of deforestation and degradation in Mexico (Table 2).

Depending on its capacity, a sensor is able to detect a complex or a simple process; this suggests (next sub-section) that definitions of deforestation should be compatible with the capacity of sensors, in order for the measurement of deforestation to be more reliable. The measurement of a complex process (e.g. forest degradation) also means higher costs than the measurement of a simpler process (forest to non-forest). Also, the use of a sensor of higher capacity (from 1 to 3) means more costs for the measurement of a given process.

Sensor discrimination capacity	Minimum detectable area (ha)	Indicative temporal resolution of sensor set	Access/ indicative cost in Mexico	Processing cost for an area of 200x200km (indicative)
Capacity 1: Resolution 250-1000m (e.g. MODIS, AVHRR)	6-100	Daily	Free	1 day person
Capacity 2: Resolution10-30m (e.g. Landsat, ASTER, SPOT XS multispectral)	0.05-0.30	3 days	$US 250 per scene of 180x180 km² /Free*	3 days person
Capacity 3: Resolution 0.5-5m (ej. Quickbird, GeoEye)	0.01	3 days	$US 20 / km²	30 days person

* Free under governmental agreement for government agencies, higher education and research institutions.

Table 2. Grouping of sensors according to their discrimination capacity for deforestation and degradation processes.

4.2 Forest cover change definitions

The definition of deforestation stems from the FAO definition of forest and refers to the forest – non forest change in a 0.5 ha surface or more. We will name this definition as the 'FAO deforestation'. Symmetrically we define as 'consolidated reforestation' the change from non forest to forest in a 0.5 ha surface or more. These processes can be detected by medium (Capability 2) to high (Capability 3) resolution sensors (Couturier et al., 2010). It is further proposed to attach the 'degradation' process to a physiognomic concept of forest compatible with its detection by remote sensors. Our proposal is to define *forest degradation* as the permanence of forest with a loss of more than 30% of its canopy cover (e.g. a canopy cover of 70% becomes a canopy cover of 40%). It is thought that this process might be detected by medium resolution (Capability 2) sensors, but should be preferably detected (with much higher reliability) by high resolution (Capability 3) sensors.

Because of the difficulties associated with a necessary 0.5 ha minimum mapping unit (or less) for the measurement of FAO deforestation (for this reason, deforestation rates in FAO reports are not 'FAO deforestation'), and because of the attractive characteristics of capacity 1 sensors (daily availability of data), we propose to consider other forest definitions as well.

In the first place, we propose the notion of '*extensive deforestation*', which would refer to the removal of forest in a convex area of 6 ha, and would be associated to a process susceptible of being detected by Capacity 1 sensors. Obviously, the measurement of FAO deforestation is more costly than the measurement of extensive deforestation.

Finally, forested land is ecologically considered to have many life forms in mega-diverse Mexico (Table 3), a few of which are not included in the FAO definition of forest (e.g. some low tropical forests, sub-tropical shrublands, Chaparrales, open oak forests). Many are defined as 'forested vegetation' in Mexico (according to forest law LGDFS) and included in the FAO category of Other Forested Land (OFL). To report the removal of this forested vegetation is very relevant in the case of Mexico.

Formation	Vegetation Type
Temperate Forest	1. Cedar forest, 2. Fir forest, 3. Pine forest, 4. Conifer scrubland, 5. Douglas fir forest, 6. Pine-oak woodland, 7. Pine-oak forest, 8. Oak-pine forest, 9. Oak forest, 10. Mountain cloud forest, 11. Gallery forest.
Tropical forest	*Humid/ evergreen & sub-evergreen tropical forests:* 12. Tropical evergreen forest, 13. Tropical sub-evergreen forest, 14. Tropical evergreen forest (medium height), 15. Tropical sub-evergreen forest (medium height), 16. Tropical evergreen forest (low height), 17. Tropical sub-evergreen forest (low height), 18. Gallery forest.
	Deciduous & sub-deciduous forests: 19. Tropical sub-deciduous forest (medium height), 20. Tropical deciduous forest (medium height), 21. Tropical sub-deciduous forest (low height), 22. Tropical deciduous forest (low height), 23. Tropical forest of thorns.
Scrubland	24. Sub-montane scrubland, 25. Spiny Tamaulipecan scrubland, 26. Cactus-dominated scrubland 27. Succulent-dominated scrubland, 28. Succulent-cactus-dominated scrubland, 29. Sub-tropical scrubland, 30. Chaparral, 31. Xerophytic scrubland, 32. Succulent-cactus-dominated cloud scrubland, 33. Rosetophyllic scrubland, 34. Desertic xerophytic rosetophyllic scrubland, 35. Desertic xerophytic microphyllic scrubland, 36 *Prosopis* spp.-dominated, 37. *Acacia* spp.-dominated, 38. Vegetation of sandy desert.
Grassland	39. Natural grassland, 40. Grassland-huizachal, 41. Halophilous grassland, 42. Savannah, 43. Alpine bunchgrassland, 44. Gypsophilous grassland.
Hygrophilous vegetation	45. Mangrove, 46. Popal-Tular (Hydrophilous grassland), 47. Riparian vegetation.
Other vegetation Types	48. Coastal dune vegetation, 49. Halophilous vegetation.

Table 3. Classification scheme of the INEGI land use land cover cartography (only natural land cover categories are indicated):

For this reason we propose the definition of *total forest cover* that encompass woody plants of low size and shrubs which are not secondary vegetation, at the intersection of the notions of 'Forested vegetation' in Mexico and Other Forested Land in the FAO nomenclature. The notion of '*FAO-Mexico deforestation*' is then defined as the change of total forest cover (Forest + Other Forested Land for FAO) to non-forested cover in a 0.5 ha area. It is noteworthy to mention that Capacity 1 sensors are not likely to detect change in low size forest cover with reasonably high accuracy, even for areas of more than 6 ha (see Couturier, 2010).

In synthesis, we define three notions of deforestation, one of them detectable by Capacity 1 sensors and two by Capacity 2 and 3 sensors, as showed in Tables 4 and 5. The accuracy assessment of the process implies the use of a sensor with more capacity than the sensor involved in the map production process.

Type of Detectable Process	Definition
Extensive deforestation	Change from forest to non-forest in a convex area of 6 ha.
Consolidated extensive reforestation	Change from non-forest to forest in a convex area of 6 ha.
Extensive permanence of forest	Permanence of forest in a convex area of 6 ha.
Extensive permanence of non-forest	Permanence of non-forest in a convex area of 6 ha.

Table 4. Processes detectable by capacity 1 sensors (low spatial resolution)

Type of Detectable Process	Definition
Deforestation (FAO or FAO-Mexico)	Change from forest to non-forest in an area of 0.5 ha.
Consolidated reforestation (FAO or FAO-Mexico)	Change from non-forest to forest in an area of 0.5 ha.
Degradation (FAO or FAO-Mexico)	Permanence of forest but with a decrease of more than 30% of canopy cover (e.g. A canopy cover of 70% decreases to 40%).
Regeneration (FAO or FAO-Mexico)	Permanence of forest but with an increase of more than 30% of canopy cover (e.g. A canopy cover of 40% increases to 70%).
Forest permanence (FAO or FAO-Mexico)	Permanence of forest in an area of 0.5 ha.
Non-forest Permanence (FAO or FAO-Mexico)	Permanence of non-forest in an area of 0.5 ha.

Table 5. Processes detectable by capacity 2 and 3 sensors (medium to high spatial resolution)

4.3 Sampling design for LULC change classes

The method comprises a sampling design that efficiently controls the spatial distribution of samples for all classes of the forest cover change map, including sparsely distributed (or ' rare') change classes. Previous assessments have relied on two-stage sampling schemes, where simple random or stratified by class random sampling was employed in the first stage. Couturier et al. (2007) demonstrated that these strategies fail in the case of sparsely distributed (rare) classes. This research proposes a two-stage hybrid scheme where proportional stratified sampling is employed for the rare change classes.

The first stage of the sampling design consists in the selection of two subsets of Primary Sampling Units (PSUs). The first subset of PSUs is obtained with a simple random selection and shall be used for the assessment of non-change classes. The second subset of PSUs is obtained with a proportional random selection of PSUs, and shall be used for the assessment of change classes. In the latter selection, the probability of selection attributed to each PSU is proportional to the abundance of the change class in that PSU, as described in Stehman et al. (2000, further discussed via personal communication); this mode of selection is retained as an appropriate way for including all classes, in the sample while maintaining a low complexity level of statistics (i.e. standard stratified random formulae to compute estimators of accuracy).

According to this scheme, the PSU selection process is made independently for each change class and a given PSU can be potentially selected multiple times (for rare classes as well as for common classes). This hybrid selection scheme, differentiated according to non-change (a common class) and change (a rare class), was proposed and detailed in Couturier et al. (2007), where its potential advantages with respect to sampling designs formerly applied in the literature were evaluated.

Once the sample PSUs are selected, all points of the second stage grid included within these PSUs are assigned the attribute of their mapped class. The full second stage sample consists of the selection of 200 points [Secondary Sampling Units (SSUs)] for each class mapped in the area. For each non-change (common) class, the selection is a simple random sorting of points within the second stage grid in the first subset of PSUs. For change (rare) classes, the selection of points is obtained via proportional random sampling in the second subset of PSUs, this time with a probability inversely proportional to the abundance of the class. This mode of selection can preserve equal inclusion probabilities at the second stage within a rare class (see the option of proportional stratified random sampling advocated in Stehman et al. 2000. A sequence of ArcView (2010) and Excel-based simple Visual Basic routines, for easy and fast repeated use on vector attributes of each class, was specifically designed to perform this proportional selection at both stages.

5. Preparing the framework for the case of Southeastern Mexico

The Grijalva and Usumacinta rivers in Southeastern Mexico are two of the most important in Mexico and North America. In terms of stream flow, the Usumacinta river (ranks 7th worldwide) is the most important in the Gulf of Mexico after the Mississippi river. The Grijalva – Usumacinta basin, one of the major rain-laden regions in Mexico (figure 2) is characterized by a contrasted anthropogenic transformation of the landscape, ranging from a highly modified coastal plain, to two mountain chains with mainly indigenous agricultural management, to some very well conserved forested lands on the Guatemala border. This contrast reflects the level of incorporation of agricultural products to local, regional or international markets. This research first presents some results of a LULC change study in the Grijalva – Usumacinta basin, based on INEGI national level maps (sub-section 5.1). And then results of a deforestation study which approaches the FAO forest definition are obtained in the Marquéz de Comillas area (sub-section 5.2), a highly dynamic agricultural frontier within the Usumacinta watershed.

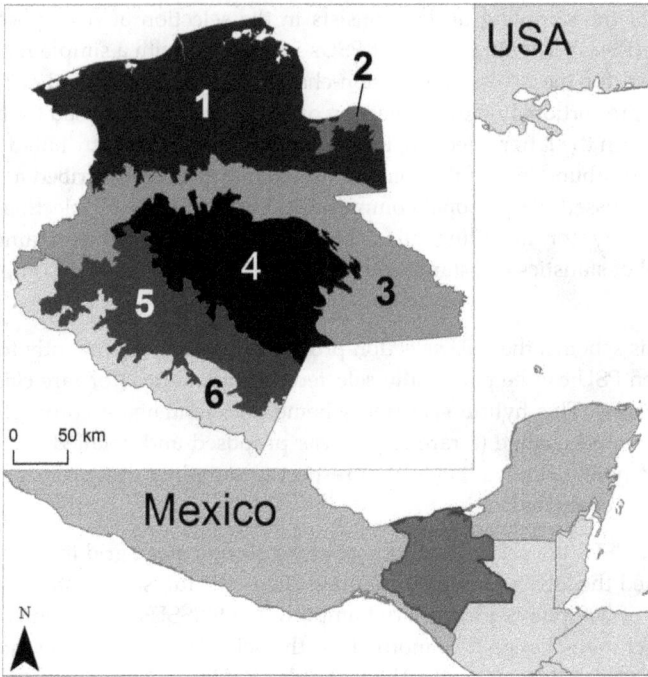

Fig. 2. The Grijalva-Usumacinta watershed and its ecoregions with the main vegetation types. 1) Gulf of Mexico Coastal Plain with Wetlands and Tropical evergreen forest, 2) Hills with High and Medium Tropical sub-evergreen forest, 3) Hills with Medium and High Tropical Evergreen Forest, 4) Chiapas Highlands with Conifer, Oak, and Mixed Forest, 5) Chiapas Depression with Low tropical deciduous and medium tropical sub-deciduous Forest, 6) Central American Sierra Madre with Conifer, Oak, and Mixed Forests.

5.1 'Extensive deforestation' measurement derived from regional maps in Southeastern Mexico

Three spatial data sets of LULC from the INEGI 1:250 000 series were used to analyse changes during the periods 1993–2002 and 2002–2007. For this purpose, the 55 original LULC classes were grouped into 18 categories (Table 6) following a hierarchical classification system developed for the INEGI maps. This system takes into account the vegetation dynamic and gives consistent results in time series analysis (Velázquez et al. 2002).

The level of anthropogenic modification of the forest cover is reflected in the appellation 'primary' versus 'secondary' in order to estimate the forest degradation. Forest degradation is understood as a forest change from a well conserved state ('primary') into a highly modified ('secondary') state. Additionally, 3 temperate and 2 tropical forest types were distinguished in order to specifically analyse LULC changes in each of these forest classes. Pastures for extensive cattle ranching and several agricultural classes were considered, as they are responsible of deforestation processes. Once possible and impossible transitions were established, thematic errors of the maps were detected and corrected with a revision of additional maps.

Formation	Vegetation and land use types	Categories of analysis	Original land use and cover classes
Temperate forests	Coniferous and broad leaved	Coniferous forests (1)	Primary *Juniperus* forests, primary fir forests, primary pine forests, primary pine-oak forests
		Broad-leaved forests (2)	Primary oak forests, primary oak-pine forests
		Montane cloud forests (3)	Primary montane cloud forests
Tropical forests	Rain	Rain forests (4)	Primary evergreen forests (tall, medium and low), primary sub-evergreen forests (medium), primary sub-evergreen forest of thorns (low)
	Dry	Dry forests (5)	Primary deciduous forests (low), primary sub-deciduous (medium)
Hydrophilic vegetation	Mangrove forests, reed, halophilic vegetation	Mangrove forests, reed (6)	Primary and secondary mangrove forests, reed, primary and secondary halophilic vegetation, primary halophilic grasslands
Secondary vegetation	Temperate forests	Secondary coniferous forests (7)	Secondary *Juniperus* forests, secondary fir forests, secondary pine forests, secondary pine-oak forests
		Secondary broad-leaved forests (8)	Secondary oak forests, secondary oak-pine forests
		Secondary montane cloud forests (9)	Secondary montane cloud forests
	Tropical forests	Secondary rain forests (10)	Secondary evergreen forests (tall, medium and low), secondary semi evergreen forests (medium), secondary sub evergreen forest of thorns (low)
		Secondary dry forests (11)	Secondary deciduous forests (low), secondary sub deciduous forests (medium)
Pastures	Pastures	Pastures (12)	Cultivated and induced grasslands, savanna
Cultivated areas	Agriculture	Irrigated agricultura (13)	Irrigated, eventually irrigated, suspended irrigation
		Permanent crops (14)	Permanent and semi-permanent
		Rain fed agricultura (15)	Annual crops
	Plantations	Forest plantations (16)	Forest plantations
Others	Urban areas	Urban areas (17)	Urban areas
	Other vegetation types	Othervegetation types (18)	Primary palm forests, induced palm forests, bare, primary and secondary riparian vegetation and forests

Table 6. Land use land cover (LULC) categories of analysis and classification scheme of the original LULC classes in the Grijalva-Usumacinta watershed (Southeast Mexico).

Deforestation and other changes were mapped to calculate the surface distribution and to capture the patterns of change and permanence. The proportion of change with respect to the initial extent (LULC rate of change), was calculated for each year as follows (FAO, 1996):

$$R = [(1-(A_1-A_2)/A_1)^{1/t})-1]*100 \tag{1}$$

where 'R' is the annual change rate in percentage, 'A1' is the area at 't1' , 'A2' the area at 't2' and 't' the number of years in the period. For deforestation rates, primary and secondary forest classes were aggregated and the results were multiplied by –1 to obtain positive numbers for negative change rates.

One way to determine LULC change dynamics is to establish the major change processes resulting from observed changes; these were defined as:

- deforestation, the conversion of forest into land use classes,
- forest degradation, a process leading to a temporary or permanent deterioration in the density or structure of the vegetation cover,
- transitions, change between different land use classes, and
- regeneration, the transitions of any land use into secondary vegetation.

The land use change processes were identified based on annualized change probabilities calculated with Markov chain properties based on area change matrices with the software package DINAMICA-EGO (Soares-Filho et al. 2009). Afterwards, the transitions with a probability greater than 0.00 were used for an analysis of the major change processes and the related dynamics by subsuming them into principal change processes:

$$P^t = M * V^{1/t} * M^{-1} \tag{2}$$

where 'P' is the annualised probability of change, 'M' the Eigen values of the matrix, 'V' the associated Eigenvectors and 't' the number of time steps within a time period.

The detailed LULC change data revealed that from 1993 to 2007, the major land cover losses were in tropical rain forests, temperate coniferous forests (both >300 000 ha) and secondary tropical dry forests (128 000 ha). For other land cover categories, the loss was smaller and mainly between 1993 and 2002. The primary tropical dry forests had the lowest cover loss (4000 ha). Secondary vegetation increased in almost all forest types, though most gain belonged to secondary coniferous forests (227 000 ha). Among land use classes, the extent of pasture increased most (392 000 ha) followed by rain-fed agriculture (264 000 ha).

However, for the reasons developed in this chapter, the results presented in tables 7 & 8 should be read with caution. The INEGI cartography, a key input of this study, lacks error margins, and does not permit a rigorous assessment of deforestation, forest degradation or regeneration rates in Mexico. Indeed, the partial accuracy assessment of the 2000 NFI map (Couturier et al., 2010) called for prudence in interpreting land cover change from Landsat-based INEGI-like maps, especially in the case of degradation studies. In contrast with the relatively high levels of accuracy of vegetation cover with little modification (classes labeled as 'primary' in the INEGI legend), many errors were reported for classes of highly modified vegetation cover (classes labeled as 'secondary vegetation'). For instance, in the Cuitzeo watershed, the accuracy of sub-tropical scrubland (78%), oak-pine forest (97%), pine forest

	1993-2002	2002-2007	1993-2007
Forest type	Deforestation rate (%)	Deforestation rate (%)	Deforestation rate (%)
Coniferous forests	0.77	0.55	0.69
Broad-leaved forests	1.33	1.31	1.33
Montane cloud forests	0.38	0.18	0.31
Tropical rain forests	0.91	0.68	0.83
Tropical dry forests	2.62	1.55	2.24
Total	1.02	0.70	0.90

Table 7. Deforestation rates for different forest types (Kolb and Galicia, 2011).

	1993-2007		1993-2002		2002-2007	
Land cover and land use classes	Δ Area (ha)	Change rate	Δ Area (%)*	Change rate	Δ Area (%)*	Change rate
Coniferous forests	-316,258	-6.08	79	-6.61	21	-5.13
Broad-leaved forests	-96,317	-6.16	85	-7.39	15	-3.91
Montane cloud forests	-118,878	-3.34	87	-4.34	13	-1.5
Tropical rain forests	-305,440	-2.41	78	-2.79	22	-1.72
Tropical dry forests	-4,263	-3.31	88	-4.36	12	-1.41
Hydrophilic vegetation	1,798	0.02	248	0.07	-148	-0.08
Secondary coniferous forests	227,023	3.09	81	4.06	19	1.37
Secondary broad-leaved forests	50,772	2.89	101	4.56	-1	-0.04
Secondary montane cloud forests	93,881	2.14	89	3.03	11	0.57
Secondary tropical rain forests	93,596	0.74	91	1.05	9	0.18
Secondary tropical dry forests	-127,520	-2.22	78	-2.59	22	-1.55
Pastures	391,513	1.12	69	1.24	31	0.91
Irrigated agriculture	18,294	1.14	99	1.76	1	0.03
Permanent agriculture	-151,041	-3.46	56	-2.69	44	-4.82
Rain-fed agriculture	263,653	2.11	60	2.08	40	2.17
Forest plantations	2,132		10		90	59.29
Other vegetation types	3,816	0.06	7	0.01	93	0.17
Urban areas	15,038	2.94	41	2.09	59	4.49

Table 8. Areas of land use and land cover (LULC) change and change rates for each category and period (Kolb and Galicia, 2011). Δ Area is the difference in area for the different LULC classes for 1993-2007. Δ Area in percentage for 1993-2002 and 2002-2007 is relative to the total change area for 1993-2007.

(79%) and fir forest (76%) contrast with the accuracy of highly modified oak forest (46%), highly modified pine forest (12%) and highly modified mixed forest (45%). From both the taxonomical and landscape points of view, a class of highly modified vegetation cover is close to a wide set of land use classes as well as low modification vegetation cover classes, which makes it prone to more confusions than a class of low modification vegetation cover. These low accuracy levels, however, appear as a real challenge for improving the quality of future forest cover cartography because degradation estimates are probably characterized by very poor reliability, and yet degradation studies are an important part of the REDD - based forest management.

In this research, the accuracy assessment method proposed for the case of *extensive deforestation* measurement with regional LULC cartography consists in a multi-spectral SPOT coverage for reference data and a sampling design defined in section 4.3 where SPOT frames are the PSUs. In the extent of the Grijalva – Usumacinta region, a total of 5 SPOT images per change class and 7 SPOT images for non change classes is thought to achieve a good spatial distribution of the sample. SSUs should be constructed as squared frames centered on the points of the periodic INFyS, and the amount of SSUs should be selected so as to achieve a sampling intensity of at least 4% for all classes.

5.2 'FAO deforestation' measurement in the Marquéz de Comillas area

The forest monitoring program over the Marquéz de Comillas area is an instrument to measure the impact of conservation programs around the Montes Azules Biosphere Reserve (figure 3). The main objective is the measurement, via remote sensing, of the forest cover at the landscape scale. Part of the challenge is to establish deforestation estimates with error margins at a scale approaching the FAO forest definition.

For this purpose, forest cover was defined as: "Areas densely covered by tree vegetation, photosynthetically active at the evaluation season, and canopy cover of more than 30% of the observation area". This definition makes no reference to forest use (e.g., plantations, forest area under management), successional stages (secondary forest or 'acahual', vs low modification or 'primary' forest), or seasonal conditions (sub-evergreen forests). The purpose of this definition of forest cover is to provide a general framework on the dynamics of forest cover.

Landsat TM, ETM + and SPOT HRVIR multispectral images from three different years were used to develop forest cover maps (Table 9). Values were sampled in homogeneous reflectance areas, which were used to search for patterns from a number of independent variables containing spectral and spatial information on the forest cover. From these patterns, a pixel-based probability of ownership to the forest class was derived. We then used a multivariate logistic regression model, in which different spectral and spatial transformations (e.g., vegetation indices and topographic information obtained from a Digital Elevation Model) were the independent variables. The accuracy of the map was measured at every date of study (1990, 2000, 2010), using a stratified random sampling, the visual appraisal of colour composites of Landsat/ SPOT original data and auxiliary ground data.

The forest cover data was derived with accuracies of 91% and Kappa coefficient (K) of 0.7055 for 1990, 88% and K = 0.7540 for 2000 and finally in 2010 the accuracy was estimated at 88% and Kappa coefficient of 0.7660. The estimated annual deforestation rate was -2.1% for the entire period, showing a net loss of 88 098 hectares. In 1990, 95% of the study area was forest cover, while in 2000 forest cover had decreased to 78% and, finally, by 2010 the forest cover declined to 61%. The results show a loss of 4,557 ha/year for the period from 1990 to 2000, down to 4,252 ha/year for the period 2000 to 2010 (Figure 4).

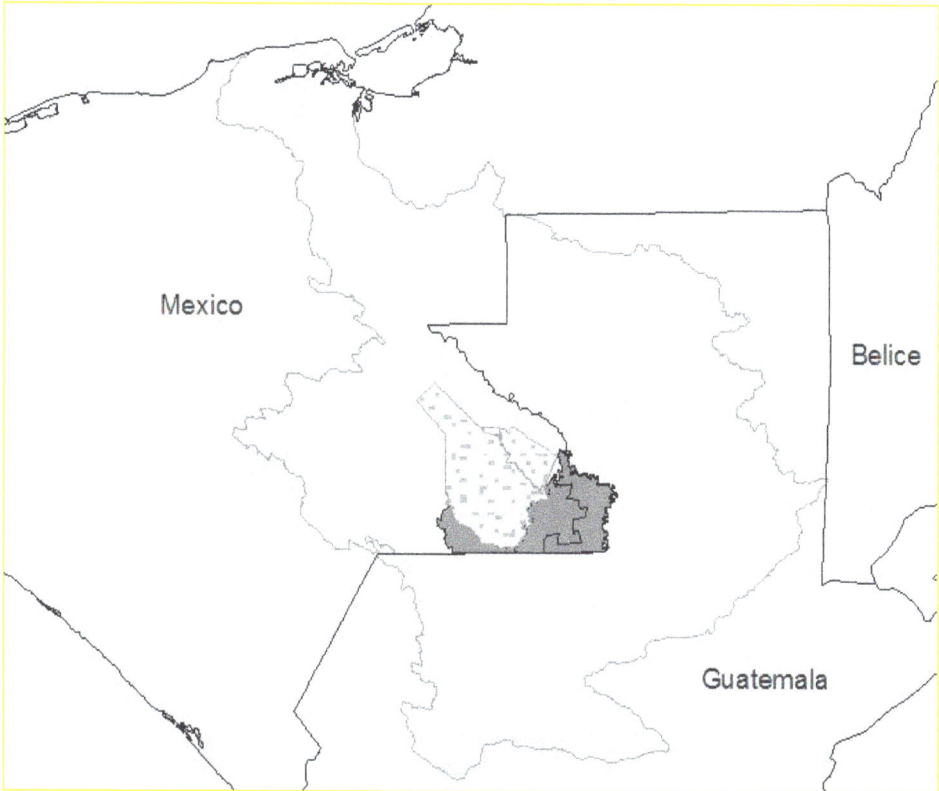

Fig. 3. The 'Marquéz de Comillas' study area is located between Montes Azules Biosphere Reserve (grey dots) and the Mexico-Guatemala border. The Usumacinta watershed was delineated in grey. The study area includes the Mexican municipalities of Maravilla Tenejapa, Marquéz de Comillas and Benemérito de las Americas, in the state of Chiapas.

Platforms & Sensors	Number of scenes	Spectral bands	Pixel size (m)	Year
Landsat 5 TM	2	7	30	1990
Landsat 7 ETM+	2	7	30	2000
SPOT 5 HRVIR	4	4	10	2010

Table 9. Principal characteristics of Landsat TM, ETM + and SPOT HRVIR used for forest cover mapping.

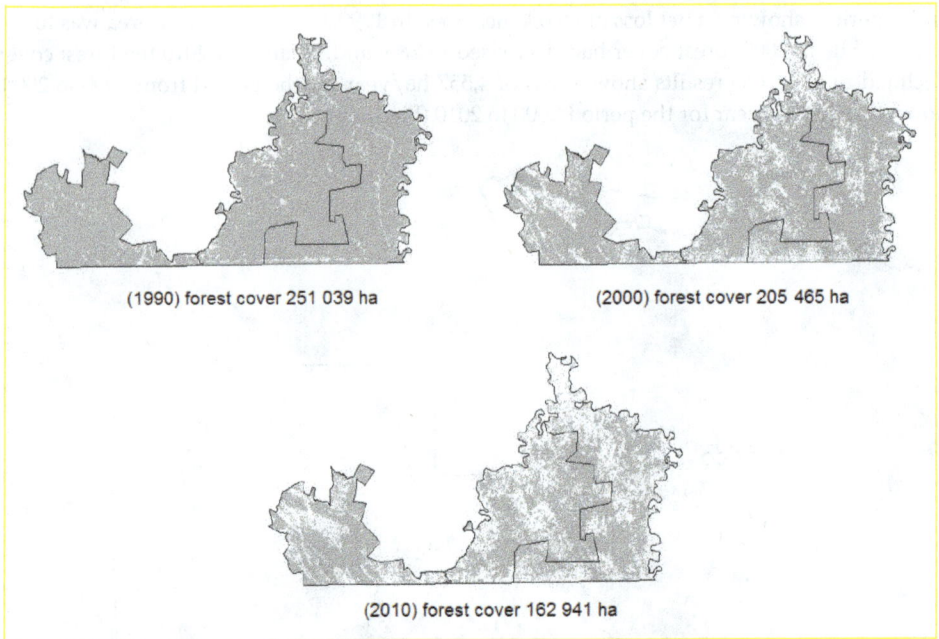

(1990) forest cover 251 039 ha

(2000) forest cover 205 465 ha

(2010) forest cover 162 941 ha

Fig. 4. Forest cover data 1990, 2000 and 2010 for the study area in Chiapas, Mexico.

The accuracy indices correspond to each individual forest cover map in 1990, 2000, and 2010, but the accuracy of the deforestation rate, still cannot be derived. The tools presented in this research will provide grounds for the measurement of *total deforestation* (because the forest definition of this study approaches the FAO definition) in this area, and more generally in the region surrounding the Biosphere Reserve.

For the sake of comparison, we identified studies specific to the region and with forest definitions similar to ours. The rate of change obtained in this study (2.13%) is quite comparable with those reported by Velazquez et al. (2002) about forests in the period 1993 to 2000 (2.06%) and almost equal to that reported for the Lacandona rainforest in the period 1984 to 1991 (2.14%), after 1978 when the Montes Azules Biosphere Reserve was officially decreeted (Mendoza and Dirzo, 1999).

Our deforestation figures are also above the national figures in the same period. According to official FAO reports for the period 1990 to 2010 in Mexico, about 298 000 hectares was lost (FAO, 2010). If we were to compare local with national figures, annual forest loss was estimated in this area at about 4,463 hectares, representing 1.5% of the national loss, in just 0.1% of the country. However, as said earlier, the deforestation definitions are not compatible, by at least two aspects: in the first place the Minimum Mapping Unit of forest is much smaller in our study than in the INEGI national cartography (source of the FAO 2010 Mexico report) and also the percentage of canopy cover is stricter in our study and would not encompass a large variety of forest covers at the national level. In any case, this study perhaps highlights the importance of spatial variability in the dynamics of forest cover throughout the country, and illustrates at what point a national average of *extensive deforestation* hides the magnitude of the fragmentation of forests and of spatial differences throughout the country.

In this research, the accuracy assessment method proposed for the case of local forest - non forest cartography consists in a panchromatic SPOT coverage for reference data and a sampling design defined in section 4.3 where SPOT frames are the PSUs. In the extent of the Marquéz de Comillas area, a total of 2 SPOT images per change class and 3 SPOT images for non change classes are thought to achieve a good spatial distribution of the sample. SSUs should be constructed as squared frames centered on a regular grid of the area, and the amount of SSUs should be selected so as to achieve a sampling intensity of at least 4% for all four classes.

6. Conclusion

International schemes such as REDD related to tropical forest monitoring, critically depend on the reliability of forest cover maps and tropical deforestation rates. In contrast with the poor (almost null) information on this reliability in the world, this chapter provides much evidence from previous studies and experiences that high uncertainty and imprecision still characterize the cartography, remote sensing data, and the forest definitions from which these rates are produced. For example, the FAO reports, which include per-nation tropical deforestation rates (flowchart on figure 5a), are based on national level cartography, provided by sub-tropical agencies, which understandably produce cartography that do NOT (and maybe CANNOT) correspond to the FAO definition of forest, in the first place because this definition requires very fine scale mapping (0.5 hectares Minimum Mapping Unit). Additionally, for the overwhelming majority of governmental agencies in the world, the quality of the cartography is easily confounded with the spatial resolution, or temporality of the satellite imagery used in the map production process. Confusions between thematic classes on the imagery that lead to errors on the map are simply ignored, so that the derived deforestation rates, forest extent baselines, etc. are quantities without error margins, therefore without statistical support.

A rigorous accuracy assessment scheme with appropriate forest definitions and adapted remote sensing data is thus a pending challenge in sub-tropical countries where the baseline cartography is essentially produced. This research proposes a novel deforestation assessment framework, adapted to typical materials and cartography in sub-tropical countries and suitable for REDD schemes. This framework comprises two features. The first feature consists in considering a set of three definitions of forest cover change based on the FAO definitions of forests as well as the Mexican standards on the forest cover definitions. This set of forest cover change definitions permits different levels of deforestation assessment ('*FAO deforestation*' which would reflect *total* deforestation and corresponds to flowchart illustrated in figure 5b, and only '*extensive deforestation*') and considers the need for reporting change of a diversity of vegetation types in Mexico. Accordingly, remote sensors with low or high discrimination capacity are suited to different definitions of deforestation/degradation. The second feature, derived from recent theoretical advances made by the geo-science community, consists in a sampling design that efficiently controls the spatial distribution of samples for all classes, including non-change classes.

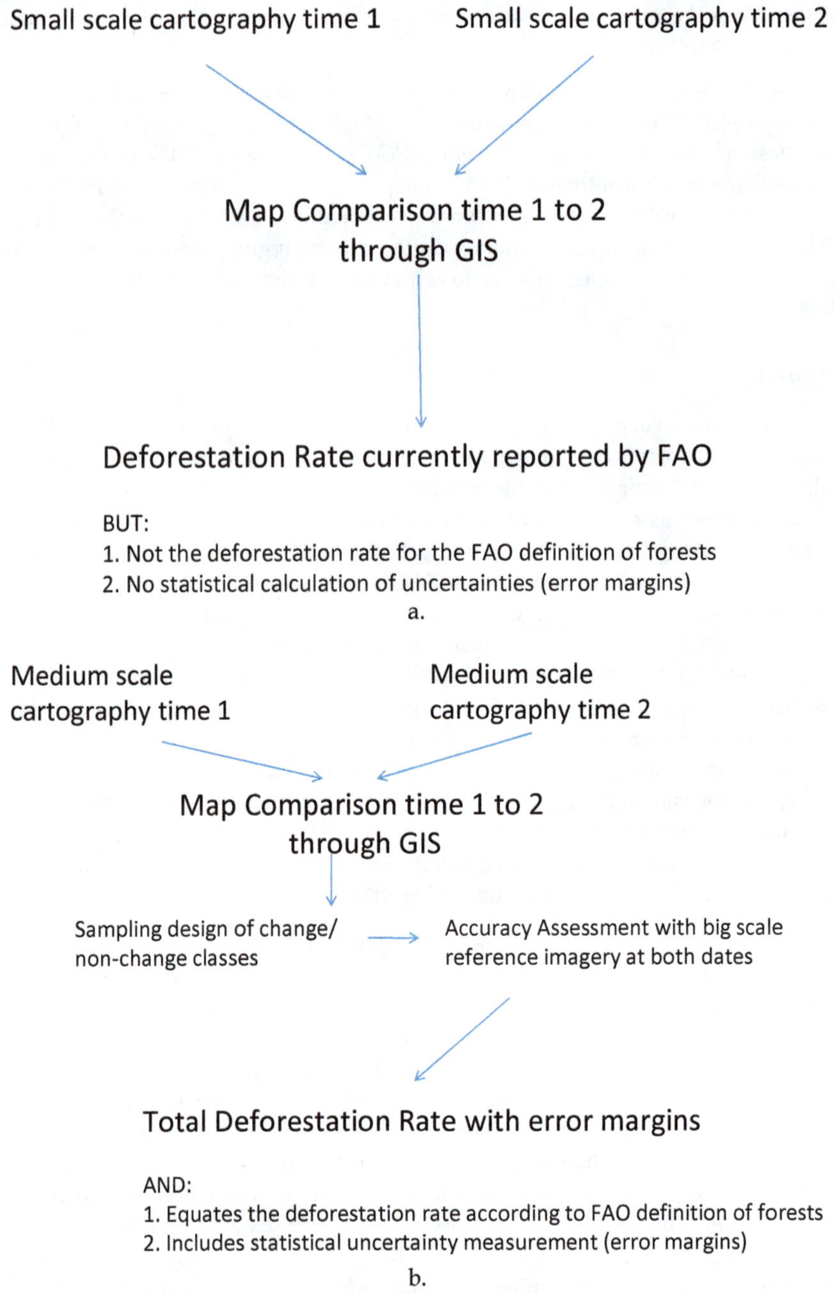

Small scale cartography time 1   Small scale cartography time 2

Map Comparison time 1 to 2
through GIS

Deforestation Rate currently reported by FAO

BUT:
1. Not the deforestation rate for the FAO definition of forests
2. No statistical calculation of uncertainties (error margins)

a.

Medium scale   Medium scale
cartography time 1   cartography time 2

Map Comparison time 1 to 2
through GIS

Sampling design of change/   ⟶   Accuracy Assessment with big scale
non-change classes   reference imagery at both dates

Total Deforestation Rate with error margins

AND:
1. Equates the deforestation rate according to FAO definition of forests
2. Includes statistical uncertainty measurement (error margins)

b.

Fig. 5. Sequence of GIS for deforestation calculation: a. Traditional sequence (which leads to FAO deforestation figures currently), b: Proposed sequence (which is compatible with FAO definition of forests and includes error margins)

This chapter provides the planning for future application of this framework on two cases of ongoing deforestation measurement and analysis, at regional and landscape scales, in biodiverse Southeast Mexico. The first case is the use of typical national level (INEGI) cartography in the Grijalva – Usumacinta basin and the second case is a more optimal use of medium resolution imagery for the measurement of the deforestation in accordance with the FAO definition of forest, in the highly dynamic edge of a National Biosphere Reserve (Montes Azules). In addition, since 2003, the monitoring of deforestation in Mexico is partly ensured using the MODIS sensor (CONAFOR, 2008), which is comparable with the SPOT-VEGETATION sensor used by Stibig et al. (2007) in Asia. We recommend the method presented here be extended to the national level for comprehensive accuracy assessment of these SEMARNAT vegetation cover annual maps. This method would ensure very reasonable costs and would contribute to solve the polemical discussions on the reliability of deforestation rates and land use change rates in the country. We conclude that the work presented here contributes to set grounds for the quantitative accuracy assessment of forest cover change cartography in the context of the REDD programme.

7. Acknowledgments

This research was jointly funded by CONACyT and DGAPA-UNAM. The CONACyT funded Project is entitled "Desarrollo de redes para la Gestión Territorial del Corredor Biológico Mesoamericano – México". The funding was obtained through the resource allocation agreement No. 143289, between the Fomento regional para el Desarrollo Científico Tecnológico y de Innovación (FORDECyT), and the Centro de Investigación en Geografía y Geomática Ingeniero Jorge L. Tamayo A.C (CentroGeo). The two DGAPA-UNAM projects (PAPIIT IN307410 and IACOD IC300111) are related to the construction of a Territorial Observatory for Environmental Risk Assessment in Mexico.

8. Abreviations

ASTER	Advanced Spaceborne Thermal Emission and Reflection Radiometer
AVHRR	Advanced Very High Resolution Radiometer
CLC	CORINE Land Cover Program of Europe
CONAFOR	the Mexican National Commission of Forests (Comisión Nacional Forestal)
COP	Conference Of Parties
CORINE	European 'Coordination of Information on the Environment' Program
EOSD	Earth Observation Sustainable Development Program of Canada
ESA	European Space Agency
FAO	Food and Agriculture Organization
FRA	Forest Resources Assessment
GAP	Gap Analysis Project in the USA
GFA	Global Forest Assessment of the United Nations
GIS	Geographic Information Science
HRVIR	High Resolution Visible and Infra Red
INEGI	National Institute of Statistics, Geography and Informatics in Mexico
INFyS	National Inventory of Forests and Soils in Mexico
IPCC	Intergovernmental Panel on Climate Change

JRC   Joint Research Council
Landsat GLC   Global Land Cover program of the Landsat satellite
LULC   Land-Use and Land-Cover
MODIS   Moderate Resolution Imaging Spectroradiometer
NASA   National Aeronautics and Space Administration of the USA
NFI   Mexican National Forest Inventory
NLCD   National Land Cover Data of the USA
OFL   Other Forested Land
PSU   Primary Sampling Unit
REDD   Reduction of Emissions from Deforestation and Degradation
SPOT   French Satellite for Earth Observation (Systeme Pour l'Observation de la
                     Terre')
SSU   Secondary Sampling Unit
SEMARNAT   National Environmental Agency (Secretaría del Medio Ambiente y
                     Recursos Naturales)
UNFCCC   United Nations Framework Convention on Climate Change
USA   United States of America
USM   United States of Mexico

9. References

Achard, F., Stibig, H.-J., Eva, H.D., Lindquist, E.J., Bouvet, A., Arino, O., & Mayaux, P.
    (2010). Estimating tropical deforestation from Earth observation data. *Carbon
    Management*, Vol. 1, No.2, pp. 271-287.
Arcview (2010). ArcGIS for Desktop Basic, ESRI, Retrieved from:
    http://www.esri.com/software/arcgis/arcgis-for-desktop/index.html
Broich, M., Stehman, S.V., Hansen, M.C., Potapov, P., & Shimabukuro, Y.E. (2009). A
    comparison of sampling designs for estimating deforestation from Landsat
    imagery: a case study of the Brazilian legal Amazon. *Remote Sensing of Environment*,
    Vol. 113, No 11, pp. 2448-2454.
Burrough, P.A. (1994). Accuracy and error in GIS, In *The AGI Sourcebook for Geographic
    Information Systems 1995*, Green, D.R. & D Rix (eds.), pp. 87-91, AGI, London.
Büttner, G., & Maucha, G. (2006). The thematic accuracy of CORINE Land Cover 2000:
    Assessment using LUCAS, EEA Technical Report/No7/2006,
    http://reports.eea.europa.eu/ , accessed 04/2007.
Card, A. (1982). Using known map category marginal frequencies to improve estimates of
    thematic map accuracy, *Photogrammetric Engineering and Remote Sensing*, Vol. 48,
    No. 3, pp. 431-439.
Castillo, P. E. P., P. Lehtonen, M. Simula, V. Sosa and R. Escobar (1989). Proyecciones de los
    principales indicadores forestales de México a largo plazo (1988-2012). Reporte
    interno, Subsecretaría Forestal, Cooperación México-Finlandia, SARH, México
Cochran, W.G. (1977). *Sampling Techniques* (3rd ed.), John Wiley and Sons, New York.
CONAFOR (2004). Baja el índice de deforestación en México. Retrieved from :
    http://fox.presidencia.gob.mx/buenasnoticias/   ?contenido=16205&pagina=308.
    Accessed 12/2008.
CONAFOR (2008). Cartografía de cobertura vegetal y usos de suelo en línea. Retrieved
    from: http://www.cnf.gob.mx:81/emapas/. Accessed 03/2008.

CONAFOR (2010). Mexico's REDD+ Vision. National Forestry Commission. Retrieved from: http://www.conafor.gob.mx :8080/documentos/ver.aspx?grupo=35&articulo=2520

Congalton, R.G. (1988). Comparison of sampling scheme use in generating error matrices for assessing the accuracy of maps generated from remotely sensed data, *Photogrammetric Engineering and Remote Sensing*, Vol. 54 , No. 5, pp. 593-600.

Couturier, S., Mas, J.-F., Vega, A., & Tapia, V. (2007). Accuracy assessment of land cover maps in sub-tropical countries: a sampling design for the Mexican National Forest Inventory map. *Online Journal of Earth Sciences*, Vol. 1, No. 3, pp. 127-135.

Couturier, S., Mas, J.-F., Cuevas, G., Benítez, J., Vega-Guzmán, A., & Coria-Tapia, V. (2009). An accuracy index with positional and thematic fuzzy bounds for land-use/ land-cover maps. *Photogrammetric Engineering and Remote Sensing*, Vol. 75, No. 7, pp. 789-805.

Couturier, S., Mas, J.-F., López, E., Benítez, J., Coria-Tapia, V., & Vega-Guzmán, A. (2010). Accuracy Assessment of the Mexican National Forest Inventory map: a study in four eco-geographical areas, *Singapore Journal of Tropical Geography*, Vol. 31, No. 2, pp. 163-179.

Couturier, S. (2010). A fuzzy-based method for the regional validation of global maps: the case of MODIS-derived phenological classes in a mega-diverse zone. *International Journal of Remote Sensing*, Vol. 31, No. 22, pp. 5797-5811.

Díaz-Gallegos, J.R., Mas, J.-F., & Velázquez, A. (2010). Trends of tropical deforestation in Southeast Mexico. Singapore *Journal of Tropical Geography*, Vol. 31, pp. 180-196.

Duveiller, G., Defourny, P., Desclée, B., & Mayaux, P. (2008). Deforestation in Central Africa: estimates at regional, national and landscape levels by advanced processing of systematically distributed Landsat extracts. *Remote Sensing of Environment*, Vol. 112, No. 5, pp. 1969-1981.

Eva, H.D., Carboni, S., & Achard, F., Stach, N., Durieux, L., Faure, J.-F., & Mollicone, D. (2010). Monitoring forest areas from continental to territorial levels using a sample of medium spatial resolution satellite imagery. *ISPRS Journal of Photogrammetry and Remote Sensing*, Vol. 65, pp. 191-197.

FAO (1988). An interim report on the state of forest resources in the developing countries, Forest Resource Division, Forestry Department, , 40pp.

FAO (1995). Forest Resources Assessment 1990. Global synthesis, Forestry paper No 124, Rome, Retrieved from: http://www.fao.org/docrep/007/v5695e/v5695e00.htm

FAO (1996). Forest Resources Assessment 1990. Survey of tropical forest cover and study of change processes. n° 130. Rome, Retrieved from: http://www.fao.org/docrep/007/w0015e/w0015e00.htm

FAO (1997). State of the world forests 1997. FAO, Retrieved from: http://www.fao.org/docrep/ w4345e/w4345e00.htm

FAO (2004). State of the world forests 2004. FAO, Rome.

FAO (2010). Global Forest Resources Assessment 2010. Main Report, Forestry paper no 163, Rome, Italy. 340pp

Foody, G.M. (2002). Status of land cover classification accuracy assessment, *Remote Sensing of Environment*, Vol. 80, pp. 185-201.

Fritz, S., McCallum, I., Schill, C., Perger, C., Grillmayer, R., Achard, F., Kraxner, F., & Obersteiner, M. (2009). Geo-wiki.org: the use of crowdsourcing to improve global land cover. *Remote Sensor*, Vol. 1, No. 3, pp. 345-354.

Grainger A. (1984). Rates of deforestation in the humid tropics: overcoming current limitations, *Journal of Forestry Resources Management*, Vol. 1: pp. 3-63.

Green, D.R., & Hartley, W. (2000). Integrating photo-interpretation and GIS for vegetation mapping: some issues of error, In: *Vegetation Mapping from Patch to Planet*, Alexander, R. & Millington, A.C. (editors), John Wiley & Sons Ltd., pp. 103-134.

Hagen, A. (2003). Fuzzy set approach to assessing similarity of categorical maps. *International Journal of Geographical Information Science*, Vol. 17, No. 3, pp. 235-249.

Hansen, M.C., Stehman, S.V., Potapov, P.V., Loveland, T.R., Townshend, J.R.G., DeFries, R.S., Pittman, K.W., Arunarwati, B. Stolle, F., Steininger, M.K., Carroll, M. & DiMiceli, C. (2008). Humid tropical forest clearing from 2000 to 2005 quantified by using multitemporal and multiresolution remotely sensed data. *Proceedings of the National Academy of Science of the USA*, Vol. 105, No. 27, pp. 9439-9444.

INEGI (1980). Sistema de Clasificación de Tipos de Agricultura y Tipos de Vegetación de México para la Carta de Uso del Suelo y Vegetación del INEGI, escala 1:125 000. Instituto Nacional de Estadística, Geografía e Informática, Aguascalientes, Ags, Mexico.

Kolb, M. and Galicia, L. (2011). Challenging the linear forestation narrative in the Neo-tropics: regional patterns and processes of deforestation and regeneration in Southern Mexico. *The Geographical Journal*. Online version.

Laba, M., Gregory, S.K., Braden, J., Ogurcak, D., Hill, E., Fegraus, E., Fiore, J.& DeGloria, S.D. (2002). Conventional and fuzzy accuracy assessment of the New York Gap Analysis Project land cover map. *Remote Sensing of Environment* Vol. 81, pp. 443-455.

Lambin, E., Geist, H.J., & Lepers, E. (2003). Dynamics of Land-Use and Land-Cover change in tropical regions. *Annual Review of Environment and Resources*, Vol. 28, pp. 205-241.

Mas, J.-F, Velázquez, A., Palacio-Prieto, J.L., Bocco, G., Peralta, A., & Prado, J. (2002). Assessing forest resources in Mexico: Wall-to-wall land use/ cover mapping. *Photogrammetric Engineering and Remote Sensing*, Vol. 68, No. 10, pp. 966-969.

Mas, J.-F., Velázquez, A., Díaz-Gallegos, J.R., Mayorga-Saucedo, R., Alcántara, C., Bocco, G., Castro, R., Fernández, T. & Pérez-Vega, A. (2004). Assessing land use/cover changes: a nationwide multidate spatial database for Mexico, *International Journal of Applied Earth Observation and Geoinformation*, Vol. 5, No.4:249-261.

Masera, O. R., Ordóñez, M. J, &Dirzo, R. (1997). Carbon emissions from Mexican forests: current situation and long-term scenarios. *Climatic Change*, Vol. 35,pp. 265-295.

Mendoza, E., & Dirzo, R. (1999). Deforestation in Lacandonia (southeast Mexico): evidence for the declaration of the northernmost tropical hot-spot. *Biodiversity and Conservation*. Vol. 8, pps. 1624-1641.

Myers, N. (1989). *Deforestation rates in tropical forests and their climate implications*. Friends of the Hearth, Great Britain.

PRODES (2010). Monitoramento da Floresta Amazônica Brasileira por Satelite, Projeto PRODES. Instituto Nacional de Pesquisas Espaciais (INPE), Sao Jose dos Campos, Brazil.

Repetto, R. (1998). *The forests for the trees? Government policies and the misuse of forest resources*, World Resources Institute, Washington D.C., USA.

Rockström, J., Steffen, W., Noone, K., Persson, A., Chapin F.S., Lambin E., et al. (2009). Planetary boundaries: exploring the safe operating space for humanity. *Ecology and Society*, Vol. 14, No.2, 32 pp. (online)

Sánchez Colón, S., Flores Martínez, A., Cruz-Leyva, I.A. & Velázquez, A. (2008). Estado y transformación de los ecosistemas terrestres por causas humanas. II Estudio de país. CONABIO, Mexico.

SARH (1992). Inventario Forestal nacional de Gran Visión. Reporte principal. Secretaría de Agricultura y Recursos Hidraulico, Subsecretaría Forestal y de fauna Silvestre, México, 49pp.

SARH (1994). Inventario Forestal Nacional Periódico, México 94, Memoria Nacional Secretaría de Agricultura y Recursos Hidráulicos, Subsecretaría Forestal y de Fauna Silvestre, Mexico. 81 pp.

SEMARNAT (2006) Inventarios forestales y tasas de deforestación. Available in: http://app1.semarnat.gob.mx/dgeia/ informe_04/02_vegetacion/recuadros/c_rec3_02.htm. Accessed 08/2008.

Soares-Filho, B.S., Rodrigues, H.O., & Costa, W.L.S. (2009). Modelamiento de Dinámica Ambiental con Dinámica EGO. Centro de Sensoriamento Remoto/ Universidade Federal de Minas Gerais. Belo Horizonte, Brasil. Retrieved from: http://www.csr.ufmg.br/dinamica/. Accessed 01/2011.

Solomon, S., Qin, D., Manning, M. et al. (2007), Technical summary. In: *Climate Change 2007: The Physical Science Basis*. Contribution of Working Group I to the Fourth Assessment Report of the Intergovernmental Panel on Climate Change. Solomon S., Qin D., Manning M. et al. (Eds). Cambridge University Press, Cambridge, UK.

Stehman, S.V., & Czaplewski, R.L. (1998). Design and analysis for thematic map accuracy assessment: fundamental principles. *Remote Sensing of Environment*, Vol. 64, pp. 331-344.

Stehman, S.V. (2001). Statistical rigor and practical utility in thematic map accuracy assessment, *Photogrammetric Engineering and Remote Sensing*, Vol. 67, pp. 727-734.

Stehman, S.V., Wickham, J.D., Smith, J.H., & Yang, L. (2003). Thematic accuracy of the 1992 National Land-Cover Data for the eastern United-States: Statistical methodology and regional results. Remote Sensing of Environment 86, pp. 500-516.

Stehman, S.V., Wickham, J.D., Yang, L., &Smith, J.H. (2000). Assessing the Accuracy of Large-Area Land Cover Maps: Experiences from the Multi-Resolution Land-Cover Characteristics (MRLC) Project, 4th International Symposium on Spatial Accuracy Assessment in Natural Resources and Environmental Sciences, Amsterdam, pp. 601-608.

Stibig, H. J., Belward, A. S., Roy, P. S., Rosalina-Wasrin, U., Agrawal, S., Joshi, P.K., Hildanus, Beuchie, R.), Fritz, S., Mubareka, S., & Giri, C.2007). A land-cover map for South and Southeast Asia derived from SPOT- VEGETATION data. *Journal of Biogeography*, Vol. 34, pp. 625-637.

Toledo, V.M. (1989). *Bio-economic costs of transforming tropical forest to pastures in Latinoamerica. En: S. Hecht (editor). Cattle ranching and tropical deforestation in Latinoamerica.* Westview Press, Boulder, Colorado, USA.

Torres Rojo (2004). Estudio de tendencias y perspectivas del sector forestal en América Latina al año 2020. Informe Nacional México. Juan Manuel Torres Rojo. FAO, Roma.

van der Werf, G.R., Morton, D.C., DeFries, R.S., Olivier, J.G.J., Kasibhatla, P.S., Jackson, R.B., Collatz, G.J., & Randerson, J.T. (2009), CO2 emissions from forest loss. *Nature Geosciences*, Vol. 2, pp. 737-738.

Velázquez, A., Mas J.-F., Díaz, J.R., Mayorga-Saucedo, R., Alcántara, P.C., Castro, R., Fernández, T., Bocco, G., Escurra, E., & Palacio, J.L. (2002). Patrones y tasas de cambio de uso del suelo en México, Gaceta Ecológica, INE-SEMARNAT, Vol. 62, pp. 21-37.

Wickham, J.D., Stehman, S.V., Smith, J.H., &Yang, L. (2004). Thematic accuracy of the 1992 National Land-Cover Data for the western United-States, *Remote Sensing of Environment*, Vol. 91, pp. 452-468.

Wulder, M.A., Franklin, S.F., White, J.C., Linke, J., &Magnussen, S. (2006). An accuracy assessment framework for large-area land cover classification products derived from medium-resolution satellite data, *International Journal of Remote Sensing*, Vol. 27, No. 4, pp. 663-68.

Wulder, M. A., White, J.C., Magnussen, S., & McDonald, S. (2007). Validation of a large area land cover product using purpose-acquired airborne video. *Remote Sensing of Environment*, Vol. 106, pp. 480-491.

Zhu, Z., Yang, L., Stehman, S.V., & Czaplewski, R.L. (2000), Accuracy Assessment for the U.S. Geological Survey Regional Land-Cover Mapping Program: New York and New Jersey Region, *Photogrammetric Engineering & Remote Sensing*, Vol. 66, pp. 1425-1435.

16

Exchange of Carbon Between the Atmosphere and the Tropical Amazon Rainforest

Julio Tóta[1], David Roy Fitzjarrald[2] and Maria A. F. da Silva Dias[3]

[1]*Universidade do Estado do Amazonas,*
[2]*State University of New York,*
[3]*Universidade de São Paulo,*
[1,3]*Brasil*
[2]*USA*

1. Introduction

The terrestrial biosphere is an important component of the global carbon system. The long term exchanges estimates of terrestrial biosphere is a challenge and has resulted in ongoing debate [*Baldocchi*, 2008; *Aubinet*, 2008]. For monitoring long-term net ecosystem exchange (NEE) of carbon dioxide, energy and water in terrestrial ecosystems, tower-based eddy-covariance (EC) techniques have been established worldwide [*Baldocchi*, 2008].

It is now recognized that the EC technique has serious restrictions for application over complex terrain and under calm and stable nighttime conditions with low turbulence or limited turbulent mixing of air [*Goulden et al.*, 1996; *Black et al.*, 1996; *Baldocchi et al.*, 2001; *Massman and Lee*, 2002; *Loescher et al.*, 2006; *Aubinet*, 2008, *Tóta et al.*, 2008]. To overcome this problem, the friction velocity (u*)-filtering approach has been formalized by the FLUXNET committee for the estimation of annual carbon balances [*Baldocchi et al.*, 2001; *Gu et al.*, 2005]. This approach simply discarded calm night's flux data (often an appreciable fraction of all nights) and replaced them with ecosystem respiration rates found on windy nights [*Miller et al.*, 2004]. *Papale et al.*, [2006] pointed out that this approach itself must be applied with caution and the friction velocity (u*) corrections threshold is subject to considerable concerns and is very site specific. *Miller et al.*, [2004] reported that depending on the u* threshold value used to correct the flux tower data at Santarem LBA site (Easterly Amazon Region – Brazil), the area can change from carbon sink to neutral or carbon source to the atmosphere.

The transport of CO_2 by advection process has been suggested by several studies as the principle reason for the "missing" CO_2 at night [*Lee*, 1998; *Finnigan*, 1999; *Paw U et al.*, 2000; *Aubinet et al.*, 2003; *Feigenwinter et al.*, 2004; *Staebler and Fitzjarrald*, 2004]. The search for this missing CO_2 has spurred a great deal of research with the goal of explicitly estimating advective fluxes in field experiments during the last decade, in order to correct the NEE bias over single tower eddy covariance measurements (*Aubinet et al.*, 2003, 2005; *Staebler and Fitzjarrald*, 2004, 2005; *Feigenwinter et al.*, 2004; *Marcolla et al.*, 2005; *Sun et al.*, 2007; *Leuning et al.*, 2008; *Tóta et al.*, 2008; *Yi et al.*, 2008; *Feigenwinter et al.*, 2009a, b).

The complexity of topography and the presence of the valley close to the eddy flux tower have increased the importance to investigating if subcanopy drainage flow account for the underestimation of CO_2 respiration as past studies have asserted [*Froelich and Schmid*, 2006]. The Manaus LBA site (Central Amazon Region – Brazil), is an example of moderately complex terrain covered by dense tropical forest. The NEE bias is reported by preview works [*Kruijt et al.*, 2004; *de Araújo*, 2009; and references there in], and a possible explanation for this is that advection process is happening in that site. This work examines subcanopy flow dynamics and local micro-circulation features and how they relate to spatial and temporal distribution of CO_2 on the Manaus LBA Project site. The contribution of exchange of carbon between the atmosphere and the tropical Amazon Rainforest is discussed and correlated with the present work.

2. Material and methods

2.1 Site description

The study site is located in the Cuieiras Biological Reserve (54° 58′W, 2° 51′S), controlled by National Institute for Amazon Research (INPA), about 100 km northeast from Manaus city. At this site, named K34, was implemented a flux tower with 65m height to monitor long term microclimate, energy, water and carbon exchanges (*Araújo et al.*, 2002), and various studies have been conducted in its vicinity. The measurements are part of the Large-Scale Biosphere-Atmosphere experiment in Amazonia (LBA). Figure 1 presents the study site location including the topographical patterns where the maximum elevation is 120m and the total area (upper panel) is 97.26 km2, with distribution of the 31% of plateau, 26% of slope and 43% of valley [*Rennó et al.*, 2008]. The site area is formed by a topographical feature with moderately complex terrain including a landscape with mosaics of plateau, valley and slopes, with elevation differences about 50m (Figure 1), and with distinct vegetation cover (Figure 2). The eddy flux tower at Manaus K34 site has footprints that encompass this plateau-valley mosaic.

The vegetation cover on the plateau and slope areas is composed by tall and dense terra firme (non-flood) tropical forest with height varying between 30 to 40m, maximum surface area density of 0.35 m^2m^{-3} (Figure 2b, see also *Parker et al.*, [2004]), and average biomass of 215 to 492 ton.ha^{-1} [*Laurance et al.*, 1999; *Castilho*, 2004].

On the valley area, the vegetation is open and smaller with heights from 15 to 25 m, but with significant surface area density more than the 0.35 m^2m^{-3} (Figure 2b). The soil type on the plateau and slopes area is mainly formed by Oxisols (USDA taxonomy) or clay-rich ferrasols ultisols (FAO soil taxonomy), while on the valley area, waterlogged podzols (FAO)/spodosols (USDA) with sand soil low drained predominates. Also, in the valley area the presence of small patchy of *Campinarana* typical open vegetation with low biomass is also common [*Luizão et al.*, 2004].

The precipitation regime on the site show wet (December to April) and dry (June to September – less than 100 mm.month^{-1}) periods. The total annual rainfall is about 2400 mm and the average daily temperature is from 26 (April) to 28°C (September). For more detailed information about the meteorology and hydrology of this site see *Waterloo et al.* [2006] and *Cuartas et al.* [2007].

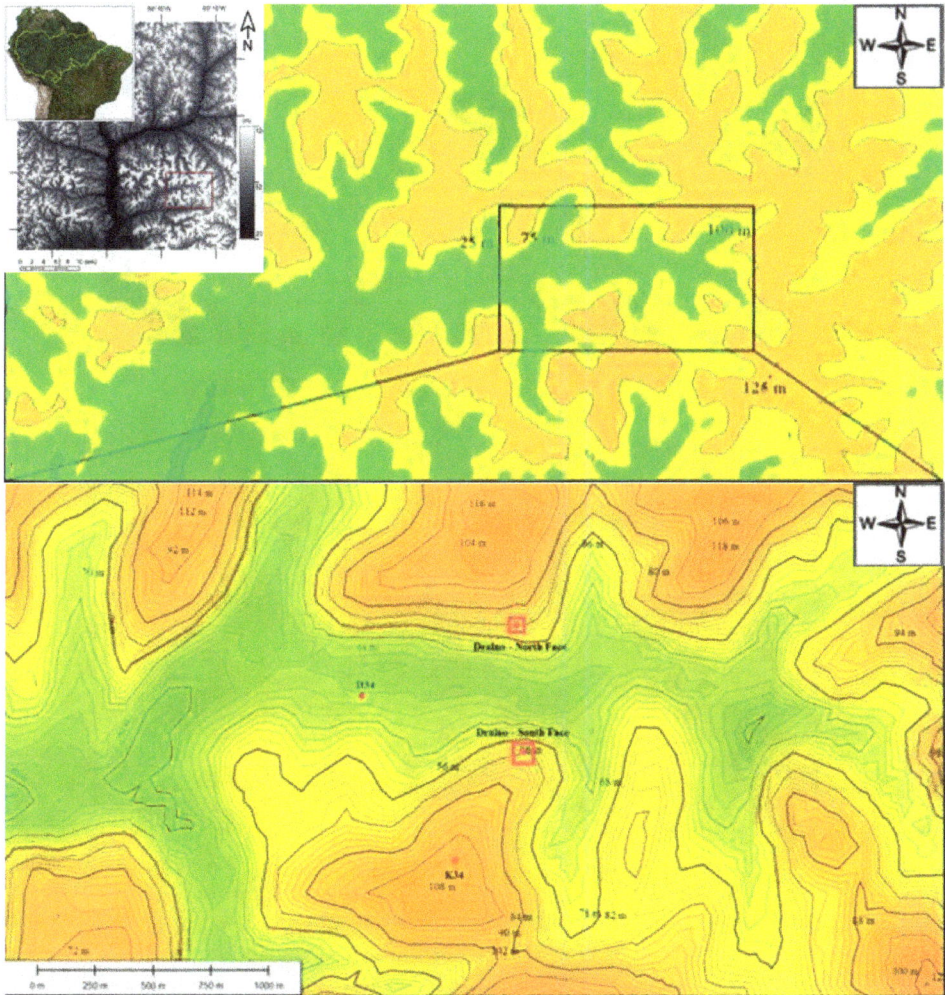

Fig. 1. Detailed measurements towers's view in the ZF-2 Açu catchment (East-West valley orientation) from SRTM-DEM datasets. The large view in the above panel and below panel the points of measurements (B34 – Valley, K34 – Plateau, and subcanopy Draino system measurements over slopes in south and north faces (red square).

a)

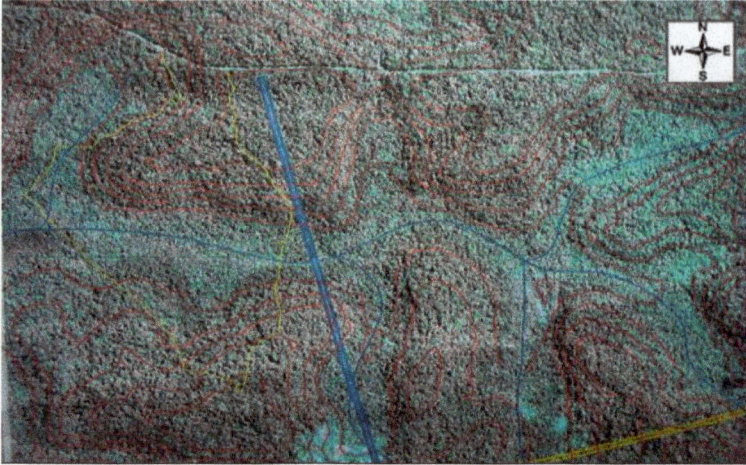

b)

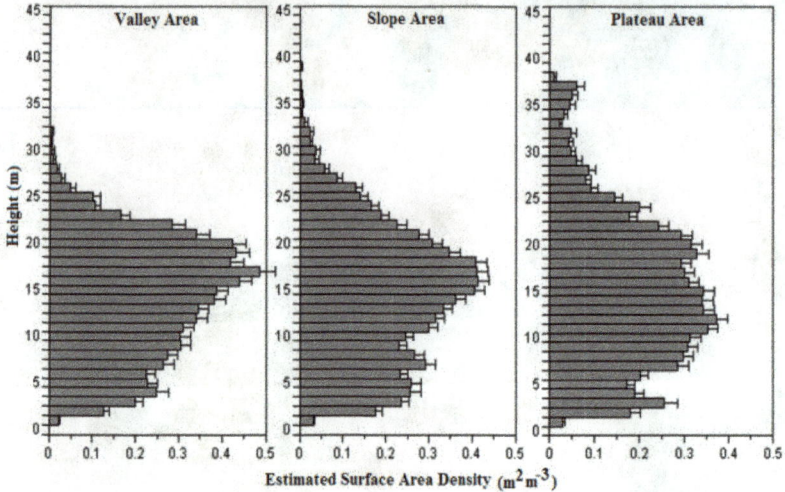

Fig. 2. (a) IKONOS's image of the site at Açu Cachment with level terrain cotes and vegetation cover and (b) vegetation structure measured from LIDAR sensor over yellow transect in (a). From (a) the valley vegetation (blue color) and vegetation transition to plateau areas (red colors).

2.2 Measurements and instrumentation

The datasets used in this study include a measurement system to monitor airflow above and below the forest, horizontal gradients of CO_2, and the thermal structure of the air below the canopy, named "DRAINO System" [see, *Tóta et al.*, 2008]. The data used in this study were collected during the wet season (DOY 1-151) and the dry season (DOY 152-250) of the year 2006. Complementary information was used from flux tower K34 (LBA tower) on the plateau, and sonic anemometer data collected in the valley flux tower (B34, see *de Araújo*

[2009] for details). The flux tower K34 includes turbulent EC flux and meteorological observations of the vertical profiles of the air temperature, humidity and CO_2/H_2O concentrations, and vertical profile of wind speed, as well as radiation measurements. The fast response eddy flux data were sampled at 10 Hz and slow response (air temperature and wind profiles) at 30 min average [see *Araújo et. al.*, [2002] for details information].

- DRAINO measurement System – Manaus LBA ZF2 site

The Draino measurement system used in Manaus LBA Site was similar to that developed by State University of New York, under supervision of Dr. David Fitzjarrald, and applied at Santarem LBA Site, including the same methodological procedures and sampling rates [see, *Tóta et al.*, 2008]. However, due to the terrain complexity, it was modified for Manaus forest conditions including a long distance power line and duplication of CO_2 observations for different slopes areas (Figure 4). The Draino measurement system used in Manaus LBA Site was mounted in an open, naturally ventilated wooden house (Figure 3).

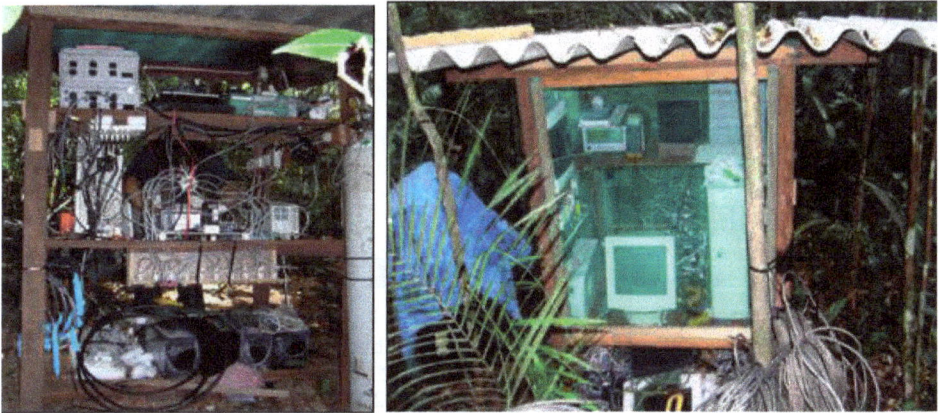

Fig. 3. Draino measurement system used in Manaus LBA (South Face, see also Figure 4).

The system and sensors were deployed (Figure 4) with measurements of air, temperature and humidity (red points), CO_2 concentration (green points), and wind speed and direction (blue points), for both south and north faces. The observations of the 3-D sonic anemometer were sampled at 10 Hz and all the other parameters (CO_2, H2O, air temperature and humidity) were sampled at 1 Hz (Figure 4).

The acquisition system developed at ASRC was employed (Staebler and Fitzjarrald, 2005). It consists of a PC operating with Linux, an outboard Cyclades multiple serial port (CYCLOM-16YeP/DB25) collecting and merging serial data streams from all instruments in real time, the data being archived into 12-hour ASCII files. At Manaus LBA Site two systems in the both south and north valley slope faces were mounted (Figure 3 and 4).

For each slope face, a single LI-7000 Infrared Gas Analyzer (LI-COR inc., Lincoln, Nebraska, USA) was used. A multi-position valve (Vici Valco Instrument Co., Inc.) controlled by a CR23x Micrologger (Campbell Scientific, Inc., Logan, Utah, USA), which also monitored flow rates was also used. This procedure minimizes the potential for systematic concentration errors to obtain the horizontal and vertical profiles. Following *Staebler and*

Fitzjarrald [2004] and *Tóta et al.* [2008] a similar field calibration was performed during the observations at Manaus LBA Site, including initial instrument intercomparison.

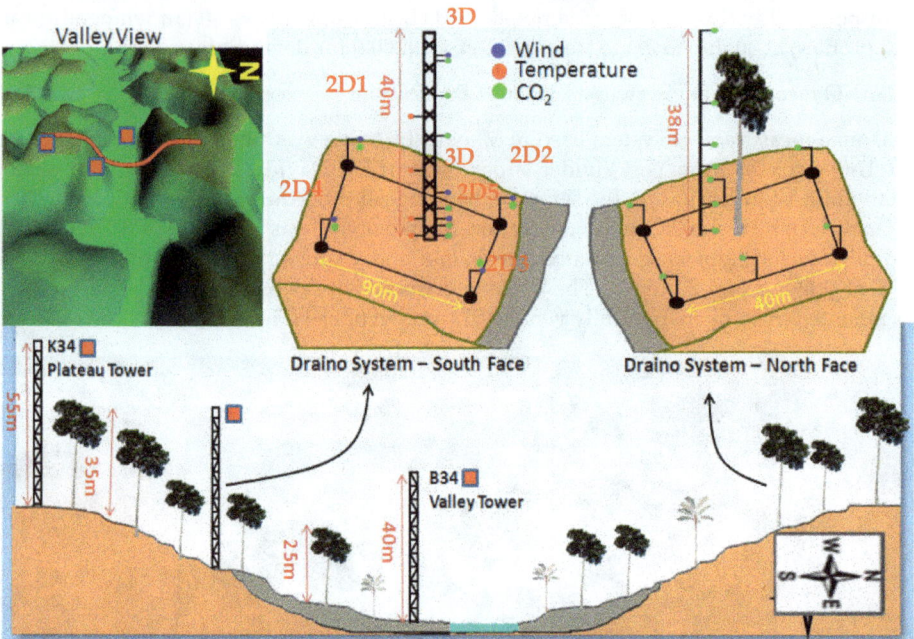

Fig. 4. Draino measurement system (South and North Slope face) implemented at Manaus LBA Site, including topographic view and instrumentation deployed.

The result was similar to that obtained by *Tóta et al.* [2008], with CO_2 mean standard error was < 0.05 ppm and mean standard error of about 0.005 ms-1 for wind speed measurements. After intercomparison, the sonic anemometers and the CO_2 inlet tubes were deployed as shown in Figure 4.

On the south face, the instrument network array (Figure 4 and Table 1) consisted of 6 subcanopy sonic anemometers, one 3-D ATI (Applied Technologies Inc., CO, USA) at 2m elevation in the center of the grid (named 3-D ATI), and 5 SPAS/2Y (Applied Technologies Inc., CO, USA), 2-component anemometers (1 sonic at 6m in the grid center and 4 sonic along the periphery at 2m, see Figure 4), with a resolution of 0.01 m s-1. Also, a Gill HS (Gill Instruments Ltd., Lymington, UK) 3-component sonic anemometer was installed above the canopy (38 m). The horizontal gradients of CO_2/H_2O were measured in the array at 2 m above ground, by sampling sequentially from 4 horizontal points surrounding the main tower location at distances of 70-90m, and from points at 6 levels on the main Draino south face tower, performing a 3 minute cycle. On the north face, similar CO_2 measurements were mounted including a 6 level vertical profile and 6 points in the array at 2 m above ground, performing a 3 minute cycle.

On both slope faces the air was pumped continuously through 0.9 mm Dekoron tube (Synflex 1300, Saint-Gobain Performance Plastics, Wayne, NJ, USA) tubes from meshed inlets to a manifold in a centralized box. A baseline air flow of 4 LPM from the inlets to a central manifold was maintained in all lines at all times to ensure relatively "fresh" air was being sampled. The air was pumped for 20 seconds from each inlet, across filters to limit moisture effects. The delay time for sampling was five seconds and the first 10 seconds of data were discarded. At the manifold, one line at a time was then sampled using an infrared gas analyzer (LI-7000, Licor, Inc.). To minimize instrument problems, only one LI-7000 gas analyzer sensor, for each slope face, was used to perform vertical and horizontal gradients of the CO_2.

Level (m)	Parameter	Instrument
38	u' v' w' T'	Gill 3D sonic anemometers
2	u' v' w' T'	ATI 3D sonic anemometer
6,2	u' v' w' T'	CATI/2 2D sonic anemometers
2	CO_2 Concentration (horizontal array)	LI-7000 CO_2/H_2O analyzer
38,26,15,3,2,1	CO_2, H_2O Profile (Sourth face)	LI-7000 CO_2/H_2O analyzer
35,20,15,11,6,1	CO_2, H_2O Profile (North face)	LI-7000 CO_2/H_2O analyzer
18,10,2,1	Air Temperature and Humidity	Aspirated thermocouples

u', v', w': wind components and T', air temperature fluctuation

Table 1. DRAINO system Sensors at ZF2 LBA Manaus Site

3. Results and discussion

The datasets analyzed in this study were obtained during the periods defined by dry (DOY 1-150 January to June) and wet (DOY 152-250 July to October) seasons of 2006. Figure 5 presents an example of the datasets cover, with 10 days composite statistic, for CO_2 concentration and air temperature at south face area of the DRAINO system and the total precipitation on the plateau K34 tower measurements.

The measurements covered almost the entire year of 2006, including dry, wet and the transition from wet to dry season. The air temperature amplitude above canopy on the slope area of the DRAINO System was higher, as expected, in the dry season. A good relationship is observed between CO_2 concentration and air temperature with much large amplitudes in the dry season than in the wet season. It is probably associates with less vertical mixing during dry than wet season producing much higher subcanopy CO_2 concentration and vertical gradient along the forest.

3.1 Air temperature field

3.1.1 Plateau K34 tower

The vertical profiles of air temperature from plateau K34 tower show a very different pattern from that on the slope area, probably due to canopy structure differences (Figure 2b, *Parker et al.*, [2004]). The canopy structure is important for characterizing its thermal regime as it can be seen in Figure 6. The mean canopy layer stores large quantity of heat during the daytime and distributes it downward and upward throughout the nighttime (Figure 6, 7).

Above canopy layer, over plateau area, the neutral or unstable conditions were predominant during the daytime for both seasons (Figure 6a, c). During the nighttime, stable conditions dominates during dry period (Figure 6b) and neutral to stable conditions for the wet period

(Figure 6d). Similar pattern has been reported elsewhere for plateau forests in the Amazonia (*Fitzjarrald et al.,* [1990]; *Fitzjarrald and Moore,* [1990]; *Kruijt et al.,* [2000]; *Goulden et al.,* [2006]).

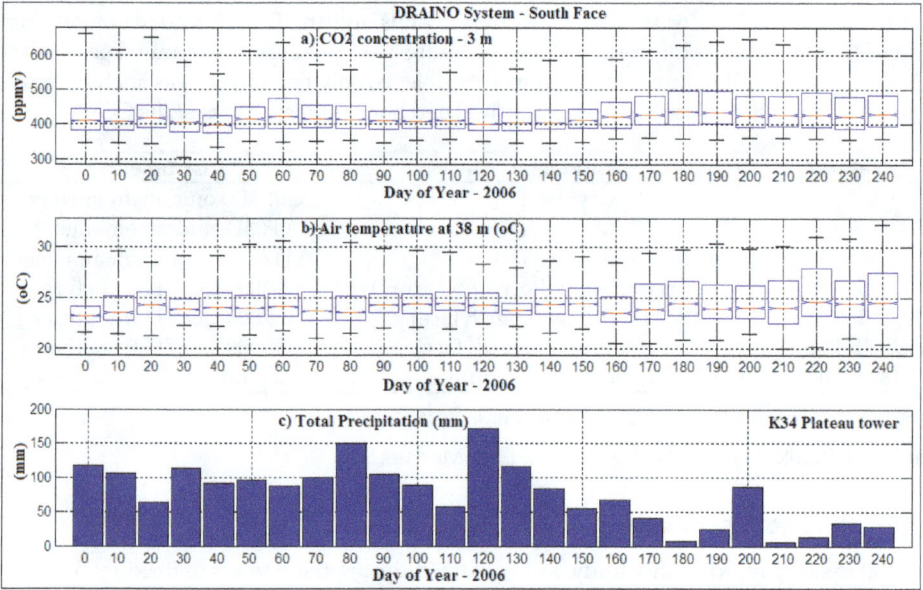

Fig. 5. 10 days time series of the CO_2 concentration (a), air temperature (b) (DRAINO System) and total precipitation (c) (plateau tower).

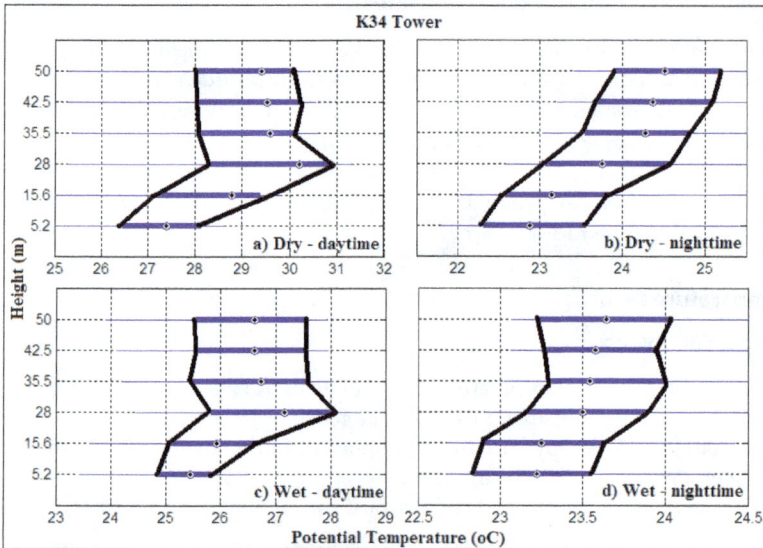

Fig. 6. Boxplot of the virtual potential temperature vertical profile for dry (a, b) and wet periods (c, d) of the 2006 during night (b, d) and daytime (a, c), on the plateau K34 tower.

The below-canopy layer of ambient air on the plateau area was stable at all times (Figure 6a, b, c, d), indicating that this layer is stable where the cold air concentrated in the lower part of the canopy air space as shown in Figure 7.

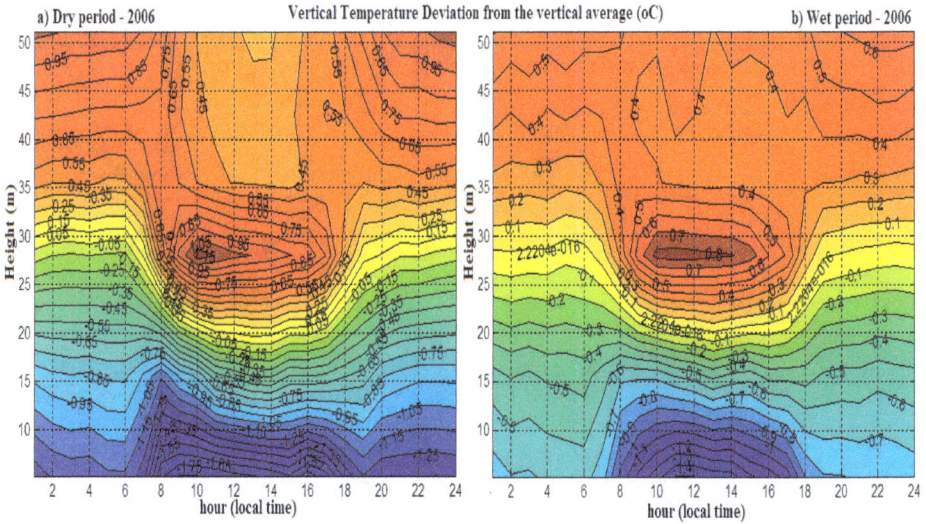

Fig. 7. Daily course of the vertical deviation of the virtual potential temperature ($[\theta'_{v=}\theta_v(z) - \overline{\theta_v(z)}_{5.2}^{55}]$), during dry (a) and wet (b) periods of the 2006, over plateau K34 tower.

The Figure 7 presents daily course of the vertical deviation of the virtual potential temperature, e.g., ($[\theta'_{v=}\theta_v(z) - \overline{\theta_v(z)}_{5.2}^{55}]$), the temperature differences from each level in relation to the vertical average profile. The subcanopy air space was relatively colder during both dry and wet season, showing a similar feature of strong inversion. The same pattern was reported by *Kruijt et al.* [2000] measured over a tower located 11 km northeast of our site with a similar forest composition.

Note that a very interesting length scale can be extracted from the observation when the deviation is about zero. The vertical length scale has mean value of about 30 m during nighttime and 20 m during daytime (yellow color in the Figure 7a, b). Those values are comparable with above canopy hydrodynamic instability length scale used in most averaged wind profile models [*Raupach et al.*, [1996]; *Pachêco*, [2001]; *Sá e Pachêco*, [2006]; *Harman and Finnigan*, [2007]].

3.1.2 DRAINO system slope tower

On the slope area south face (see Figure 2) air temperature at 5 levels underneath the canopy (heights 17, 10, 3, 2, and 1 m) was measured. The observations of the air temperature profile inside canopy are used to monitor the possible cold or warm air layer that generates drainage flow on the slope area. Figure 8 presents observations of the virtual potential temperature vertical profile for both dry and wet periods, during both day and nighttime.

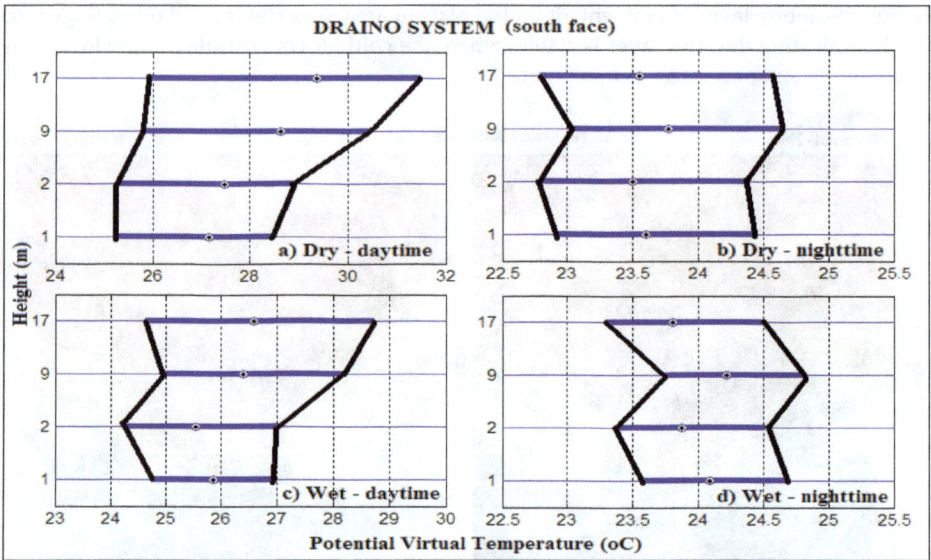

Fig. 8. Boxplot of the virtual potential temperature vertical profile for dry (a, b) and wet periods (c, d) of the 2006 during night (b, d) and daytime (a, c), on the slope area DRAINO System tower (south face, see Figure 2).

The pattern on the slope area is clearly very different when compared with that on the plateau K34 area (Figure 6), except in dry period during daytime when the air was stable inside the canopy. During nighttime (wet and dry periods) a very stable layer predominates with inversion at about 9 m. These can likely be interpreted as a stable layer between two convective layers is associated with cold air (Figure 8). Yi [2008] hypothesized about a similar "*super stable layer*" developing during the night in sloping terrain at the Niwot Ridge AmeriFlux site. This hypothesis suggests that above this layer, vertical exchange is most important (vertical exchange zone) and below it horizontal air flow predominates (longitudinal exchange zone). The relationship between subcanopy thermal structure and the dynamic of the airflow on the slope area will be discussed in next section.

Figure 9 presents a daily cycle composite of the virtual potential temperature deviation from the vertical average ($[\theta_v(z) - \overline{\theta_v}(z)_1^{18}]$). There is persistent cold air entering during nighttime for both dry and wet periods, a characteristic pattern observed on the slope area. It is a very different vertical thermal structure from that of the plateau area.

The cold air in the subcanopy upper layer is probably associated with top canopy radiative cooling, while the cold air just above floor layer is associated with upslope wind from the valley area (as discussed later in the next section).

The average of the vertical gradient virtual potential temperature was negative during nighttime and positive during daytime for both periods dry and wet (Figure 9). This observation shows that during the daytime a relative cooler subcanopy air layer predominates creating inversion conditions. In contrast, a relative hotter subcanopy air layer generates a lapse conditions during nighttime. In general that is not a classical thermal condition found on the sloping open areas without dense vegetation. This general pattern was present at several specific study cases not show here due limited size paper. A similar pattern was reported by *Froelich and Schmid* [2006] during "leaf on" season.

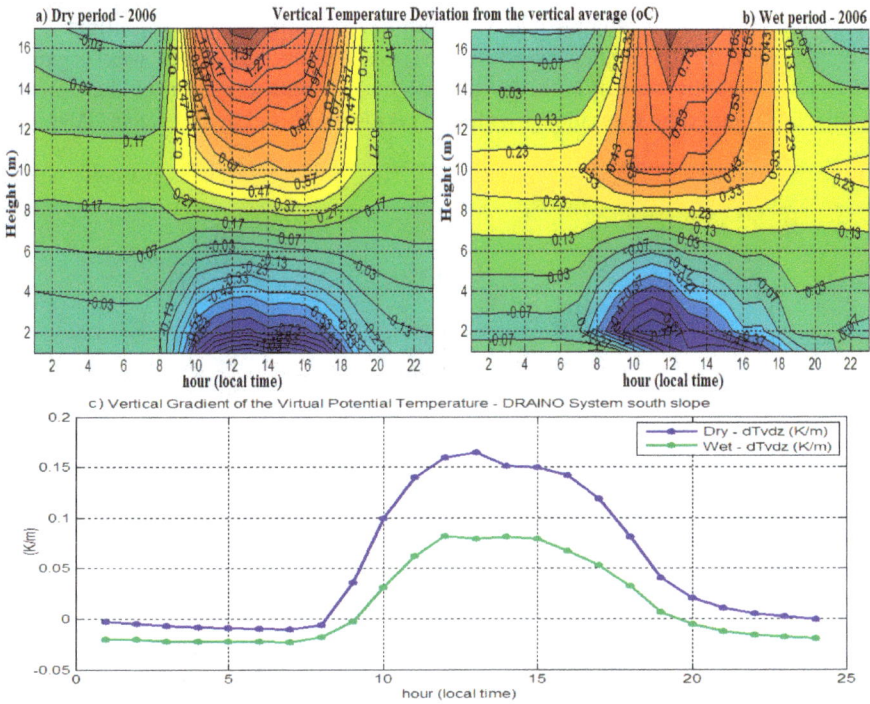

Fig. 9. Daily course of the vertical deviation of the virtual potential temperature for dry (a) and wet (b) periods of the year 2006, and the virtual potential temperature vertical gradient (c), over slope area DRAINO System tower.

3.2 Wind field

The LBA Manaus Site has moderately complex terrain when compared with the Santarem LBA Site (Figure 1, 2). This complexity generates a wind airflow regime much complex to be captured by standard measurement system like a single tower. At the Manaus LBA site, we implemented a complementary measurement system on the slope area to support the plateau K34 tower and better understand how the airflow above and below the canopy interact and also to describe how the valley flow influences the slope airflow regimes. It is important to note that the valley in the microbasin is oriented from East to West (Figure 2, 4).

3.2.1 Horizontal wind regime - Above canopy

3.2.1.1 Plateau K34 tower

Above the canopy (55m above ground level – a.g.l.) on the plateau area K34 tower, the wind regime was strongest (most above 2 m.s^{-1}) during daytime for both dry and wet periods of

2006, with direction varying mostly from southeast and northeast for dry and wet period, respectively (Figure 10). During nighttime, the wind regime was slower (most below 3 m.s⁻¹) and with same direction variation from northeast to southeast (Figure 10). As reported by *de Araújo* [2009], the above canopy valley area wind speed and direction was different from that of the plateau area, suggesting a decoupling mainly during nighttime. A clear channeling effect on the valley wind regime was observed; which was oriented by microbasin topography during both day and nighttime, with direction of the flow in the valley area determined by the valley orientation [as also reported by *de Araújo*, 2009].

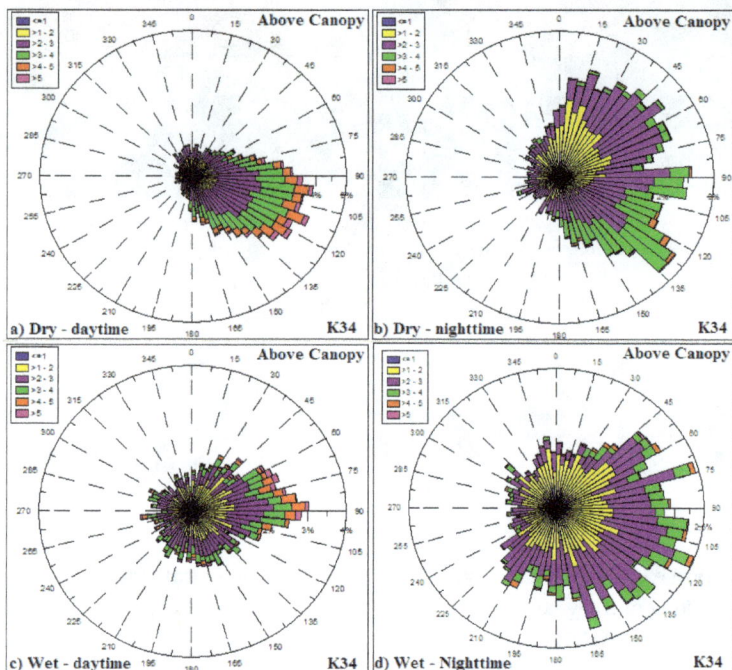

Fig. 10. Frequency distribution of the wind speed and direction. For dry (a, b) and wet (c, d) periods from the year 2006 during day (a, c) and nighttime (b, d), on the plateau K34 tower.

3.2.1.2 DRAINO System slope tower

The above canopy (38 m above ground level – a.g.l.) on the slope area DRAINO system south face (see Figure 4, 3D sonic), the wind regime was very persistent from east quadrant direction during day and nighttime in both dry and wet periods of the 2006 (Figure 11). The daytime wind speed during the dry season was between 1 to 3 m s⁻¹ and much stronger during the wet period with values up to 4 m s⁻¹.

During the nighttime the wind speed was slower than 2 m s⁻¹, except from northeast during the wet period. The wind direction pattern was similar to that on the plateau K34 tower (Figure 10) prevailing from northeast to southeast. This observation indicates that the airflow above the canopy on the slope area is related to how the synoptic flow enters in the eastern part of the microbasin (see Figure 2, 4).

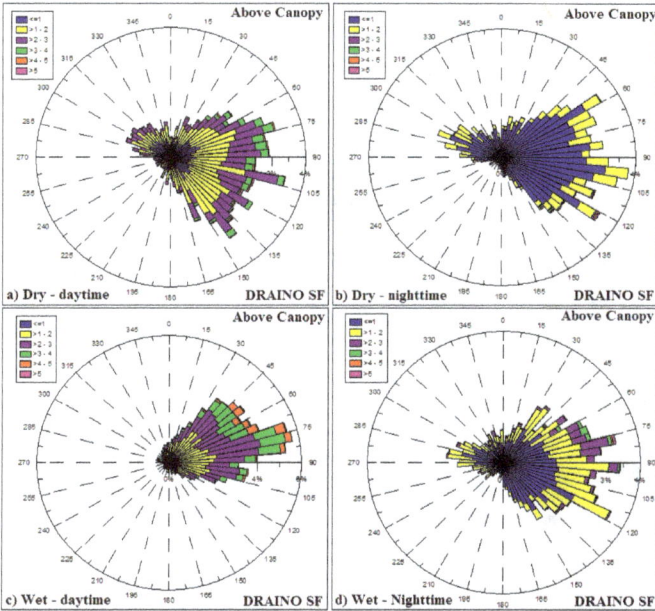

Fig. 11. Frequency distribution of the wind speed and direction above canopy (38 m above ground level – a.g.l). For dry (a, b) and wet (c, d) periods from the year 2006 during day (a, c) and nighttime (b, d), on the slope area at DRAINO system tower.

3.2.2 Horizontal wind regime – Subcanopy array measurements (2 m a.g.l)

In Figure 12, the subcanopy array frequency distribution of the wind speed and directions is shown for both dry and wet periods of the year 2006, during both day and nighttime. The observations show that the airflow in the subcanopy is very persistent and with similar pattern during both dry and wet periods of the year 2006. It is important to observe that the south slope area in the DRAINO System (see Figure 4) is *downslope* from **south** and *upslope* from **north** quadrants.

Subcanopy **daytime** wind regime

During daytime, in both dry (Figure 12a-c) and wet periods (Figure 12g-i), the wind direction prevailed from south-southeast (190-150 degrees) on the three slope regions [Figure 12, Top (a, d, g, j), Middle (b, e, h, k) and Low slope part (c, f, i, l)]. The airflow in the subcanopy was decoupled from the wind regime above the canopy (Figure 11) most of the time. The wind direction in the subcanopy airflow was dominated by a daytime downslope regime during the majority of the period of study, suggesting a systematic daytime katabatic wind pattern.

The wind speed in the subcanopy during the daytime was mostly from 0.1 to 0.4 m/s, and strongest at middle slope region (Figure 12b, e, h, k) about 0.3 to 0.5 m/s or above. A similar daytime katabatic wind regime was reported by *Froelich and Schmid* [2006] during "leaf on" season in Morgan-Monroe State Forest (MMSF), Indiana USA.

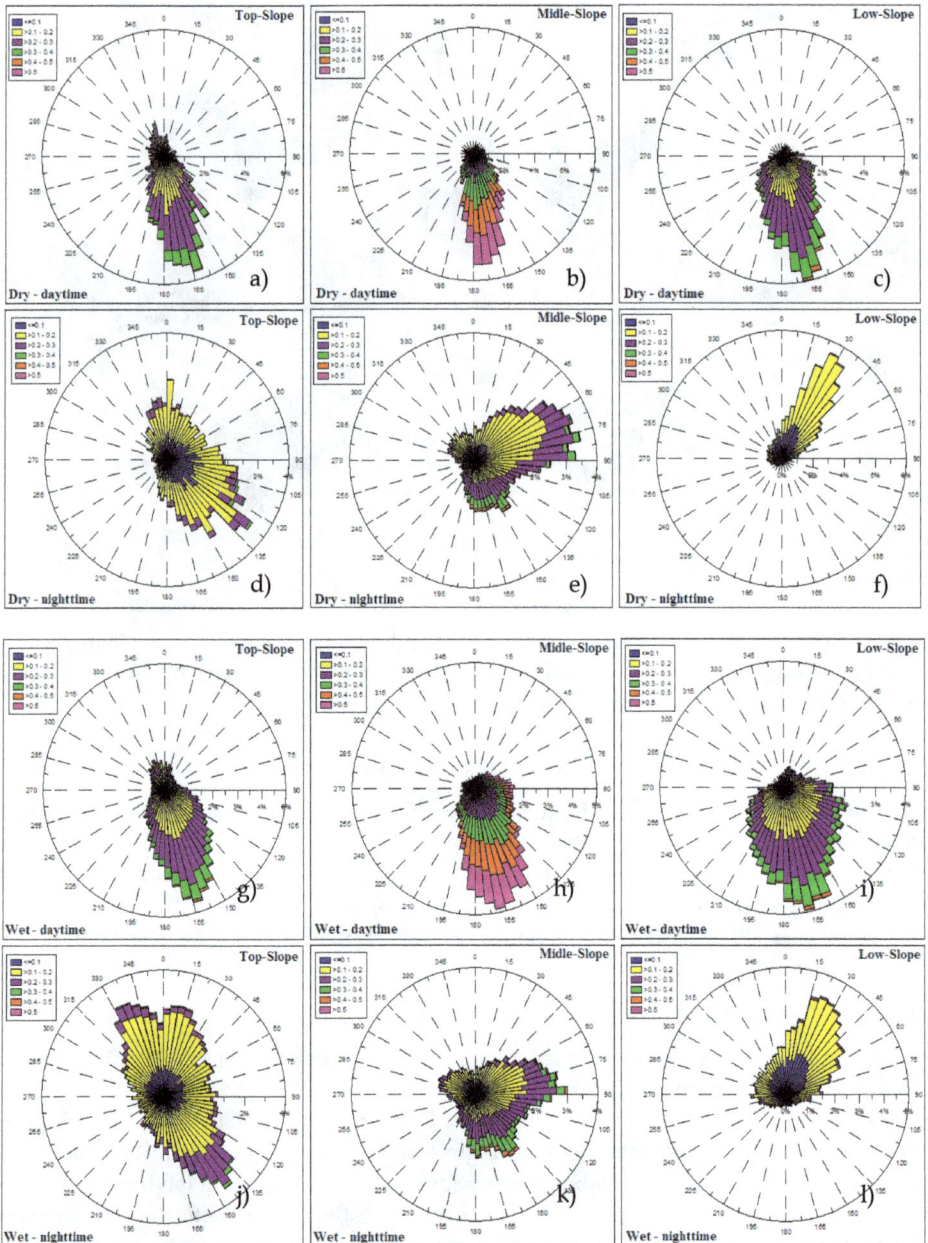

Fig. 12. Frequency distribution of the wind speed and direction in the subcanopy array (2 m above ground level – a.g.l) on the microbasin south face slope area at DRAINO horizontal array system (see Figure 4). For dry (a-f) and wet (g-l) periods from the year 2006, during day (a, b, c, g, h, i) and nighttime (d, e, f, j, k, l).

The daytime downslope wind was also supported by the subcanopy thermal structure (Figure 9), where the air was cooling along the day by inversion of the virtual potential temperature profile with a positive vertical gradient (Figure 9c). This results shows that subcanopy flows in a sloping dense tropical rainforest are opposite to the classical diurnal patterns of slope flows studied elsewhere in the literature [e.g.; *Manins and Sawford*, 1979; *Sturman* (1987); *Papadopoulos and Helmis*, 1999; *Kossmann and Fiedler*, 2000]. It is important to note that few studies have been done in forested terrain and it is unclear why similar reversed diurnal patterns have not been observed in studies at other forested sites [*Aubinet et al.*, 2003; *Staebler and Fitzjarrald*, 2004; *Yi et al.*, 2005], except by a single point subcanopy measurement observed by *Froelich and Schmid* [2006].

Subcanopy **nighttime** wind regime

The nighttime subcanopy wind regime on the slope area (see the terrain on Figure 4) was very complex and differentiates from that one above the canopy vegetation. It was observed that, on the up-slope part, the nighttime airflow was southeast downsloping direction (130°-170°) and northeast-northwest (45°-340°) uphill direction (Figure 12d, j). In the middle-part of slope area, the wind moved uphill (from northeast; 30°-90°) and also downsloping wind direction from southeast (Figure 12e, k), and with lightly higher wind speed. On the lower-part of the slope area (Figure 12f, l) the wind direction prevailed from the northeast (10°-70°), indicating upsloping pattern (anabatic). It is interesting to note that, on the up-slope area, the wind direction regime (northeast-northwest, 45°-340°) suggest a reversal lee side airflow (re-circulation or separation zone) probably in response to the above canopy wind (see Figure 11b, d). It is has been suggest by *Staebler* [2003] and reported by simulations using fluid dynamic models [*Katul and Finnigan*, 2003; *Poggi et al.*, 2008].

The upsloping subcanopy flows pattern, on the lower-part the slope area, is supported by subcanopy relative heat air layer along the slope during the night, as observed by lapse rate condition of the virtual potential temperature negative vertical gradient (Figure 9c). This observation does not follow the classical concept of nighttime slope flow pattern, as commented previously (section 3.1.2), this is a example of non-classical microscale slope flow. *Froelich and Schmid* [2006], has reported similar feature where they found anabatic wind regime during nighttime in their seasonal forest study area. Figure 13 presents the frequency distribution of the subcanopy wind direction on the south face slope area at DRAINO horizontal array system during upsloping (from north quadrant) and downsloping (from south quadrant) events.

3.2.3 Mean vertical wind velocity – Subcanopy and above canopy

Several correction methods have been proposed to calculate the mean vertical velocity, e.g. linear regression method [*Lee*, 1998], coordinate rotation [*Finnigan et al.*, 2003] and the planar fit method [*Wilczak et al.*, 2001]. We use the linear regression method by *Lee* [1998] to determine the "true" mean vertical velocity: $\bar{w} = w - a(\alpha_i) - b(\alpha_i)u$, where '$a$' and '$b$' are coefficients to be determine, for each α_i (10° azimuthal wind direction), by a linear regression of measured mean vertical velocity (w) and horizontal velocity (u) in the instrument coordinate system. Figure 14a presents the original and the correction results by method application of the mean vertical velocity as function of wind direction. In Figure 14b, the results of the hourly mean vertical velocities for plateau K34, DRAINO system (above and below canopy) and valley B34 towers. As expected, not only low but non-zero values were observed for all points of measurements.

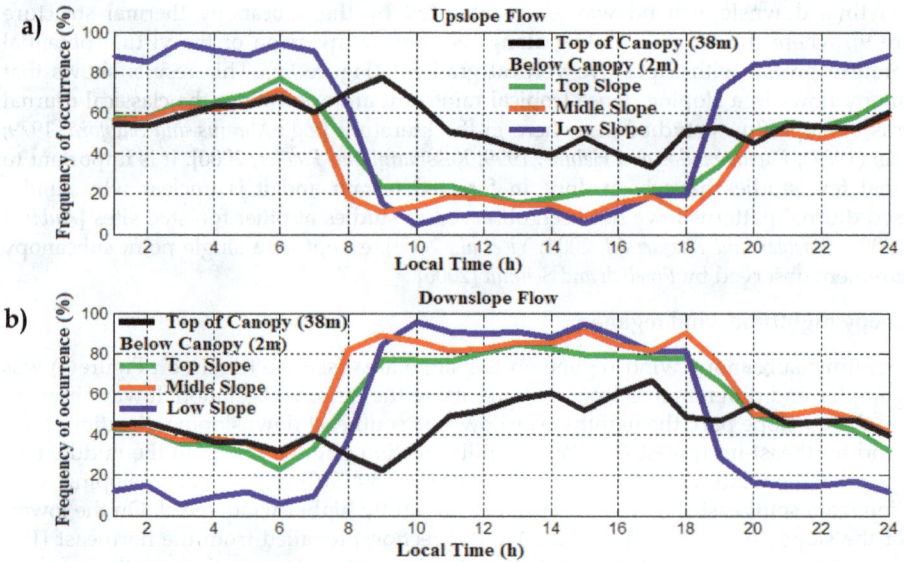

Fig. 13. Frequency distribution of the subcanopy wind direction (a) upsloping (from north quadrant) and (b) downsloping (from south quadrant) on the south face slope area at DRAINO horizontal array system (see Figure 4).

On the plateau area, the mean vertical velocity was always positive indicating upward motion or vertical convergence at the top of the hill during night and daytime. In the valley area during nighttime, negative or zero values were observed, indicating a suppression of vertical motion (mixing) in the valley, as also reported by *de Araújo* [2009].

However, during the daytime a transition is observed, where beginning in the morning, downward motion is observed, changing after mid-morning to upward motion (Figure 14b). This suggests that probably the cold air pooled during night moved downslope and started to warm, resulting in a breakdown the inversion over the valley (see *de Araújo* [2009], for detailed description and references there in for this process). The mechanism of the breakdown, the inversion process over the valley is consistent with positive vertical velocity observed above canopy at slope area by the DRAINO system tower during daytime (Figure 14b).

The subcanopy diurnal pattern of the mean vertical velocity observed shows positive values during nighttime and negative during daytime, consistent with observed up and downsloping flow regime, respectively (Figure 13 a, b). This is consistent with thermal vertical virtual potential temperature gradient on the slope (see Figure 9c), where during nighttime (daytime) an unstable (inversion) condition is associated with upward (downward) mean vertical velocity (see Figure 9c).

3.3 Phenomenology of the local circulations: Summary

The Figure 15 shows a schematic cartoon of local flow circulation from the previews sections observations.

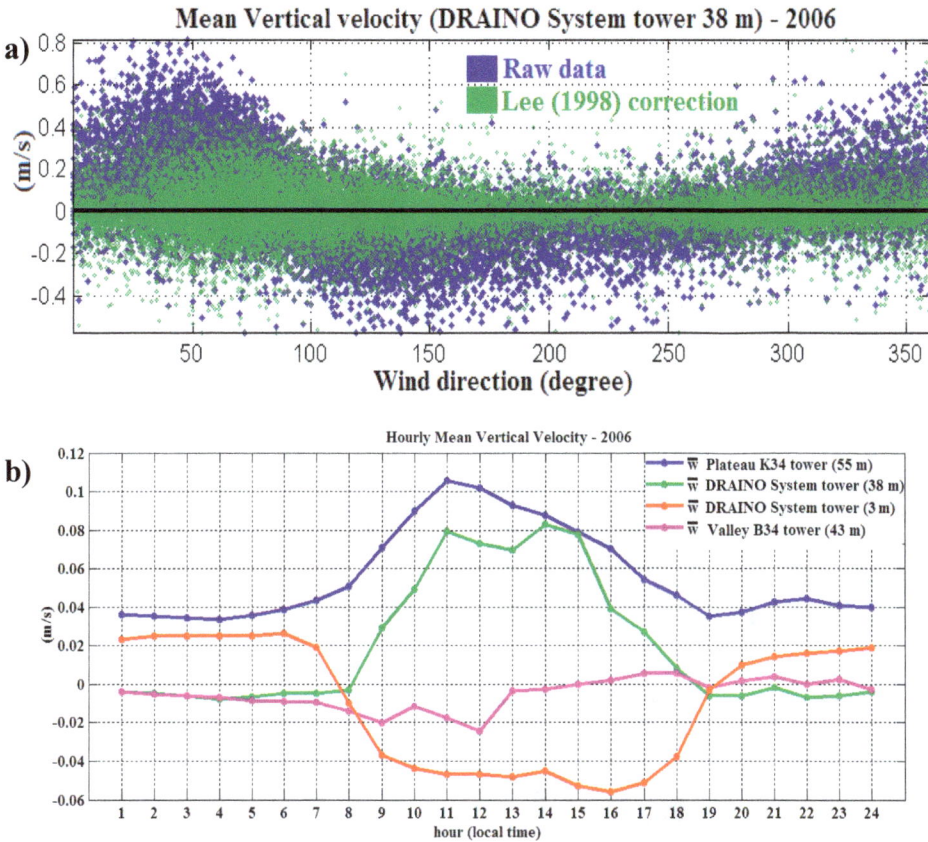

Fig. 14. Mean vertical velocity raw and correct vertical velocity (a) for DRAINO system slope tower (38 m), and hourly mean vertical velocity (b) for: plateau K34 tower (55 m), DRAINO system slope tower (above canopy - 38 m and subcanopy - 3 m) and for valley B34 (43 m) towers (see Figure 4, for details).

In Figure 15a is show the above canopy airflow over valley space (red arrow) and correspondent (induced) the most probable airflow above canopy over slope areas (blue arrow). In the same figure is show the main physical mechanisms (pressure gradient force) producing that micro-scale circulations. The observations result from preview sessions suggests that the balance of the buoyancy and pressure gradient forces generates the airflow or microcirculations patterns in the site studied.

During nighttime (Figure 15b), in the subcanopy, there is an upslope flow reaching about 10 m height above the ground, associated with positive mean vertical velocity (indicating upward movement). Also, above canopy, there is a downslope flow associated with negative mean vertical velocity, with downward convergence above the canopy. The microcirculation along the plateau-slope-valley is promoted by an feedback mechanism of accumulation of cold air drainage above canopy into the valley center (Figure 15b), creating the forcing needed to sustain nighttime pattern. The air temperature structure above canopy in the valley (see de

Araújo, 2009) is a good indication of cold air pool in the center of the valley. Maybe the local pressure gradient force due to the cold air accumulation promotes the upward airflow in both the slopes of the valley. During daytime periods an inverse pattern is found (not show), indicating that this microcirculation is a systematic pattern in the site.

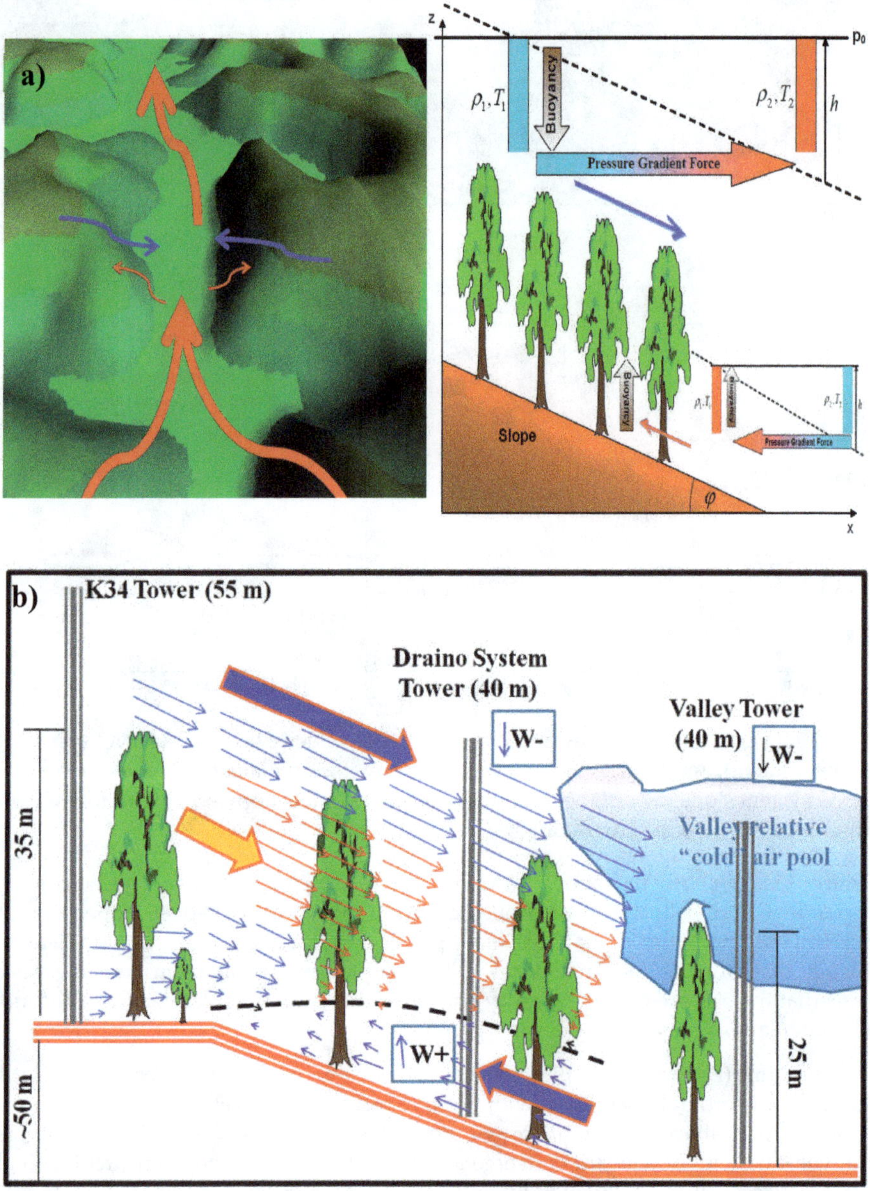

Fig. 15. Schematic local circulations in the site studied, valley and slopes flow (a), 2D view from suggested below and above canopy airflow (b).

3.4 CO_2 concentration and subcanopy horizontal wind field

The CO_2 concentration was measured by DRAINO system on the south face slope area for dry and wet periods of the year 2006, and on the north face slope during the dry period (Figure 4). The Figure 16 presents an example, for midnight (local time), of the horizontal wind field and spatial CO_2 concentration over the DRAINO System south face domain.

The wind field was interpolated from the blue points onto a 10 m grid. Similar procedures have been reported in the literature (*Sun et al.*, 2007; *Feigenwinter et al.*, 2008). The horizontal wind regime plays an important role in modulating the horizontal spatial distribution of CO_2 concentration (Figure 16).

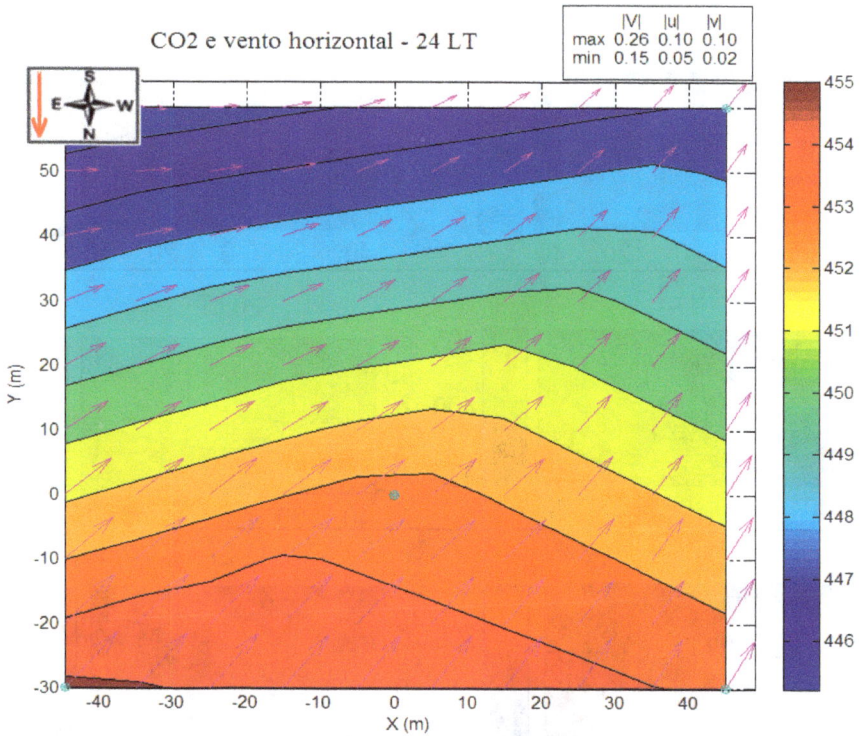

Fig.16. Example at midnight (local time) of the horizontal CO_2 concentration (ppmv) over the DRAINO System south face domain including an interpolated horizontal wind field (10 m grid). Note the geographic orientation and the red arrow indicating slope inclination (see Figure 4).

In Figure 17 (a, b, c) the typical pattern observed is shown for both dry and wet periods of the year 2006 measured by the DRAINO system on the south-facing slope area. During the daytime (Figure 17c), the wind prevailed downslope inducing a strong horizontal gradient of CO_2 in the slope area (about 0.2 ppmv m^{-1}). In the evening, periods of changes of the horizontal wind pattern (as described in section 3.1) show an upsloping regime in the lower-part and downsloping in the upper-part of the slope areas (Figure 17b). The wind regimes produce direct responses in the spatial feature of the horizontal gradient of CO_2 concentration. Later during the night, the upsloping regime is well established and also the horizontal gradient of CO_2 is growing from lower part of the slope to the top (Figure 17a).

These observations suggest a subcanopy drainage flow and its influence on the scalar spatial distribution. Therefore, as discussed in the previews sections, the flow above the canopy indicates a reverse pattern of downward motion (negative mean vertical velocity, see section 3.2.3) that suggests vertical convergence and possible horizontally divergent flow during nighttime. The report by *Froelich and Schmid* [2006] and more recently *Feigenwinter et al.,* [2009a, b] describing similar features of the airflow interaction between above and below canopy.

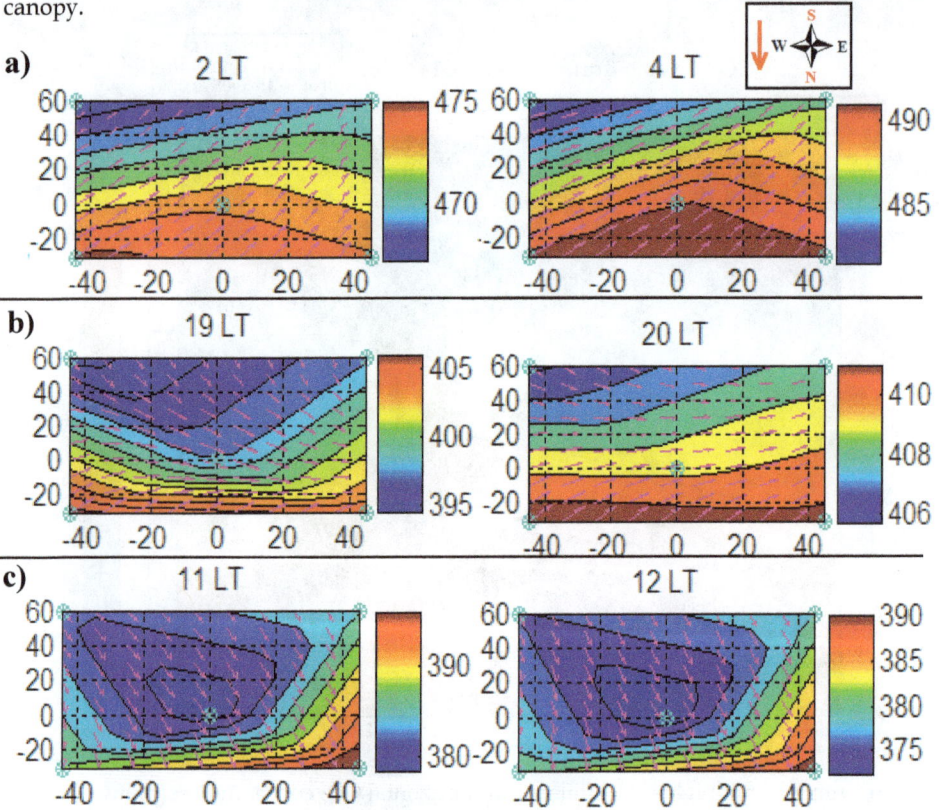

Fig. 17. Hourly average of the subcanopy (2 m) CO_2 concentration and horizontal wind speed over DRAINO System south face area during dry period of the year 2006, note the geographic orientation and the red arrow indicating slope inclination (see Figure 4). The axis represents distances from center of the main tower. Daytime (a), transition period - evening (b), established nighttime (c).

The spatial distribution of the horizontal CO_2 concentration, Figure 18 along the north face, shows a similar pattern than the south face described previously. Despite there being no wind information in that area, if one assumes the same spatial correlation between horizontal wind and CO_2 concentration, it is possible predict that the wind should present an inverse pattern from the south face suggesting that during daytime the downslope wind direction should be from the northeast (Figure 18c, from blue to red color).

During evening period (Figure 18b), it should be indicating downslope (from northeast) in the upper part of the north face slope and upslope (from southeast) in the lower part of the slope, an inverse feature from Figure 17b. Finally, later in the night, on the north face slope, the wind pattern should present an upslope wind direction regime from southeast, an inverse regime that one from Figure 17a on the south face slope.

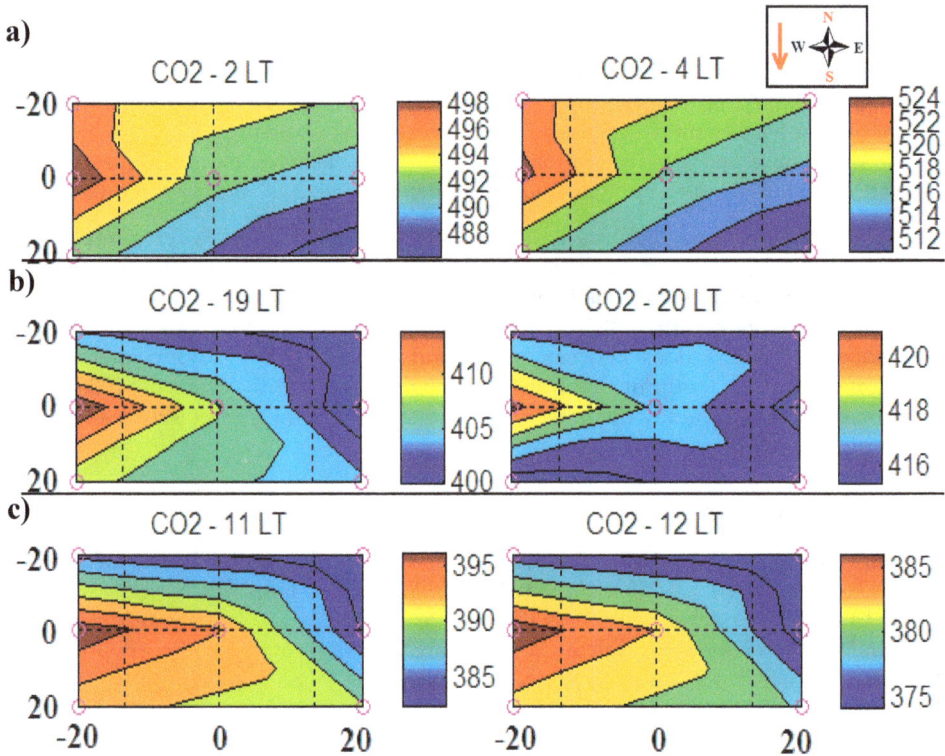

Fig. 18. Hourly average of the subcanopy (2 m) CO_2 concentration on the DRAINO System north face area during dry period of the year 2006. Note the geographic orientation and the red arrow indicating slope inclination (see Figure 4). The axis represents distances from center of the main tower. Daytime (a), transition period - evening (b), established nighttime (c).

One possible explanation to this subcanopy slopes wind regime and spatial distribution of CO_2 concentration, is the valley wind channeling effect and how it is meandering when it enters in the valley topography [as described by de Araújo, 2009]. This valley wind pattern, probably causes oscillations as those observed on the CO_2 concentration along the day (Figures 17, 18), the known "Seiche phenomena" (*Spigel and Imberger*, 1980).

4. Summary and conclusions

The main objective of this study was to measure and understand the local circulation over a dense forest site in Manaus with moderately complex terrain and to verify the existence of the drainage flow regimes on slope and valley areas. The main pattern of the airflow above and below the canopy in dense tropical forest in Amazonia was captured by a relative simple measure system, as also has been done by more sophisticated measurements system as those described recently by *Feigenwinter et al.*, [2009a, b]. As described and discussed in preview sections, it was identified as drainage flow in both day and nighttime periods in the site studied. Evidence of the drainage current above canopy was suggested by *Goulden et al*, (2006) similar to the one observed here. The study highlighted that the local micro-circulation was complicated and presented tri-dimensional nature. Where to estimate the advection flux at this site seems uncertain and not possible with the limited measurement system employed. As reported recently by *Feigenwinter et al.*, [2009a, b], even using a more sophisticated measurement design, the level of uncertainties is still high and some processes are not yet known and need more exploration perhaps using a more complete spatial observation network or even applying model resources (*Foken*, 2008; *Belcher et al.*, 2008).

In summary, the drainage flow exists and is observed at K34 LBA site. Very large carbon uptake estimates reported previously should be questioned [*Kruijt et al.*, 2004; *Araújo et al.*, 2002] and more research is warranted. The use of nighttime $u*$ correction to avoid estimating canopy storage is inappropriate. One cannot get by using only above canopy turbulence information. The interactions between motions above and below canopy question the foundations of the footprint analysis [*Schuepp et al.*, 1990; *Schimid*, 2006]. The representativeness of the eddy flux tower is most in question for complex terrain, especially on calm nights).

5. Acknowledgments

This work is part of the LBA-ECO project, supported by the NASA Terrestrial Ecology Branch under grants NCC5-283 and NNG-06GE09A to the Atmospheric Sciences Research Center, University at Albany, SUNY, LBA-ECO team CD-03. The latter grant included a subcontract to the Fundação Djalma Batista, Manaus. Julio Tóta was also supported by FAPEAM RH-POSGRAD and by MCT-INPA - Fundacao Djalma Batista during 2006–2007, with a fellowship. Thanks to LBA Manaus team for support and during the field work.

6. References

Araújo, A. C., et al., (2002), Comparative measurements of carbon dioxide fluxes from two nearby towers in a central Amazonian rainforest: The Manaus LBA site, *Journal of Geophysical Research*, 107(D20), 8090, doi:10.1029/2001JD000676.

de Araújo, A. C., (2009), Spatial variaton of CO_2 fluxes and lateral transport in an area of terra firme forest in central Amazonia. PhD Thesis, Vrije Universiteit Amsterdam, VU, Holanda.

Aubinet M, P. Berbigier, C. H. Bernhofer, A. Cescatti, C. Feigenwinter, A. Granier, T. H. Grunwald, K. Havrankova, B. Heinesch, B. Longdoz, B. Marcolla, L. Montagnani, P. Sedlak (2005), Comparing CO_2 storage and advection conditions at night at different Carboeuroflux sites. *Boundary Layer Meteorol., 116*: 63-94.

Aubinet, M., B. Heinesch, and M. Yernaux (2003), Horizontal and vertical CO_2 advection in a sloping forest, *Boundary Layer Meteorol.*, *108*(3), 397–417.

Aubinet, Marc. (2008), Eddy Covariance CO2 flux measurements in nocturnal conditions: An analysis of the problem. Ecological Applications, 18:1368–1378. [doi: http://dx.doi.org/10.1890/06-1336.1]

Baldocchi, D., et al. (2001), FLUXNET: A new tool to study the temporal and spatial variability of ecosystem-scale carbon dioxide, water vapor, and energy flux densities, *Bull. Am. Meteorol. Soc.*, *82*(11), 2415 – 2434.

Baldocchi, D.D . (2008). 'Breathing' of the Terrestrial Biosphere: Lessons Learned from a Global Network of Carbon Dioxide Flux Measurement Systems. Australian Journal of Botany. 56, 1-26.

Belcher, S. E., J. J. Finnigan, and I. N. Harman. (2008). Flows through forest canopies in complex terrain. Ecological Applications, 18:1436–1453. [doi: http://dx.doi.org/10.1890/06-1894.1]

Black, T. A., et al. (1996), Annual cycles of water vapour and carbon dioxide fluxes in and above a boreal aspen forest, *Global Change Biol.*, *2*(3), 219–229.

Castilho, C., (2004), Variação espacial e temporal da biomassa arbórea viva em64 km2 de floresta de terra-firme na Amazônia Central. 87p. Doctoral Thesis, ecology, Instituto Nacional de Pesquisa da Amazônia, Manaus-AM.

Cuartas, L.A. et al., (2007), Interception water-partitioning dynamics for a pristine rainforest in Central Amazonia: Marked differences between normal and dry years. *Agricultural and Forest Meteorology*, 145(1-2): 69-83.

Feigenwinter, C., C. Bernhofer, and R. Vogt, (2004), The influence of advection on the short term CO_2-budget in and above a forest canopy, *Boundary Layer Meteorol.*, *113*(2), 201–224.

Feigenwinter, C., C. et al., (2008), Comparison of horizontal and vertical advective CO_2 fluxes at three forest sites, *Agric. Forest. Meteorol.*, 145,1–21.

Feigenwinter, C., et al. (2009a), Spatiotemporal evolution of CO_2 concentration, temperature, and wind field during stable nights at the Norunda forest site. *Agric. Forest Meteorol.*, doi:10.1016/j.agrformet.2009.08.005

Feigenwinter, C., Montagnani, L., Aubinet, M. (2009b), Plot-scale vertical and horizontal transport of CO_2 modified by a persistent slope wind system in and above an alpine forest. *Agric. Forest Meteorol.*, doi:10.1016/j.agrformet. 2009.05.009.

Finnigan, J.J., R. Clement, Y. Malhi, R. Leuning and H.A. Cleugh, (2003), A reevaluation of long-term flux measurement techniques. Part I: Averaging and coordinate rotation. Bound.-Layer Meteorol., 107, 1-48.

Finnigan JJ. 1999. A comment on the paper by Lee (1998): On micrometeorological observations of surface-air exchange over tall vegetation. Agricultural and Forest Meteorology. 97: 55-64.

Fitzjarrald, D. R., and K. E. Moore, (1990), Mechanisms of nocturnal exchange between the rain-forest and the atmosphere, *Journal of Geophysical Research*, *95*(D10), 16,839–16,850.

Fitzjarrald, D. R., K. E. Moore, O. M. R. Cabral, J. Scolar, A. O. Manzi, and L. D. D. Sa (1990), Daytime turbulent exchange between the Amazon Forest and the atmosphere, *Journal of Geophysical Research*, *95*(D10), 16,825 – 16,838.

Foken, T. (2008), The energy balance closure problem: An overview. *Ecological Applications*: Vol. 18, No. 6, pp. 1351-1367.

Froelich, N.J. and Schmid, H.P., (2006), Flow divergence and density flows above and below a deciduous forest Part II. Below-canopy thermotopographic flows. *Agricultural and Forest Meteorology*, Volume 138, Issues 1-4.

Goulden, M. L., J. W. Munger, S. M. Fan, B. C. Daube, and S. C. Wofsy (1996), Measurements of carbon sequestration by long-term eddy covariance: Methods and a critical evaluation of accuracy, *Global Change Biol.*, 2(3), 169– 182.

Goulden, M.L., Miller, S.D. and da Rocha, H.R., (2006), Nocturnal cold air drainage and pooling in a tropical forest. *Journal of Geophysical Research-Atmospheres*, 111(D8), 10.1029/2005JD006037.

Gu L., E. Falge, T. Boden, D. D. Baldocchi, T. A. Black, S. R. Saleska, T. Suni, T. Vesala, S. Wofsy, L. Xu (2005), Observing threshold determination for nighttime eddy flux filtering, Agric. For. Meteorol., 128:179–197.

Harman, I.N. and Finnigan, J.J., (2007), A simple unified theory for flow in the canopy and roughness sublayer. Boundary-Layer Meteorology, 123, 339-363.

Katul, G. G., Finnigan, J. J., Poggi, D., Leuning, R. and Belcher, S. E. (2006) The influence of hilly terrain on canopy-atmosphere carbon dioxide exchange. Boundary-Layer Meteorology, 118 (1). pp. 189-216. ISSN 0006-8314.

Kossmann, M., and F. Fiedler, (2000) Diurnal momentum budget analysis of thermally induced slope winds. Meteor. Atmos. Phys., 75, 195–215.

Kruijt, B. J., A. Elbers, C. von Randow, A. C. Arau'jo, P. J. Oliveira,A. Culf, A. O. Manzi, A. D. Nobre, P. Kabat, and E. J. Moors (2004), The robustness of eddy correlation fluxes for Amazon rain forest conditions, Ecol. Appl., 14, suppl. S, S101–S113.

Kruijt, B., et al., (2000), Turbulence Statistics Above and Within Two Amazon Rain Forest Canopies. Boundary-Layer Meteorology, v. 94, n. 2, p. 297-331.

Laurance, et al., (1999), Relationship between soils and Amazon Forest biomass: a landscape-scale study. Forest Ecology Management, 118:127-138.

Lee, X. H. (1998), On micrometeorological observations of surface-air exchange over tall vegetation, Agric. For. Meteorol., 91(1– 2), 39–49.

Leuning, R., Zegelin, S. J., Jones, K., Keith, H., Hughes, D., (2008), Measurement of horizontal and vertical advection of CO2 within a forest canopy. Agricultural and Forest Meteorology, Volume 148, Issue 11, Pages 1777-1797.

Loescher, H.W., Law, B.E., Mahrt, L., Hollinger, D.Y., Campbell, J. and Wofsy, S.C. (2006). Uncertainties in, and interpretation of, carbon flux estimates using the eddy covariance technique. Journal of Geophysical Research 111(D21): doi: 10.1029/2005JD006932. issn: 0148-0227.

Luizão, R.C.C., Luizão, F.J., Paiva, R.Q., Monteiro, T.F., Sousa, L.S., Kruijt, B., (2004), Variation of carbon and nitrogen cycling processes along a topographic gradient in a central Amazonian forest. Global Change Biology, 10: 592-600.

Manins, P. C., and B. L. Sawford, (1979) Katabatic winds: A field case study. Quart. J. Roy. Meteor. Soc., 105, 1011–1025.

Marcolla, B, A. Cescatti, L. Montagnani, G. Manca, G. Kerschbaumer and S. Minerbi, (2005), Importance of advection in the atmospheric CO2 exchanges of an alpine forest. Agric. For. Meteorol., 130, 193-206.

Massman, W.J.; Lee, X. (2002) Eddy covariance flux corrections and uncertainties in long-term studies of carbon and energy exchanges. Agricultural and Forest Meteorology. 113: 121-144.

Miller, S. D., M. L. Goulden, M. C. Menton, H. R. da Rocha, H. C. de Freitas, A. M. E. S. Figueira, and C. A. D. de Sousa (2004), Biometric and micrometeorological measurements of tropical forest carbon balance, Ecol. Appl., 14(4), suppl. S, S114–S126.

Pachêco , V. B., (2001), Algumas Características do Acoplamento entre o Escoamento Acima e Abaixo da Copa da Floresta Amazônica em Rondônia. 2001 109f. Dissertação (Mestrado em Meteorologia) - Instituto Nacional de Pesquisas Espaciais, São José dos Campos.

Papadopoulos, K. H. and Helmis, C. G.: (1999) Evening andMorning Transition of Katabatic Flows, Boundary-Layer Meteorol. 92, 195–227.

Papale D., Reichstein M., Aubinet M., Canfora E., Bernhofer C., Longdoz B., Kutsch W.,Rambal S., Valentini R., Vesala T., Yakir D. (2006). Towards a standardized processing of Net Ecosystem Exchange measured with eddy covariance technique: algorithms and uncertainty estimation. Biogeosciences, 3, 571-583.

Parker, G., and D. R. Fitzjarrald (2004), Canopy structure and radiation environment metrics indicate forest developmental stage, disturbance, and certain ecosystem functions, paper presented at III LBA Scientific Conference, Braz. Minist. of Sci. and Technol., Brasilia, Brazil, July.

Paw U. K. T., D. D. Baldocchi, T. P. Meyers, and K. B. Wilson (2000), Correction of eddy covariance measurements incorporating both advective effects and density fluxes. Boundary Layer Meteorol., 97, 487-511.

Poggi, D., Katul, G., Finnigan,J. J., Belcher,S. E., (2008) Analytical models for the mean flow inside dense canopies on gentle hilly terrain Q. J. R. Meteorol. Soc. 134: 1095–1112.

Raupach, M. R., J. J. Finnigan, and Y. Brunet., (1996), Coherent eddies and turbulence in vegetation canopies: The mixing-layer analogy. Boundary-Layer Meteorology 78: 351-382.

Rennó, C. D., et al., (2008), HAND, a new terrain descriptor using SRTM-DEM: Mapping terra-firme rainforest environments in Amazonia. Remote Sensing of Environment, doi:10.1016/j.rse.2008.03.018

Sá, L. D. A., Pachêco, V.B., (2006), Wind velocity above and inside Amazonian Rain Forest in Rondônia. Revista Brasileira de Meteorologia, v.21, n.3a, 50-58, 2006.

Schmid, H.P. (2006): On the "Dos" and "Don't"s of footprint analysis in difficult conditions. iLEAPS SpecialistWorkshop Flux Measurement in Difficult Conditions. 26-28 January, 2006, Boulder, USA.

Schuepp, P. H., Leclerc, M. Y., MacPherson, J. I. and Desjardins, R. L.: 1990, Footprint Prediction of Scalar Fluxes from Analytical Solutions of the Diffusion Equation', Boundary-Layer Meteorol. 50, 355-373.

Spiegel, R. H., and J. Imberger. (1980) The classification of mixed layer dynamics in lakes of small to medium size. J. Phys. Oceanogr. 10: 1104–1121.

Staebler R.M., and Fitzjarrald D.R. (2005), Measuring canopy structure and kinematics of subcanopy flows in two forests. J. Appl. Meteor., 44, 1161-1179.

Staebler, R. M., and D. R. Fitzjarrald (2004), Observing subcanopy CO_2 advection, Agric. For. Meteorol., 122(3– 4), 139– 156.

Staebler, R.M., 2003. Forest subcanopy flows and micro-scale advection of carbon dioxide. Ph.D. Dissertation, SUNY Albany.

Sturman, A.P., 1987. Thermal influences on airflow in mountainous terrain. Prog. Phys. Geog. 11, 183–206.

Sun J., S. P. Burns, A. C. Delany, S. P. Oncley, A. A. Turnipseed, B.B. Stephens, D. H. Lenschow, M. A. LeMone, R. K. Monson, D. E Anderson (2007), CO2 transport over complex terrain. Agric. Forest. Meteorol., 145,1–21.

Tóta J., Fitzjarrald, D.R., Staebler, R.M., Sakai, R.K., Moraes, O.M.M., Acevedo, O. C., Wofsy, S.C., Manzi, A.O., (2008), Amazon rain Forest subcanopy flow and the carbon budget: Santarém LBA-ECO site, Journal Geophysical Research - Biogeosciences, 113, G00B02, doi:10.1029/2007JG000597.

Waterloo, M.J. et al., (2006), Export of organic carbon in run-off from an Amazonian rainforest blackwater catchment. Hydrological Processes, 20(12): 2581-2597.

Wilczak, J.M., S.P. Oncley, and S.A. Stage (2001), Sonic anemometer tilt correction algorithms. Bound.-Layer Meteorol.,99, 127-150.

Yi, C., (2008), Momentum transfer within canopies. Journal of Applied Meteorology and Climatology, 47, 262-275, doi:10.1175/2007JAMC1667.1.

Yi, C., R. K. Monson, Z. Zhai, D. E. Anderson, B. Lamb, G. Allwine, A. A. Turnipseed, and S. P. Burns (2005), Modeling and measuring the nocturnal drainage flow in a high-elevation, subalpine forest with complex terrain, Journal of Geophysical Research, 110, (D22)303, doi:10.1029/2005JD006282.

Part 6

Tropical Forest Protection and Process

17

Patterns of Tree Mortality in a Monodominant Tropical Forest

Patrick J. Hart
Department of Biology, University of Hawaii at Hilo, Hilo, Hawaii

1. Introduction

Tree death in forests is an important process at many ecological levels. The mortality rates of trees affect carbon and nutrient cycling, stand structure, community composition, and successional processes (van Mantgem et al. 2009). Tree death substantially increases resources such as light, nutrients, water, and energy available to other organisms. Dead standing trees (snags) provide important habitat for wildlife and dead fallen trees are often critical to seedling establishment as nurse logs (Franklin et al 1987). Despite their importance, the mechanisms that drive tree mortality are often unclear, especially in tropical forests (Swaine et al. 1987). There is particular lack of information on how both mortality and longevity varies with the species, individual size, and competition with the other trees. In order to predict the various impacts of global change on the dynamics of tropical forests, it is essential to understand the role these factors play on tree mortality at both the individual and landscape scale.

While most tropical forests have very high species richness, there are some in which a single species dominates and richness is fairly low. In many of these forests, monodominance may be perpetuated through periodic mortality as a result of massive, landscape-level disturbances such as volcanic eruptions, hurricanes, landslides, and fire (Connell and Lowman 1989, Hart 1990, Read et al. 1995). Like their more diverse tropical counterparts, the basic life-history characteristics of the species that comprise monodominant forests, particularly those related to mortality patterns, remain poorly understood.

Forests on oceanic islands in the tropics are often dominated by one or a few tree species (Mueller-Dombois and Fosberg 1998). The native, wet forests of the Hawaiian archipelago are generally dominated by a single endemic tree species, ohia (*Metrosideros polymorpha*). This tree forms monodominant stands from the earliest pioneer stages on recent lava flows to rain forest on substrates millions of years old (Atkinson 1970, Jacobi 1989, Gagne and Cuddihy 1990). In contrast to tropical forests in continental areas where gap phase dynamics plays a dominant role in succession (Denslow 1987), *Metrosideros* forest stands often occur in cycles, (*cyclic succession*) with generations of similar-aged canopy trees lasting from 300-500 years before they experience a canopy dieback event (Mueller-Dombois 1986). Pollen cores showing 2-3 pollen depressions per 1000 years (Hotchkiss 1998), and radiocarbon dating demonstrating non-overlapping age groups of canopy trees (Hart 2010), provide additional support for this cyclic succession model. While the proximate causes of periodic cyclic mortality in *Metrosideros* are likely variable across the landscape, this pattern of mortality itself, where many, if not all of the trees in the canopy die at once, may help perpetuate monodominance in most Hawaiian forests.

Here, I use information from a long term study in permanent vegetation plots within 200 ha of monodominant Hawaiian wet forest to address the following questions: How does tree mortality vary with respect to species, size, position in the canopy (crown class), and geographic location? What is the age of trees in this forest? To what extent can patterns of mortality provide evidence for succession in this forest?

2. Methods

2.1 Study area and data collection

Fieldwork was conducted from January 1996-August 2005 within the 13,246 hectare Hakalau Forest National Wildlife Refuge on the eastern slope of Mauna Kea volcano, island of Hawaii (approx. 19°50N, 155° 20W). The Koa/'Ohi'a (*Acacia/Metrosideros*) Montane Wet Forest (Gagne and Cuddihy 1990) on this refuge is remarkably intact and is dominated by *Metrosideros* and *Acacia* trees that regularly reach 1m or more in diameter and heights greater than 30m (please see Hart 2010 for a further description of the study area). Like most other montane wet forests in Hawaii, federal and/or state laws prohibit logging, however invasive plant and animal species pose significant threats to native biodiversity. The substrate at Hakalau is generally comprised of silty clay (Akaka soil association; Sato et al. 1973) that developed in 10,000 – 30,000 year BP Tephra deposits (Wolfe and Morris 1996). Road cuts indicate that the soil is quite deep (greater than 3m in many areas) and relatively free of rock. Mean annual rainfall was approximately 2250mm at Hakalau during the study period. In 1996, two 100 ha study sites separated by approximately 4 km of continuous forest, were established at elevations between 1750 and 1900 m. Within each study site, 10 one km long transects were established with permanent 30m diameter circular plots (stations) placed at 100m intervals (200 plots total). From 1996-1998, all stems within each plot greater than 5 cm diameter at breast height (DBH) were marked with an aluminum tag secured to the tree with an aluminum nail at approximately 1.3 meters above the ground. The diameter of each tree (both live and dead) was measured with a synthetic fabric diameter tape at a point approximately 3 cm above the tag. In addition, all trees were classified as belonging to one of four distinct crown classes within the canopy stratum: dominant, co-dominant, intermediate, and suppressed (Smith 1986). Dominant trees generally had full sun exposure to the crown throughout the day. Co-dominant trees had full exposure to the top of the crown, but were partially shaded on the sides by other co-dominant or dominant trees. Intermediate trees were those in which the upper portion of the crown was shaded by neighboring trees for a portion of the day. Suppressed trees were generally completely shaded by neighboring trees throughout the day. From 2004- 2005, all tagged trees within each of the 200 survey plots were re-measured 3 cm above the aluminum tag, and changes in growth state (live to live vs. live to dead) were determined. Trees that had died between surveys were further characterized as "dead-standing" or "dead-fallen" to provide further evidence for cause of mortality.

2.2 Mortality rates

Annual mortality (m) was calculated following equation 6 of Sheil et al. (1995) as:

$$m = 1 - (N_1/N_0)^{1/t},$$

where 'N_0' and 'N_1' are population counts at the beginning and end of the measurement interval, and 't' is the mean number of years between measurements. Bootstrapped standard error estimates of annual mortality rates for each species were calculated using 1000 bootstrap replicates from the binomial distribution, where 'p' = observed annual probability of mortality and 'n' = sample size (Chernick 2008).

To model survival, I used a logistic ANCOVA (Crawley 2007) with growth state (live vs. dead) as a response variable and DBH and crown class, along with their interaction, as predictors. Four different models were constructed and ranked with Akaike's Information Criterion (AIC:Akaike 1973) using R software (version 2.12.1; The R Foundation for Statistical Computing). This model selection criterion is based upon the principle of parsimony and represents a tradeoff between model fit and the number of parameters in the model (Burnham and Anderson 2002). Deletion tests (Crawley 2007) were then used to cross validate the model selection procedure by assessing the significance of the increase in deviance that resulted when a particular term was removed from the full model. Deletion tests were run on model pairs using the *anova* function in R.

If mortality is associated with stand-level dieback in *Metrosideros*, it might be expected that there are discernable spatial patterns of mortality across the landscape. To address this idea, I used the proportion of trees that died during the study period at each station at the Pua Akala study area to calculate a Getis-Ord Gi* statistic (also known as a z score) using the Hot Spot Analysis tool in ArcGIS 9.2 (ESRI 2010). This statistic helps identify those clusters of points with values higher in magnitude than would be expected by chance, with 'z' scores near zero providing evidence that, in this case, there is little clustering of mortality across the landscape.

Mortality rates for *Metrosideros* were used to evaluate longevity estimates based on growth rates and radiocarbon dating (Hart 2010). Using plot data, I estimated that the overall density of large, living trees (> 50cm DBH) within the study areas is approximately 27 trees Ha-1, then used a conservative, pooled, annual mortality estimate for these trees of 1.1% to construct mortality curves to predict the number of years that individuals in this cohort of large trees will persist. This value was then added to the median, minimum, and maximum age estimates for the smallest trees in this cohort (Hart 2010) to produce tree longevity estimates.

3. Results

A total of 480 of the original 6173 trees of all species greater than 5 cm DBH died over the course of the study. The mean number of years between measurements was 6.75. Total mortality (number of trees that died during the study divided by total number of trees) ranged from 6.4% for *Ilex anomala* and 7.1% for *Metrosideros* to 33.3% for *Coprosma ochracea*. Annual mortality (Sheil et al. 1995) varied with species, ranging from 0.98%/yr for *Ilex* and 1.1% for *Metrosideros* to 5.82 %/yr for *Coprosma* (Fig. 1). The high annual mortality for *Coprosma* may be an over-estimate due to small sample size for this species (n = 21).

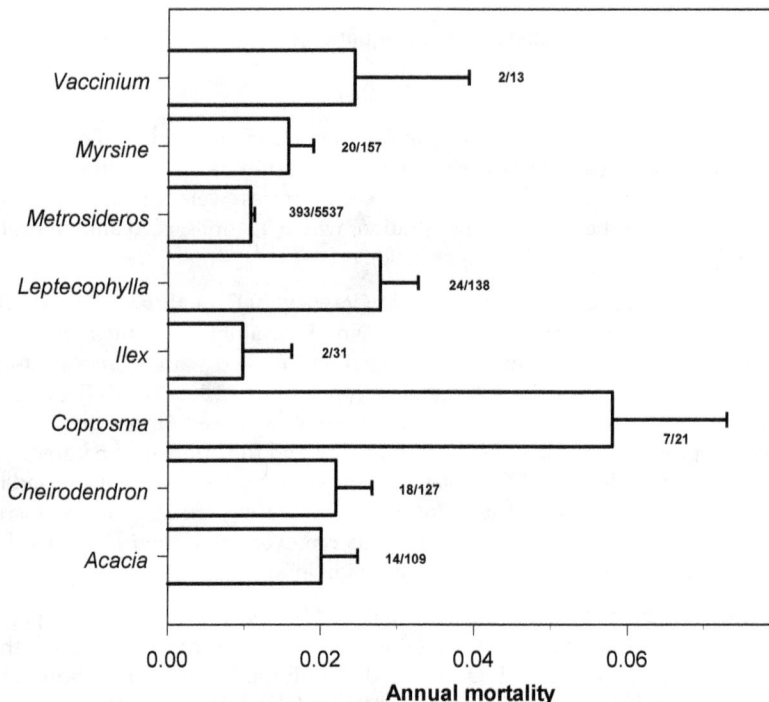

Fig. 1. Annual mortality for the eight most abundant tree species at Hakalau Forest NWR (*Acacia koa, Cheirodendron trigynum, Coprosma ochracea, Ilex anomala, Leptecophylla tameiameiae, Metrosideros polymorpha, Myrsine lessertiana,* and *Vaccinium calycinum*). Values above bars represent the total number of trees that died during the course of study vs. total trees examined for each species.

There was no evidence for any sort of clumped spatial patterns in *Metrosideros* mortality (Getis-Ord Z = 0.418, P = 0.675). The full logistic model that included crown class, tree size (DBH) and their interaction provided the best fit to the data for predicting whether a *Metrosideros* tree lived or died during the study (DBH: df = 1, F = 117.04, P < 0.0001; Crown Class: df = 3, F=143.5, P < 0.0001; interaction: df = 3, F = 20.1, P < 0.0001). This model had the lowest AIC value, and deletion tests of the model with vs. without the interaction term were highly significant, providing further evidence for the importance of the interaction term (Fig. 2).

For *Metrosideros*, percent annual mortality varied with size, ranging from 0.33% for trees 40-45cm DBH to 1.8% for trees between 10-15 cm DBH (Fig. 3). Annual mortality decreased with size between approximately 15cm – 45cm DBH, and again increased in the larger size classes (Fig. 3). When viewed by crown class category, suppressed trees accounted for much of the overall mortality, dying at a rate of approximately 3.8% per year (Fig.4). A Chi-square test on the frequency of mortality for each crown class category shown in Fig. 4 was highly significant (χ^2 = 298.2, df = 2, P < 0.001).

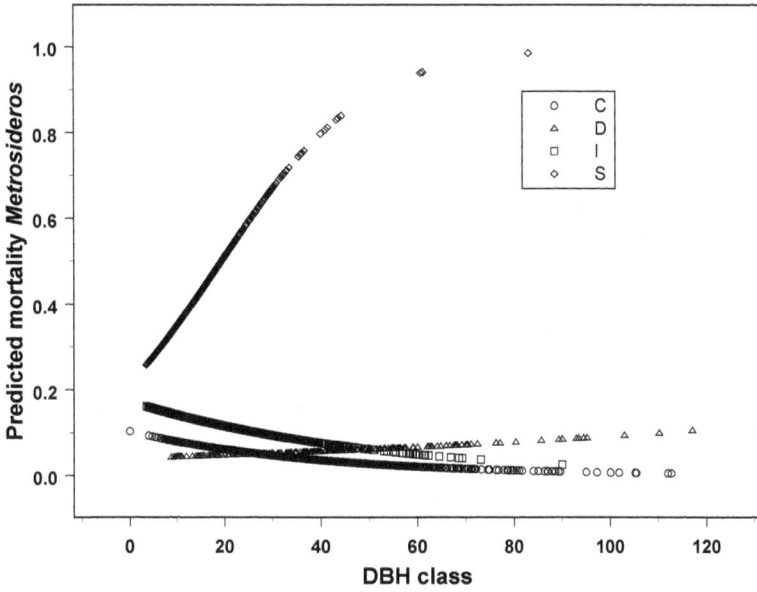

Fig. 2. Predicted mortality of *Metrosideros* from the full logistic model illustrating the interaction between Crown Class (particularly suppressed trees) and DBH. C = Co-dominant, D = Dominant, I = Intermediate, and S = Suppressed.

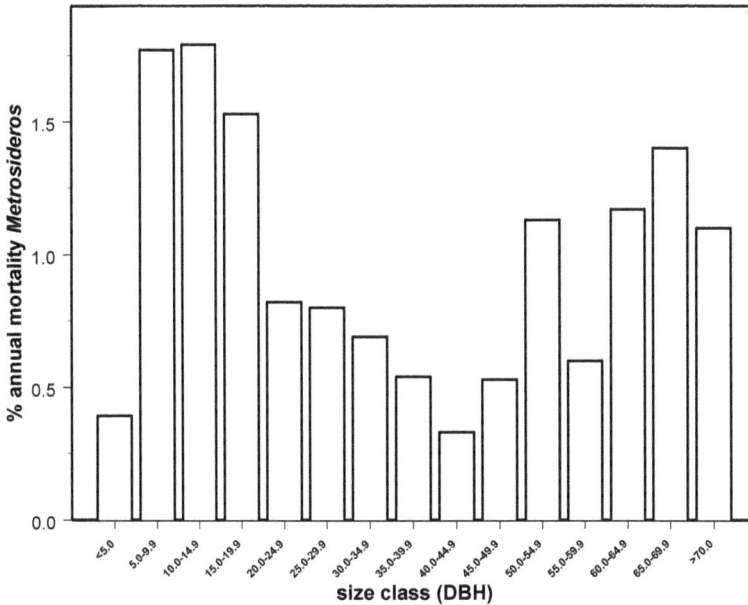

Fig. 3. Percent annual mortality for *Metrosideros* as a function of size class (cm DBH) at Hakalau Forest NWR.

Mortality for *A. koa* showed rather different trends with size. Total mortality during the course of the study for *A. koa* trees in smaller size classes (5-59.9 cm DBH) was 7.6 % (3 out of 39 trees) vs. 15.7 % (11 out of 70) for trees in the larger (> 60 cm) DBH size classes, however this difference was not significant (two-sided Fisher's exact test; P = 0.370). Annual mortality for all size classes of *A. koa* combined was 2.01%; annual mortality of trees below 60 cm DBH was 1.2% vs. 2.4% for those greater than 60 cm DBH.

Of 315 ohia trees between 5-29.9 cm DBH (generally those in the suppressed or intermediate categories) that died during the study period, 55.5% were "dead standing" and presumably victims of competition with neighboring trees, In contrast, of the 78 trees > 30cm DBH that died, only 33.3% were "dead standing. A chi-square test demonstrated a difference in presumed cause of mortality between the large and small size classes (χ^2 = 12.35, df = 1, P < 0.001).

Tree age – Approximately one *Metrosideros* tree per ha reaches the age of 650 yrs (range = 590-800 yrs) at Hakalau (Fig. 5). This estimate assumes that the median age of any tree greater than 50cm DBH is 360 yrs, with a range of approximately 300-500 years based on upper and lower 95% Confidence Bounds (Hart 2010) and that the annual mortality rate for these large trees is 1.1% per year.

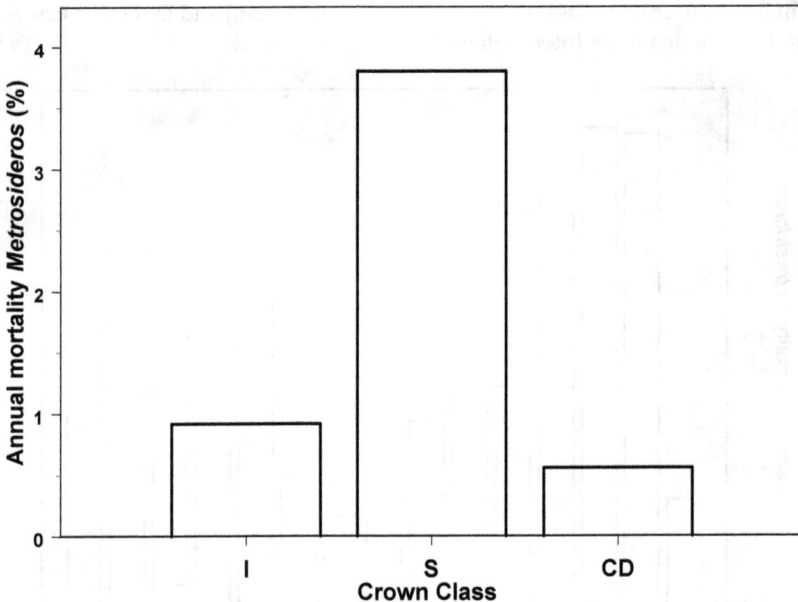

Fig. 4. Percent annual mortality for *Metrosideros* as a function of crown class at Hakalau Forest NWR. Co-dominant and dominant crown classes are combined due to small sample size.

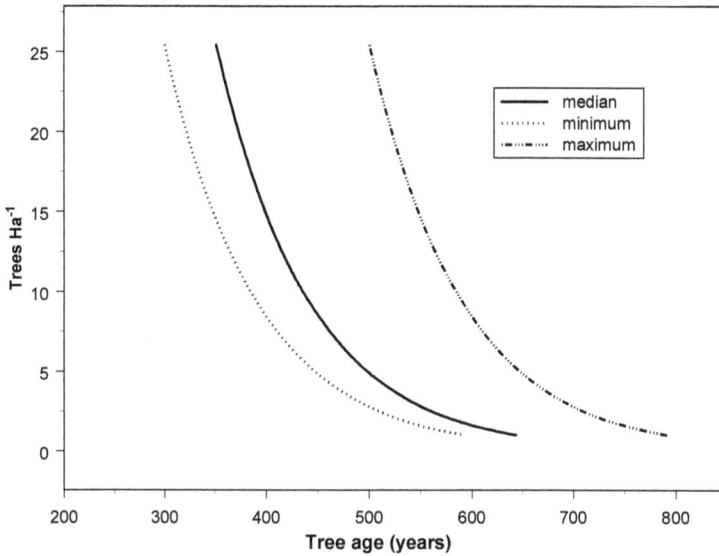

Fig. 5. Prediction curve for tree density (Tree ha^{-1}) vs. age for *Metrosideros* > 50 cm DBH at Hakalau Forest NWR. Assuming an annual mortality of 1.1%, it takes approximately 290 years for the current cohort of 27.1 trees ha-1 >50 cm DBH [with a median age of 360 yrs (range = 300-500 yrs)] to reach a density of 1 tree ha-1.

4. Discussion

Tree mortality patterns in this monodominant forest were somewhat similar to those in far more diverse continental tropical forests. The *forest-wide* annual mortality rate of 1.1% for *Metrosideros* at Hakalau is on the low end of the approximate 1.0 - 2.0 % annual mortality reported for most tropical forests around the globe (Swaine et al. 1987). This mortality rate may be viewed as relatively low especially if one takes into account that *Metrosideros* is dominant across all successional levels, from primary colonizer to late successional "climax" forests (apparently like Hakalau). In most continental forests, primary colonizers have far higher annual mortality, ranging up to 10% and more (Condit et al. 1995). The relatively U-shaped relationship between size and mortality, where intermediate-sized *Metrosideros* individuals exhibited much lower mortality rates than both the smallest and to a certain degree the largest size classes (Fig. 3) also appears to be a common feature of forests, including those from tropical (Korning and Balslev 1994), temperate (Coomes and Allen 2007, Lines et al. 2010), and even sub-boreal (Umeki 2002) regions. However, many other long-term studies in tropical mixed forests (Lieberman and Lieberman 1987; Manokoran and Kochumen 1987; Swaine et al. 1987; Carey et al. 1994) as well as monodominant forests (Nascimento et al. 2007) have reported mortality to be generally independent of size class for trees > 10 cm DBH (but see Clark and Clark 1992).

While it is difficult to disentangle the effects of size vs. crown class on mortality rates due to autocorrelation, examining mortality as a function of crown class is especially revealing. Trees in the suppressed category (those that are competitively inhibited by taller neighbors) were able to grow slowly, but they suffered by far the highest annual mortality rates (Fig. 4). Conversely, the low annual mortality of 0.56% demonstrated here for the co-dominant and dominant size classes combined indicates that once trees become established in the canopy, mortality is relatively rare. Within this crown class category, the smaller size classes experience mortality rates as low as 0.33% per year, and the largest size classes up to 1.40% per year, possibly due to senescence.

With the exception of studies within *Metrosideros* dieback areas on younger substrates (Gerrish et al. 1988), there has been little previous work on mortality rates of *Metrosideros* trees greater than sapling size in Hawaii. Burton and Mueller –Dombois (1984) found high mortality (57%/yr) for *Metrosideros* seedlings growing in < 5% irradiance, but that annual mortality decreased to 13% for seedlings growing at > 5% irradiance. For trees species other than *Metrosideros*, the mortality data presented here should probably be treated with caution because of low sample sizes. Also, the understory and mid-canopy of the higher elevation sections of the study sites are currently regenerating following disturbance by cattle which should presumably lead to low estimates of mortality for those species that predominate in these layers.

Tree mortality, at both the individual and stand level, is often the result of a combination of abiotic (drought, high winds) and biotic (predation, competition, disease, and senescence) factors. The way in which a tree dies suggests possible factors that contributed to its death (Carey et al. 1994). In this study, there was a significantly higher proportion of trees in the smaller size classes (< 30 cm DBH) than the larger size classes that were classified as "dead standing". These smaller (often intermediate or suppressed) trees that died standing were assumed to be victims of light competition with larger neighbors, with the remainder primarily being killed by neighboring tree falls. This latter process may be important to successional processes in this forest in terms of gap formation. In contrast, large dominant or co-dominant trees that died standing were generally assumed to have senesced. Why might there be a lower proportion of large trees that are dead standing? The answer may be that the heartwood of the largest (>60cm DBH) *Metrosideros* individuals at Hakalau decomposes within a few meters from the ground, presumably predisposing them to treefall as live trees during wind storms, which are common at Hakalau during winter (pers. obs.).

Implications for succession in montane Hawaiian wet forests- The patterns of annual mortality of *Metrosideros* at Hakalau are primarily a function of size and crown class, with suppressed and, to a lesser extent, intermediate trees experiencing much higher mortality rates than co-dominant or dominants. This is the opposite of the majority of Hawaiian wet forests described so far, where the dominant and co-dominant ohia trees that comprise the canopy belong to a similar-aged cohort, with mortality occurring primarily over a relatively short time period as "stand level dieback". As expected, there was also a slight overall increase in mortality as *Metrosideros* reach the largest size classes (Fig. 3), but this senescence occurred on the individual, rather than stand level. With most mortality occurring in the smaller, suppressed classes, these results do not support the cyclic succession model for this forest. Instead, montane wet forests such as the one at Hakalau may be similar to the "Type I" monodominant forests described by Connell and Lowman (1989), in which the dominant species persists through multiple generations. Gap phase dynamics, where large tree falls create openings in the canopy that allow the establishment of seedlings (primarily

Metrosideros) may be an important ecological process that maintains monodominance in large-statured montane wet forests in Hawaii.

5. References

Akaike, H. 1973. Information theory and an extension of the maximum likelihood principle. Pages 267-281. *In* International symposium on information theory. Second edition. (B. N. Petran and F. Csaki Eds.). Akademiai Kiado, Budapest Hungary.

Atkinson, I.A.E. 1970. Successional trends in the coastal and lowland forest of Mauna Loa and Kilauea volcanoes, Hawaii. Pacific Science 24: 387-400.

Burnham, K. P., and D. R. Anderson. 2002. Model selection and multimodel inference: a practical information-theoretic approach. Springer-Verlag, New York, USA.

Burton, P.J. and D. Mueller-Dombois. 1984. Response of *Metrosideros polymorpha* seedlings to experimental canopy opening. Ecology 65: 779-791.

Carey, E.V., S. Brown, A.J.R. Gillespie, and A.E. Lugo. 1994. Tree mortality in mature lowland tropical moist and tropical lower montane moist forests of Venezuela. Biotropica 26: 255-265.

Chernick, M.R. 2008. Bootstrap methods: a guide for practitioners and researchers. John Wiley & Sons, Inc. Hoboken, New Jersey, USA.

Clark, D.A. and D.B. Clark. 1992. Life history diversity of canopy and emergent trees in a neotropical rain forest. Ecological Monographs 62: 315-344.

Condit, R., S.P. Hubbell, and R.B. Foster. 1995. Mortality rates of 205 neotropical tree and shrub species and the impact of a severe drought. Ecological Monographs 65:419-439.

Connell, J.H. and M.D. Lowman. 1989. Low-diversity tropical rain forests: some possible mechanisms for their existence. The American Naturalist 134: 88-119.

Crawley, M.J. 2007. The R Book. John Wiley & Sons Ltd, West Sussex, England.

Coomes, D. A. and R. B. Allen. 2007. Mortality and tree-size distributions in natural mixed-age forests. Journal of Ecology 95: 27-40.

Denslow, J.S. 1987. Tropical rain forest gaps and tree species diversity. Annual Review of Ecology and Systematics 18: 431-451.

ESRI. 2010. Copyright ESRI. All rights reserved. ArcGIS 9.2 is a registered trademark of ESRI in the United States, the European Community, and other jurisdictions.

Franklin, J. F., H. H. Shugart, and M. E. Harmon. 1987. Tree death as an ecological process. BioScience 37: 550-556.

Gagne, W.C., and L.W. Cuddihy. 1990. Vegetation. Pages 45-114 *in* W.L. Wagner, D.H. Herbst, and S.H. Sohmer, editors. Manual of the flowering plants of Hawaii. Bernice P. Bishop Museum, Honolulu, Hawaii, USA.

Gerrish, G., D. Mueller-Dombois, and K.W. Bridges. 1988. Nutrient limitation and *Metrosideros* forest dieback in Hawaii. Ecology 69: 723-727.

Hart, P. J. 2010. Tree growth and age in an ancient Hawaiian wet forest: vegetation dynamics at two spatial scales. Journal of Tropical Ecology 26: 1-11.

Hart, T.B. 1990. Monospecific dominance in tropical rain forests. Trends in Ecology and Evolution 5: 6-11.

Hotchkiss, S.C. 1998. Quaternary vegetation and climate of Hawai`i. Ph.D. Dissertation, University of Minnesota, MN, USA

Jacobi, J.D. 1989. Vegetation maps of the upland plant communities on the islands of Hawai`i, Maui, Moloka`i, and Lana`i. Technical Report Number 61. Cooperative National Park Resources Study Unit, University of Hawaii at Manoa, Honolulu, Hawaii, USA.

Korning, J. and H. Balslev. 1994. Growth rates and mortality patterns of tropical lowland tree species and the relation to forest structure in Amazonian Ecuador. Journal of Tropical Ecology 10: 151-166.

Lieberman, D. and M. Lieberman. 1987. Forest tree growth and dynamics at La Selva Costa Rica, (1969-1982). Journal of Tropical Ecology 3: 347-358.

Lines, E. R., D. A. Coomes, and D. W. Purves. 2010. Influences of forest structure, climate, and species composition on tree mortality across the Eastern US. PLoS ONE 5: 1-12.

Manokaran, N., and K.M. Kochummen. 1987. Recruitment, growth and mortality of tree species in a lowland *Dipterocarp* forest in Peninsular Malaysia. Journal of Tropical Ecology 3: 315-330.

Mueller-Dombois, D. 1986. Perspectives for an etiology of stand-level dieback. Annual Review of Ecology and Systematics 17: 221-243.

Mueller-Dombois, D. and F.R. Fosberg. 1998. Vegetation of the tropical Pacific Islands. Springer-Verlag New York, Inc.

Nascimento, M. T., R. I. Barbosa, D. M. Villela, and J. Proctor. 2007. Above-ground biomass changes over an 11-year period in an Amazon monodominant forest and two other lowland forests. Plant Ecology 192: 181-191.

Read, J., P. Hallam, and J. Cherrier. 1995. The anomaly of monodominant tropical rainforests: some preliminary observations in the *Nothofagus*-dominated rainforests of New Caledonia. Journal of Tropical Ecology 11: 359-389.

Sato, H.H., W. Ikeda, P. Paeth, R. Smythe, and M. Takehiro Jr. 1973. Soil survey of the Island of Hawaii, State of Hawaii. United States Department of Agriculture, Soil Conservation Service, University of Hawaii Agricultural Experiment Station. Honolulu. 115pp.

Sheil, D.F., R.P. Burslem, and D. Alder. 1995. The interpretation and misinterpretation of mortality rate measures. Journal of Ecology 83: 331-333.

Smith, D.M. 1986. The practice of silviculture. Eighth edition. John Wiley and Sons. New York, New York, USA.

Swaine, M.D., D. Lieberman, and F.E. Putz. 1987. The dynamics of tree populations in tropical forest: a review. Journal of Tropical Ecology 3: 359-366.

Umeki, K. 2002. Tree mortality of five major species on Hokkaido Island, northern Japan. Ecological Research 17: 575-589.

Van Mantgem, P. J., N. L. Stephenson, J.C. Byrne, L. D. Daniels, J. F. Franklin, P. Z. Fulé, M. E. Harmon, A. J. Larson, J. M. Smith, A. H. Taylor, and T. T. Veblen. 2009. Widespread increase of tree mortality rates in the western United States. Science 323: 521-524.

Wolfe, E.W., and J. Morris. 1996. *Geologic Map of the Island of Hawai`i*. USGS Misc. Investigations Series Map i-2524-A. Washington, D.C.: U.S. Geological Survey.

18

Direct Sowing: An Alternative to the Restoration of Ecosystems of Tropical Forests

Robério Anastácio Ferreira and Paula Luíza Santos
Federal University of Sergipe, Department of Forest Sciences,
Brazil

1. Introduction

In tropical regions, under certain situations, we can employ a nature-oriented regeneration, as opposed to the artificial one. This is possible when there still is some resilience in the environment which can enable its own regeneration. The nature-oriented regeneration is a simple method of low cost, which may promote the formation of productive forests in the areas previously degraded (Shono et al., 2007). However, for the areas that have been very fragmented and with a more severe environmental degradation promoted by human disturbances, there are physical, chemical and biological barriers that reduce the resilience of these areas and prevent natural regeneration processes (Shono et al., 2007).

In these situations, it is necessary that humans (foresters, ecologists and farmers) intervene, through artificial means (seedlings plantation or direct sowing) to start processes of soil covering and vegetation restoration. The use of direct sowing to restore degraded ecosystems has become a viable alternative for this purpose in the tropical forest regions, as opposed to the conventional method of planting seedlings (Ferreira et al., 2007).

In this context, this chapter aims to analyze ecological, technical, socio-economic and forestry aspects involved in this process. It also highlights that several experiments conducted in the tropical regions around the world may contribute to broaden the perspectives and enhance methodologies, so that the ecological restoration can be used more widely.

Herein some already undertaken research findings will be reported which will bring to light some experiences that may contribute to the decision making on the choice of direct sowing for restoration of degraded ecosystems.

2. Importance of restoration of tropical forest ecosystems

In the tropical regions, natural resources are of great economic importance to the countries therein, especially to communities that depend exclusively on these resources for their livelihood and survival (Lamb & Gilmour, 2003).

Among natural resources, exploitation of native forests, either for obtaining direct or indirect benefits, has always been associated with the development of human communities. However, as an immediate consequence of continuous exploitation, the environment degradation has become increasingly apparent. This is resulting from the use of development models based on super-exploitation, and in most of cases the latter is done without adequate planning and without any commitment to the future generations. Thus, it is globally observed in a significant reduction framework of forested areas.

The highest rates of change in relation to the use of soil are recorded in the tropics. According to Lamb & Gilmour (2003), it is estimated that approximately 17 million of hectares of forests are annually changed from their natural conditions to other forms of use. It is also estimated that half of this area is in the range of wet tropical forest.

Even if we consider all the richness of biodiversity of the plants, animals, and microorganisms of the tropical forests, besides the high rates of rainfalls, high incidence of sunlight, large amount of biomass, and intense recycling of organic matter, these are still considered as fragile environments (Ambasht, 1993). One should also consider that as a result of poor edaphic conditions in most of the areas, forests may be easily converted and degraded, leading to savannization and desertification (Ambasht, 1993).

In an attempt to reverse the accelerated environmental degradation condition, practices of recovering degraded ecosystems are very old, and can find examples of its use in the history of different peoples, times and regions (Rodrigues & Gandolfi, 2004). Nevertheless, until recently, recovery processes of degraded zones were more characterized as silvicultural practices without close ties to the theory, and usually performed as a practice of seedlings plantation with specific purposes, such as erosion control, slope stabilization, and landscape improvement (Rodrigues & Gandolfi, 2004).

Knowledge broadening concerning processes related to the dynamics of natural vegetation formations has promoted meaningful changes in the orientation of environmental recovery programs (Rodrigues & Gandolfi, 2004). In this case, the actions of recovery no longer have the connotation of sheer application of silvicultrural practices that aim only at the restoration of forest species in a given environment in order to fulfill the difficult task of trying to recover the ecological processes. As result, it implies in the attempt to restore the complex interactions of vegetal communities having intrinsic characteristics to assure the perpetuation and evolution of the natural ecosystems (Rodrigues & Gandolfi, 2004).

Overall, it is possible to note that most of the experiences in restoration of degraded areas in tropical forests involve large investments. However, according to Jesus & Rolim (2005), the main and most current trends to recover degraded areas are related to an appropriate species selection, implementation of more effective plantation models, and researches are always focusing on the reduction of implementation costs. As stated by Ferreira (2002), the development of technologies aiming at the recovery of degraded areas at lower costs is indispensable, since in tropical forests most areas are in possession of small owners, who have limited or no available resource to be employed in the recovery process. In this context, despite very effective experiences and already consolidated by means of seedlings plantation, the employment of direct sowing in the field, similar to agricultural areas, may also be feasible in silvicultural, ecological, social, and mainly economic aspects promoting restoration of ecosystems at lower costs.

It is observed that the existing imbalance between the use and maintenance of forest resources is caused principally by their excessive exploitation, when in most situations it even happens without the awareness of required mechanisms in order to have an efficient regeneration of these environments (Maury-Lechon, 1993).

In the tropics, to recover degraded areas, it is necessary that the socio-economic conditions of the local communities are taken into consideration, since in many areas, these factors may be more important to select suitable methods to be employed for this purpose than the ecological and silvicultural factors themselves (Lamb & Lawrence, 1993). In this case, according to the aforesaid authors, human communities that exploit forests directly for their own survival are of great importance in the ecological processes and their restoration.

According to Ketler (2001), this interdependence may be observed in different countries where their economy is strongly related to efficient use of the land. For instance, some experiments conducted in New Zealand, Mexico, and Philippines show that when restoration is planned, besides the ecological processes, the social and cultural values of the communities also need to be observed and taken into consideration (Ketler, 2001).

3. Systemic approach to the restoration of forest ecosystems

Environmental systems are complex and dynamic, consisting of a large number of interconnected elements, capable of exchanging information with their surroundings, and able to adapt their inner structure as a result of interactions among their elements (Cristofoletti, 2004). Although they are organizationally closed, but open in terms of flows of matter and energy, environmental systems receive inputs, transforming them and generating products. In view of their model, it is necessary to evaluate the flow and transformation of some inputs, such as water, sediments, light, raw matter, nourishment, and others (Cristofoletti, 2004).

In the case of forest ecosystems, organisms and their communities (flora and fauna) reproduce hierarchically complex levels of structural organizations that function in closed form, and where each group fulfills its own ecological purpose. That is, plants absorb light energy and conduct photosynthesis, herbivores eat plants, carnivores eat other animals, fungi recycle nutrients, and so forth, with the whole system working and self-producing within the limits of thermodynamics laws (Odum, 1988; Ricklefs, 2003).

According to Aumond (2003), this model assumes irreversibility of the process, in which matter and environment energy flow continuously considering the whole structure of the ecosystem, and keeping it beyond the state of equilibrium. The instability of this process, associated with the mechanisms and techniques of restoration that allow the internalization of part of the matter and energy flow, leads to a self-organization resulting from the emergence of new structures that work as attractions of a growing complexity.

In degraded systems, the absence of some ecological variables hinders the development and improvement of biodiversity, and the organization model will be open, with high entropy, resulting in progressively greater and irreversible losses (Aumond & Comin, 2008). The water seeps out of the system, consequently dragging particles, such as organic matter, macro and micronutrients, further impoverishing the soil of the degraded area. In areas lacking vegetation, water retention will always be lower than in areas with vegetation cover. Direct sunlight on the soil surface leads to extreme temperature fluctuations. Solar heat transfer to the environment by means of conduction, radiation, and convection, produces considerable temperature fluctuations in the soil followed by great losses to the atmosphere. Therefore, in open systems, there are appreciable losses of matter and energy, and hence it is necessary to internalize the ecological processes; that is, to enable the organizational closing of the degraded zone, inducting into the introspection of environmental variables, aiming at the increase of flow of matter and internal energy in the system (Aumond & Comin, 2008).

The sustainability of ecological systems is supported by three pillars: biodiversity, nutrient cycling, and energy flow (Franco & Campelo, 2005). Also, in the opinion of these authors, to maintain the soil any system need to include as many vegetal species as possible, sustain high levels of organic matter in conjunction with the soil microbiota, besides being most effective in the use of water, light, and nutrients.

Several authors consider the soil, vegetation, water, fauna (invertebrates), microclimate, and relief rugosity as being the main components to be taken into account when recovering a degraded area (Odum, 1988; Ricklefs, 2003; Aumond, 2007). Considering the environment and degraded system, with the main components, by means of an influence diagram, one may observe that the presence of water flow in the surface may indicate an increase of erosion in the soil; the presence of a vegetal layer corresponds to greater microclimatic softening due to the decrease of solar radiation, which results in lower environment temperatures and higher relative humidity of air; roughness such as relief changes may indicate more water, sediments, and propagules retained in the degraded system; water affects the microclimate through the mitigation of its variables, and so on.

Upon studying an integrative model to recover areas degraded by mining activities, Aumond (2007) developed an ecological model, considering the area as a complex and dynamic system, sensitive to initial conditions for the site preparation, with the application of rugosity techniques to trigger over time emergent properties that accelerate the recovery process. Over a period of time of 26 months, the author assessed the evolution of the components (soil, vegetation, fauna, water, microclimate, and rugosities), concluding terrain variations helped retaining and minimizing the volume and flowing coefficient of the water in the terrain, with an impressive decrease of erosive processes. The author also confirmed the development of the *Mimosa scabrela* species (through the use of seedlings plantation) and the height of spontaneous vegetation had better growth in irregular areas, contributing to the changes of solar radiation, temperature and relative humidity of air which resulted in the death of herbaceous plants and shrubs, standing out the trees, confirming an advanced stage of succession in irregular areas.

Thus, the restoration process for degraded zones starts from its reforestation by using methods that make ecological succession feasible, covering the exposed soil and stimulating the development of new vegetal species. When this process is successful, not only the development of new species in the environment occurs, it also resumes the condition of self-sustaining, and the reconduction of ecological relations enables the integration of the area recovered with areas preserved in its surroundings (Van den Berg et al., 2008).

When it comes to the restoration of permanent preservation zones, programs will only be successful if the land owners consider the actions attractive and think that they can bring benefits or payments in the form of ecological goods and services, such as: improvement of quality and increase in the quantity of water produced by the hydro resources, carbon sequestration and conservation of biodiversity (Lamb et al., 2005). However, according to

Miller & Hobbs (2007), defining programs and priority actions to restore landscapes based only on types of characteristic habitats or forms of vegetal cover, from metric standards, such as the index of fragmentation of a specific zone or yet, considering very broad objectives like conservation of biodiversity, may constitute large setbacks to reach more effective and well succeeded restoration.

In several countries, it is observed that the most widely employed method for such purpose is by means of seedlings plantation of different ecological groups with mixed populations. Nevertheless, the use of direct sowing presents satisfactory results both from the ecological and economic-silvicultural points of view by using native species, making it feasible to recover the degraded zones.

4. Direct sowing system for tropical forests restoration

Upon starting a forest restoration program by using native tree species, it is essential to be aware of some aspects of the seed technology of the species to be used, mainly related to germination. As stated by Figliolia et al. (1993), awareness of favorable conditions for seed germination, principally temperature and lighting, is a determining factor in the germination process, since both the factors are directly interconnected with ecological characteristics of the species.

Seeds of pioneer species need to have conditions of high luminosity and temperature to germinate, and seedlings are intolerant to shade, besides having dormancy and forming a seed bank in the soil. On the other hand, seeds of climax species are not demanding of lighting to germinate and present little or no dormancy, constituting a seedling bank in the soil (Budowski, 1965).

Direct sowing consists of introduction of specific forest seeds directly into the soil of the zone to be forested. In principle, it is a technique recommended only for some initial pioneer and secondary species in areas lacking vegetation, and also for late secondary and climax species, for the enrichment of secondary forests (Kageyama & Gandara, 2004).

It is an inexpensive and versatile reforestation technique, which can be used on most sites, with a greater focus on situations where natural regeneration or seedlings plantation cannot be performed (Mattei, 1995a). Besides that, it shows favorable results in degraded areas of difficult access and steep terrain slopes (Barnett and Baker, 1991).

The methods most commonly employed to perform direct sowing with the purpose of restoring degraded zones are: spread of seeds throughout the area, sowing through grooves or plantation lines, and sowing through ditches (Barnett & Baker, 1991). According to Groot & Adams (1994), this method is very efficient whenever one desires to regenerate areas at very low costs, where clear cutting operations are conducted (Groot & Adams, 1994).

In Figure 1, one may find the steps required for the forest restoration process by using direct sowing. At first, one shall make a diagnosis of the area to be restored by means of a climatic and soil analysis to check whether there is or not the possibility to use this technique; followed by selection of vegetal species aiming at making the environment as close as possible to the original condition with the intent to facilitate the adaptation of the species.

Primarily, it is recommended to employ the species that naturally occurs in the region, or native species that may be associated with different ecological groups, and also the socio-economic importance of the species to the human communities within the areas to be restored. However, as claimed by Maury-Lechon (1993), the employment of species of the genus *Pinus, Eucalyptus,* and *Acacia* may be recommended for many countries where they do not occur naturally, since they are associated with their use with positive socio-economic impacts, are well adapted to the climatic conditions, and that do not compromise the local ecological balance. One must bear in mind that it is necessary to do careful planning to bring in their gradual replacement and assure the development of native species. In this case, from report of Durigan (1996), the exotic species may be initially recommended, playing the role of pioneer species. According to the author, this is desired in areas where the degradation degree is high and that may require a long-term recovery because they may provide a better environment to the native species of later successional stages.

It is also important to know the basics of germination, since the seeds of some species need to have pre-germination treatments to increase the percentage and standardize the germination. By means of tests originally carried out in the laboratories (Figure 2), it is possible to assess the physiological quality of the batches of seeds and establish the amount that will be used to perform the direct sowing, without wasting them.

The use of direct sowing is related to show some advantages, such as reduction of deformations of root system of field plants. It may also stimulate better development of seedlings, and yet, eliminates the cost of seedlings production during the nursery phase. On the other hand, it is observed that one of the major difficulties related to its use, to a large extent, is that it is still necessary to do a good analysis, because in most cases the results are inconsistent concerning emergence, survival, and growth of plants (Winsa & Bergsten, 1994).

Thus, some factors may be associated with the difficulty in applying this method to large areas because of high diversity of vegetal species in the tropical forests. Little is known of the species demands in the environment as to the essential factors to the emergence and development of seedlings, such as: lighting, temperature, and required humidity. These factors are associated directly with the features of the areas where the seeds are sown. In addition, another difficulty is related to the definition of the relation between the survival of seedlings and quantity of seeds that should be sown in order to assure a good recovery and good development of species.

When it comes to sowing density there is little information in literature. But, considering an immense diversity of vegetal species in the tropical forests, it is practically impossible to define the one that could be considered as ideal. However, Burton et al. (2006) recommend that a density that should be considered desirable shall be that one representing at least 50% of the soil cover; minimum density shall be related to maximum production, without observing an increment even with an increase of density and when it achieves a balance in demography, without a decrease or failure in the final stand.

One observes that for the restoration of degraded zones in the tropical forest ecosystems, the success of the use of direct sowing is related to the degree of degradation in the zone, that is to say, the possibility of success of the development of plants decreases with the increase in the degree of degradation. In wet tropic forest areas of Australia, the employment of direct sowing is recommended only before verifying a state of more intensive degradation (Sun et al., 1995).

Steps for the implementation of direct sowing in the forest restoration process.

I – Zone diagnosis;

II – Soil preparation;

III – Seed plantation, e.g. *Erythrina velutina* Wild;

IV – Emergence of seedling of *Erythrina velutina* 8 days after direct sowing;

V – Seedling of *Erythrina velutina* 30 days after the implementation of direct sowing;

VI – Plant of *Erythrina velutina*, 240 days after direct sowing.

Fig. 1. Illustration of steps required for the implementation of direct sowing in the process of restoration of degraded and riparian zones (Pictures of Santos, 2010).

Fig. 2. Analysis of viability of seeds lots in the laboratory by testing them in a germination camera (BOD), at constant temperature and under continuous lighting, before the implementation of direct sowing in the field (A); Germination of the seed lots of *Erythrina velutina* Wild (B) and *Sapindus saponaria* L (C).

In Brazil, some experiments were carried out in an attempt to make the direct sowing technique viable in the ecological and silvicultural terms, both in restoration of ecosystems and implantation of populations for economic purposes (Ferreira et al., 2007). Several experiments presented good results in the implementation of the species *Pinus sp.* (Mattei, 1997; Brum et al., 1999; Mattei et al., 2001; Finger et al., 2003), recovery of degraded slopes (Pompéia et al., 1989), and in the implementation of riparian forests (Santos Jr et al., 2004; Almeida, 2004; Klein, 2005; Ferreira et al, 2007, 2009 and Santos, 2010).

As claimed by Kageyama & Gandara (2004), the direct plantation of forest seeds may be used both for implementation of pioneer species in areas without forest covering, and for the implementation of slow-growing species (late secondary and climax species) in the enrichment of secondary forests.

However, in both direct sowing and work with seed banks, the emergence of seedlings of native species is irregular with predominance of few species, and pioneer ones prevailing, requiring the replacement of seeds in places where there were failures in germination. This way, the need for selected species for this purpose to present rapid emergence of seedlings in order to quickly obtain an effective soil recovery rate was observed. In order to have the acceleration of the germination process of the seeds and promotion of rapid development of seedlings, the use of treatments to break seed dormancy may be required, since the intent is to promote the fast soil covering (Winsa & Bergstein, 1994; Aerts et al., 2006; Ferreira et al., 2007; 2009; Santos, 2010).

The quality of the seeds, as assessed by the germination and vigor of each lot, is indispensable to ensure the germination in the field. Seeds with low vigor are unable to germinate in adverse conditions, and when they germinate, in most cases, they do not generate vigorous seedlings (Botelho & Davide, 2002).

Many authors identify the preparation of the site as an essential factor in the development of the seeds in the field (Smith, 1986, Winsa & Bergstein, 1994; Fleming & Mossa, 1994; Falck, 2005; Andrade, 2008; Santos, 2010), since in degraded areas, the exposure of soil to weathering results in the change of its physical, chemical and biological characteristics, retarding or making the development of any species impossible. Therefore, it is necessary to prepare the soil, prior to sowing, minimizing the physical difficulties to be found by the seedlings, and increasing the absorption of water by the soil and providing nutrients located in the lower layers of soil, besides other factors (Santos Jr, 2000).

In an experiment carried out by Santos (2010), with direct sowing, in the municipality of São Cristovão, Sergipe, Brazil, in two subsystems with different types of soil occupation (agriculture and pasture), it was found that the soil slips by means of plowing and harrowing may have contributed to the loss of seeds, mainly in the portions without a physical protector, due to the soil movement, leading to a landfill or seed dragging. This fact was evidenced in the pasture area, where the seeds of the species *Guazuma ulmifolia*, *Machaerium aculeatum*, *Lonchocarpus sericeus* and *Bowdichia virgilioides* presented low emergence in comparison to the portions with physical protectors. In an agricultural area, this fact was evidenced for the seeds of the species *B. virgilioides* and *M. aculeatum* (Table 1).

Species	Seedling Emergence (%)			
	Subsystem 1 – Pasture		Subsystem 2 – Agriculture	
	WP	WOP	WP	WOP
Erythrina velutina	57.14 aA	78.09 aA	99.05 aA	84.76 aA
Bowdichia virgilioides	16.19 bA	0.00 cA	96.19 aA	39.05 bB
Sapindus saponaria	40.00 aA	42.86 bA	79.04 aA	64.76 aA
Guazuma ulmifolia	26.67 bA	7.14 cA	52.38 bA	26.19 bA
Lonchocarpus sericeus	50.00 aA	6.19 cB	44.76 bA	44.28 bA
Machaerium aculeatum	7.14 bA	0.95 cA	14.76 cA	1.91 cA
Average	32.86A	22.54A	64.37A	43.49B

Averages followed by the same letter do not differ between them as per the Scott-Knott test at 5% probability (Ferreira, 2006).
Small letters in columns compare species for each protector (WP/WOP) in subsystem 1 and 2.
Capital letters in lines compare protectors, for each species, in subsystem 1 and 2.

Table 1. Seedling emergence of forest species (%), until 90 days after sowing in the absence and presence of physical protectors in two subsystems placed in the Rural Campus of Federal University of Sergipe, municipality of São Cristóvão, Sergipe. WP – with physical protectors; WOP – without physical protectors. (Source: Santos, 2010).

Sun et al. (1995) stated that the competition with grasses and the lack of soil fertility are the factors that also affect the seedling survival, once weeds have certain aggressiveness in the field. This characteristic makes them exceptional competitors, since in few months they colonize the area, interfering with the forest species, mainly in the development of climax species.

Weeds also present effective mechanisms for survival in the environment, as strategies to withstand adverse conditions, for instance: high reproductive capacity, effective dispersion mechanisms, and large seed longevity. These mechanisms are of great importance to ensure their success and win the competition with other species. The strategies of development in the environment noticed in weeds are similar to those mentioned by Budowski (1965) and Swaine & Whitmore (1988) for the pioneer or colonizing tree species.

The dry season, burial of seeds by torrential rains, and the severe cold are considered as the main climatic factors that cause damage to direct sowing (Mattei 1995b). It is also mentioned as a failure in direct sowing due to lack of contact of the seed with the mineral soil, displacement of the seed after sowing, flooding or excessive moisture close to the seed, and losses resulting from the attack of birds and ants.

Santos (2010) states that the presence of laminar erosion in the soil during the rainy season may affect the emergence of species due to dragging and burial of seeds that are planted without any physical protectors. And yet, even the portions with physical protectors may be affected by burial, and the presence of water in some places (Figure 3). Therefore, there is clear need to evaluate and monitor the area where the seeds will be directly sown.

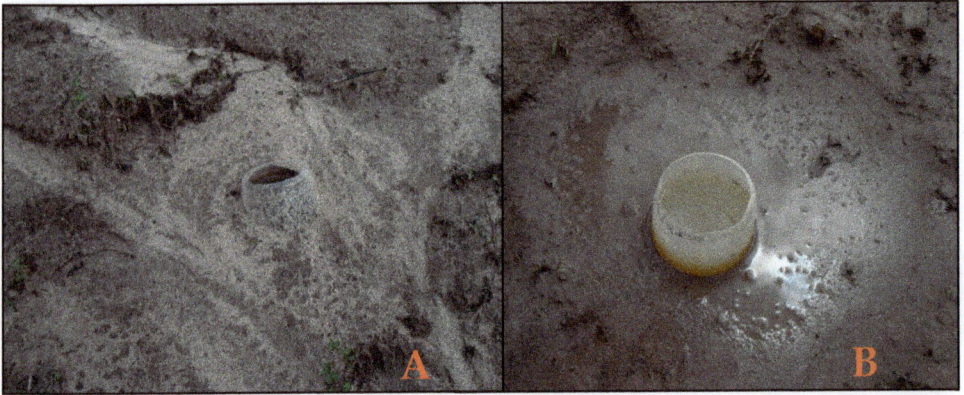

Fig. 3. Presence of water in the physical protector (A) and burial through laminar erosion (B) after rainy season in a pasture area located in the municipality of São Cristovão, Sergipe, Brazil, after the implementation of direct sowing (Source: Santos, 2010).

One of the first experiments performed by using direct sowing in Brazil is mentioned by Silva Filho (1988) where tests were conducted in an attempt to recover zones in severe conditions of degradation located in Serra do Mar (Hill), São Paulo. Considering the high slope steepness, *Brachiaria* seeds were sown, but due to the rains, they were dragged to the bottom of the hill, promoting great emergence of seedlings in this area.

Upon evaluating the direct sowing of *Pinua taeda* and *Cedrela fissilis* through the use of two techniques of soil preparation (without preparation and plowing and harrowing), Mattei (1995b) concluded that the sandy soil proved to be inappropriate for the direct plantation of seeds in the rainy season because of the movement of its particle and rapid infiltration.

With the intent to make the direct sowing process possible, several authors tested the use of physical protectors aiming at the reduction of herbivory rate and at increasing the temperature and moisture of the superficial layer of soil (Mattei, 1997; Santos Jr. et al., 2004; Falck, 2005; Klein, 2005; Ferreira et al., 2007; Andrade, 2008; Santos, 2010).

According to Mattei (1997), the objective of use of physical protectors is to promote the improvement of germination of seeds and survival of seedlings in the field, creating a microenvironment for the development of young plants (Figure 4). Besides, it prevents the soil slips in conjunction with the seeds in different times of heavy rains, preserving the depth of sowing, and facilitating the emergence and hindering the attack of natural enemies (Mattei, 1995b).

Predation by ants and birds, considered as one of the major problems in the implementation of direct sowing, has been reduced by using physical protectors, which showed a significant decrease in predation (Schneider et al., 1999; Mattei & Rosenthal, 2002). As claimed by Serpa & Mattei (1999), the use of physical protectors may help in retaining moisture at the sowing point, as the presence of water is one of the essential factors in emergence, survival and development of plants.

The use of physical protectors in direct sowing cannot be recommended as a standard methodology for all species and situations, since its effectiveness cannot always be proven.

According to Ferreira et al. (2007), the use of plastic protectors was not effective either for the emergence or for the survival of seedlings, for the pioneer species: *Senna multijuga, Senna macranthera, Solanum granuloso-leprosum* and *Trema micrantha* (Table 2). On the other hand, the use of physical protectors, similar to those employed by Ferreira et al. (2007), was effective for the survival of the seedlings of the climax species such as *Cedrella fissilis, Copaifera langsdorffii, Enterolobium contortisiliquum, Piptadenia gonoacantha* and *Tabebuia serratifolia*, mainly preventing the attack of ants as observed by Santos Jr et al. (2004).

Fig. 4. Direct sowing of the species *Erythrina velutina* Wild (A) and *Bowdichia virgilioides* Kunth (B) through the use of physical protectors for germination (Source: Santos, 2010).

Species	Seedlings emergence (%)		Seedlings survival (%)	
	WP	WOP	WP	WOP
Senna multijuga	38.24 a	40.81 a	86.75 a	87.40 a
Senna macranthera	15.55 a	15.71 a	91.25 a	90.91 a
Solanum granuolo-leprosum	22.72 a	18.26 a	52.50 a	33.67 a
Trema micrantha	15.29 a	17.58 a	94.72 a	95.36 a

Tukey Test conducted at level 5% of probability
Comparisons of averages between treatments – same letters in lines do not differ between them.

Table 2. Seedling emergence and tree species survival 3 months after sowing under field conditions (WP – with physical protectors; WOP – without physical protectors). Adapted from Ferreira et al. (2007).

Physical protectors to improve emergence and/or survival of seedlings may be fabricated by using different materials like wood laminates, plastic cups, cardboard and plastic bottles. However, the latest trend is to use biodegradable materials to reduce the costs of implementation and which do not jeopardize the environment. It should be emphasized that whenever plastic containers are used it is necessary to remove them after accomplishing their purpose.

To Smith (1986), the success of direct sowing is to create a microenvironment with favorable conditions for rapid seedling emergence, with enough moisture during the period of emergence of seedlings and in the following phase, since the plants that germinate and grow in the field have restrictive protection in relation to the numerous lethal agents, which may be controlled in nurseries. Consequently, there are more risks of survival to be lower by using direct sowing than with seedling plantation. However, it is one of the most promising techniques in the recovery process for the degraded areas, especially when one of the objectives is to reduce the implementation costs (Santos Jr, 2000).

In tropical regions, with great environmental heterogeneity, it is possible to observe a significant variation of temperature moisture. In most studies, this range is from 20° to 30° C (Longman & Jenik, 1974). One of the ways mentioned to make the environment more appropriate for emergence is through the use of physical protectors, once they foster a more suitable microenvironment (Santos, 2010). On the other hand, in other studies the use of physical protectors for this purpose would be unnecessary in this phase (Santos Jr et al., 2004; Ferreira et al., 2007).

Another relevant factor in the use of direct sowing, is related to the size of the seed, as this characteristic may significantly influence both emergence and development of the seedlings in the degraded zones (Doust et al., 2006).

In this context, some information about tropical forest species were provided by Sautu et al (2006) for a group of 100 species that compose the Panama Canal Basin, which can be useful in coping with direct sowing.

Ferreira et al (2009) observed that both emergence and development of seedlings have been influenced by the seed density in the study carried out by using *Caesalpinia leiostachya* (0.154g), *Schinus terebinthifolius* (0.013g), *Cassia grandis* (0.753g), *Himenea courbaril* (4.994g), and *Enterolobium contortisiliquum* (0.664g). As claimed by the authors, the species that had bigger seeds, with higher density, tended to show faster and higher seedling emergence.

This fact was also verified by Doust et al. (2006) in a study conducted in Queensland, Australia, with 16 species for restoration of tropical forests. The authors noted that species with bigger seeds (>5.0g) presented better development in relation to small seeds (0.01g – 0.099g), and those considered as intermediate (0.1g – 4.99g).

This technique has a high potential to recover degraded areas provided that in the forest formation the main form of regeneration, both in gaps and in the expansion of the remaining, takes place by means of natural sowing (Botelho & Davide, 2002) which, under favorable conditions, fosters good germination of seeds.

5. Final considerations

Direct sowing, in view of experiments conducted and results obtained by several researchers, offers itself as a feasible alternative to restoration of degraded ecosystems in the tropical forests (Santos Jr et al., 2004; Ferreira et al., 2007; Ferreira et al., 2009; Santos 2010).

For its employment, some essential factors need to be taken into consideration: the degradation state of the area, selection of species with best capability for this purpose, site characteristics (terrain steepness, fertility, temperature, moisture, etc.), presence of weed, a thorough analysis of the need for physical protectors as they can foster a more favorable microenvironment to the emergence of seedlings, and protection against agents that predate seeds and seedlings, mainly ants and birds. On the other hand, it is necessary to understand that the use of physical protectors cannot be recommended as a standard method as a result of great interaction observed between them, the environment conditions, and the species employed. The results are not always favorable and, in many cases, their use is unnecessary.

One of the main favorable factors in the use of direct sowing to restore degraded forest ecosystems is the possibility to reduce implementation costs. Nevertheless, one needs to understand that, in addition to the ecological and silvicultural aspects of the selected species, it is also necessary to observe the socio-economic conditions of the local communities where the restoration will be carried out. In many regions, communities are dependent on natural resources for their living, and they need to be involved in the process.

From broader knowledge of ecology of ecosystems, searching understanding about the diverse interactions and its functioning may stimulate better uses of direct sowing, as it may represent an alternative to recreate or restore degraded environments in the tropical regions with characteristics more similar to the original natural environments.

6. References

Aerts, R., Maes, W., November, E., Negussie, A., Hermy, M. & Muys, A. (2006). Restoring dry afromontane forest using bird and nurse plant effects: direct sowing of *Olea europaea* ssp. *cuspidate* seeds. *Forest Ecology and Management*, Oxford, v.30, p.23-31. ISSN: 0378-1128

Almeida, N.O. (2004). *Implantação de matas ciliares por plantio direto utilizando-se sementes peletizadas*. Lavras: UFLA, 269p. (Doctoral Thesis).

Ambasht, R.S. (1993). Conservation of some disturbed Indian tropical forest ecosystems. In: Lieth, H. & Lohmann, M. (eds.). *Restoration of tropical forest ecosystems*. Netherlands: Kluver Academic Publishers, p.203-208. ISBN: 978-90-481-4198-2

Andrade, A.P.A. (2008). *Avaliação da utilização de protetor físico de germinação e semeadura direta das espécies Copaifera langsdorffii Desf. e Enterolobium contortisiliquum (Vell.) Morong. em área degradada pela mineração*. Brasília: UNB, 2008. 90p. (Master Thesis).

Aumond, J.J. (2003). Teoria dos sistemas: uma nova abordagem para recuperação e restauração ambiental. In: Simpósio Brasileiro de Engenharia Ambiental, 2., Itajaí. *Anais...* Itajaí: UNIVALI/CTT, p. 43-49.

Aumond, J.J. (2007). Adoção de uma nova abordagem para a recuperação de área degradada pela mineração. Florianópolis: UFSC, 2007. 265p. (Doctoral Thesis).

Aumond, J.J., Comin, J.J (2008). Abordagem sistêmica e o uso de modelos para recuperação de áreas degradadas. In: Simpósio Nacional sobre Recuperação de Áreas Degradadas 7, Curitiba. *Anais...* Curitiba: FUPEF, p.03-34.

Barnett, J.P. & Baker, J.B. (1991). Regeneration methods. In: Duryea, M.L. & Dougherty, P.M. (eds.). *Forest regeneration manual*. Dordrecht: Kluver Academic Publishers, p.35-50. ISBN: 978-07-923-0959-8

Botelho, S.A. & Davide, A.C. (2002). Métodos silviculturais para recuperação de nascentes e recomposição de matas ciliares. In: Simpósio Nacional sobre Recuperação de Áreas Degradadas 5, Belo Horizonte. *Anais...* Belo Horizonte, p.123-145.

Brum, E.S.; Mattei, V.L. & Machado, A.A. (1999). Emergência e sobrevivência de *Pinus taeda* L. em semeadura direta a diferentes profundidades. *Revista Brasileira de Agrociência*, Pelotas, v.5, n.3, p. 190-194. ISSN: 0104-8996

Budowski, G. (1965). Distribution of tropical American rainforest species in the light of sucessional process. *Turrialba*, Costa Rica, v.15, n.1, p.40-42. ISSN: 0041-4360

Burton, C.M., Burton, P.J., Hebda, R. & Turner, N.J. (2006). Determining the optimal sowing density for a mixture of native plants used to revegetate degraded ecosystems. *Restoration Ecology*, Oxford, v.14, n.3, p.379-390. ISSN: 1526-100X

Cristofoletti, A. (2004). Sistemas dinâmicos: as abordagens da teoria do caos e da geometria fractal em geografia. In: Guerra, A.J.T. & Vitte, A.C. (Orgs). *Reflexões sobre a geografia física no Brasil*. Rio de Janeiro: Bertrand Brasil, p.89-110. ISBN: 978-85-286-1049-9

Doust, S.J., Erskine, P.D. & Lamb, D. (2006). Direct seeding to restore rainforest species: microsites effects on the early establishment and growth of rainforest tree seedlings on degraded land in the wet tropics of Australia. *Forest Ecology and Management*, Amsterdam, v.234, p.333-343. ISSN: 0378-1128

Durigan, G. (1996). Revegetação em áreas de Cerrado. In: Simpósio IPEF, 6, Piracicaba, 1996. *Anais*...Piracicaba: ESALQ, v.1, p.23-26.

Falck, G.L. (2005) *Recobrimento de sementes de Pinus elliottii Engelm como alternativa para semeadura direta em campo*. Pelotas: UFPEL. 58p. (Master Thesis).

Ferreira, D.F. (2006). Sisvar - Sistema de Análise de Variância. Lavras: UFLA.

Ferreira, R.A. (2002). *Estudo da semeadura direta visando à implantação de matas ciliares*. Lavras: UFLA, 138p. (Doctoral Thesis).

Ferreira, R.A., Davide, A.C., Bearzoti, E. & Motta, M.S. (2007). Semeadura direta com espécies arbóreas para recuperação de ecossistemas florestais. *Cerne*, Lavras, v.13, n.3, p.21-279, ISSN: 0104-7760

Ferreira, R.A., Santos, P.L., Aragão, A.G., Santos, T.I.S., Santos Neto, E.M. & Rezende, A.M.S. (2009). Semeadura direta com espécies florestais na implantação de mata ciliar no Baixo São Francisco em Sergipe. *Scientia Forestalis*, Piracicaba, v.37, n.81, p.37-46, ISSN: 1413-9324

Figliolia, M.B., Oliveira, E.C. & Piña-Rodrigues, F.C.M. (1993). Análise de sementes. In: Aguiar, I.B.; Piña-Rodrigues, F.C.M. & Figliolia, M.B. (Coords). *Sementes florestais tropicais*. Brasília: ABRATES, p.137-174.

Finger, C.A.G., Schneider, P.R., Garlet, A., Eleotério, J.R. & Berger, R. (2003). Estabelecimento de povoamentos de *Pinus elliottii* Engelm pela semeadura direta a campo. *Ciência Florestal*, Santa Maria, v.13, n.1, p.107-113. ISSN: 0103-9954

Fleming, R.L. & Mossa, D.S. (1994). Direct seeding of black spruce in northwestern Ontario: seedbed relationships. *The Forestry Chronicle*, Ottawa, v.70, n.2, p.151-158. ISSN: 0015-7546

Franco, A.A. & Campello, E.F.C. (2005). Manejo nutricional integrado na recuperação de áreas degradadas e na sustentabilidade dos sistemas produtivos utilizando a fixação biológica de nitrogênio como fonte de fósforo. In: Aquino, A.M. & Assis, R.L. *Processos biológicos no sistema solo-planta: ferramentas para uma agricultura sustentável*. Brasília: Embrapa Informação Tecnológica. p.201-220. ISBN: 85-7383-304-1

Groot, A. & Adams, M.J. (1994). Direct seeding black spruce on peatlands: fifth-year results. *The Forestry Chronicle*, Ottawa, v.70, n.5, p.585-592, Sep./Oct. ISSN: 0015-7546

Jesus, R.M. & Rolim, S.G. (2005). Experiências relevantes na restauração da Mata Atlântica. In: Galvão, A.P.M. & Porfírio-da-Silva, V. *Restauração florestal: fundamentos e estudos de caso*. Colombo: Embrapa Florestas. p.59-86. ISBN: 85-892-8104-3

Kageyama, P.Y. & Gandara, F.B. (2004). Recuperação de áreas ciliares. In: Rodrigues, R.R. & Leitão Filho, H.F. (eds.). *Matas ciliares: conservação e recuperação*. São Paulo: EDUSP/FAPESP, 2004. p.249-269. ISBN: 85-314-0567-X

Ketler, J.S. (2001). A dependence on people: examples of ecological restoration and land-based economies from three countries. *Ecological Restoration*, Madison, v.19, n.1, p.27-33. ISSN: 1543-4060

Klein, J. (2005). *Utilização de protetores físicos na semeadura direta de timburi e canafístula na revegetação de matas ciliares*. Marechal Cândido Rondon: UNIOESTE. 80p. (Master Thesis).

Lamb, D. & Gilmour, D. (2003). *Rehabilitation and restoration of degraded forests*. Issues in Forest Conservation, IUCN. 110p. ISBN: 28-327-0668-8

Lamb, D., Erskine, P.D. & Parrota, J.A. (2005). Restoration of degraded tropical rain forest landscapes. *Science*, v.310, p.1628-1632. ISSN: 0036-8075

Lamb, D. & Lawrence, P. (1993). Mixed plantations using high value rainforest trees in Australia. In: Lieth, H. & Lohmann, M. (eds.). *Restoration of tropical forest ecosystems*. Netherlands: kluver Academic Publishers. p.101-108. ISBN: 978-07-923-1945-0

Longman, K.A. & Jenik, J. (1974). Tropical forest and its environment. London: Lowe & Brydone Ltd. ISBN: 0-582-44045-9

Mattei, V.L. (1995a). Importância de um protetor físico em pontos de semeadura de *Pinus taeda* L. diretamente no campo. *Revista Árvore*, Viçosa, v.19, n.3, p.277-285. ISSN: 0100-6762

Mattei, V.L. (1995b). Preparo do solo e uso de protetor físico, na implantação de *Cedrela fissilis* V. e *Pinus taeda* L., por semeadura direta. *Revista Brasileira de Agrociência*, Pelotas, v.1, n.3, p.127-132. ISSN: 0104-8996

Mattei, V.L. (1997). Avaliação de protetores físicos em semeadura direta de *Pinus taeda* L. *Ciência Florestal*, Santa Maria, v.7, n.1, p.91-100. ISSN: 0103-9954

Mattei, V.L., Romano, C.M. & Teixeira, M.C.C. (2001). Protetores físicos para semeadura direta de *Pinus elliottii* Engelm. *Ciência Rural*, Santa Maria, v.31, n.5, p.775-780. ISSN: 0103-8478

Mattei, V.L. & Rosenthal, M.D.A. (2002). Semeadura direta de canafístula (*Peltophorum dubium* (Spreng.) Taub.) no enriquecimento de capoeiras. *Revista Árvore*, Viçosa, v.26, n.6, p.649-654. ISSN: 0100-6762

Maury-Lechon, G. (1993). Biological and plasticity of juvenile trees stages to restorate degraded tropical forests. In: Lieth, H. & Lohmann, M. (eds.). *Restoration of tropical ecosystems*. Netherlands: Kluver Academic Publishers. p.37-46. ISBN: 978-07-923-1945-0

Miller, J.R. & Hobbs, R.J. (2007). Habitat restoration – do we know what we're doing? *Restoration. Ecology*, v.15, n.3, p.382-390. ISSN: 0012-9658

Odum, E. (1988). Ecologia. Rio de Janeiro: Guanabara Koogan, 446 p. ISBN: 85-201-0249-2

Pompéia, S.L., Pradella, D.Z.A., Martins, S.E., Santos, R.C. & Diniz, K.M. (1989). A semeadura aérea na Serra do Mar em Cubatão. *Ambiente*, São Paulo, v.3, n.1, p.13-19.

Ricklefs, E. A. (2003). Economia da Natureza. 5ª ed. Rio de Janeiro: Guanabara Koogan, 470p. ISBN: 85-277-0798-5.

Rodrigues, R.R. & Gandolfi, S. (2004). Conceitos, tendências e ações para a recuperação de florestas ciliares. In: Rodrigues, R.R. & Leitão Filho, H.F. (eds.). *Matas ciliares: conservação e recuperação.* São Paulo: EDUSP/FAPESP, p.235-247. ISBN: 85-314-0567-X

Santos, P.L. (2010). *Semeadura direta com espécies florestais nativas para recuperação de agroecossistemas degradados.* São Cristóvão: UFS. 69p. (Master Thesis).

Santos Júnior, N.A. (2000). *Estabelecimento inicial de espécies florestais nativas em sistemas de semeadura direta.* Lavras: UFLA. 96p. (Master Thesis).

Santos Júnior, N.A., Botelho, S.A. & Davide, A.C. (2004). Estudo da germinação e sobrevivência de espécies arbóreas em sistema de semeadura direta, visando à recomposição de mata ciliar, *Cerne,* Lavras, v.10, n.1, p.103-11. ISSN: 0104-7760

Sautu, A., Baskin, J.M.; Baskin, C.C.& Condit, R. (2006). Studies on seed biology of 100 native species of trees in a seasonal moist tropical forest, Panama, Central America. *Forest Ecology and Management,* Amsterdam, v.234, p.254-263. ISSN: 0378-1128

Schneider, P.R.; Finger, C.A.G.; Schneider, P.S.P. (1999). Implantação de povoamentos de *Dodonaea viscosa* (L.) Jacq. com mudas e semeadura direta. Revista Ciência Florestal, Santa Maria, v.9, n.1, p.29-33. ISSN: 0103-9954

Serpa, M.R. & Mattei, V.L. (1999). Avaliação de diferentes materiais de cobertura e de um protetor físico, no estabelecimento de plantas de *Pinus taeda* L. por semeadura direta no campo. Revista Ciência Florestal, Santa Maria, v.9, n.2, p.93-101.

Shono, K.; Cadaweng, E.A.; Durst, P.B. (2007). Application of assisted natural regeneration to restore degraded tropical forestlands. *Restoration Ecology,* v.15, n.4, p.620-626. ISSN: 1061-2971

Silva Filho, N.L. (1988). Recomposição da cobertura vegetal de um trecho degradado da Serra do Mar, Cubatão SP. Campinas: Fundação Cargil, 53p.

Smith, D.M. (1986*). The practice of silviculture.* New York, John Wiley & Sons. 527p. ISBN: 978-04-718-0020-0

Sun, D., Dickinson, G.R. & Bragg, A.L. (1995). Direct seeding of *Alphitonia petriei* (Rhamnaceae) for gully revegetation in tropical northern Australia. *Forest Ecology and Management,* Amsterdam, v.73, n.73, p.249-257. ISSN: 0378-1128.

Swaine, M.D. & Whitmore, T.C. (1988). On the definition of ecological species groups in tropical rain forests. Vegetatio, Dordrecht, v.75, p.81-868.

Van den Berg, E., Guimarães, J.C.C., Silva, A.C. & Nunes, M.H. (2008). Avaliação da recuperação da vegetação em área minerada para explotação de bauxita, no planalto de Poços de Caldas, MG, e métodos atuais para recuperação ecológica. In: Simpósio Nacional sobre Recuperação de Áreas Degradadas, 7. Curitiba. Anais... Curitiba: FUPEF, 2008. p.67-84.

Winsa, H. & Bergstein, U. (1994). Direct seeding of *Pinus sylvestris* using microsite preparation and invigorated seed lots of different quality: 2-year results. *Canadian Journal of Forest Research,* Ottawa, v.24, n.1, p.77-86. ISSN: 0045-5067

19

Conservation, Management and Expansion of Protected and Non-Protected Tropical Forest Remnants Through Population Density Estimation, Ecology and Natural History of Top Predators; Case Studies of Birds of Prey (*Spizaetus* taxon)

M. Canuto, G. Zorzin, E. P. M. Carvalho-Filho, C.E.A. Carvalho,
G.D.M. Carvalho and C.E.R.T Benfica
S.O.S. Falconiformes,
Centro de Pesquisa Para a Conservação das Aves de Rapina Neotropicais,
Brazil

1. Introduction

Among the Accipitridae family, the genus *Spizaetus* is included in the group of the Booted Eagles (Griffiths et al., 2007; Haring et al., 2007), which according to old taxonomic classifications, was the most diverse, with 22 species that inhabited the tropical forests of the New and the Old World (Brown and Amadon 1968, del Hoyo et al. 1994, Ferguson-Lees and Christie 2001). Most specimens of this taxon are medium sized (1,100-1,600g), but there are some exceptions which are smaller (*S.nanus*, 510-610g) or larger (*S.nipalensis*, 2,500-3,500g) (del Hoyo et al. 1994, Ferguson-Lees and Christie 2001). The first description of the genus *Spizaetus* was introduced by Vieillot in 1816, when he described the taxon *S. ornatus* (Haring et al. 2007). This species, along with *S.tyrannus*, is found in the neotropical region, and both are divided into two subspecies (Brown and Amadon 1968, Ferguson-Lees and Christie 2001). Only one species occurs in the African continent, while most of the species of this genus are distributed throughout Southeast Asia (Haring et al. 2007). However, the grouping of this genus based on comparisons of the external morphology and feather patterns, according to traditional taxonomy, was refuted by a recent classification based on genetic studies (Haring et al. 2007). These authors consider the genus as paraphyletic, segregating the taxon into three distinct groups (Neotropical, African and Asian): while the neotropical representatives maintained the genus name due to Vieillot's first classification (*Spizaetus ornatus*, 1816); the African representative was inserted in the genus *Aquila* and the Asian ones renamed as *Nisaetus*.

Haring et al. (2007) and Griffiths et al. (2007) also evaluated the phylogenetic relationship of the genus with two other neotropical taxons, *Spizastur melanoleucus* and *Oroaetus isidori*. Although indicating that the archetypes from South America would not form a monophyletic clade, the authors adopted a more conservative classification, adding these two species to the genus *Spizaetus* (Helbig et al. 2005, Haring et al. 2007, Griffiths et al. 2007), which is the same taxonomy adopted by the AOU (American Ornithologist's Union). Thus, the genus would be represented by *S. isidori*, with distribution restricted to the Andes strip extending from Venezuela to the far north of Argentina; *S.tyrannus*, *S.ornatus* and *S.melanoleucus*, which are much more widely distributed through South America, including almost all of Brazil (Sick 1997) and part of Central America (Blake 1977, Ferguson-Lees and Christie 2001).

Even with phylogenetic uncertainties, these species have in common many ecological characteristics which are important in terms of biodiversity conservation. They demand large areas of well conserved forests (Jullien and Thiollay 1996, Thiollay 1989a,b, 2007, Bierregard 1998, Canuto 2009), which characterizes them as "umbrella" species for their environment (Newton 1979), when populations from other species, with smaller home ranges are preserved in locations that support viable populations of these predators (Sergio et al. 2006). These birds are also paramount for ecological balance, regulating populations of their eventual prey (Newton 1979, Bierregard 1998, Watson 1998).

The loss of top predators, such as these hawk-eagles, may entail a large increase in meso-predator release, affecting the populations of their prey and may even lead them to extinction (Soulé et al. 1988, Terborgh 1992). Like other raptors, they are also good indicators of environmental change (Newton 1979, Thiollay 2007, Seavy and Apodaca 2002, Carrascal et al. 2008a, b, Carvalho and Marini 2007), as they are sensitive to anthropic disturbances, including hunting and selective logging (Thiollay 1989a, 2007; Stotz et al. 1996).

Deforestation and fragmentation of tropical forests are this species' main threats (Bierregard 1998, Bildstein et al. 1998, Thiollay 2007, IUCN 2010). Considering the rates of tropical deforestation in the last few decades (Whitmore 1997), some taxons have become especially vulnerable to local extinction.

The neotropical species, *S. melanoleucus*, *S. ornatus* and *S. tyrannus*, are much more widely distributed throughout all of South America and parts of Central America (Blake 1977, Ferguson-Lees and Christie 2001). These species have large populations in the Hylean Amazon, thus contributing to the fact that they were not considered endangered in Brazil or worldwide (Machado et al. 2008, IUCN 2010). However, they have become extremely rare outside the Amazon domain (Sick and Teixeira 1979, Sick 1997", Drummond et al. 2008). The regional decrease of these taxons, due to the reduction of forest environments in Southeastern Brazil, is made clear with their presence in state lists of endangered species in Southeastern and Southern regions of Brazil (Bergallo et al. 2000, Silveira et al. 2009, Fontana et al. 2003, Mikich and Bérnilis 2004, Simon 2007). As well as the deforestation of the Atlantic rainforest "(SOS Mata Altântica e INPE 2010), which is identified as these species' biggest threat, trapping and hunting are secondary contributors to the decrease of their population status (Albuquerque 1995, Machado et al. 1998, Ribon et al. 2003).

Despite recent records in many localities of the Atlantic domain and in other areas of the Cerrado (Kirwan et al. 2001, Azevedo et al. 2003, Manõsa et al. 2003, Ribon et al. 2003, Carlos and Girão 2006, Joenck 2006, Olmos et al. 2006, Roda and Pereira 2006, Zorzin et al. 2006, Dias and Rodrigues 2008, Carvalho Filho et al. 2009, Faria et al. 2009, Salvador-Jr. and Silva

2009, Salvador-Jr. et al. 2011), there is no evidence about the reproduction of these species in most areas where they were found. Also, there is a complete lack of substantial data on their local density as well.

The best efforts of data collection on the biology and ecology of these taxons were concentrated in areas of continuous forests in Guatemala (Flatten 1990, Julio et al. 1991, Sixto, 1992, Montenegro 1992), the French Guyana (Thiollay 1985, 1989a,b Jullien and Thiollay 1996, 2007) and in large forest fragments in Brazil (Canuto 2009, Manosa et al. 2002, 2003). Other works regarding territoriality and estimates of genus density usually include collection of complementary data with the use of radio telemetry equipment (Manõsa and Pedrochi 1997, van Balen et al. 2001 and Gjersshaug et al. 2004).

In some studies, such as Thiollay (1989a,b), Preulethner and Gamauf (1998), Whitacre et al. (1992a) van Balen et al. (2001) and Canuto (2009), the authors estimated the density of pairs and territoriality by direct observation of pairs and individuals performing flight circuits and aerial displays. Generally speaking, population census in preserved habitats of many species of Accipitriformes – especially those that are more selective such as the individuals of the *Spizaetus* genus – enable the prediction of the distribution of a specific population in a mosaic of modified environments. This is how the regional or local status of the species may be evaluated (Thiollay 2007). However, the lack of specific census that aim to evaluate more precisely the density of these species in altered environments, does not permit further inference on their local state of conservation in most of their current areas of occurrence. Therefore, there is less information available on the dynamics of these species' populations in fragmented areas, such as the Southeast and Center-West regions of Brazil.

In the current chapter we describe the results of five studies conducted in the country by analyzing the incidence of specimens of the genus *Spizaetus* in areas with different fragmentation histories and considering the different population and reproductive ecological aspects of these taxons collected at each locality. By analyzing each case study we promote a reflection on the perspectives of local and punctual conservation of these species, according to their ecological requirements. Thus, we used these species as "flags" to point out the problems involving conservation of top predators, which present small density but demand a large area, in fragmented and continuous areas.

2. Methods and study areas

For the study sites we present the different localities, including different regions of the Cerrado and Atlantic Rainforest and their respective transition areas where different data collections occurred. Later, these results are discussed in a regional context according to previous results by the authors obtained from the Amazon biome and a second tropical region (Central America).

In each case study specific methods were employed for census, mapping and monitoring of these species according to Whitacre and Turley (1990), Thiollay (1989), Bibby et al. (2000), Manõsa et al. (2003), Canuto (2009) and Granzinelli and Motta Junior (2010). These taxons require an ample and specific sample effort, as they are highly territorial birds that occur in low density and occupy extensive territories. However, these methods were applied differently according to the focus of each study by different authors, thus involving different sampling parameters according to the environment, conditions of region preservation and the logistics involved. The methods are described below:

i.   **Observations through point counts:** This method consists of recording species from a fixed observation point during a pre-established period, over a large time-scale, involving different seasons. It was used to evaluate the density, frequency of occurrence and incidence of pairs and individuals of the focus species in the sampling areas. For the density estimates, we used a fixed observation radius, and for calculating the frequency of occurrence, all individuals accurately identified within the range of observation were considered. Each point should allow an ample field of vision, and observations were made with the aid of binoculars 10x42 Nikon and a spotting scope (20-60x), during a specific period of time that varied between 4 and 5 hours, beginning first hours of the day.

ii. **Data collection by routes:** This technique was used to estimate the frequency of occurrence of the species, as well as to identify new areas of use and nesting. Surveys were conducted in roads, slowly driving a vehicle, as well as in rivers with a motorboat or boat, or on foot in paths in the surrounding and interior areas of the remaining forests.

iii. **Spot mapping:** According to Bibby et al. (2000) and Sutherland (2004), this technique consists of mapping the movements of the individuals from observations made from fixed points, including use of climbing techiniques for point counts within the forest canopy. All the movements and routes of individuals of these species were recorded, followed by the representation of these movements and activities on a digital map of the area. After the identification of active territories or breeding sites, various observation points were distributed in the area defended by the resident individuals. The more repetitions performed within these territories, the more accurate are the estimates of displacement, therefore confirming which sites were occupied or defended. The birds observed were usually identified by the imperfections in the flight feathers (remiges and rectrices) and patterns of plumage (adult, sub-adults or youngsters). Other points in the surrounding area of the occupied territories and adjacent ones were scouted with the objective of verifying the overlap of territories or unoccupied areas.

As previously described, differences occurred in the use of these methods according to the intrinsic characteristics of each area, logistics, and the objectives of the following studies:

• Case Study 1 "Density of the representatives of the genus *Spizaetus* in remaining continuous Atlantic Rainforest in the east of the state of Minas Gerais, Brazil"

• Case Study 2 "Estimates of territoriality, incidence of pairs and reproduction sites of *Spizaetus tyrannus* in the south of the state of Minas Gerais, Brazil"

• Case Study 3 "Estimates of territoriality, incidence of pairs and reproduction sites of *Spizaetus melanoleucus* in Atlantic Rainforest fragmented areas in the east of the state of Minas Gerais"

• Case Study 4 "Density and territoriality of *Spizaetus ornatus* in the Araguari river valley, Triângulo Mineiro Region"

• Case Study 5 "The effects of fragmentation of the Atlantic Rainforest on the richness and abundance of Accipitriformes in the region of Viçosa, Minas Gerais"

3. Results

3.1 Case study 1

"Density of the representatives of the genus *Spizaetus* in remaining continuous Atlantic Rainforest in the east of the state of Minas Gerais, Brazil"

The Doce River State Park (Parque Estadual Rio Doce - PERD) is the largest Atlantic Rainforest park in Minas Gerais, with an area of 35.976,43 ha, and 120 km in perimeter. The reserve is located in the Doce River Valley, eastern part of the state, between the 42o 38'W and 48o 28'W meridians and the 19o 45'S and 19o 30'S parallels, with an altitude varying between 230 and 515m. The reserve has as important landmarks, the limits of the Doce and Piracicaba rivers, to the east and northeast respectively (Hirsch 2003a).

The PERD vegetation belongs to the lower montane Atlantic Rainforest formation (Rizzini 1979) and is composed by a mosaic of primary and secondary forests (Ferreira and Marques 1998) classified as Semi Deciduous Forest (Viana 2001). The reserve has a freshwater system of 38 to 44 lagoons which amounts to 3,529.7 ha (approximately 6% of the total area).

For the *Spizaetus* density estimates (Table 1), the spot mapping method was used registering the routes of the three species of this genus. The results were obtained according to the occupation of the species within the 12 sampling plots, a total of 21,400 hectares. These records attested to the presence of pairs in reproductive activity or that were actively nesting. It was collected 126 samples between 2006 and 2009, concentrated between the months of July and January. There were 1 to 4 fixed observation points in each plot, inside the reserve and border. These points occurred on the canopy of emerging trees (n=23 / 83 sample repetitions), with the use of climbing techniques, and also in open areas throughout the border of the reserve (N=10 / 43 repetitions). All 33 points had a mean distance of 831m (±514m), the minimum distance being 100m and the maximum 2 km between the points within a plot. These sets of fixed points were sampled from 07:30 to 12:00. The mean distance between plots, each one with 1-4 point counts, was 5.3 km (±2.2 km), and varied between 2.2 and 9 km. All points had a wide viewing angle, between 120^0 to 360^0, mean 260^0 (± 64), which enabled an evaluation of the occupation or defense of the plots by aerial and territorial displaying individuals. This was determined by the recordings of pairs engaged in reproductive behavior or in defense of the plots, thus generating a density estimate of the minimum number of pairs inside the reserve's boundaries. The plot was considered occupied when a minimum of two records on pairs took place. This estimate does not include subordinate birds, immatures, wandering individuals with no defined territory, and individuals not reproductively active, as predicted by Meunier's "reservation population" (1960). Methods i, ii and iii (see methods and study area sections) were applied during the study, with a total of 567 hours of direct observation.

For the *Spizaetus tyrannus* species, records of territories occupied by pairs were inexistent, so clear patterns of territoriality were not obtained. However, the species' incidence in the plots is still considered because of the random records of individuals presenting the territorial behavior.

A minimum number (Table 1) of 4 pairs of the Ornate Hawk-eagle (*S. ornatus*) in 21,400 ha, and 6.7 pairs in 36,000 hectares of the PERD was estimated, with a density of approximately 1 pair per 5,373 hectares. For the Black-and-white Hawk-eagle (*S. melanoleucus*) five plots had records of matched pairs in 21,400 ha (12 sampling plots), generating a minimum estimate of 5 pairs in 21,400 ha, and 8.4 pairs of the species in 36,000 ha, with an approximate density of 1 pair per 4,285 hectares. For the Black Hawk-eagle (*S. tyrannus*), 4 plots had individuals recorded, being those characterized by territorial defense behavior, as soaring flights along with vocalizations specific to the species. This generated an estimate of a minimum of 4 pairs in 21,400 hectares, and 6.7 pairs in 36,000 hectares of the PERD, with a density of approximately 1 pair per 5,373 hectares, similar to the first species. Therefore, *S. melanoleucus* was considered the most abundant species in this Atlantic Rainforest reserve of semideciduous seasonal lowland of 36,000 hectares, more than 40 lake formations, and three watersheds, as well as the Doce River, along the eastern limits of the reserve, and the Piracicaba River, along the northern limits.

3.2 Case study 2

"Estimates of territoriality, incidence of pairs and reproduction sites of *Spizaetus tyrannus* in the south of the state of Minas Gerais, Brazil"

For this study, three sites were evaluated (Table 1), two located between the municipalities of Ibertioga, Juiz de Fora and Itutinga (Study Sites 1 and 2), and one in the municipality of Capela Nova (Study Site 3), all of which are in the state of Minas Gerais (20⁰ 49'S - 21⁰ 34'S - 44⁰ 13W 43⁰ 24'W). The region has remnants of Atlantic Rainforest, which are concentrated on the slopes and hilltops. It also has tracts of Cerrado vegetation, with modifications due to agriculture. The region's altitude varies between 780 to 1,200 meters ASL. It is characterized by seasonal, semi-deciduous forests associated to Cerrado vegetation, with field areas with anthropogenic interference and areas of aquatic environment. The municipality of Capela Nova is characterized by a mosaic of forest fragments on hill tops and depressions, pasture areas, rural properties and river valleys, as well as rocky formations on forested slopes.

Photo 1. Black hawk-eagle nest with young. Photo credit: Eduardo Pio Carvalho

The three sites are located to the south of the state capital, Belo Horizonte, and are near or part of the Serra da Mantiqueira mountain chain.

For case study 2, the methods included direct observation of the species using fixed points of variable radius in accordance with Whitacre and Turley (1990), Whitacre et al. (1992a) and Canuto (2009), in landscape vantage points such as hill crests (sites 1 and 2). An observation platform was also used in a tree canopy (site 3), with ascension techniques for the monitoring of a nest (Photo 1). To avoid influences on the individual's behavior, a *blind tent* was used, in such a way that the nesting pair could not see the observers most of the time. The different areas were monitored by at least quarterly visits for collecting data on the reproductive biology, monitoring of the reproductive sites or active nesting and, consequently, evaluating the patterns of territoriality. In the nesting area monthly visits were carried out.

In sites 1 and 2, two pairs were monitored, 23 km apart, through methods i, ii, and iii. Both pairs presented display or territorial defense behavior in both areas in the years of 2007 and 2008. In both cases the forests were considered of late secondary succession and had distinct strata. Movements and territoriality patterns were characterized in displacements of up to 1.6 km. On one occasion, 3 individuals were sighted in one of the areas, which could have been a pair and a youngster. Both areas are characterized by forested slopes or hilltops on rugged terrain.

In site 3 a nest was identified (Charles Moreira, personal communication) inside a ravine in a small drainage basin at an altitude of 776 meters, in a secondary fragment in intermediate stage of regeneration. Starting at this point, specimens used a route along the ravine valley to exit and enter the nest as well as other points. The patch is approximately 175 hectares in size and is inserted in a mosaic of larger and smaller forest patches in a region of river valleys with some local agropastoral pressure. The nest was on an Angico tree (*Anadenanthera sp.*) at a height lower than 20 meters. The tree is located 400 meters from the Piranga River and 700 meters from a stream. Its circumference at breast height (CBH) is 1.40m.

3.3 Case study 3

"Estimates of territoriality, incidence of pairs and reproduction sites of *Spizaetus melanoleucus* in Atlantic Rainforest fragmented areas in the east of the state of Minas Gerais".

Eastern Minas Gerais, precisely the middle Doce river basin, is located between parallels 18°45' and 21°15' S and meridians 39°55' and 43°45' W, comprising a drainage area of about 83,400 sq. km, of which 86% belongs to the state of Minas Gerais and the remaining area to Espírito Santo. In its southern border is the Paranaíba do Sul river basin, and in the west the São Francisco river basin, as well as some of the Grande river. In the north, it borders the Jequitinhonha and Mucuri river basins and in the northeast the São Mateus river basin. The average annual precipitation varies from 1,500 mm, in the springs located in the Mantiqueira and Espinhaço mountain ranges, to 900 mm, in the region of the city of Aimorés, increasing towards the coast.

The study areas were the Doce River State Park and surroundings (Case study 1), the Private Reserve of the Feliciano Miguel Abdala Natural Patrimony in Caratinga, the municipalities of São José do Mantimento and São Pedro do Suaçuí, and a district near the city of Governador Valadares, Baguari, all of which are in the middle Doce river watershed, in the state of Minas Gerais. In these areas, the presence of active or inactive nests of *S.*

melanoleucus, or incident pair, were documented. Patterns of territoriality and distance estimates (Table 1) were obtained according to the spot mapping method. Also, data was collected on the reproductive biology of the species. Methods i, ii and iii were used.

In the first area, located in the Doce River State Park and surroundings (see Case study 1), two nests were found and tracked. One inside the reserve (Canuto 2008b), located on a pink Jequitiba tree (*Cairina legalis* – CBH 291cm) at a height of 37 meters and at a distance of about 800 meters from the border. It was located at the base of the hill among different ravines or drainage systems, near a ridge of approximately 250 meters in altitude and about 770 meters away from the Turvo stream.

A second nest (Fabiano R. de Melo, personal communication) was studied around the same location in a hillside forest patch in a dry valley of approximately 1,700 hectares, which is located 2 km away from the Mumbaça stream, border of the PERD. This nest was 13 km away from the first nest. The patch was isolated from the reserve by *Eucalyptus* crops and open areas. The nest was located on an Arara Nut Tree (*Joannesia* sp.) of approximately 35 meters in height. Displacements up to 3.5 km were mapped in both areas.

In the second area, a nest which was already known (Fernanada Tabacow, personal communication) was visited and found to be active (Photo 2) in the municipality of Caratinga, which has protected semidecidual forests of 1,000 hectares in size. The nest was on a *Zeyheria* cf. *tuberculosa* of 51 meters in height and CBH 2.7 meters. This nesting tree is located at an altitude of 700 meters, also along forested ridges and valley depressions. The location of the nest provides a broad view of the depression. The tree is 1.7 km away from the Manhuaçu river. Displacements of up to 1.5 km were mapped. This area is located at a distance of 75 km from the first two nests mentioned.

Photo 2. Black-and-white Hawk-eagle nest with young. Photo credit: Marcus Canuto

In the third area, near the city of Governador Valadares, the nest was on an emerging tree (≥30 meters) located at the crest of a hill, 280 meters in altitude, near many slopes and depressions, and at a distance of 1.73 km from the Doce river. The tree was beside a small portion of native vegetation on top of the hill, even though the environment is dominated by *Eucalyptus* sp. plantations (approximately >60%). The tree was not measured. The species had been previously recorded by different observers on an island with mature trees (>30m), in a nearby area, 8 km away from the nest, which had not yet been discovered at the time. After the island was flooded by a hydroelectric company, the pair was recorded on a monthly basis around the nest area for a period of 8 months before evidence of nesting was confirmed. Displacements of up to 4 km were mapped. The area used by the specimens throughout the months amounts to 1,293 hectares of *Eucalyptus*, streams, native vegetations and environments inhabited by man.

Another nest of the same species was found in 2001 (Marcelo Vasconcelos personal communication) in the northern region of the Doce river basin, in the municipality of São Pedro do Suaçuí, a more dry and arid region. This nest was located 900 meters away from the margins of the Suaçuí Grande river. The area was revisited in 2010, where an individual was recorded 7.5 km away from the previous nesting tree; however, the structure of twigs was no longer present. At the same location, an adult individual of the species *S. tyrannus* was recorded. This area is located at approximately 95 km northwest of the Governador Valadares nest.

To the east of the Doce river basin, a pair was sighted in São José do Mantimento, showing courtship or territorial defense behaviors, as observed by the vocal emissions and diving or dashing against each other over the course of two days. This was recorded in a 966 hectares forest patch that is located at the margins of the José Pedro river. This site is approximately 30 km southeast of the Caratinga nest.

3.4 Case study 4

"Density and territoriality of *Spizaetus ornatus* in the Araguari river valley, Triângulo Mineiro Region"

The Araguari river source is located in the Serra da Canastra national park in the municipality of São Roque de Minas and covers an area of approximately 2185 ha, with an extension of 475 km. The vegetation above the river gorge, such as the remnants of the Cerrado *latu sensu*, were heavily modified due to the creation of extensive pastures and the modernization of agricultural techniques of edaphic correction, geared towards the production and exporting of grains (Ferreria et al. 2005). Enclaves of Atlantic Rainforest are concentrated along the valley, where the rugged topography has limited the advance of agropastoral activities, with some patches remaining somewhat isolated on steeper slopes. However, the implementation of the hydroelectric power plants of Nova Ponte, Miranda and Amador Aguiar I and II eradicated the riparian vegetation and contributed to the disfigurement of the remaining vegetation physiognomy, such as the deciduous forests.

Estimates of vegetation coverage and the use of the soil in the valley's area point to few areas occupied by deciduous forests, riparian vegetation and savannahs (19%), all in secondary stages, with the largest part of the area used for pastures (50%) and other annual cultures (31%) (Rosa et al. 2006). The region's climate is of the Aw type, according to Köppen's classification, which is characterized by a dry period, from the months of April to September, and a humid period between the months of October to March (Rosa et al. 2006).

The valley's altitude varies between 500 and 950m, and the slope percentage varies between 20 and 80%.

To protect the few remaining areas of native vegetation, the Pau Furado State Park was created, located between the municipalities of Araguari and Uberlândia, and established as a compensatory measure for the Amador Aguiar I and Amador Aguiar II hydroelectric power plants. The area includes 2.200 ha, covered by remnant forests in different stages of succession. It's the first conservation unit of this category located in the Triângulo Mineiro Region, and one of the areas in the state that has been most affected by deforestation throughout the years.

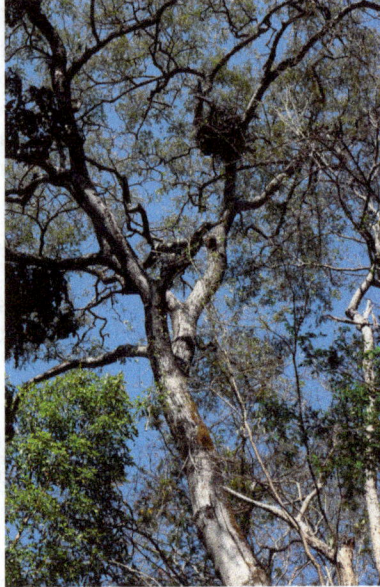

Photo 3. Ornate Hawk-eagle nest Photo credit: Eduardo Pio Carvalho

After the year 2005, when a few sporadic contacts with S. ornatus and S. tyrannus occurred in the Araguari river valley, another two years of efforts were carried out with the objective of finding the reproductive sites of these species and mapping the distribution of these taxons in an area of approximately 700 sq. km, including part of the valley and adjacent areas.

Some contacts occurred randomly during field work done in the site in 2005, while other records were obtained during surveys carried out in the area in the following two years, when survey methods by transit and fixed point observations and spot mapping were used. Several recordings occurred during car trips and motorboat trips on the Araguari river. Many observations were randomly performed throughout the valley, with an average time of 4 hours of observation (08 am – 12 pm). The selected sites had a field view of at least 150 degrees and were located on hilltops with an ample view of the valley. Through direct observations, the routes of individuals or matched pairs were mapped when it was possible to distinguish the individuals by imperfections in their flight feathers and describe distinct territories. Additional observations were performed after the identification of the occupied areas to better map the individuals. Other points were established among the identified territories in order to identify possible overlapping territories or unoccupied areas.

Through mapping, pairs of *S. ornatus* were recorded in an area of 70.000ha. Their activities were concentrated around the two largest remaining forest areas in the region, of approximately 1.000 ha, and a distance of 21 km between them (Table 1). In one of these areas we found an active nest in 2006. The area is characterized by secondary vegetation with some patches in initial stages and others more advanced, and with a few emerging trees, including patches of Cerrado. The nest was built on a Copaiba (*Copaifera langsdorffii*) with approximately 23m in height and a CBH of 2.07m. The tree was located 1.9 km from the Araguari river, 40m from a small stream and 27m from a pasture in the valley. During the reproduction period we observed displacements of at least 4.5 km from the nest. In the other forest fragment no nests were found; however, a pair was observed performing courting flights and vocalizations. The records of *S. tyrannus* in the valley can be summarized by two isolated contacts with adult individuals.

3.5 Case study 5

"The effects of fragmentation of the Atlantic Rainforest on the richness and abundance of Accipitriformes in the region of Viçosa, Minas Gerais"

This effort was focused between the municipalities of Viçosa, Paula Cândido and Cajuri, in the mesoregion of the "Zona da Mata" in Minas Gerais, which covers an extension of 100 sq. km, between the parallels 20° 42′ 30″and 20° 50′ 00″ S and 42° 48′ 45″and 42° 56′ 15″W. The original vegetation is part of the Sub-montane Semideciduous Forest formation (Oliveira-Filho and Fontes 2000), characterized by the seasonal loss of 20 to 50% of tree leaves, and by an altitude of 300 to 700m.

In 1817, the region was covered by primary forests, with small secondary patches due to subsistence farming of indigenous peoples and isolated farming activities (Ribon et al. 2003). The effective occupation of the Forest Zone began with the introduction of coffee plantations in the Parnaíba river basin in Minas Gerais, around mid-19th century, with the occurrence of widespread destruction and fragmentation of the landscape, caused by the creation of extensive plantations (Valverde 1958). Following the coffee plantations, cattle ranching expanded as the crop became unproductive, and the plantations became pasture grass (*Melinisminutiflora*) (Resende and Resende 1996).

The coffee cycle ended in the first decades of the 20th century, which probably led to the regeneration of various forest fragments from old abandoned crops (Ribon et al. 2003). Thus, many fragments in the region of Viçosa have between 20 and 60 years of regeneration (Ribon et al., 2003). According to the latest expectations, the area´s forest cover reaches 20% (Zorzin 2011), with fragments concentrated on slopes and hilltops, rarely on lowlands and caves, and practically none with watercourses in their interior (Ribon 1998). Most of these remaining fragments are no larger than 50ha and few cover an area superior to 300ha. Even with the regeneration of many fragments, they are still exposed to distinct anthropic disturbances, such as fire, invasion by domestic cattle and selective logging (Ribon et al. 2003, Silva Júnior et al. 2004), while many also have their area reduced by the implantation of small cultures (personal communication).

Currently, the landscape matrix has different uses, such as coffee plantations of various extensions, other crops to a lesser extent (such as corn, rice and beans), small eucalyptus (*Eucalyptus spp.*) plantations, pastures and secondary forests (Ribon et al. 2003).

For the recording of medium and large sized Accipitriformes, we used fixed observation points and trajectory surveys. The selection of points was based on a PRISM/ALOS satellite image with a 2.5m resolution. We selected 25 points that had distances longer than 800m from other points and offered a different field view from these. The shortest distance between selected points was 820m and the average minimum distance was 1.714 km.

These points were selected during two seasons, according to the Brazilian species' reproductive cycle (Sick 1997, Carvalho-Filho et al. 1998, 2005, 2006, 2009, Ferguson-Lees and Christie 2001, Zorzin et al. 2004, Azevedo et al. 2006, Canuto 2008b, Canuto2009). The first sampling was concentrated between December of 2009 and February of 2010, which corresponds to the end of the reproductive period, when it was possible to record the dispersal of the youngsters; the second period occurred between the months of August and November of 2010, which includes reproductive activity such as pairing of mates, courting flights and incubation.

Observations at the points occurred for five hour periods, from 7 am to 12 pm. The transit observations were conducted in the interior and outskirts of the region's 35 remaining forest fragments.

After reviewing Ribon et al (2003), Zorzin (2011) attested to the local extinction of *S.ornatus* and *S.melanoleucus*. These species were previously considered as critically endangered in the region by Ribon et al. (2003). The only species observed by Zorzin (2011) was *S.tyrannus*, when a minimum number of 3 adults, identified by imperfections in the remigi, were estimated in an area of 10,000ha (Table 1). One of these birds was recorded at points distanced by 10 km. Zorzin (2011) also analyzed the landscape's influence, in different scales, on the abundance of *S. tyrannus* in the region, analyzing the presence of this species in 35 fragments. This author found significant correlations (p<0.01) between the species abundance and the sizes of the remaining forest areas. Using buffers of 500, 1000, 1500 and 2000m, this author also found significant correlations between the abundance of the taxon and the percentage of forest cover around the fragments, the amount of nuclear area and the average size of the remaining fragments of the landscape. On the broadest scale adopted, the 2000m radius, the abundance of the species showed significant negative correlation (p<0.05) to the average isolation of the surrounding fragments.

4. Discussion

For the case studies conducted in the Atlantic Rainforest and transition areas, according to territoriality estimates for the genus corresponding to Table 1, we were able to evaluate the incidence of the group in forest patches of at least 179 ha for *S. tyrannus* nests, and of 840-1.000 hectares for the incidence of pairs and nests of *S. melanoleucus* and *S. ornatus* in areas with intense history of fragmentation. It is clear that their protection should be a priority, as well as of small, medium and large Atlantic Rainforest fragments, to create wildlife corridors which guarantee the existence of refuges and source or dispersal areas (see Laurance et al. 1997). However, most of what is left of the Atlantic Rainforest is found in small forest fragments <100ha (Ranta et al. 1998; Ribeiro et al. 2009) which are isolated from each other and composed of portions of secondary forests in recent or medium regeneration stages (Viana et al. 1997, Metzger 2000, Metzger et al. 2009 In Ribeiro et al. 2009). Therefore, the pressure caused by deforestation of the genus in the Atlantic Rainforest is still present, whereas the process or history of deforestation in the biome is prior to the 20th century.

Minimum Density Estimate, Distance among Nests, Territory and Movement for *Spizaetus* in the Neotropics					
Bibliography	Estimates	S. tyrannus	S.melanoleucus	S. ornatus
Study case 1 (Canuto 2009)	Dist. among Terriotries	x	6.9±1.5 Km	6.9±1.5 Km
	Minimum Movement	x	4,7 km	6.7 km
	Density Estimate in PERD	1 pair/5373 ha	1 pair/4285 ha	1 pair/5373 ha
	Distance between nests	x	13 km	x
Study case 2	Minimum Defended Territory	729 ha	x	x
	Minimum movement	1.6 km	x	x
	Dist. among Terriotries	23 km	x	x
Study case 3	Forest site occupied	x	840 ha	x
	Dist. among Nests	x	13 km	x
	Minimum Movement	x	4 km	x
Study case 4	Forest site occupied	x	x	1000 ha
	Dist. among Terriotries	x	x	21 km
	Movement	x	x	4.5 km
	Density Estimate	x	x	2 pair/70000 ha
Study case 5	Minimum Movement	10 km	x	x
	Density Estimate	3 individuals /10000 ha	x	x
Thiollay (1989 a)	Territory	x	925 to 1090 ha	1375 ha
	Density Estimate	x	2 pair/4200 ha	4 pair/10000 ha
	Maximum territory size	x	4200 to 5800 m	3000 to 6900 m
Flatten (1990)	Dist. between nests	x	x	3.5 km
	Territory	x	x	977 to 1315 ha
	Max. movement from nest	x	x	5 km
	Nest density	x	x	1/787 ha
Montenegro (1992)	Male territory	1090 ha	x	x
	Female territory	1450 to 2470 ha	x	x
Sixto (1992)	Max. Movements from nest by female	3 km	x	x
Julio (1991)	Female territory	x	x	256 to 2100 ha
	Max. Movements from nest by female	x	x	5 km
	Max. Movements from nest by male	x	x	3.3 km
Joenck (2006)	Density estimate	2 indiv./4500 ha	x	x
Canuto (2008)	Density Estimate	1 pair/2960 ha	x	1 pair/4440 ha
	Dist. among Terriotries	5.7 km	x	7.9 km

Table 1. Case studies and literature on estimates and patterns of nesting and individuals densities, specimen movements, telemetry gear data and territoriality among *Spizaetus*.

In the study areas, the use of large trees for nesting by the species *S. melanoleucus* (N=4) and *S. ornatus* (N=3) is evident. The location of the reproduction sites for the three species of the genus also coincided with areas of rugged terrain, valleys and forested ravines, which are favorable for isolation or areas with difficult access for human activities, according to Silva et al. (2007) (Figure 1, Photo 2 and 3).

In areas with patches of 6,000 to 12,000-36,000 ha, in contiguous or protected remnants, it is possible to perceive the incidence of the three different species in the same fragment. It is also possible to observe population patterns, regarding the defense of different neighboring areas, through mapping and reproductive behavior. These may be considered to be refuges for reproductively active populations, although they are isolated at a regional level. Only large fragments with mature forests are capable of preserving sensitive species, especially those with large area requirements (Ferraz et al. 2007) or with strict habitat requirements, whose survival is particularly problematic in the present fragmented state (Aleixo 1999; Ribeiro et al. 2009). Therefore, these species, though present in extensive or contiguous forest environments, have low density and are naturally rare in the landscape, according to characteristics that are sparse in the environment, (for example Nesting areas; photo 4, Newton 1979). Thus, these ecological parameters make them even more susceptible to fragmentation (Laurance et al. 1997, Henle et al. 2004).

Photo 4. Black-and-white hawk-eagle nesting site. Photo credit: Marcus Canuto

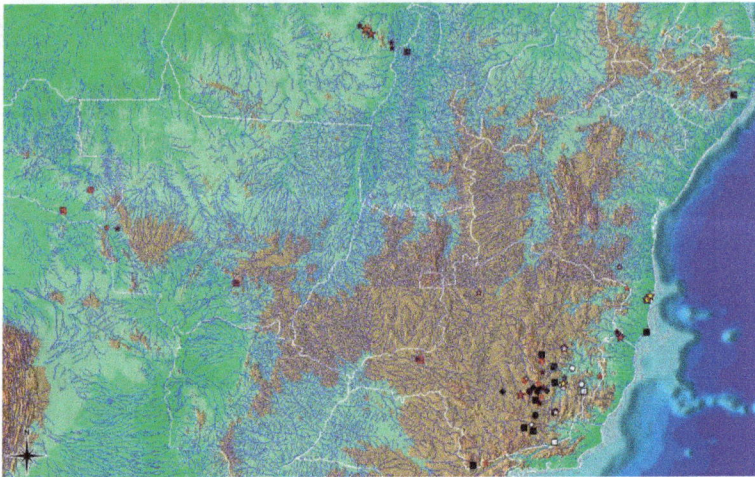

Fig. 1. Map of *Spizaetus* records in the Atlantic Rainforest and Amazon; squares indicate records of pairs, circles indicate nests and diamonds indicate records from cited literature. Big stars indicate the incidence of the three species of *Spizaetus* (yellow), or of two, *S. ornatus* and *S. tyrannus* (brown with black), and *S. tyrannus* and *S. malanoleucus* (white with black). Small stars indicate records of species in this study. In general, white is for *S. malanoleucus*, black is for *S. tyrannus* and brown is for *S. ornatus*. Scale: 1:10685101.

The distance of 25 km from most remaining fragments of Atlantic Rainforest to its largest reserves (Ribeiro et al. 2009) presents serious conservation problems even for medium sized birds according to the patterns and estimates of movement and density presented (Table 1).

According to the biometrics of the nesting trees previously discussed, we observe a tendency of the species *S. melanoleucus* in using emergent or mature trees in inland Atlantic Forest and transition areas, as does *S. ornatus*. For *S. tyrannus*, according to Sixto (1992) and the nest monitored, the species used medium sized trees (20-25m) for nesting. According to case study 3, in the highly fragmented region of Governador Valadares, an active *Spizaetus melanoleucus'* nest was located in an area mostly covered by *Eucalyptus sp.* The nest was located in a fragment of approximately 1,293 hectares, with an area roughly evaluated to be 70% covered by *Eucalyptus*. However, the presence of an emerging tree for nesting, isolated from the rest of the vegetation, in a drainage system of difficult access, near extensive bodies of water and riparian vegetation, are probably decisive factors in the occupation of this territory. The population maintenance of these species depends on creating or increasing wildlife corridors, due to the large life area required by the taxon. A better evaluation is needed on the dispersion pattern of adults and youngsters throughout the various fragments in a mosaic of altered environments, as well as the selection of nesting areas (see Table 2).

Characterization on areas of nests and records on pairs						
Study case	Species	Record	Site	Forest site of the record	Landscape Features	Habitat
1 and 3	S. melanoleucus	Nest	Baixa Verde MG	1700 ha	Forested slope, valley and lakes	Tall protected forest
1 and 3	S. melanoleucus	Nest	PE Rio Doce MG	36000 ha	Lowland forest, small valleys and lakes	Tall protected forest
2	S. tyrannus	Pair	Ibertioga MG	x	Slope, valleys and forested hill tops	Late seconday growth
2	S. tyrannus	Pair	Ibertioga MG	x	Slope and forested hill tops	Late seconday growth
2	S. tyrannus	Nest	Caranaiba MG	179 ha	Valleys and forested hill tops	Early secondary growth
3	S. melanoleucus	Nest	Governador Valadares MG	1293 ha	Lowland forest, Eucalyptus plantations and valleys	Early secondary growth
3	S. melanoleucus	Nest	São Pedro do Suaçuí MG	x	Valleys and forested hill tops	Early secondary growth
3	S. melanoleucus	Nest	RPPN Miguel Abdala, Caratinga MG	1000 ha	Valleys and forested hill tops	Tall protected forest
5	S. ornatus	Nest	Uberlândia MG	1000 ha	Forested valley	Early secondary growth
5	S. ornatus	Pair	Uberlândia MG	1000 ha	Forested valley	Early secondary growth
6	all 3 Spizaetus	Incidence	PN do Pau Brasil, Porto Seguro BA	12000 ha	Lowland forest with valley formations	Tall protected forest
6	all 3 Spizaetus	Incidence	Estação Veracel	6000 ha	Lowland forest with valley formations	Tall protected forest
6	S. melanoleucus	Pair	PN do Pau Brasil, Porto Seguro BA	12000 ha	Lowland forest with valley formations	Tall protected forest
Pers. comm. 1	S. tyrannus	Pair	Antônio Dias	x	Slope, valleys and forested hill tops	Late seconday growth
Pers. comm. 2	S. tyrannus	Pair	São Bartolomeu/Ouro Preto MG	2846 ha	Slope, valleys and forested hill tops	Early secondary growth
Pers. comm. 3	S. tyrannus	Pair	Serro MG	x	Cerrado of altitude, deciduous and semideciduous forest	Early secondary growth
x	S. melanoleucus	Pair	Além Paraíba MG	840 ha	Valleys and forested hill tops	Late seconday growth
x	S. tyrannus	Pair	Camanducaia MG	410 ha	Montane forest and slope	Late seconday growth
x	S. tyrannus	Young and adults	Nova Lima, MG	343 ha	Moantne forest, valleys and forested hill tops	Late seconday growth
x	S. melanoleucus	Pair	São José do Mantimento MG	966 ha	Lowland forest and valleys	Early secondary growth
x	S. ornatus	Nest	Matozinhos MG	900 ha	Cerrado, deciduous and semideciduous forest	Early secondary growth
Forest site of the record: it was only considered the fragment of the nest or from where individuals appeared, and not the surrounding ones.						

Table 2. Characteristics of some of the areas, and records used in the case studies to illustrate area sizes in hectares whereas records of the three species took place. Personal communication 1 (Gustavo Pedersoli); personal communication 2 (Frederico Pereira de Castro Andrade); personal communication 3 (Estefane N. L. Siqueira).

For *S. ornatus* and *S. tyrannus*, case studies 2 and 4, the density estimation is lower in fragmented areas. There are two pairs with approximately 20 km between them. In extensive fragments, there are two pairs every 5.3-13 km, according to mapping and estimations for the three species (case study 1; Canuto 2008a). According to Ribeiro et al. (2009) the average isolation between forest fragments along the Atlantic Rainforest extension is 1.4 km, varying from a few meters to dozens of kilometers. By removing fragments smaller than 50 hectares from the equation, the isolation between fragments increases to 3.5 km, and by removing fragments up to 200 hectares, this average isolation becomes 8 km.

Therefore, the existing and continuous fragmentation of the biome has isolated populations and individuals, and is still continuously depleting prey resources and top predator populations (according to Newton 1979, Thiollay 1989a, 2007). Jullien and Thiollay (1996), in the French Guiana, after extensive sampling with various habitats and gradients of anthropogenic degradation, described the decline and loss of forest species and their substitution by more generalist and open habitat species.

According to Harestad and Bunnell (1979), Schoenerand and Schoener (1982), and Kennedy et al (1994) cited in Perry (2000), the interspecific variations in the life size areas, which were not evaluated in this study, have also been correlated to age, sex and seasons. Therefore, according to Table 1, the differences between values derived from the different methods, environments, ecological patterns and specimens. However, we emphasize that, regardless of the differences, there is evidence of great dislocations in kilometers in fragmented and continuous environments, low density of the species in the area and reproduction sites in all the studies. There is also incidence of the species in environments or sites that have characteristics such as isolation, difficult access (rugged topography), and the presence of bodies of water and mature trees for nesting. Also, certain local populations are present along extensive protected forest areas, that are however isolated from the Atlantic Rainforest.

For Salvador-Jr. et al (2011), 16 records of the species *S. tyrannus* were obtained in an area of 720,000 hectares, including 37 municipalities, known as the "Quadrilátero Ferrífero" (Iron Quadrilateral) in the mountain formations in the state of Minas Gerais. According to these authors, the region has been under great pressure from mining and deforestation throughout its colonization history. In the previous and present studies, 23 sites were contemplated for the species, with the farthest points covering a range of approximately 200sq. km in fragmented environments. The records show an average distance of 18.8+_11, varying between 40 km and 8 km. Thus, we point out that the "Quadrilátero Ferrífero" is of extreme importance for conservation of potential reproductive populations of the species.

In Mendonça-Lima et al. (2006) several records were made of *S. ornatus* and its respective geographical locations in the south of the country generally associated with preserved forest portions. However, the authors suggest more efforts are needed to conserve this species in the south of the country, with the species having been sub-sampled, which is partly due to their refuge in environments of difficult access. Still regarding the southern Atlantic Rainforest, Joenck (2006) recorded the species *Spizaetus tyrannus* in several occasions in a preservation park, estimating two individuals in 4,500 hectares (Table 1).

In the northeast, the Endemism Center of Pernambuco, the species *S. tyrannus* was recorded as shown in Figure 1. According to Carlos and Girão (2006), still in the northeastern Atlantic Rainforest, the species *S. ornatus* may never have occurred, such as *S. melanoleucus* that were never recorded in the region. However, we emphasize the sub-sampling problems, inadequate methods, prior to habitat fragmentation, which may have made this/these species extinct in the state, according to previously cited studies.

In the case of nest records or pairs incidence for the genus in the Brazilian Amazon (Figure 1), these records were associated to extensive forests or mosaics of continuous environments, but also to fragmented areas. The presence of extensive source areas may complicate an evaluation on territoriality or density in the Amazon Forest, or fragments of the forest, according to the recruiting of individuals. *S. tyrannus* was recorded in extremely fragmented areas, with records of pairs in areas distanced between 22 and 67 km, with open and altered environmental mosaics among these, as well as in portions of continuous forest, with records of two specimens exhibiting territorial behavior distanced in 28 km. However, possible gaps or occupied territories among these were not evaluated in the continuous areas. *S. ornatus*, in the Amazon, was only recorded in continuous or extensive fragments, forests with a minimum of dozens of kilometers in extension.

5. Conclusion

Alvarez et al. (1996) obtained throughout five years in Venezuela, during hike and drives, one record of *S. melanoelucus*, two records of *S. ornatus* and four records of *S. tyrannus*. These species were seen in areas that suffer from selective logging, monoculture plantations, clearings and other altered environments. The authors attested that methods for censuses of birds of prey must be complemented with observations above the forest canopy. This warranted the evaluation in this study conducted with appropriate techniques for the visualization of these species, presenting results of low frequency occurrence and low density for *S. melanoleucus* and *S. ornatus* in the fragmented areas, which shows a greater sensibility to these environmental changes for both species.

Some authors consider *Spizaetus tyrannus* less sensitive to fragmentation than its congeners because it is observed more frequently in fragmented areas (Table 2) and altered forests (Brown and Amadon, 1968; Jullien and Thiollay, 1996; Thiollay, 2007). In addition, there are recent records of their occurrence in fragments near major urban centers (Azevedo et al. 2003, Zorzin et al. 2006, Canuto 2008, Salvador-Jr et al. 2011).

We obtained sparse or isolated records of its congeners *S. melanoleucus* and *S. ornatus* in the different case studies. This loss of sympatry (presence of sister species in the same place), previously cited and evident in extensive or continuous areas (case study 1), shows a lesser sensibility of the species *S. tyrannus* to fragmentation when compared to its congeners in the Atlantic Rainforest, where the species' natural distribution would be throughout all of the biome (Brown and Amadon 1968, del Hoyo et al. 1994, Ferguson-Lees and Christie 2001). According to Canuto (2008a), in the lowland tropical rainforest of Panama, the density of the group was higher than that of case study 1 (semideciduous Atlantic Forest), with the use of quite different methods in each study, random hikes and direct observation of the canopy respectively, but with the same sampling effort (approximately 550 hours). Thus, for the group, both methods are essential and complementary, with the need of continuous efforts and several counts or annual campaigns. Therefore, according to these density differences we can infer, primarily, that Hawk-eagles in less productive areas should be spaced further apart than those in more productive areas (Hustler and Howells, 1988). According to Lyon and Kuhnigk (1985), a region with higher primary productivity should support a higher density of Hawk-eagles. This was attested here for *S. tyrannus* and *S. ornatus*, according to Canuto (2008a) study but not for the few unclear results showed here for the Amazon forest, in comparison with study case 1.

Also it's important to show that populations, in Amazon forest or other tropical or subtropical forets (Canuto 2008a, Carvalho-Filho 2009, Canuto 2009, Thiollay 1989a, 2007), on large or protected areas, includes 2-3 *Spizaetus* species at the same sites or portions of mature forest. Also according to the cited studies the species are of low densitie even at large portions of mature forest. At Minas Gerais state, those, earlier sympatric species, are being separated or already are, through local extinction, on several regions by deforestation, with exception of one extensive reserve in Minas Gerais as PERD (study case 1), or probably just a few other areas.

Also, as evaluated by Carvalho and Marini (2007), urbanization led to a loss of one-third of diurnal raptors at the surrounding areas of the Minas Gerais state capital, Belo Horizonte, including all three *Spizaetus*, that became extremely rare or probably extinct.

Azevedo et al (2003) recorded several times two individuals together of *S. tyrannus* at an island of approximately 42300ha, apart from the main land approximately by 2-10 km, at Santa Catarina state, on the Atlantic rainforest domain south of Brazil. No other taxon from *Spizaetus* was recorded or estimates on density were made.

Therefore, the local decline and extinction of forest raptors in the Atlantic Rainforest, with their posterior substitution by countryside generalist taxons, due to deforestation of tropical forests, is very worrying and documented in several studies (Thiollay 1985, Albuquerque 1986, Ellis et al. 1990, Jullien and Thiollay 1996, Sick 1997, Seavy and Apodaca 2002, Salvador-Jr. and Silva 2009, Zorzin 2011).

Along with the loss of habitats, the local preservation of the genus *Spizaetus* may be affected by selective logging, since these species use large trees to build their nests (Brown and Amadon 1968, Lyon and Kuhnigk 1985, Montenegro et al. 1992, Canuto 2008), and these evidences of suppression of emergent trees was evidentiated on several occasions at the Minas Gerais state (personal communication by the authors), furthermore, these trees and nesting sites are sparse or rare in already fragmented habitats.

Due to the extinction of those species on several locations of the Atlantic rainforest its fundamental the creation of a bridge between technical-scientific knowledge and public opinion through environmental education and publicizing results and conservation programs, as well as the creation of wildlife corridors. The examples of pressure of deforestation of areas surrounding conservation parks and along the Atlantic Rainforest due to mining, selective logging and fragmentation, as well as changes in the Brazilian Forest Code, and hunting and poaching of the group's species, are all directly linked to the public opinion in rural communities and environmental sectors of the municipal, state and federal governments. This consequently influences the preservation of the Atlantic Rainforest habitat, along with the *Spizaetus* group and other indicator or rare species.

According to Figure 2, the central east, northeast, south and west of the state of Minas Gerais and the south of the state of Bahia are extremely important for the preservation and study of the genus. Including areas of natural incidence or occurrence, these portions include the middle basin of the Doce river, the south-central section of the Espinhaço mountain chain, the Mantiqueira mountain chain, MG transition *cerrado* areas in the Triângulo Mineiro Region and the northeast and northwest (insufficient data) part of the state, and the boundarie between the states of Minas Gerais and Bahia.

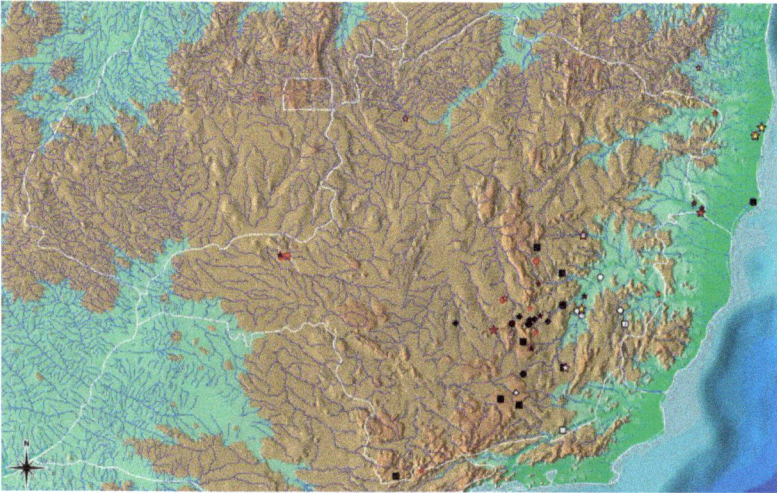

Fig. 2. Map of *Spizaetus* records in the Atlantic Rainforest; squares indicate records of pairs, circles indicate nests and diamonds indicate records from cited literature. Big stars indicate the incidence of the three species of *Spizaetus* (yellow), or of two, *S. ornatus* and *S. tyrannus* (brown with black), and *S. tyrannus* and *S. melanoleucus* (white with black). Small stars indicate records of species in this study. In general, white is for *S. melanoleucus*, black is for *S. tyrannus* and brown is for *S. ornatus*. Scale: 1:5036552.

Several unpublished records of most of individuals, from different observers, were not used on the assessments. This work focus in population status at some regions, pairs, young and nests, than distribution or occurrence. Therefore according to the pattern of specimen movements, several kilometers, and due to human persecution or deforestation pressure, these could lead certain species to wander throughout the regions, mistaken the actual active breeding population with isolated records of individuals in dispersal or other.

We also have to realize that the few or absent records on fledged youngs, meaning reproductive success, at least in Minas Gerais state, are a crucial alert to lack of active breeding populations, yet to be evaluated for future conservation plans for the medium or large birds of prey. As any charismatic or umbrela species, that figures as conservation flags for any threatened environments, predators generally uses extensive portions of mature, or protected, or in need to protection, forest habitats, and therefore could provide and serve as conservation flags for our natural resources and those of local communities.

6. Acknowledgments

We would like to thank Biocev Serviços de Meio Ambiente and staff, the Veracel Station and staff, the administration of PARNA do Pau Brasil and staff, Tânia Sanaiotti (INPA), the Instituto Estadual de Florestas (State Forest Institute), the staff and all the surrounding communities of the "Parque Estadual do Rio Doce", The Peregrine Fund, the Federal University of Ouro Preto staff, teachers, students, city community and CAPES, and the Federal University of Viçosa. Also the collaborators Fernanda Pedreira Tabacow, Breno G. M. da Silva, Carla de Borba Possamai and staff of the RPPN Miguel Feliciano Abdala-FMA,

Charles Lopes Moreira and Djalma de Carvalho Junior, Geraldo dos Santos Adriano (Canela) Fabiano R. Melo, Marcelo Ferreira Vasconcelos, Frederico Pereira de Castro Andrade, Gustavo Pedersoli, Luiz Gabriel Mazzoni, Alyne Perillo, and Estefane N. L. Siqueira for their valuable contribution in recording data.

Photo 5. Point count in the canopy (Study case 1). Photo credit: Marcus Canuto

Photo 6. Galery forest, Cerrado-Amazon forest, Mato Grosso state, site of incidence of *S. ornatus*. Photo credit: Eduardo Pio Carvalho

Photo 7. Climbing techiniques to access Black-and-white Hawk-eagle nest for measurements. Study case 3. Photo credit: Breno G M da Silva

Photo 8. Adult individual of *S. tyrannus* on flight. Photo credit: Eduardo Pio Carvalho

Photo 9. Adult individual os *S. ornatus*. Photo credit: Eduardo Pio de Carvalho

Photo 10. Fledgling of *S. melanoelucus* in the nest. Photo credit: Carlos Eduardo Carvalho

Photo 11. Selective logging, Viçosa, Minas Gerais. Photo credit: Giancarlo Zorzin

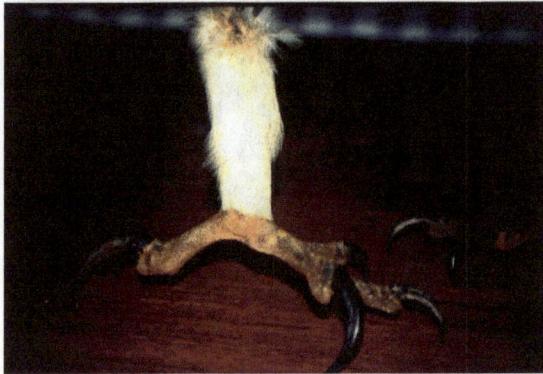

Photo 12. Tallon of hunted *Spizaetus* specime. Photo credit: Gustav Specht

7. Filliation universities and institutions

Some of the data were presented as part of master degree programs at Universidade Federal de Ouro Preto – PPG Ecologia de Biomas Tropicais, and Universidade Federal de Viçosa - Museu de Zoologia João Moojen. Those dissertations were supported by the scientific scholarship CAPES and The Pregrine Fund, Boise state, Idaho, USA.

8. References

Albuquerque, J.L.B. 1995. Observations of rare raptors in southern Atlantic rainforest of Brazil.Journalof Field Ornithology 66: 363-369.
Alvarez, E., D. H. Ellis, D. G. Smith, and C. T. Larue. 1996. Diurnal raptors in the fragmented rain forest of the Sierra Imataca, Venezuela. Pages 263-273

ANA, 2001.Proposta de Instituição do Comitê de Bacia Hidrográfica do Rio Doce, donforme Resolução No 5, de 10 de Abril de 2000, do Conselho Nacional De Recursos Hídricos.

Azevedo, M.A.G., Machado, D.A. e Albuquerque, J.L.B. 2003. Aves de rapina na Ilha de Santa Catarina, SC: composição, freqüência de ocorrência, uso do habitat e conservação. Ararajuba, 11 (1):75-81.

Bergallo, H. G., C. F. D. Rocha, M. A. S. Alves and M. Vansluys.2000. A fauna ameaçada de extinção do Estado do Rio de Janeiro. . 1vols, p. 166. Rio de Janeiro: EdUERJ.

Bibby, C.J., Burgess, N.D., Hill, D.A., Mustoe, S.H., 2000.Bird Census Techniques, 2nd ed. Academic Press, London.

Bierregaard, R.O., Jr. 1998. Conservation status of birds of prey in the South American tropics. Journal of Raptor Research 32: 19-27

Bildstein, K. L. 1998. Linking raptor migration science to mainstream ecology and conservation: an ambitious agenda for the 21 stcentury.Pages 583-602 in Holarctic birds of prey (B.-U.Meyburg, R. D. Chancellor, and J. J. Ferrero, Eds.).ADENEX, Merida, Spain; and World Working Group for Birds of Prey and Owls, Berlin, Germany.

Blake, E.R.1977. Manual of neotropical birds. The University of Chicago Press, Chicago and London.

Brown, L. and D. Amadon. 1968. Eagles, hawks, and falcons of the world. v.1 and 2. Country Life Books, London.

Burnham, W. A., D. F. Whitacre, and J. P. Jenny, (eds). 1990. Progress report III, 1990: Maya Project--use of raptors as environmental indices for design and management of protected areas and for building local capacity for conservation in Latin America. The Peregrine Fund, Inc., Boise, Idaho.201 pp.

Burnham, W. A., D. F. Whitacre, and J. P. Jenny. 1994. The Maya Project: use of raptors as tools for conservation and ecological monitoring of biological diversity. pp. 257-264 in B.-U.Meyburg and R. D. Chancellor, eds., Raptor conservation today.World Working Group on Birds of Prey, The Pica Press.

Canuto, M. 2008a. Observations of two hawk eagle species in a humid lowland tropical forest reserve in Central Panama. Journal of Raptor Research 42 (4): 287-292.

Canuto, M. 2008b. First description of the nest of the black-and-white hawk eagle (Spizaetusmelanoleucus) in the Brazilian Atlantic Rainforest, southeast Brazil.Neotropical Ornithology 19: 607-610.

Canuto, M. (2009). Ecologia de comunidades de aves de rapina (Cathartidae, Accipitridae e Falconidae) em fragmento de Mata Atlântica na região do médio Rio Doce, MG. Dissertação de Mestrado. Ouro Preto: Universidade Federal de Ouro Preto.

Carlos, C.J. and W. Girão. 2006. [The history of the Ornate Hawk-Eagle, Spizaetusornatus, in the Atlantic forest of northeast Brazil]. Revista Brasileira de Ornitologia 14:405-409.

Carrascal, L.M.; Palomino, D. Polo, V. 2008a. Patrones de distribución, abundancia y riqueza de especies de la avifauna terrestre de la isla de La Palma (islas Canarias). *Graellsia* 64: 209-232.

Carrascal, L.M.; Seoane, J.; Palomino, D.; Polo, V. 2008b. Explanations for bird species range size: ecological correlates and phylogenetic effects in the Canary Islands. JournalofBiogeography35:2061–2073.

Carvalho, C. E. A. & M. Â. Marini. 2007. Distributional patterns of raptors in open and forested habitats in Southeast Brazil and effects of urbanisation. Bird Conservation International 17: 367-380.

Carvalho-Filho, E. P. M., Carvalho, C. E. A. e Carvalho, G. D. M. 1998. Descrição de ninho e ovos de Micrastur semitorquatus (Falconidae) no interior de habitação rural, no município de sete lagoas - MG. Atualidades Ornitológicas, 86: 12.

Carvalho Filho, E. P. M., Carvalho, G. D. M. e Carvalho, C.E.A. 2005. Observations of nesting Gray-Headed Kites (Leptodon cayanensis) in southeasters Brazil. Journal Raptor Research, 39 (1): 91-94.

Carvalho Filho, E.P.M., Canuto, M. e Zorzin, G. 2006. Biologia reprodutiva do gavião preto (Buteogallus u. urubitinga: Accipitridae) no sudeste do Brasil. Revista Brasileira de Ornitologia, 14 (4): 445-448

Carvalho Filho, E.P.M., Zorzin, G., Canuto, M., Carvalho, C.E.A. & Carvalho, G.D.M. 2008. Aves de rapina diurnas do Parque Estadual do Rio Doce, Minas Gerais, Brasil. Instituto Estadual de Florestas - MG.Biota v.1.n.5. ISSN 1983-3678.

Carvalho Filho, E. P. M., Zorzin, G., Canuto, M., Carvalho, C. E. A. e Carvalho, G. D. M. 2009. Aves de Rapina Diurnas do Parque Estadual do Rio Doce. MG Biota, 1(5): 04-43.

CETEC. 1981. Vegetação do Parque Florestal do Rio Doce. Programa de Pesquisas Ecológicas no Parque Florestal Estadual do Rio Doce. Relatório Final, Vol. 2. Fundação Centro Tecnológico de Minas Gerais (CETEC), Belo Horizonte. 277pp.

del Hoyo, Josep, Andrew Elliott, JordiSargatal, et al. 1994. Handbook of the birds of the world: Vol. 2. New World Vultures to GuineafowlLynx Edicions. Barcelona. 638 pg.

Dias, D F ; Rodrigues, M. . Registro do gavião-pombo-pequeno Leucopternis lacernulatus em Belo Horizonte, Minas Gerais. Atualidades Ornitológicas, v. 147, p. 20-21, 2009.

Drummond, G. M.; Machado, A. B. M.; Martins, C. S.; Mendonça, M. P.; Stehmann, J. R. Listas vermelhas das espécies de fauna e flora ameaçadas de extinção em Minas Gerais. 2 ed. Belo Horizonte, MG: Fundação Biodiversitas, 2008.

Faria, L.C.P., L.A. Carrara, F.Q. Amaral, M.F. Vasconcelos, M.G. Diniz, C.D. Encarnação, D. Hoffmann, H.B. Gomes, L.E. Lopes and M. Rodrigues. 2009. The birds of Fazenda Brejão: a conservation priority area of cerrado in northwestern Minas Gerais, Brazil. Biota Neotropica 9(3): 223-240.

Ferguson-Lees, J. & Christie, D.A. (2001) Raptors of the world. Christopher Helm, London.

Ferreira, R. L. and M. M. G. S. M. Marques. 1998. A Fauna de Artrópodes de Serapilheira de Áreas de Monocultura com Eucalyptussp e Mata Secundária Heterogênea. Anais da Sociedade Entomológica do Brasil 27:395–403.

Flatten, C.J. 1990. Biology of the Ornate Hawk-Eagle (Spizaetusornatus). Pages 129– 143 in W.A. Burnham, D.F. Whitacre, and J.P. Jenny [EDS.], Maya Project: use of raptors as environmental indices for design and management of protected areas for building local capacity for conservation in Latin America, progress report III. The Peregrine Fund Inc., Boise, ID U.S.A.

Fontana, C. S., Bencke, G. A. e Reis, R. E. (orgs.). 2003. Livro vermelho da fauna ameaçada de extinção no Rio Grande do Sul. Porto Alegre: EDIPUCRS.

Francischetti CN. 2007. Ephemeroptera (Insecta) do Parque Estadual do Rio Doce, Minas Gerais, Brasil: Biodiversidade e distribuição espacial. D.S. Thesis, Universidade Federal de Viçosa. 97 pp.

Helbig, A.J., A. Kocum, I. Seibold, and M.J. Braun. 2005. A multi-gene phylogeny of aquiline eagles (Aves: Accipitriformes) reveals extensiveparaphyly at the genus level. Molecular Phylogenetics and Evolution 35:147-164.

Gilhuis, J.P. 1986. Vegetationsurveyofthe Parque Florestal Estadual do Rio Doce, MG, Brazil. Tese de Mestrado, Agricultural University of Wageningen, Netherlands.Viçosa, Universidade Federal de Viçosa/AgriculturalUniversityof Wageningen. 86 pp

Gjershaug, J. O., N. Røv, T. Nygård, D. M Prawiradilaga, M. Y. Afianto and Hapsoro, and A. Supriatna. 2004. Home-range size of the Javanhawk-eagle (Spizaetusbartelsi) estimated from direct observations and radiotelemetry. Journal of Raptor Research 38:343-349.

Granzinolli, Marco AntonioMonteiro ; Motta-junior, J. C. . Aves de rapina: levantamento, seleção de habitat e dieta. In: Von Matter, S.; Straube, F.; Accordi, I.; Piacentini, V.; Cândido Jr., F. J.. (Org.). Ornitologia e conservação: ciência aplicada, técnicas de pesquisa e levantamento. 1 ed. Rio de Janeiro: Technical Books editora, 2010, v., p. 169-187

Griffiths, C.S., G.F. Barrowclough, J.G. Groth and L.A. Mertz. 2007. Abstract: Phylogeny, diversity, and classification of the Accipitridae based on DNA sequences of the RAG-1 exon. pp.22 In Program and abstracts: Kettling on the Kittatinny 12-16 September 2007, Holiday Inn Conference Center, Lehigh Valley, Fogelsville, Pennsylvania. Raptor Research Foundation and Hawk Migration Association of North America, Fogelsville, PA.

Haring, E. et al., 2007.Convergent evolution and paraphyly of the hawk-eagles of the genus Spizaetus [Aves, Accipitridae] – phylogenetic analyses based on mitochondrial markers. Journal of Zoological Systematics and Evolutionary Research 45: 353-365).

Hirsch, A. 2003a. Avaliação da Fragmentação do Habitat e Seleção de Áreas Prioritárias para a Conservação dos Primatas da Bacia do Rio Doce, Minas Gerais, Através da Aplicação de um Sistema de Informações Geográficas. Tese de Doutorado. Programa de Pós-Graduação em Ecologia, Conservação e Manejo de Vida Silvestre, ICB / UFMG, Belo Horizonte. 227pp + Anexos.

Hirsch, A. 2003b.Habitat fragmentation and priority areas for primate conservation in the Rio Doce Basin, Minas Gerais.Neotropical Primates, 11(3):195-196.

Hustler, K. and Howells, W.W. 1988. The effect of primary production on breeding success and habitat selection in the African Hawk-Eagle. Condor 90:583--587

IBGE. 2003. Base Cartográfica Integrada do Brasil ao Milionésimo Digital - bCIMd. Versão 1.0. Fundação Instituto Brasileiro de Geografia e Estatística, Rio de Janeiro. 8pp.

Joenck, C.M. 2006. [Records oftheSpizaetustyrannus (Acciptridae) in the Centro de Pesquisa e Conservação da NaturezaPró-Mata (CPCN Pró-Mata) in northeast of the state of Rio Grande do Sul, Brazil] .RevistaBrasileira de Ornitologia 14:427-428

Jullien, M. and J.M. Thiollay. 1996. Effects of rain forest disturbance and fragmentation: comparative changes of the raptor community along natural and human-made gradients in French Guiana. Journal of Biogeography 23(1): 7-25.

Julio, A., Madrid, M., Héctor, D., Sixto H., Funes, A., Lopez, J., Botzoc, G.R. and Ramos, A. 1991. Reproductive biology and behavior of the Ornate Hawk-Eagle (Spizaetus ornatus) in Tikal National Park. Pages 93–13 in W.A. Burnham, D.F. Whitacre, and J.P. Jenny [EDS.], Maya Project: use of raptors and other fauna as environmental indices for design and management of protected areas for building local capacity for conservation in Latin America, progress report IV. The Peregrine Fund Inc., Boise, ID U.S.A.

Kirwan, G.M., J. M Barnett. e J. Minns (2001) Significant Ornitological observations from the Rio Sao Francisco Valley, Minas Gerais, Brazil, with notes on conservation and biogeography. Ararajuba 9(2): 145-161.

Laurance, W. F. & Bierregaard-Jr., R.O. (eds.). 1997. Tropical Forest Remnants.Ecology, Management, and Conservation of Fragmented Communities.University of Chicago, Chicago, Illinois.

Lins, L.V. (2001) Diagnóstico ornitológico do Parque Estadual do. Rio Doce. Instituto Estadual de Florestas, Belo Horizonte.

Lyon, B. and K.A. Kuhnigk. 1985. Observations on nesting Ornate Hawk-Eagles in Guatemala. Wilson Bull. 97:141--147.

Machado, A.B.M., G A.B. Fonseca, R. B. Macahado, L. M. S. Aguiar e L. V. Lins. 1998. Livro vermelho das espécies ameaçadas de extinção da fauna de Minas Gerais. Belo Horizonte: FundaçãoBiodiversitas.

Madrid, J. A.; Madrid, H. D.; Funes, S. H. A.; Avila, J. A.; Botzoc, R. G. and Ramos, A. Reproductive biology and behavior Ornate Hawk-eagle in Tikal National Park, p. 92-113 In: WHITACRE, D. F; BURNHAM, W. A. and JENNY, J. P. (eds.) Maya Project: Use of raptos as environmeental indices for design and management of protect area sand for building local capacity for conservation in Latin America. The Peregrine Fund, Inc., Boise, Idaho, Progress Report 4, 1991.

Mañosa, S. and V. Pedrocchi. 1997. A raptor survey in the Brazilian Atalntic rain forest. Journal of Raptor Research 31: 203-207.

Mañosa, S.; Mateos, E.; Pedrocchi, V. and Martins, F.C. (2002). Birds of Prey Survey (Aves: Cathartiformes and Accipitriformes) in the Paranapiacaba Forest Fragment. Pp. 165-179. In: Mateos, E.; Guix, J.C.; Serra, A. and Pisciotta, K. (eds.). Censuses of Vertebrates in a Brazilian Atlantic Rainforest Area: The Paranapiacaba Fragment. Barcelona: Universitat de Barcelona.

Mañosa, S., E. Mateos, and V. Pedrocchi. 2003. Abundance of soaring raptors in the Brazilian Atlantic rainforest. Journal of Raptor Research 37:19-30.

Mendonça-Lima, A, Zilio, F., Joenck, C. M. and Barcellos, A. 2006. Novos registros de Spizaetusornatus (Accipitridae) no sul do Brasil. Revista Brasileira de Ornitologia 14 (3) 279-282

Meunier, K. 1960. GrundsltzlicheszurPopulationsdynamik der Vogel.Z. wiss. Zool., 163 :397-445.

Metzger, J.P., 2000. Tree functional group richness and landscape structure in a Brazilian tropical fragmented landscape. Ecological Applications 10, 1147– 1161.

Metzger, J.P., Martensen, A.C., Dixo, M., Bernacci, L.C., Ribeiro, M.C., Teixeira, A.M.G, Pardini, R., 2009.Time-lag in biological responses to landscape changes in a highly dynamic Atlantic forest region. Biological Conservation 142, 1166– 1177.

Mikich, S. B. e Bérnils, R. S. Livro vermelho da fauna ameaçada no Estado do Paraná. Curitiba: Instituto Ambiental do Paraná, 2004.

Montenegro, H.D.M. 1992. Behavior and breeding biology of the Ornate Hawk-Eagle.Pages 179-191 in W.A. Burnham, D.F. Whitacre, and J.P. Jenny [EDS.], Maya Project: use of raptors and other fauna as environmental indices for design and management of protected areas for building local capacity for conservation in Latin America, progress report V. The Peregrine Fund Inc., Boise, ID U.S.A.

Newton, I. 1979. Population ecology of raptors.Buteo Books, Vermillion, SD.

Oliveira-Filho, A. T. e M. A. Fontes. 2000. Patterns of floristic differentiation among Atlantic forests in southeastern Brazil and the influence of climate. Biotropica, 32: 793–810

Olmos, F., J.F. Pacheco, and L.F. Silveira. 2006. [Notes on Brazilian birds of prey]. RevistaBrasileira de Ornitologia 14:401-404.

Perry, M.Z. 2000.Factors affecting interspecies variation in home-range size of raptors. The Auk 117 (2): 511-517.

Preleuthner M., Gamauf A 1998: A possible new subspecies of the Philippine Hawk- Eagle (Spizaetusphilippensis) and its future prospect. J. Raptor Res. 32:126-135.

Prugh, L.R., Stoner,C.J., Epps, C.W, Bean, W.T., Ripple, W.J. Laliberte, A.S. e. Brashares, J.S.2009. The Rise of the Mesopredator. BioScience 59(9): 779-790.

Ranta, P., Blom, T., Niemelä, J., Joensuu, E., Siitonen, M., 1998. The fragmented Atlantic rain forest of Brazil: size, shape and distribution of forest fragments. BiodiversityandConservation 7, 385–403.

Resende, S. B. e Resende, M. 1996. Solos dos Mares de Morros: ocupação e uso. In: Alvares, V . H. V. (Org.). O solo nos grandes domínios morfoclimaticos do Brasil e o desenvolvimento sustentado. SBCS, Viçosa: 261-288.

Ribeiro, MC. et al., 2009. Brazilian Atlantic forest: how much is left and how is the remaining forest distributed? Implications for conservation. Biological Conservation, 142: 1141-1153.

Ribon, R. 1998. Fatores que influenciam a distribuição da avifauna em fragmentos de Mata Atlântica nas Montanhas de Minas Gerais. Dissertação de mestrado do Programa de Pós-Graduação em Ecologia, Conservação e Manejo da Vida Silvestre da UFMG, Belo Horizonte

Ribon, R., Simon, J.E. & Mattos, J.E. 2003. Bird extinctions in Atlantic Forest fragments on the Viçosa region, southeastern Brazil. Conserv. Biol. 17:1827-1839.

Rizzini CT, 1979, Tratado de fitogeografia do Brasil. Aspectos sociologicos e florísticos. HUCITEC, Sao Paulo, 2: 374.

Roda, S. A. e Pereira, G.A. 2006. Distribuição recente e conservação das aves de rapina florestais do Centro Pernambuco. Revista Brasileira de Ornitologia 14 (4) 331-344.

Ryall, K.L. and L. Fahrig. 2006. Response of predators to loss and fragmentation of prey habitat: A review of theory. Ecology 87: 1086-1093.

Salvador-Jr., L.F. & Silva, F.A. 2009. Rapinantes diurnos em uma paisagem fragmentada de Mata Atlântica no alto rio Doce, Minas Gerais, Brasil. Boletim do Museu de Biologia Mello Leitão (N. Sér.)25: 53-65.

Salvador-Jr., L.F.; Canuto, M.; Carvalho, C.E.A &Zorzin, G. 2011. Aves, Accipitridae, Spizaetustyrannus (Wied, 1820): New records in the QuadriláteroFerrífero region, Minas Gerais, Brazil. Check List 7(1): 32-36

Sergio, F., I. Newton, and L. MarchesI. 2006. Ecologically justified charisma: preservation of top predators delivers biodiversity conservation. J. Appl. Ecol. 43:1049– 1055.

Seavy, N.E., and C.K. Apodaca. 2002. Raptor abundance and habitat use in a highly-disturbed -forest landscape in western Uganda. Journal of Raptor Research 36:51-57.

Sick, H. & Teixeira, D.M. 1979. Notas sobre aves brasileiras raras ou ameaçadas de extinção. Publ.Avuls.Mus.Nac. 62:1-39.

Sick, H. 1997. Ornitologia Brasileira, uma introdução. Nova Fronteira, Rio de Janeiro, pp. 492-503.

Silva, W.G.S., Metzger, J.P., Simões, S., Simonetti, C., 2007. Relief influence on the spatial distribution of the Atlantic Forest cover at the Ibiúna Plateau, SP. Brazilian Journal of Biology 67, 403–411.

Silveira, L. F. et al. 2009. Aves. in Bressan, P.M., M.C.M. Kierulff and A.M. Sugieda. 2009. Fauna ameaçada de extinção no Estado de São Paulo: Vertebrados. Fundação Parque Zoológico de São Paulo, Secretaria do Meio Ambiente. 645 p.

Simon et al. 2007. As aves ameaçadas de extinção no estado do Espírito Santo. In Passamani, M. e Mendes, S.L. 2007. Espécies da Fauna ameaçadas de Extinção no Estado do Espírito Santo. Instituto de Pesquisa da Mata Atlântica. Vitória.

Shultz, S. 2002. Population density, breeding chronology and diet of crowned eagles Stephanoaetuscoronatus in Ta‹ National Park, Ivory Coast. Ibis 144: 135-138.

Sixto, H.F. 1992. Reproductive biology, food habits, and behavior of the Black Hawk Eagle in Tikal National Park. Pages 173-178 in W.A. Burnham, D.F. Whitacre, and J.P. Jenny [EDS.], Maya Project: use of raptors and other fauna as environmental indices for design and management of protected areas for building local capacity for 67 conservation in Latin America, progress report V. The Peregrine Fund Inc., Boise, ID U.S.A.

SOS Mata Atlântica 2010.Atlas dos remanescentes florestais da mata atlântica período 2008-2010.

Soulé, M. E., Bolger, D. T., Alberts, A. C., Wright, J., Morice, M. and Hill, S. 1988. Reconstructed dynamics of rapid extinctions of chaparral-requiring birds in urban habitat islands. Conservation Biology 2: 75–92.

Stotz, D. F., J. W. Fitzpatrick, T. A. Parker III, and D. K. Moskovits.1996. Neotropical birds:
Ecology and conservation. University of Chicago Press, Chicago, Illinois.

Sutherland, W. J., Newton, I. & Green, R. E. (eds). 2004. Bird Ecology and Conservation: A
handbook of techniques. Oxford University Press, Oxford.

Terborgh, J. 1992. Maintenance of diversity in tropical forests.Biotropica 24(2b):
283— 292.

Thiollay, J. M. 1985. Falconiforms of tropical forests: a review. 155–165.in Newton, I. and R.
D. Chancellor, editors. eds. Conservation studies on raptors. ICBP Techn. Publ. No.
5. Cambridge, U.K.

Thiollay, J.M. 1989a. Area requirements for the conservation of rain forest raptors and game
birds in French Guiana. Conservation Biology 3: 128-137.

Thiollay, J.M. 1989b. Censusing of diurnal raptors in a primary rain forest: comparative
methods and species detectability. Journal of Raptor Research 23: 72-84.

Thiollay, J.M. 1990. Comparative diversity of temperate and tropical forest bird
communities: the intluence of habitat heterogeneity. ActaOecol. 11: 887-911.

Thiollay, J.M. 2007. Raptor communities in French Guiana: distribution, habitat selection,
and conservation. Journal of Raptor Research 41:90-105.

Valverde, O. 1958. Estudo regional da Zona da Mata de Minas Gerais. Revista Brasileira de
Geografia. 1: 3-82.

vanBalen, S., V. Nijman, and R. Sozer. 2001. Conservation of the endemic JavanHawk-Eagle
Spizaetusbartelsi Stresemann, 1924 (Ave: Falconiformes) density, agestructureand
population numbers. Contrib. Zool 70:161–173

Viana, L.C.S. 2001. Diagnóstico da cobertura vegetal do Parque Estadual do Rio Doce.
Contribuição ao Plano de Manejo; Projeto Doces Matas. Instituto Estadual de
Florestas (IEF).

Watson, R.T. 1998. Preface - conservation and ecology of raptors in the tropics. Journalof
Raptor Research 32: 1-2

Whitacre, D. F. and C. W. Turley . 1990. Correlations of diurnal raptor abundance with
habitat features in Tikal National Park. 71–92. In W.A. Burnham, D.F. Whitacre, J.P.
Jenny (eds.). Maya Project, progress report III. The Peregrine Fund, Inc. Boise,
Idaho.

Whitacre, D.F., Jones, L.E. and Sutter, J. 1992a. In Whitacre, D. F., and R. K. Thorstrom,
(eds.). Progress report V, 1992: Maya Project, progress report V. The Peregrine
Fund, Inc., Boise, Idaho.

Whitmore, T. C. 1997. Tropical forest disturbance, disappearance, and species loss, p. 3-12.
In W.F. Laurance& R. O. Bierregaard, Jr. (eds.) Tropical Forest Remnants. Ecology,
Management, and Conservation of Fragmented Communities.University of
Chicago, Chicago, Illinois.

Zorzin, G., Carvalho, C.E.A e Canuto, M. 2004. Dados sobre a biologia reprodutiva de
Buteogallus meridionalis (Falconiformes – Accipitridae) na APA Carste de Lagoa
Santa /MG. XII Congresso Brasileiro de Ornitologia - Blumenau /SC.

Zorzin, G., C.E.A. Carvalho, E.P.M de Carvalho Filho, and M. Canuto. 2006. [New records of
rare and threatened Falconiformes for the state of Minas Gerais]. RevistaBrasileira
de Ornitologia 14:417-421.

Zorzin, G. (2011). Os efeitos da fragmentação da Mata Atlântica sobre a riqueza e abundância de Accipitriformes e Falconiformes na Zona da Mata de Minas Gerais. Dissertação de Mestrado. Viçosa: Programa de Mestrado em Biologia Animal Universidade Federal de Viçosa.

Permissions

The contributors of this book come from diverse backgrounds, making this book a truly international effort. This book will bring forth new frontiers with its revolutionizing research information and detailed analysis of the nascent developments around the world.

We would like to thank Padmini Sudarshana, Madhugiri Nageswara-Rao and Jaya R. Soneji, for lending their expertise to make the book truly unique. They have played a crucial role in the development of this book. Without their invaluable contribution this book wouldn't have been possible. They have made vital efforts to compile up to date information on the varied aspects of this subject to make this book a valuable addition to the collection of many professionals and students.

This book was conceptualized with the vision of imparting up-to-date information and advanced data in this field. To ensure the same, a matchless editorial board was set up. Every individual on the board went through rigorous rounds of assessment to prove their worth. After which they invested a large part of their time researching and compiling the most relevant data for our readers. Conferences and sessions were held from time to time between the editorial board and the contributing authors to present the data in the most comprehensible form. The editorial team has worked tirelessly to provide valuable and valid information to help people across the globe.

Every chapter published in this book has been scrutinized by our experts. Their significance has been extensively debated. The topics covered herein carry significant findings which will fuel the growth of the discipline. They may even be implemented as practical applications or may be referred to as a beginning point for another development. Chapters in this book were first published by InTech; hereby published with permission under the Creative Commons Attribution License or equivalent.

The editorial board has been involved in producing this book since its inception. They have spent rigorous hours researching and exploring the diverse topics which have resulted in the successful publishing of this book. They have passed on their knowledge of decades through this book. To expedite this challenging task, the publisher supported the team at every step. A small team of assistant editors was also appointed to further simplify the editing procedure and attain best results for the readers.

Our editorial team has been hand-picked from every corner of the world. Their multi-ethnicity adds dynamic inputs to the discussions which result in innovative outcomes. These outcomes are then further discussed with the researchers and contributors who give their valuable feedback and opinion regarding the same. The feedback is then collaborated with the researches and they are edited in a comprehensive manner to aid the understanding of the subject.

Apart from the editorial board, the designing team has also invested a significant amount of their time in understanding the subject and creating the most relevant covers. They scrutinized every image to scout for the most suitable representation of the subject and create an appropriate cover for the book.

The publishing team has been involved in this book since its early stages. They were actively engaged in every process, be it collecting the data, connecting with the contributors or procuring relevant information. The team has been an ardent support to the editorial, designing and production team. Their endless efforts to recruit the best for this project, has resulted in the accomplishment of this book. They are a veteran in the field of academics and their pool of knowledge is as vast as their experience in printing. Their expertise and guidance has proved useful at every step. Their uncompromising quality standards have made this book an exceptional effort. Their encouragement from time to time has been an inspiration for everyone.

The publisher and the editorial board hope that this book will prove to be a valuable piece of knowledge for researchers, students, practitioners and scholars across the globe.

List of Contributors

Madhugiri Nageswara-Rao
Department of Plant Sciences, University of Tennessee, Knoxville, TN, USA
Polk State College, Department of Biological Sciences, Winter Haven, FL, USA

Jaya R. Soneji
Polk State College, Department of Biological Sciences, Winter Haven, FL, USA
University of Florida, IFAS, Citrus Research & Education Center, Lake Alfred, FL, USA

Padmini Sudarshana
Monsanto Research Center, Hebbal, Bangalore, India

Marcela Zalamea, Grizelle González and William A. Gould
International Institute of Tropical Forestry, USDA Forest Service, Puerto Rico

María Fernanda Barberena-Arias
Universidad del Turabo, School of Sciences and Technology, Gurabo, Puerto Rico

Grizelle González
International Institute of Tropical Forestry/USDA Forest Service, Río Piedras, Puerto Rico

Elvira Cuevas
University of Puerto Rico, Department of Biology, San Juan, Puerto Rico

Grizelle González and Christina M. Murphy
USDA FS International Institute of Tropical Forestry (IITF), Río Piedras, Puerto Rico

Juliana Belén
University of Puerto Rico, Mayagüez, Puerto Rico

Pablo R. Stevenson and Juliana Agudelo T.
Departamento de Ciencias Biológicas, Universidad de los Andes, Bogotá, Colombia

María Clara Castellanos
Consejo Superior de Investigaciones Científicas, Centro de Investigaciones sobre Desertificación (CSIC-UV-GV), Valencia, España

Bernard Moyersoen
University of Aberdeen, School of Biological Sciences, UK

André R. Terra Nascimento and Glein M. Araújo
Instituto de Biologia, Universidade Federal de Uberlândia, Uberlândia, MG, Brasil

Aelton B. Giroldo
Pós-Graduação em Ecologia, Universidade de Brasília, Brasília, DF, Brasil

Pedro Paulo F. Silva
Pós-Graduação em Ecologia e Conservação de Recursos Naturais, Universidade Federal de Uberlândia, Uberlândia, MG, Brasil

Tiago Gomes dos Santos
Universidade Estadual Paulista, Brazil
Universidade Federal do Pampa, Brazil

Tiago da Silveira Vasconcelos and Célio Fernando Baptista Haddad
Universidade Estadual Paulista, Brazil

Fred Babweteera
Department of Forestry, Biodiversity and Tourism, Makerere University, Kampala, Uganda
Royal Zoological Society of Scotland, Edinburgh Zoo, Edinburgh, Scotland

S. C. Sahu and N. K. Dhal
Environment and Sustainability Department, Institute of Minerals and Materials Technology (formerly RRL), CSIR, Bhubaneswar (Odisha), India

Nicole L. Michel and Thomas W. Sherry
Department of Ecology and Evolutionary Biology, Tulane University, USA

Michael Keller
Complex Systems Research Center, Morse Hall, University of New Hampshire, Durham, NH, USA
International Institute of Tropical Forestry, USDA Forest Service, Rio Piedras, PR, USA

George Hurtt
Department of Geography, University of Maryland, College Park, MD, USA

Michael Palace and Steve Frolking
Complex Systems Research Center, Morse Hall, University of New Hampshire, Durham, NH, USA

Dyah R. Panuju and Bambang H. Trisasongko
Department of Soil Sciences and Land Resource, Bogor Agricultural University, Bogor, Indonesia

Mingxia Zhang and Jianguo Zhu
Kunming Institute of Zoology, The Chinese Academy of Sciences, Kunming, Yunnan, China

Stéphane Couturier and Juan Manuel Núñez
Laboratorio de Análisis Geo-Espacial, Instituto de Geografía, UNAM, Mexico

Melanie Kolb
Comisión Nacional para el Uso y Conservación de la Biodiversidad (CONABIO), Mexico

Julio Tóta
Universidade do Estado do Amazonas, Brasil

David Roy Fitzjarrald
State University of New York, USA

Maria A. F. da Silva Dias
Universidade de São Paulo, Brasil

Patrick J. Hart
Department of Biology, University of Hawaii at Hilo, Hilo, Hawaii

Robério Anastácio Ferreira and Paula Luíza Santos
Federal University of Sergipe, Department of Forest Sciences, Brazil

M. Canuto, G. Zorzin, E. P. M. Carvalho-Filho, C.E.A. Carvalho, G.D.M. Carvalho and C.E.R.T Benfica
S.O.S. Falconiformes, Centro de Pesquisa Para a Conservação das Aves de Rapina Neotropicais, Brazil